Miomir Vukobratović

Applied Dynamics of Manipulation Robots

Modelling, Analysis and Examples

With 176 Figures

Springer-Verlag Berlin Heidelberg GmbH

Professor Miomir Vukobratović, Ph. D., D. Sc.
Corr. member of Serbian Academy of Sciences and Arts
Foreign member of Soviet Academy of Sciences
Institute Mihailo Pupin
Volgina 15
P. O. Box 15
11000 Beograd
Yugoslavia

Based on the original Primenjena Dinamika Manipulacionih Robota
published by NIRO "Tehnička Knjiga", Beograd, Yugoslavia

ISBN 978-3-642-83868-2 ISBN 978-3-642-83866-8 (eBook)
DOI 10.1007/978-3-642-83866-8

Softcover reprint of the hardcover 1st edition 1989

2161/3020 543210 – Printed on acid-free paper

Preface

During the period 1982-1985, six books of the series: Scientific Fundamentals of Robotics were published by Springer-Verlag. In chronological order these were:

Dynamics of Manipulation Robots: Theory and Application, by M. Vukobratović and V. Potkonjak, Control of Manipulation Robots: Theory and Application, by M. Vukobratović and D. Stokić, Kinematics and Trajectory Synthesis of Manipulation Robots, by M. Vukobratović and M. Kirćanski, Real-Time Dynamics of Manipulation Robots by M. Vukobratović and N. Kirćanski, Non-Adaptive and Adaptive Control of Manipulation Robots, by M. Vukobratović, D. Stokić and N. Kirćanski and Computer-Aided Design and Applied Dynamics of Manipulation Robots, by M. Vukobratović and V. Potkonjak.

Within the series, during 1989, two monographs dealing with new subjects will be published. So far, amongst the published monographs, Vol. 1 has been translated into Japanese, Volumes 2 and 5 into Russian, and Volumes 1-6 will appear in Chinese and Hungarian.

In the author's opinion, the afore mentioned monographs, in principle, cover with sufficient breadth, the topics devoted to the design of robots and their control systems, at the level of post-graduate study in robotics. However, if this material was also to apply to the study of robotics at under-graduate level, it would have to be modified so as to obtain the character of a textbook. With this in mind, it must be noted that the subject matter contained in the text cannot be simplified but can only be elaborated in more detail.

This reasoning is derived from the fact that contemporary design of robots and their control systems do not allow for different levels of approach, that is, one applicable to postgraduate and the other to undergraduate study. Thus, this textbook series will in essence contain no simplifications in comparison with the preceding series of research monographs.

This series was envisaged as bearing the contents of a course in robotics involving the study of modern techniques in the design of robots and their control systems. This would encompass the treatment of controlling robots at the lowest execution level as well as cooperative robot performance and robot control at higher levels, such as is encountered in controlling flexible manufacturing cells and lines.

The minimal, yet sufficiently autonomous part of the series, consists of two books of the following content.

The first is devoted to the study of manipulation robot dynamics and its applications. It embodies a computational procedure for the automatic generation of mathematical models of robot dynamics, comprising the linearized models of robot dynamics and the parameter sensitivity models together with a selection of problems of practical significance involving the complete models of robot dynamics operating under conditions of constrained and unconstrained work space.

The second book treats the problem of controlling manipulation robots with the use of mathematical models of their dynamics which were studied in the first book. It further presents the problems associated with the numerical complexity involved in the execution of various control tasks and in their microcomputer implementation. Special attention was focussed on the programming support to the synthesis of control laws which are based on the complete mathematical models of robotic manipulator dynamics.

It is possible to extend the scope of the books in this series to include: the study of robot kinematics involving the non-redundant and redundant mechanical configurations, synthesis of trajectories in space with and without constraints and obstacles, material relating to computer-aided design of robots, programming and robot teaching, synthesis of the multilevel control of flexible manufacturing cells, the study of dynamics and control synthesis of elastic robotic mechanisms, expert systems for robot control synthesis of adaptive algorithms etc.

As previously mentioned, this book is devoted to the study of manipulation robot dynamics and its application. It is organized into three chapters and nine appendices.

The first chapter presents definitions and systematization of robotic systems as well their features and specification.

The second chapter constitutes the central part of the book. It includes the computer-aided procedure for automatic generation of mathematical models of rigid body manipulation robots dynamics having an open kinematic chain structure and cooperative manipulation. Dynamic models of robotic manipulators with elastic links are also presented. The procedure presented is founded upon the general theorems of mechanics and describes, in chronological order, the first method for computer-aided generation of mathematical models of rigid body robotic mechanisms. This procedure was selected, among other practical and educational resons, because it adheres to the physical essence of the problem in practically all stages of mathematical model generation.

Furthermore, Chapter 2 considers mechanical vibrations of fundament and their influence on the overall accuracy of the robot, and also regards the problems involving constraints on robot gripper motion. Dynamic model of such constrained gripper motion is presented.

Chapter 3 deals with automatic procedures for forming the linearized and parameter sensitivity models. These procedures naturally belong to Book 1 since both, the linearized models and models of parameter sensitivity will be used in the forthcoming books of the series dealing with non-adaptive and adaptive dynamic control synthesis of robotic manipulators.

The distinctiveness of this book is highlighted by the relatively large number of appendices.

Appendix 1 contains the coordinate transformations between two orthogonal coordinate systems which provide for better understanding of the relationships between the local and the fixed coordinate systems.

Denavit-Hartenberg coordinates, which are thought to be best suited to the forming of kinematic models of robotic mechanisms, as well the correlation between them and Rodrigue's formula, are presented in Appendix 2.

The fundamental relations between kinematic variables of a rigid body or more precisely of a robotic mechanism's kinematic pair, are given in Appendix 3, and moment of momentum, as well as Euler's dynamic equations of a rigid body are presented in Appendix 4.

In Appendix 5, the conditions under which a link of a robotic mechanism can be approximated as a cane are derived. This leads to significant simplifications in the formation of dynamic equations of robot motion. Mathematical models of actuator units of different types and varying complexity are given in Appendix 6. On the basis of the formed mathematical models of actuator dynamics, the complete mathematical model including the dynamics of the robotic mechanism can be formed.

The most characteristic examples of "STANFORD" robot and UMS-2 cylindrical manipulator configuration are given in Appendix 7. The dynamic equations of motion of these examples were derived "by hand" in compliance with the algorithm for automatic generation of mathematical models given in Chapter 2. In this manner, the computational procedure is systematically presented using the basic (3 d.o.f.) configuration of characteristic manipulation robot types. In Appendix 8, dynamic equations of the "ASEA" mechanism basic configuration are derived. In order to provide full autonomy of the computer-aided generation of dynamic equations of an open, arbitrarily complex, kinematic configuration of the manipulation mechanism, Appendix 9 contains the programme for nominal dynamics calculation of manipulation robots based on Newton-Euler's equations, described in Paragraph 2.1. This programme is written in programming language FORTRAN-77 and it can be used on arbitrary computer system with FORTRAN compiler. In Appendix 9, the programme for automatic linearization of the dynamic model of a manipulation mechanism having an open chain configuration is presented, too. Appendix 9 presents the programme VIBRO, on the basis of which and in conjunction with the main programme, the problem of mechanical vibrations at the robot fundament or vibrations in mechanisms with a mobile first link, is solved.

Appendix 9 also presents the programming support for solving the robot dynamics in cases where dynamic constraints are imposed on the manipulator gripper motion.

This book is primarily dedicated to students of undergraduate courses in robotics as well as to the engineers whose research interests lie in the field of mathematical modelling of robotic mechanism dynamics. However, it is also of importance to post-graduate students and specially to those concerned with non-adaptive and adaptive control based on complete dynamic models of robotic mechanisms. In this age of torrential development in robotics, it would be, of course, pointless and rather conservative to approach the problem of dynamic model generation, which is the crucial information in contemporary design of robotic mechanisms, in the classical manner of constructing it "by hand". On the other hand, there should be no significant difference in the level at which research is undertaken, at undergraduate or post graduate study, especially not in as far as mathematical modelling of robotic mechanisms is concerned.

It should be underlined, too, that besides its basic textbook character, this book has also some characteristics of a monograph, concerning results on modelling of elastic robots and cooperative manipulation, presented here for the first time.

It has to be pointed out that this book was meant as lecture material at technical faculties and for engineers to whom robot dynamics will be not the aim, but the means towards the solution of problems in robot control. For this reason, it was the opinion of the author, that this book should be free of extensive presentation of alternative techniques for generating mathematical models of robot mechanism dynamics. Hence, in this book robot dynamics is presented only as direct function of the concrete application of the modelling task. The presented mathematical models of robot dynamics do not include the effects of dry friction which appear in conventional realizations of mechanical transmission. Due to the evident trend towards direct drive motor application, the exclusion of friction forces becomes justifiable and in view of the delicacy involved in calculating these effects, provides the means for considerably simplifying the mechanism model. Furthermore, in these situations where the exclusion of dry friction effects is justified, and at the same time the model retains its fidelity, its significance in the control law synthesis of robotic systems becomes evident It is also clear that tasks involving adaptive control introduce the requirement for exceptionally effective computational procedures for generating mathematical models of robot dynamics which are to be implemented on modern micro-computer systems. Such procedures, which are based on symbolic modelling concept will for obvious reasons not be dealt with in this book. Instead, it will be the subject studied in one of the forthcoming books of this series which will consider the problems associated with the numerical complexity in the derivation of control laws and their subsequent microcomputer implementation, as well as the synthesis of the general purpose digital controller. The fact remains that the programme support to the modelling of robot dynamics which was given in Appendix 9 is of general interest, since the algorithm upon which it is founded is a general one, and as such, includes all phases of the model generation process, from mechanism assembly to the problem of solving the direct and inverse dynamics. As previously mentioned, this version of the application software is suitable for modular extension as well as for specific purpose applications and as such, the author believes it, to be the very procedure naturally belonging to this type of book. More detailed involvement in the dynamics of robotic mechanisms which can be of interest to mechanical engineers and applied mathematicians whose research interests lie within this domain of technical science, can be found in the bibliography at the end of Chapter 2.

I think that this textbook will enable the reader to gain a sufficient knowledge for his further work on the problems of dynamics and dynamic analysis of robotic systems and for implementation of theoretical approaches into practice. We want the reader to develop an engineer's approach to the subject and to direct him to use computer approach to learn robot dynamics, as this approach enables efficient linking of mathematical models and the practical requirements to be realized by current robots. How well we have succeeded it remains to be judged on the basis of the use of this book as a textbook in teaching practice, as well as in the research and development units for applied robotics.

In relation to this book, and bearing in mind the contributions which have been made towards the development of computer oriented methods based on general theorems of mechanics, I wish to give some comments.

In this book chronologically the first method of a numerical-iterative type was presented. I profit of this opportunity to mention Vesna Živković, Ph.D., a senior researcher at "Mihailo Pupin" Institute for her efforts in further operationalization and realizing the basic version of the computer programme (1977), based on the algorithm presented in this book, which was provided foremostly by Y. Stepanenko (1971, 1974) and extended by the author of this book and Y. Stepanenko (1972, 1973, 1976) on the class of anthropomorphic mechanisms. Subsequent computer-oriented methods presented in several joint papers and the monograph with V. Potkonjak, Ph.D. used the same principle for mathematical models generation of robot dynamics but were based either on 2nd order Lagrange's equations or else Appel's equations and Gibb's acceleration functions.

I would also like to emphasize the activity of Nenad Kirćanski, Ph.D., senior researcher at the Robotics Laboratory of the "Mihailo Pupin" Institute, who, on the basis of the same method of general theorems, together with the author of this book developed new efficient numerical-symbolic procedure which was presented in Volume 4 of the aforementioned Springer-Verlag monographic series. The latest results in the field of modelling the dynamics of manipulation robots are the symbolic models which belong to the efficient single-step and multi-step customized algorithms, in the creation of which substantial contribution was done by A. Timčenko, junior researcher, as well.

The author wishes to express his gratitude to associates in the Laboratory for Robotics and Flexible Automation of the "Mihailo Pupin" Institute, B. Karan senior researcher, D. Katić, Miss N. Djurović, Lj. Zarić, N. Djurić, junior researchers, for elaborating the mathematical models of typical robotic mechanisms and driving units, as well as to M. Djurović for derivation of the dynamic model of "ASEA" mechanism based on adopted computer-aided procedure for the mathematical modelling of robotic mechanisms. The author is also grateful to D. Vujić Ph.D. for his extension of programme support for manipulation robot dynamics with constrained gripper motion and vibrations of fundament, as well as to junior researcher A. Rodić for his participating in testing of the mentioned software. The author expresses his thanks to D. Šurdilović, M.Sc. for his essential contribution in conceiving the text dedicated to flexible manipulation robots, as well as to Miss M. Kolarski and M. Kostić research assistants for their programming and testing of the "ASEA" dynamic model. I further extend my gratitude to D. Hristić, Ph.D. for his high professional reading of the text and useful remarks, as well as to Professor M. Mićunović who reviewed the Serbo-Croation edition of this book. Finally, my appreciation goes to Miss V. Ćosić for her excellent typing of the entire book.

December 1988,
Beograd, Yugoslavia

Author

Contents

Chapter 1
General About Robots

1.1 Dedication and Classification of Robotic Systems

Robotic systems represent in principle new technical means of complex automation production processes. Handwork can be totally eliminated by their use, both in case of basic and auxiliary tehnological operations.

In contemporary industrial production, high automation of basic technological processes is characteristic, but auxiliary operations are hereby performed manually. These operations are monotonous, primitive and often strenuous, damaging, even hazardous. Nowadays, still a large part of human work effort is spent for these operations.

Practice has demonstrated that many auxiliary manual work operations cannot be automated by traditional means. Hence the realization and broad practical application of manipulation robots, i.e. multi-link spatial mechanisms with powered degrees of freedom were necessary. By action of the robot automatic control system (controller), manipulation mechanisms often perform movements that are similar to human arm motion.

Robot control systems are easily programmed to various kinds of operations. In that way, industrial robots represent multi-purpose machines, satisfying the contemporary demands of flexible production automation and the realization of economical technology and dignified work in production plants, mines, underwater and other environments.

Concerning the general classification of robotic systems, it is possible to point out the following broader classes:

- *manipulation robotic systems;*
- *mobile robotic systems;*
- *information and control robotic systems.*

Greatest development and practical application in industry has been attained by manipulation robotic systems of various types.

Mobile robotic systems are, in fact, mobile platforms, the motion of which is controlled by an automatic system. Hereby, beside the motion trajectory programme, they also possess automatic guidance to the desired point. They can be loaded and unloaded automatically. In production plants they are applied for automatic transport of parts and tools to the working machines and from them to the magazines. Manipulation mechanisms can be placed onto such mobile systems.

In agricultural production mobile robotic systems are automatically driven devices for field and plantation works, such as autonomous tractors. Mobile systems are also needed for work on the ocean bottom (Fig. 1.1) for serving oil and gas installations. Mobile robots can also be used for exploration in other unconventional and hostile environments, where human life is directly jeoparadized [2].

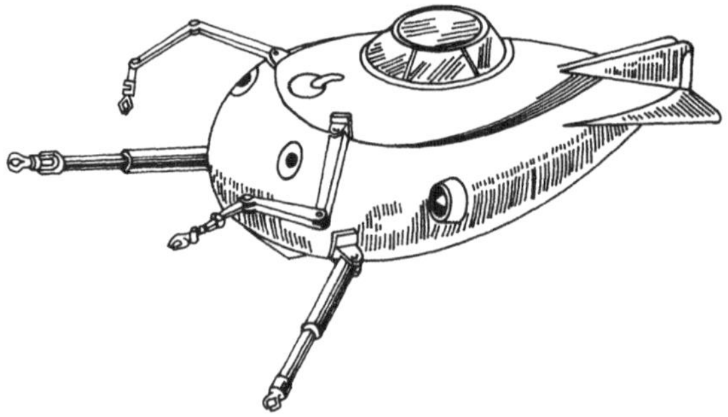

Fig. 1.1. Underwater robot

In the case of mobile robots various motion principles are used. Motion is realized by means of wheels, mechanical legs, tracks, they can be flying (pilotless aircraft), floating (crewless submarines) etc.

Information and control robotic systems represent complex measuring--informating and controlling devices, serving for acquiring, processing and transfer of data, as well as for their use in the forming of various control signals [2].

In production plants, these are automatic control systems in production processes practically without people, complexly mechanized with industrial robots used in groups (Fig. 1.2).

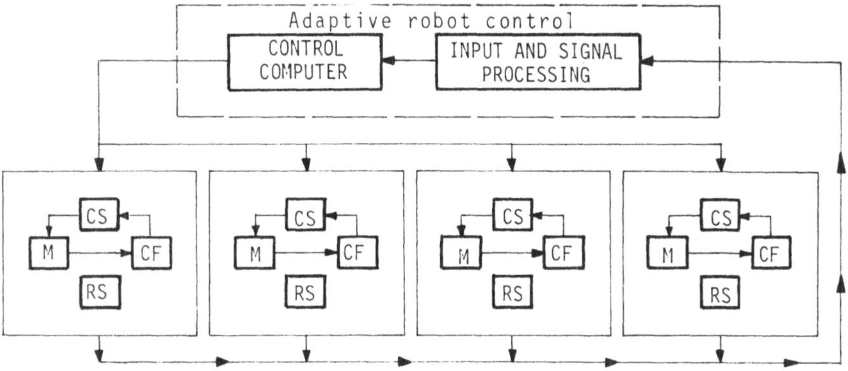

CS - Control System, M - Manipulator, CF - Control Feedback, RS - Robot Sensor

Fig. 1.2. Scheme of group robot control

In underwater conditions, these are devices equiped by measuring-informating and control systems and automatic cine-photo cameras for determining the properties of water and sea bottom (Fig. 1.3), for recognition of objects with automatic monitoring of informations, etc.

Fig. 1.3. Information underwater robot

Manipulation robot systems can be divided into three groups (Fig. 1.4):

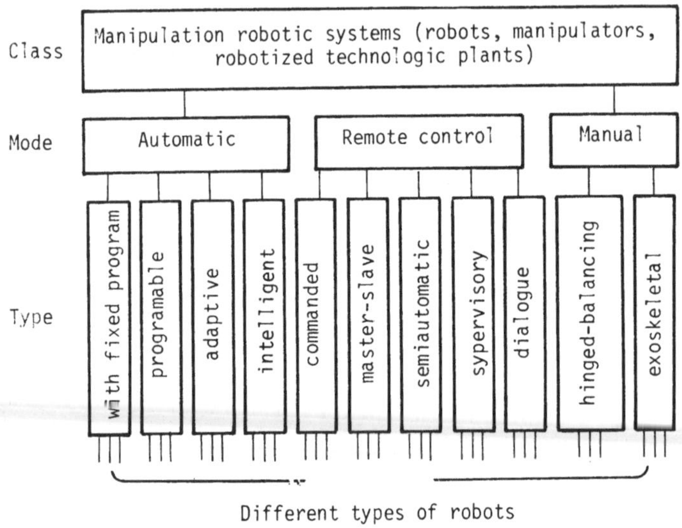

Fig. 1.4. Manipulation robots classification

1. Automatic robots, automatic manipulators and robotized technological plants;
2. Remote control robots, manipulators and technological plants;
3. Manually controlled, directly linked to the motion of arms, sometimes of legs of the human operator.

First ones are used mainly in industrial production (industrial robots and robotized complexes), and the second mainly in hostile conditions, i.e. those of nuclear radiation, air contamination, explosion danger, high or low temperatures and high pressure. Third robot group is applied for loading - unloading and heavy works [2].

Automatic manipulation devices are divided into four types: with fixed programmes, programmable, adaptive and "intelligent". Instead of the term "type", term "generation" is often used. As manipulators with fixed programmes cannot represent robots, they can be considered as the "zero" ("pre-robotic") generation. Thus the programmable robots are considered as the first, the adaptive as the second and the intelligent as the third generation. However, opposite to computer techniques, these generations do not supersede one another, but they co-exist. Hence fourth generation of robots does not exist and the artificial intelligence of the third generation can be developed in broad scopes in accordance with the development of science and technology and also with the possibilities of using newer microcomputer generations and software organization.

Let us characterize each of these generations of automatic robotic systems.

Manipulators with fixed programmes (Fig. 1.4) do not possess a programmable control system. These are mechanical arms. They are firmly linked to the technological equipment, subdueing themselves to a certain programme of the technological process. Their application is particularly characteristic for substituting manual work in mass production, e.g. on assembly lines of watch mechanisms and similar.

Programmable robots (first robot generation) have controlled driving units in all joints and their control system is easily adapted to various manual operations. However, after each adjustment these robots repeat same fixed programme in strictly defined conditions, with determined arrangement of objects. The majority of contemporary industrial robotic manipulators is of this type and they are applied for performing auxiliary operations in pressing, welding, machine tools, casting machines and similar. Such robots demand technological arrangement of the work environment and positioning of parts. This is not always feasible and, most important, "fixed" technological environment makes transferring of the robot to new operations difficult. Hence it is purposeful to make the robot control system more complex, i.e. pass over to a robot of second generation.

Second generation - adaptive robots are such robots that can orient themselves independently, to a higher or lesser degree, in the environment, which is not fully determined and to which they adapt. These robots are equiped with sensors, reacting to the situation and an information data processing system aimed at generating adaptive control signals, i.e. flexible changes in the manipulator motion programme according to the real situation. In such systems compact microprocessor systems are broadly used to-day. Adaptive industrial robots are needed in all cases, when it is difficult to ensure a strictly defined situation, when avoiding obstacles, working with parts on a moving track, in assembly operations, in arc welding, painting, applying protective layers and other operations. Second generation adaptive robots are more and more developed and used in production.

Third generation - intelligent robots possess diversified sensors with microcomputer processing of informations, recognition of situations, automatic generation of the solutions for further actions by the robot itself, in order to perform the necessary technological operations in an undetermined environment. These are robots with elements of artificial intelligence.

As already stated, robot generations do not supersede one another. Each of them is applied, where purposeful. It is natural that with new development of more reliable electronic and mechanical components the robots as complete systems will become more reliable and faster.

The general scheme of an automatic robotic system is presented in Fig. 1.5. The system can possess several manipulators and units of technological equipment, as well as transfer devices. Hereby the manipulators can be at various points of the technological plants and possess their individual control units, as well as a common control system.

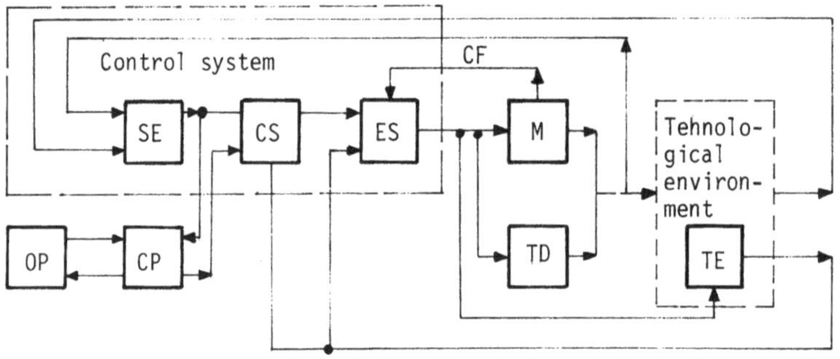

SE - Sensor Element, OP - Operator, CP - Control Panel, CS - Computer System, ES - Executive System, M - Manipulators, TD - Transfer Device, CF - Control Feedback

Fig. 1.5. General scheme of automatic robotic system

In Fig. 1.5. a basic organization of an automatic robotic system, containing the typical elements, necessary for its functioning is presented.

The second and third robot generations demand sensors for their work. With second generation robots force, tactile, location (ultra-sound) and similar sensors can be used. With third generation robots the presence of technical vision systems is characteristic, which together with the advanced microcomputer data processing forms the artificial intelligence itself, i.e. the behaviour of the robot in this case is more complete and corresponds to a certain degree to rational human behaviour in the process of working activity. Beside that, into the sensory device complex can be included means for production quality control, as well as properties of the outer world, in case required by the automatic control of the working regime.

In Fig. 1.4. the vertical lines indicate the different possible manipulation robot realizations. They differ by the principles and techniques of control system design, of the actuators in the manipulator joints, of the number of manipulator links, load mass, sensor type, mathematical programme support, etc. Here different types of industrial robots are presented: electromechanical (Fig. 1.6), electropneumatical (Fig. 1.7) and electrohydraulic (Fig. 1.8).

Fig. 1.6. Electromechanical robot UMS-1 (first Yugoslav industrial robot 1978, designed at the Mihailo Pupin Institute)

In production processes the robot, performing some determined task, is coupled into a unique system of the corresponding technological equipment. Hence, it should be regarded as an element of a complex automated

technological line. Hereby the task is usually not being solved by one robot but by a group of them in a unique technological system (flexible production cell or line).
le production cell or line).

Fig. 1.7. Electropneumatical robot

more arbitrarily with respect to the other production equipment (on the floor, ceiling-hanging, on the wall, etc). Analogous classification can be applied to robotized technological plants, too. They can be firmly programmed or adaptive to changes of external conditions. This represents the basis for the realization of flexible automated complexes, practically factories without workers.

Remotely controlled robots and manipulators are divided into five types according to the classification scheme (Fig. 1.4): direct control manipulators, master-slave manipulators, semi-automatic manipulators, robots with supervisory control, robots with dialogue (interactive) control.

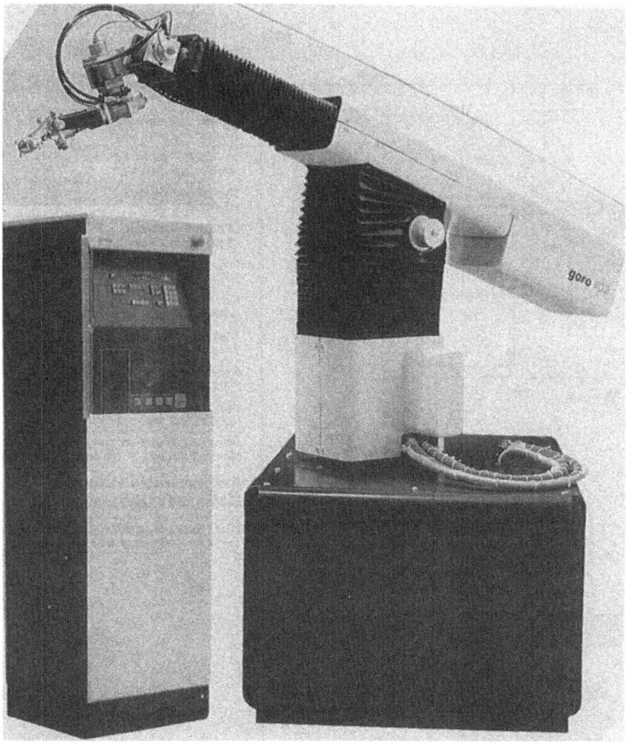

Fig. 1.8. Spray-painting electrohydraulic robot GORO-102*)

Only the last two types were denominated robots because they possess, beside remote control, fully automatic work regimes.

Direct control manipulators are defined by the fact, that the human operator remotely switches on the actuator of each manipulator joint pressing the corresponding button. By such command regime teaching of industrial robots is often performed from the command panel. The so-called teleoperators for hazardous work areas are operating analogously.

Master-slave manipulators, being in hazardous area, are controlled remotely by the human operator from a distant safe position by means of a device, which is kinematically similar to the manipulator itself (Fig. 1.9b). Hereby motion of each command manipulator joint is transferred to the corresponding joint of the working (executive) mechanism (manipulator) using the tracking system principle. Such manipulators

---

*) GORO-102 - yugoslav robot designed in cooperation between Jozef Stefan Institute (Ljubljana) and appliances enterprise GORENJE (Velenje) and Mihailo Pupin Institute (Beograd).

are used in nuclear environment, contaminated atmosphere and other hostile conditions.

Semi-automatic manipulators, compared with the master-slave, use a multifunctional joy-stick on the command panel of the operator as command mechanism, the kinematics of which can be arbitrary to suit small movements of the human hand. The electric signals from the joy-stick are transformed by means of a special-purpose computer (Fig. 1.9c) into the input command signals of the manipulator actuators. Different control algorithms are possible [2].

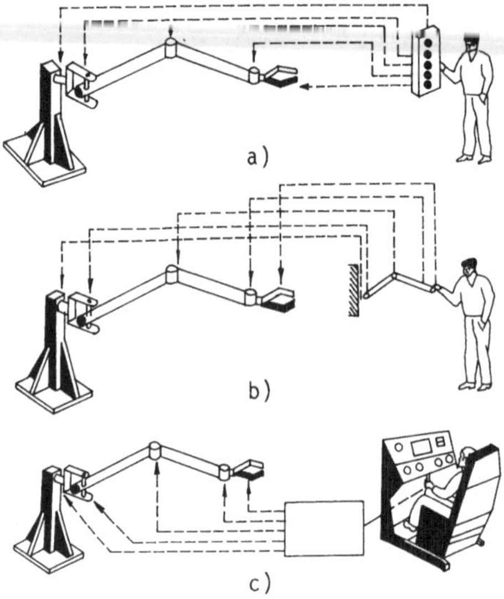

Fig. 1.9. Remotely controlled manipulator

Robots with supervisory control (Fig. 1.4) are defined by the feature that all elements of the operations they perform are pre-programmed and can be generated automatically. The human operator, supervizing from a distance the robot work, which is in hazardous zone, sends only individual command signals of the goal, according to which the individual programmes of automatic robot work are switched on. Only the function of environment (scene) recognition and decision making is left over to the human operator. After emitting the goal command by the operator, the robot performs further activities according to a particular programme. If the robot is adaptive, the human operator can give rarer and more global commands.

It should be mentioned that robots with such control mode can be encountered more and more as real means, as special vehicles operating under specific conditions, when it is necessary to withdraw the man out of the zone of direct vulnerability.

Robots with combined control are robots, with which the automatic regimes (similar to robots with supervisory control) are combined with manual-control regimes (as with semi-automatic or master-slave manipulators). They are applied in the case of underwater pilotless devices, in explosive environment, in mines without human workers, nuclear power plants, etc. Such combined control is used in the case of various teleoperator types.

Robots with dialogue (interactive) control by the rule (but not obligatorily) are intelligent and differ from the supervisory ones by the fact, that they do not receive only human commands, but they themselves participate actively in scene recognition and decision making, in that way helping the human operator [2].

Function scheme of the supervisory and dialogue (interactive) robotic systems is presented in Fig. 1.10. A joy-stick is foreseen on the command panel to enable the operator to take over the control of manipulator motion in a semi-automatic or master-slave regime if necessity arises. Such granting of the system safety by means of combinations of various remote control principles is necessary, because man himself cannot enter the hazardous zone in which the robot is working.

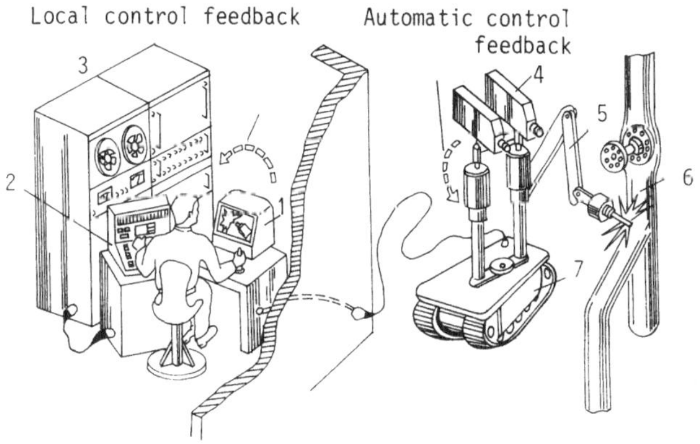

1 - display, 2 - control panel, 3 - control computer,
4 - sensors, 5 - manipulators, 6 - working object,
7 - on board computer

Fig. 1.10. Functional scheme of the interactive control systems of robots (supervisory, dialogue, combined)

Such robot functioning scheme represents the most complex control system and yields the broadest possibilities for introducing adaptive mobile systems in mostly diversified working conditions.

Finally, third mode of (manual) robotic manipulation systems, divides into hinged-balancing and exoskeletal (power amplifiers of human extremities).

The hinged-balancing manipulator (Fig. 1.11) represents a multi-segment mechanism with drives in some of the joints, which is in an equilibrium state with arbitrary load and configuration (in the scope of its possibilities). Hence the operator can move a big load easily. By moving the control stick, the operator generates control signals, whereby all load transferring work is done by the drives in the manipulator joints. Such manipulation system are very suitable for loading-unloading operations with heavier loads. Such systems are relatively simple and are widely used [2].

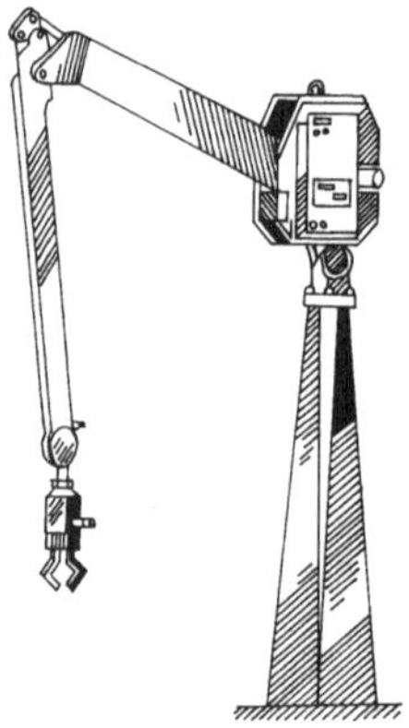

Fig. 1.11. Hinged-balancing

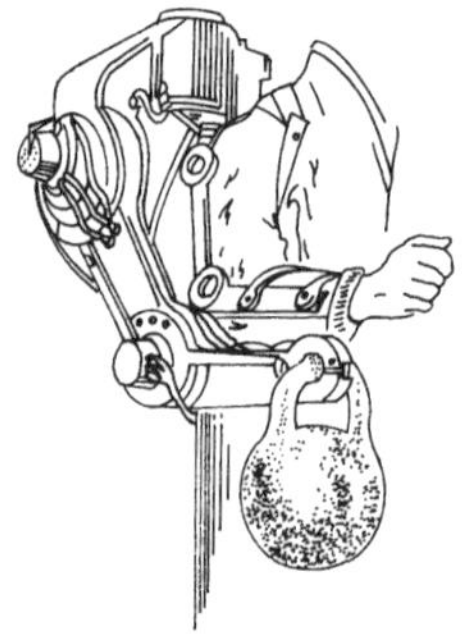

Fig. 1.12. Arm exoskeleton

Exoskeletons are multi-segment mechanisms, the segments of which are directly connected to the segments of the human arm (Fig. 1.12) or human legs. In the exoskeleton mechanism joints are also servo-controlled drives, bearing the whole workload. Movements of the human extremity only form the control signals. Such systems are applied for amplifying the power of the human extremities (and of his body), both healthy and handicapped.

In scope of the activity in the rehabilitation field, in the "Mihailo Pupin" Institute - Beograd, the first in the world exoskeletons were developed, dedicated to restoring locomotor and later also manipulation activities of the handicapped persons. A few realizations of active exoskeletons are presented here.

In Fig. 1.13. one of the first exoskeleton prototypes dedicated to paraplegics is shown, while in Fig. 1.14. a complete electrically driven exoskeleton is presented. In Fig. 1.15. the exoskeleton type electronic arm for dystrophics in advanced phase of the illness is presented.

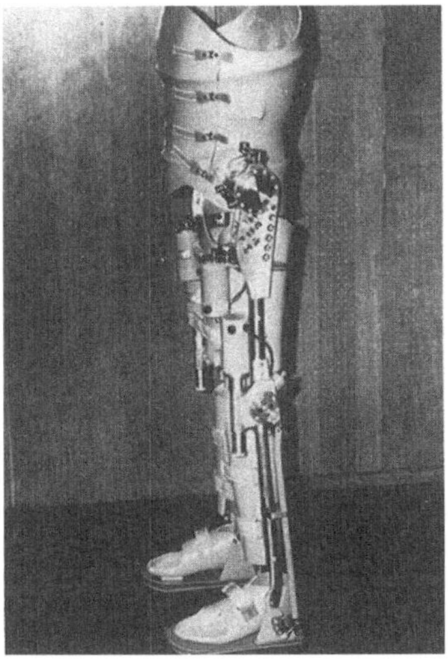

Fig. 1.14. Electrical exoskeleton for paraplegics (1974)

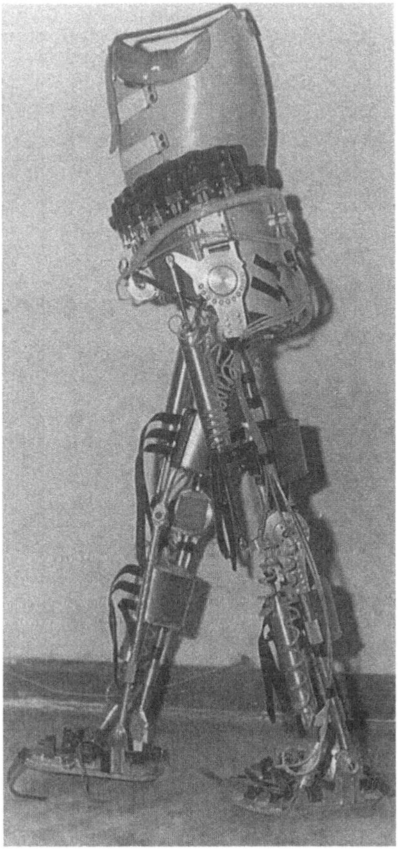

Fig. 1.13. Pneumatic exoskeleton for paraplegics (1971)

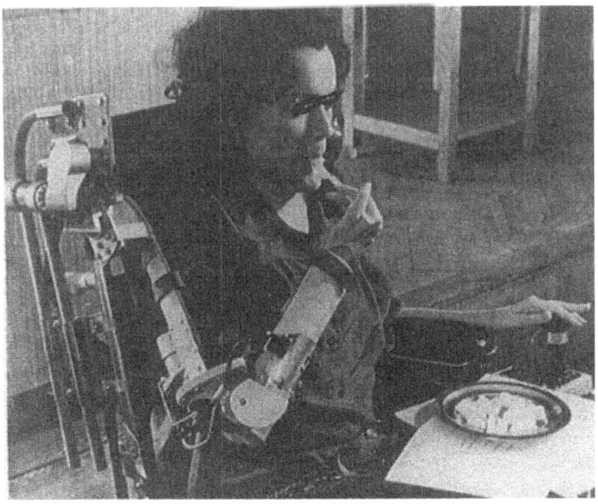

Fig. 1.15. Electronic arm for dystrophics (1981)

1.2 General Features of Robotic Mechanisms and its Classification

According to mechanism theory, active mechanisms in robotics are complex kinematic chains of variable structure, having a great number of members, some of which can be of variable length, with powered and controlled degrees of freedom.

According to control theory they are complex, nonlinear, multivariable dynamic systems. Active mechanisms can be divided according to the number of kinematic chains into:

- simple (consisting of a single kinematic chain),
- complex (comprising a number of simple chains).

According to their form, simple kinematic chains can be open or closed. Complex chains may be classified as:

- branched (comprising only simple open chains),
- combined (comprising both open and closed chains).

Depending on the kinematic constraints imposed on their end members, active mechanisms may be divided into:

- open mechanisms
- closed mechanisms (connected by kinematic pairs to the fixed base).

Members of active mechanisms are interconnected by kinematic pairs. There is no difference between kinematic pairs of active and "classical" spatial mechanisms with "kinematic drives". The difference is in the powered degrees of freedom, driven by actuators.

Kinematic pair consists of two connected members of mechanism enabling relative motion. The types of connection between mechanism members in kinematic pairs are different. The class of a kinematic pair is determined by the number of constraint conditions on the connections concerning relative motion of robot members. In the table (Fig. 1.16) kinematic pairs are arranged into five classes according to the number of relative motions (d.o.f). Kinematic pairs of the fifth class have one d.o.f. and the pairs of the first class have five d.o.f. of relative motion. Beside being partitioned into classes, kinematic pairs are divided into types, depending on the number of relative rotations within the scope of the total number of d.o.f. in the joint. Pairs of the first type allow the maximum number (3) of relative rotations, pairs of the second type two rotations, and pairs of the third type only one relative rotational motion. Beside pairs in which relative motions of members are mutually independent, there are pairs with interconnected motion. The simplest example is the screw-nut kinematic pair, in which the linear and rotational motions are linearly dependent.

Kinematic pairs of various classes are presented in the table on Fig. 1.16 [4].

In kinematic schemes, symbolic presentation for various kinematic pairs is adopted (Fig. 1.18). Kinematic pairs form kinematic chains. In Fig. 1.17. one example of spatial mechanism of the steering wheel is presented. The mechanism consists of four kinematic pairs, each link belonging to two kinematic pairs. One element of mechanism (in the fig. member 1) is fixed. This one is usually designated as the basis. The kinematic chain presented in Fig. 1.17. represents a simple closed chain. In a closed chain, each member belongs to two kinematic pairs [4].

Another example of spatial mechanism is presented in Fig. 1.19. This is a mechanism of a 5 degrees-of-freedom manipulator. The chain consists of six kinematic pairs, but the last member, differing from the previous example, belongs to one kinematic pair only. This is an example of a simple open kinematic chain [4].

In the examples considered, each mechanism member entered at most into two kinematic pairs. However there are mechanisms, in which one member can belong to several pairs. In Fig. 1.20. the scheme of an anthropomorphic robot (exoskeleton) is presented. The model contains twelve moving members connected into III class pairs of the I type. In this mechanism members 4 and 8, representing the robot body, belongs each to three kinematic pairs [4].

In the theory of machines and mechanisms, kinematic chains are classified as simple or complex, complex chains being formed by several

class of pairs	number of constr.	number of d.o.f.	TYPES OF PAIRS: I TYPE			II TYPE			III TYPE		
I	1	5	mov. number	rot.	trans.						
			allowed	3	2						
			restricted	0	1						
			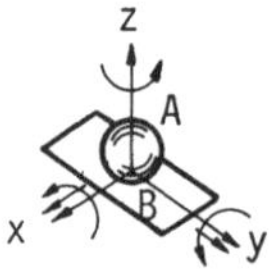								
II	2	4	mov. number	rot.	trans.	mov. number	rot.	trans.			
			allowed	0	1	allowed	2	2			
			restricted	0	2	restricted	1	1			
			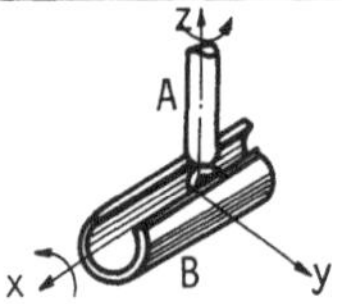			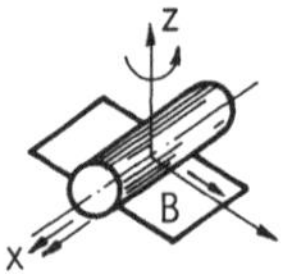					
III	3	3	mov. number	rot.	trans.	mov. number	rot.	trans.	mov. number	rot.	trans.
			allowed	3	0	allowed	2	1	allowed	1	2
			restricted	0	3	restricted	1	2	restricted	2	1
						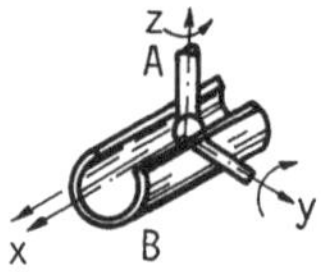			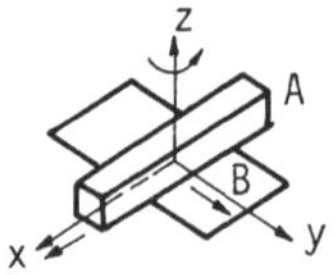		
IV	4	2	mov. number	rot.	trans.	mov. number	rot.	trans.			
			allowed	2	0	allowed	1	1			
			restricted	1	3	restricted	2	2			
			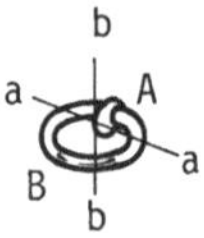			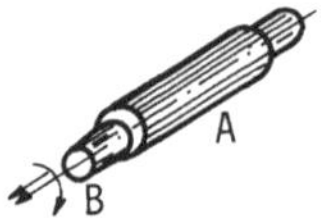					
V	5	1	mov. number	rot.	trans.	mov. number	rot.	trans.			
			allowed	1	0	allowed	0	1			
			restricted	2	3	restricted	3	2			
						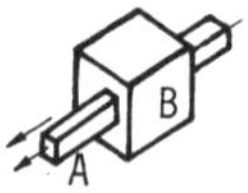					

Drawing labels:

Fig. 1.16. Table of kinematic pairs

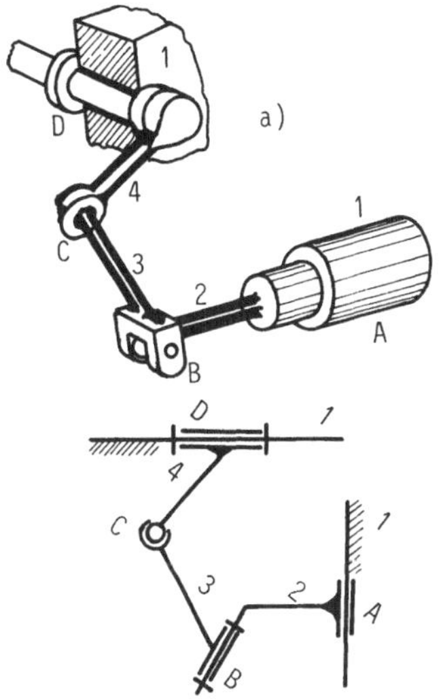

Fig. 1.17. Steering wheel mechanism

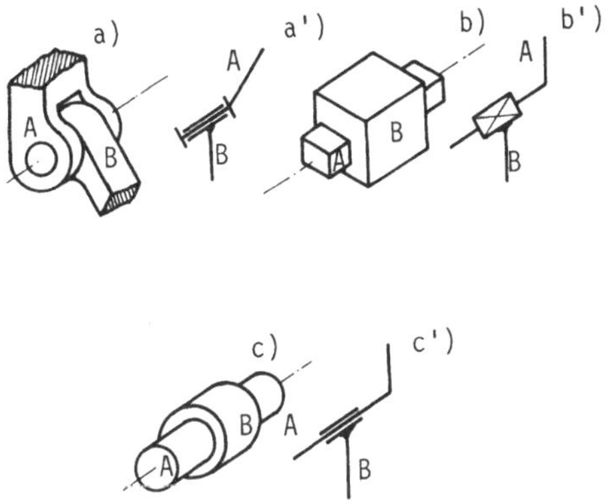

Fig. 1.18. Presentation of kinematic pairs

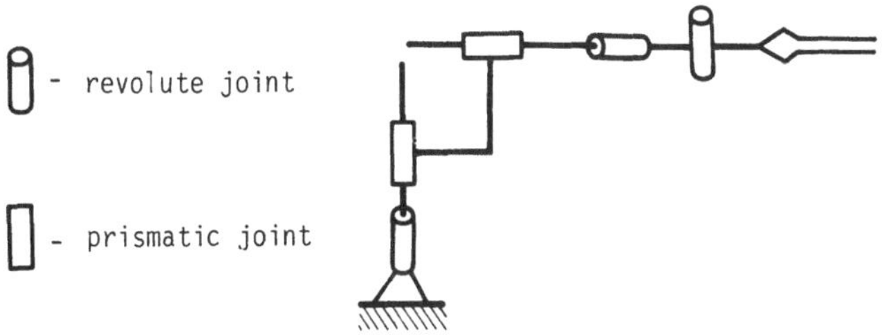

Fig. 1.19. One mechanism of "telescopic" manipulator

simple ones. Simple kinematic chains can be open or closed. In a closed chain, each member belongs to two kinematic pairs, while in an open chain, the last member belongs to one kinematic pair only. With complex kinematic chains, the individual members belong to three or more kinematic pairs. Due to more and more complex robotic mechanisms, finding their use in locomotion-manipulation mechanisms, it is necessary to consider in more detail the properties and possibilities of open and closed kinematic chains, by means of which complex spatial motions can be realized. The kinematic chain is closed when its terminal members are connected by means of kinematic pairs to one (or more) member(s), which can be: fixed (support), a member of another kinematic chain or a member of the initial chain. It should also be emphasized that in the course of working the active mechanism chains (either of manipulators or locomotion machines)

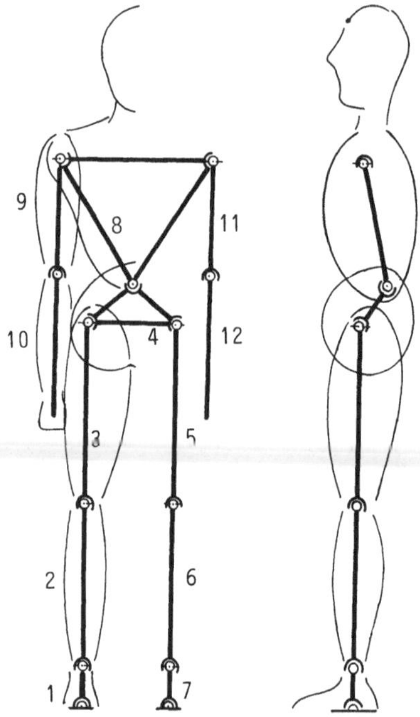

Fig. 1.20. Scheme of anthropomorphic mechanism

change their configuration once or several times from open to closed or vice versa. The kinematic chain of the manipulator is open during its motion through the working space but during execution of the operation itself (e.g. insertion or screwing in) it becomes closed. The mechanism of the locomotion biped is open during the swing phase of the step (when one foot is not on the ground) but in the double support phase becomes closed (Fig. 1.21). However, during the single support phase the anthropomorphic mechanism can also possess two configurations. The "foot" can rotate around its edges (Fig. 1.22a, b). The corresponding kinematic schemes are given in Fig. 1.22c, d. As can be seen, when the foot is supported alternately on one and then on the other foot edge, the position of hinge "O" changes abruptly.

In addition, because of permanent changes in its position, this joint cannot be equipped by a corresponding drive (actuator). On the other hand, the change of the coordinate q_o is exceptionally important because with greater values of q_o the system becomes statically unstable (it overturns). This feature of the anthropomorphic mechanism creates a special control problem because the uncontrollable degree of freedom must be controlled by means of other d.o.f. Such anthropomorphic mechanism also has the corresponding kinematic constraints at each joint in order to imitate the anthropomorphic patern. In addition, the mechanism of the locomotion biped (differing from the manipulator) is connected to the support surface by means of the frictional force only. Thus, the mechanism of an exoskeleton demonstrates a variable structure, the presence of an uncontrollable d.o.f., kinematic constraints and an essential influence of the frictional force.

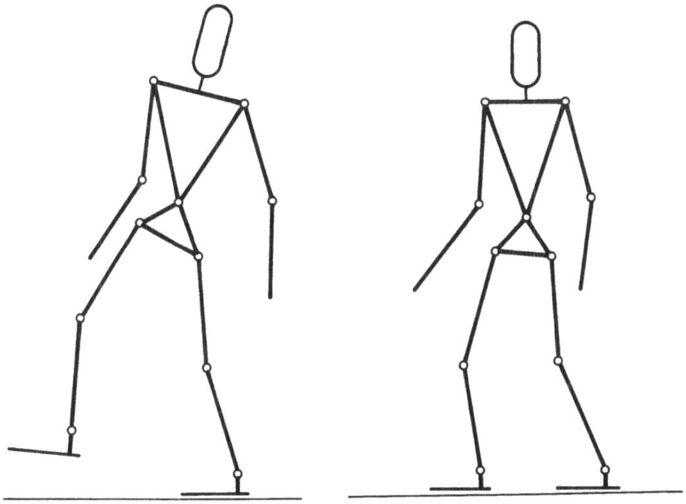

Fig. 1.21. Anthropomorphic locomotion mechanism

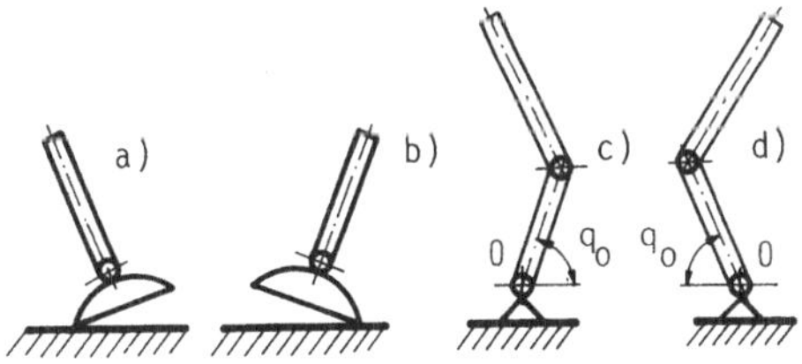

Fig. 1.22. Scheme of the uncontrollable d.o.f. of foot

The variable structure of active mechanism chains presents an essential difference from the classical spatial mechanisms, the structure of which does not change during work. A second difference is in the number of d.o.f. In spatial mechanism with driving member, number of d.o.f. is less often greater than two and the execution of the working operation is achieved in advance by synchronizing motion of the working members. The number of d.o.f. of robotic systems is considerably greater (up to ten and even more), so drives in the joints are indispensable. The desired motion is achieved by the action of the torques and forces of these motors only (during working operation). On the one hand this allows an exceptional adaptability of these mechanisms to the working environment and various tasks (which in the case of classical type mechanisms is impossible). On the other hand, it imposes exceptional difficulties in realizing control because some of the mechanism d.o.f. appear as redundant.

As already stated, robots represent active mechanisms of variable structure. For instance, one manipulator can, during its work, change the group it would belong to, according to the given classification. We will illustrate this change of structure by considering an example of an industrial manipulator in the course of inserting a cylindrical working object into a hole (Figs. 1.23. to 1.25). At first (Fig. 1.23a) the manipulator has an open kinematic scheme as in Fig. 1.23b (simple open chain). In the phase of transferring the working object (Fig. 1.24a) the kinematic chain does not change (Fig. 1.24b) but the last member (now the gripper and object together) changes its dimensions and mass, which cause the change of dynamics. Finally, in the phase of object insertion (Fig. 1.25a), the kinematic scheme of the manipulator also changes and it becomes a simple closed kinematic chain (Fig. 1.25b).

Mechanisms of legged locomotion machines are, as a rule, complex kinematic chains. Fig. 1.26. shows an arbitrary, complex kinematic chain comprising four simple chains, the first three (formed by the members 1-6) being closed and the fourth (formed by the members 7 and 8) being open. The kinematic chains connected to the support are basic chains, while the chains connected to them, but not by means of the support, are satellite chains (satellites). The procedure of separating one complex chain into a number of simple ones is hereby practically defined. The notion of an independent kinematic chain may be defined at this point. A kinematic chain is said to be independent if its motion is independent of the satellite chains, with respect to the support, e.g., if its last members are connected to the support. This means that a basic chain is independent. In that way only one autonomous chain is obtained in each complex linked mechanism with all remaining chains being satellites in respect to the autonomous one.

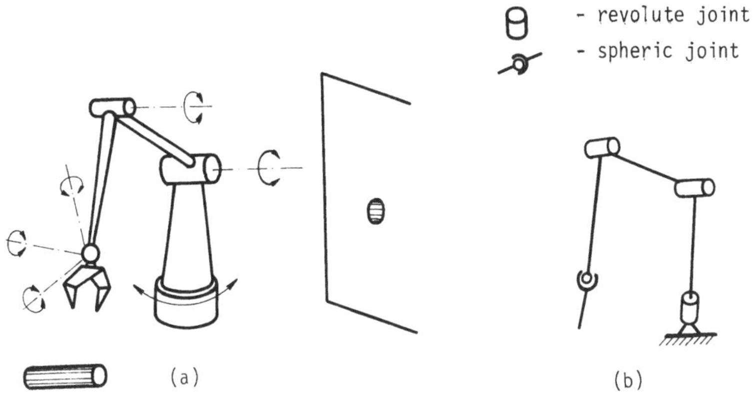

Fig. 1.23. Manipulator before grasping the object

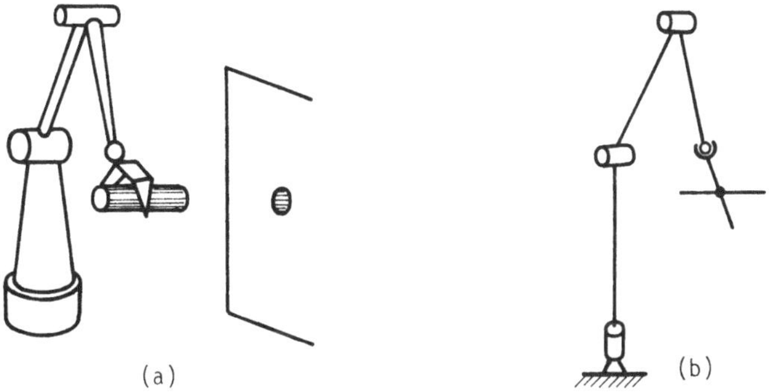

Fig. 1.24. Phase of working object transfer

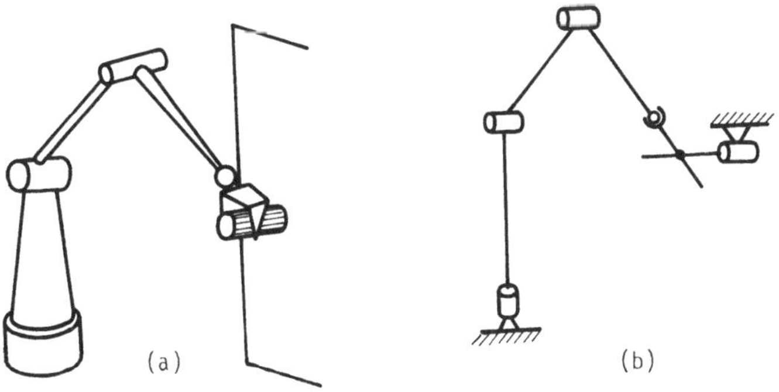

Fig. 1.25. Phase of object insertion into the hole

Taking the model of a human presented in Fig. 1.20. as an example, it is possible to isolate the chains of "legs", "body" and "arms"; here, the chain consisting of "legs" is independent since the chains of "body" and "arms" impose no kinematic constraints on it. For the mechanism presented in Fig. 1.26, if the motion of member 1 is known, the chain I is independent, all the remaining ones being satellites. The

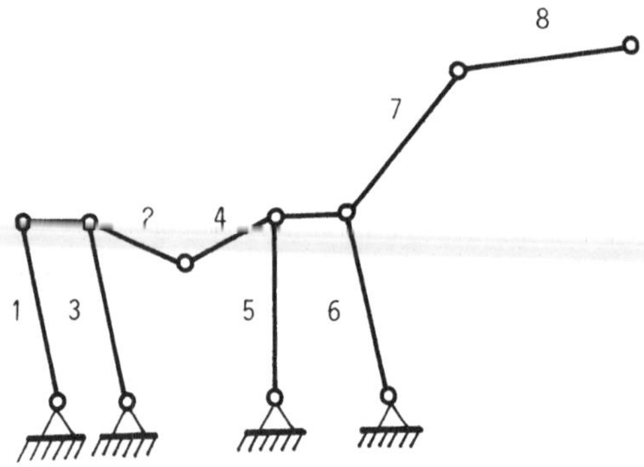

Fig. 1.26. Complex kinematic chain

sequence to be followed in performing kinematic and dynamic analyses has thus been determined; namely, the basic independent chains should be analyzed first, and the satellite (guided) chains second. Figures 1.27 and 1.28 illustrate the kinematic chains (mechanical configurations) of six-legged and four-legged walking machines, respectively,

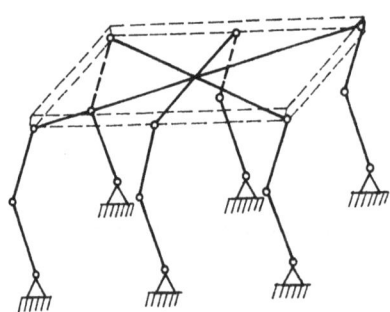

Fig. 1.27. Mechanism of a six--legged locomotion machine

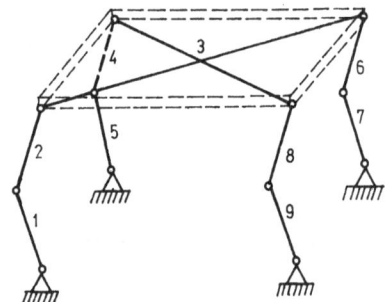

Fig. 1.28. Mechanism of a four--legged locomotion machine

without presenting the foot-to-ground connection realized by kinematic pairs. If the global kinematic constraints imposed on the mechanism and the relative motions of some mechanism members are known, it is possible to define the autonomous chain and its satellites.

1.3 Specifities of Manipulation Robots

1.3.1. Definition of position of an object in space

In order to define the position of an object in space, it is possible to fix positions of three non-colinear points of that object, or nine parameters, which are not independent, because the coordinates of these points are connected by three relations which express the invariance of the distances between them.

Consequently, definition of the position of a free object requires, in the general case, the knowledge of six independent parameters:

- three independent parameters which define the position of one point of the object (in Cartesian, cylindric, spheric, or other coordinates),

- three independent parameters which define the orientation of the object around the stated point (Euler's angles, Euler's parameters, or similar).

By definition, this object possesses six degrees of freedom.

1.3.2. Structure of an industrial manipulation robot

The mechanical structure of an industrial manipulation robot is an assembly of theoretically rigid bodies interconnected by joints. A joint (connection) exists between two bodies and permits relative motion between them (Fig. 1.16).

The number of degrees of freedom (d.o.f.) of a joint equals the minimal number of parameters which determine the position of a body B_2 in its relative motion to a body B_1 (B_1 - body No 1, B_2 - body No 2).

Class of a connection is the supplement of the number of degrees of freedom to six.

All sorts of connections are not used in industrial manipulation robots, because of the actual state of the development of actuators which perform the relative motion of B_1 and B_2.

Thus it is sufficient to consider furtheron:

- connection enabling relative rotational motion,
- connection enabling relative translatory motion.

These connections are of class 5.

In order to standardize the terms with those in international literature, we adopt (R) for rotational connections and (P) for prismatic - translational ones (the term "linear" is also used).

1.3.3. Disposition of segments and their connections

Bodies B_i and connections C_i, which form a manipulation robot, can be arranged starting from a reference body B_o, fixed or mobile, in the following manner:

- arrangement in the form of a simple chain (Fig. 1.29),
- arrangement in the form of a branched chain (Fig. 1.30),
- arrangement in the form of a complex chain (Fig. 1.31).

The last type of structure is characterized by the presence of "mechanical loops" enabling the terminal device to perform different and complex paths in the service zone of manipulation robot.

In the following text of this chapter, only manipulation robots with simple chain structure will be considered.

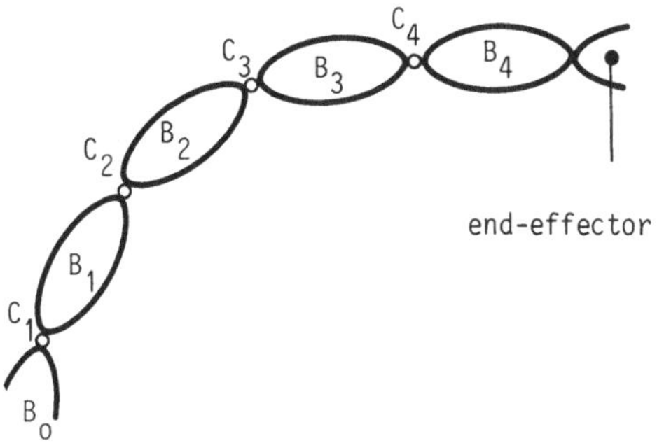

Fig. 1.29. Manipulation robot with a simple chain structure

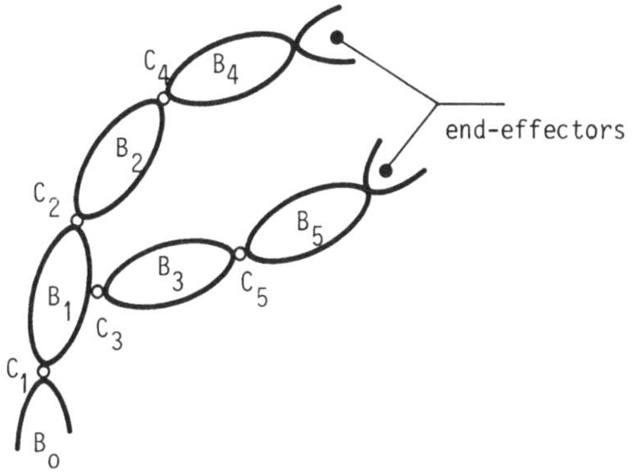

Fig. 1.30. Manipulation robot with a branched chain structure

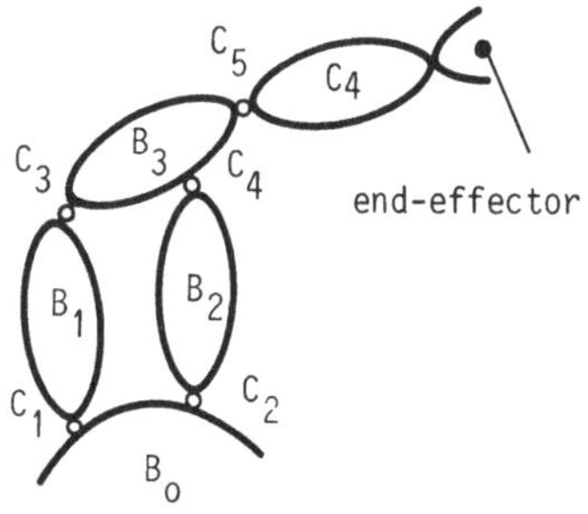

Fig. 1.31. Manipulation robot with a complex chain structure

1.3.4. Simple chain structure types

According to previous statements, the manipulation robots have a simple structure, characterized by a set of n symbols R or P which define the sequence and type of connections, starting from the reference body to the terminal device.

Table T.1. presents a few examples of structure.

Manipulation robot	Structure
Cincinati Milacron T3	R R R R R R
Unimate 4000	R R P R R R
Unimate PUMA 560	R R R R R R
ASEA IRb 60	R R R R R R
IMP*) UMS-3**)	R P P R R R

Table T.1. Some examples of structures of industrial manipulation robots

1.3.5. Mobility index and degrees of freedom of a manipulation robot

Except for the mobile robots, the reference body B_o is fixed and called the base, while bodies B_1 to B_n are mobile. Positions of these n bodies are defined by means of 6n parameters.

A class m of connection, between two bodies, defines m of these parameters. Therefrom follows, that, if the manipulation robot possesses N_m connections of class m

$$M = 6n - \sum_{m=1}^{5} m\, N_m$$

which represents the number of variable parameters, determining its configuration.

M is the *mobility index* of a manipulation robot.

Although the manipulation robots with complex chain configurations have some advantages in comparison with these with a simple or branched chain structure, e.g. greater rigidity, almost all industrial manipulation robots have a simple chain structure. Under these conditions, if all connections are rotational and/or prismatic, the general relation for the mobility index of these manipulation robots reduces to:

$$M = n$$

---

*) IMP - "Mihailo Pupin" Institute, Beograd, Yugoslavia.

**) UMS - common name for the prototype robot series, designed at the "Mihailo Pupin" Institute, Beograd, Yugoslavia.

Thus, as six parameters are necessary for the definition of the position of the terminal device in space, a manipulation robot of a simple chain structure possesses at least six rotational and/or prismatic connections.

The number of degrees of freedom of a manipulation robot (considered as a mechanical system) equals the number of independent parameters defining the position of the chain. The number of d.o.f. of the terminal device is less or equal to M.

Example 1

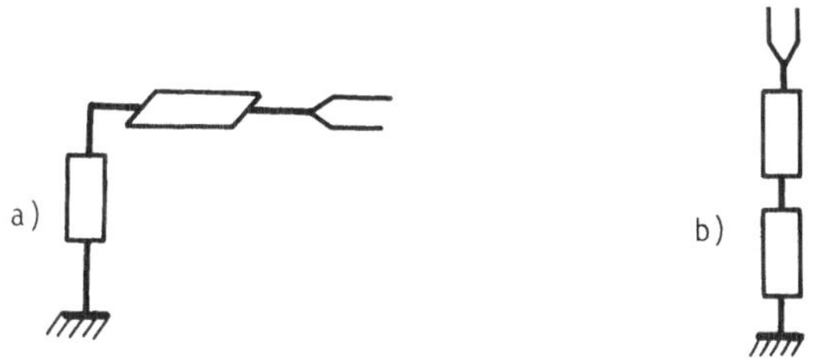

Fig. 1.32. Mobility index and degrees of freedom for two manipulation robots with two prismatic connections

The manipulation robot consists of 2 moving bodies and 2 prismatic connections:

with perpendicular axes	with parallel axes

mobility index M=2

degrees of freedom of the terminal device

d.o.f. = 2	d.o.f. = 1

because the terminal device moves

parallel to a plane	parallel to a straight line

keeping a constant orientation

Example 2

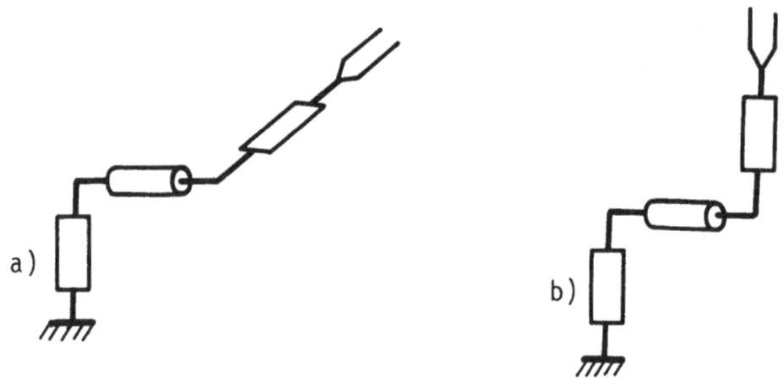

Fig. 1.33. Mobility index M and d.o.f. for a manipulation robot with three connections

The manipulation robot consists of 3 moving bodies, 1 rotational and 2 prismatic connections.

with perpendicular axes	with parallel axes
mobility index M=3	
degrees of freedom of the terminal device	
d.o.f. = 3	d.o.f. = 2
because the terminal device moves locally	
parallel to a plane with some arbitrary orientation	parallel to a plane with some imposed orientation

1.3.6. Redundancy and singularity

If number of d.o.f. of the terminal device equals M for any position of manipulation robot it is non-redundant.

This is the case of the manipulation robot from example 1(a).

In the opposite case, i.e. when d.o.f. < M, two situations are possible:

- this inequality is satisfied for all positions which the manipulation robot can take, and it is then defined as redundant (case of example 1(b)),

- this inequality is satisfied for some configurations which the manipulation robot can take, and it is then defined as locally redundant.

The corresponding configurations are called singular. This is the case of the manipulation robots from example 2(b).

1.3.7. Degrees of freedom of a task: (d.o.f.t.)

Number of degrees of freedom of a task (d.o.f.t.) equals the number of independent parameters enabling all the desired positions to be attained by the terminal device.

1.3.8. Compatibility

When d.o.f. = d.o.f.t., which is necessary, but not sufficient condition for the manipulation robot to perform some given task, the notion of compatibility expresses the possibility to find the configuration of the manipulation robot which enables it to attain the desired state of the terminal device.

1.3.9. Decoupling the orientation and the position of the terminal device

Let us consider the manipulation robot, the structure of which is presented in Fig. 1.34.

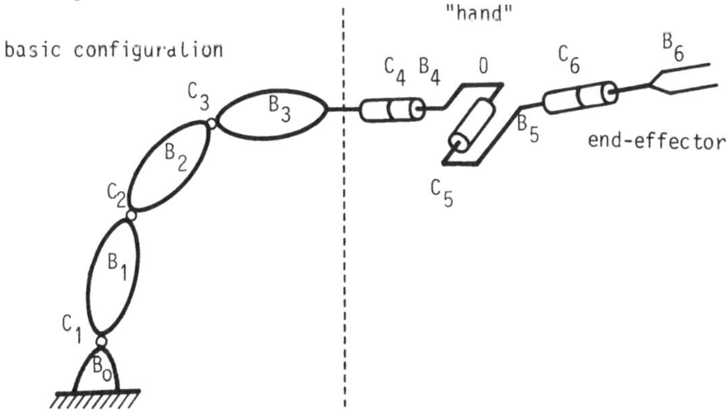

Fig. 1.34. Structure of a decoupled manipulation robot

It possesses six connections of class 5 and enables the following general situations:

- first three connections are arbitrary,

- last three connections are rotational with axes intersecting at point 0 and mutually perpendicular,

- position of point 0 depends only on the position of bodies B_1, B_2 and B_3,

- position of bodies B_4, B_5 and B_6 determines the orientation of the terminal device (last segment - gripper) with respect to point 0.

This decoupling is intended to reduce the problem of determining the six parameters of the manipulation robot configuration to two independent problems, each having three parameters only.

However, point 0 is less often the point of the terminal device that has to be positioned. Each change of its position in that way causes a change of its orientation and vice versa. Thus, the decoupling has not been solved quite generally.

1.3.10. Different minimal configurations

They are derived from different reference systems of a point in space.

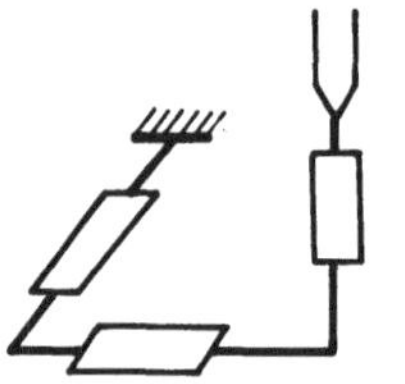

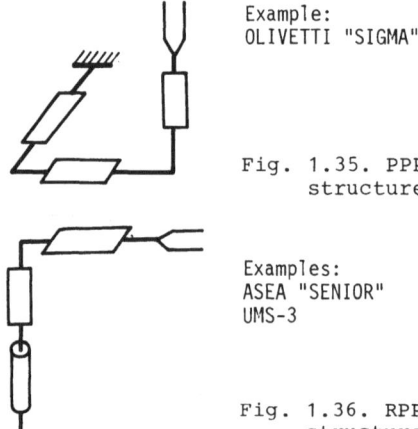

Example:
OLIVETTI "SIGMA"

Fig. 1.35. PPP - structure

This type of structure, used with approx. 14% of industrial manipulation robots, is well suited for determining the point 0 in Cartesian coordinates.

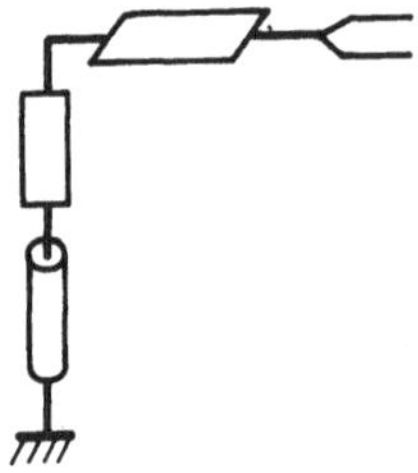

Examples:
ASEA "SENIOR"
UMS-3

Fig. 1.36. RPP - structure

This type of structure, used with approx. 45% of industrial manipulation robots, is well suited for determining the point 0 in cylindrical coordinates.

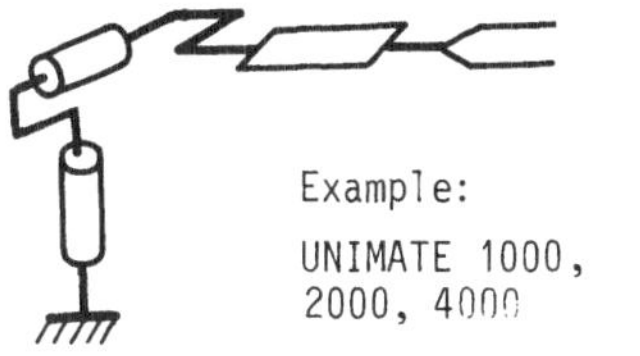

This type of structure, used with approx. 15% of industrial manipulation robots is well suited for determining the point 0 in spherical coordinates.

Fig. 1.37. RRP - structure

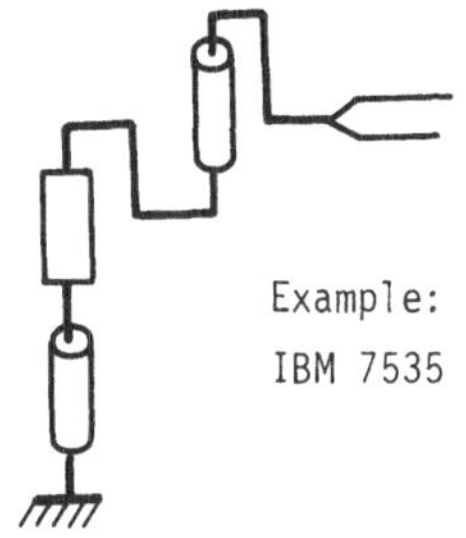

This type of structure, used with approx. 1% of industrial manipulation robots, is well suited for determining the point 0 in thorical coordinates.

Fig. 1.38. RPR - structure

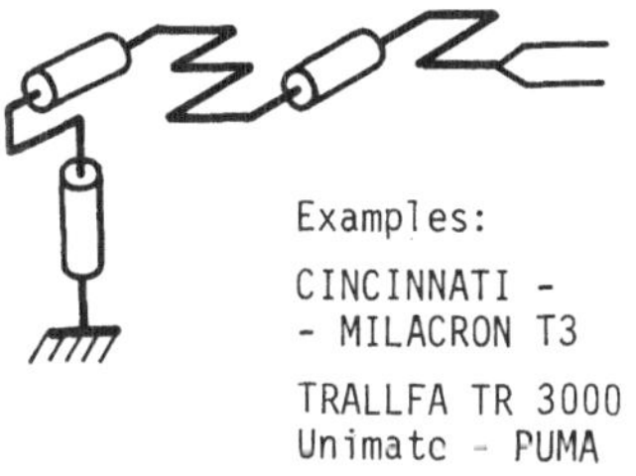

This type of structure, used with approx. 25% of industrial manipulation robots, is called arthropoid structure.

Fig. 1.39. RRR - structure

1.3.11. Workspace

Workspace of a manipulation robot is the space, physically swept by a point of the terminal device during motion of the robot configuration.

Let us note that the choice of the point mentioned above, is arbitrary. Some producers assume it as the center 0 of the wrist and thus obtain the nominal workspace. Others take the point on the tip of the terminal device.

In addition, the orientation of the terminal device does not appear in this workspace. But, although this is only an approximative characteristic of the manipulation robot performance, it anyhow permits comparison of different basic structures (minimal robotic configurations).

1.3.12. Comparison of the workspaces of different minimal configurations

We will make the following assumptions:

- permissible rotation of each rotational connection is $\approx 360^{\circ}$
- translation of each prismatic connection equals L,
- the "principal" (greatest) dimension of each segment of the manipulation robot equals L.

The workspaces illustrated suppose some arbitrary wrist, the center of which, 0, is the reference point.

(a) PPP structure

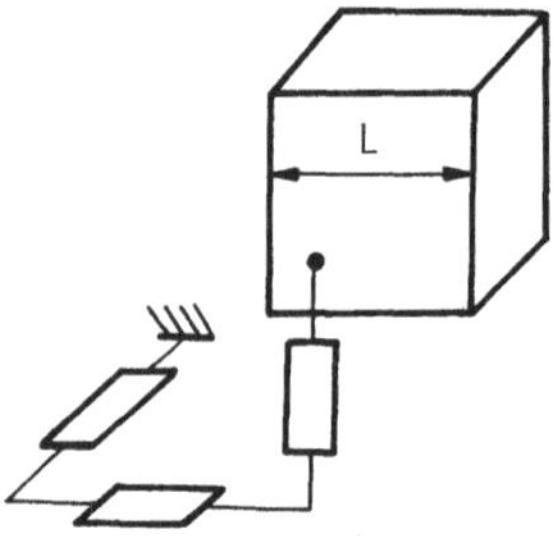

Fig. 1.40. Workspace of PPP - structure

Workspace is a cube of side L

$$V = L^3$$

(b) RPP - structure (or PRP)

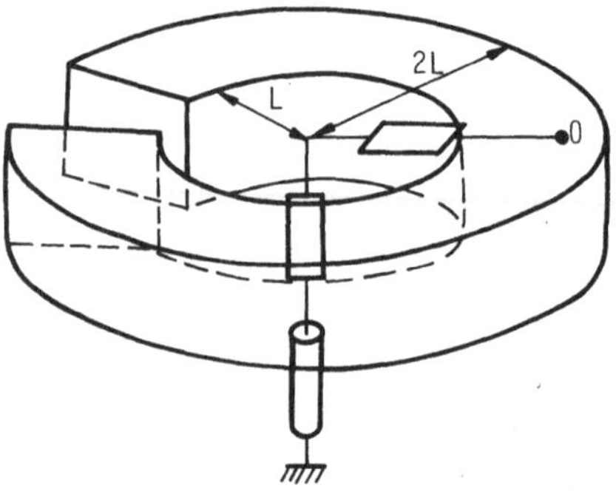

Fig. 1.41. Workspace of RPP - structure

Workspace is a hollow cylinder of square cross-section of internal radius L and external radius 2L

$$V \approx 3\pi L^3 \approx 9L^3$$

(c) RRP - structure

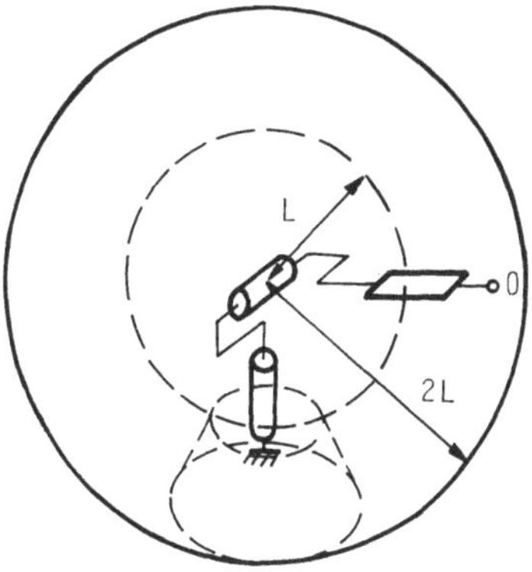

Fig. 1.42. Workspace of RRP - structure

Workspace is a hollow sphere of interior radius L and external radius 2L

$$V \approx \frac{28}{3}\pi L^3 \approx 29L^3$$

(d) RPR - structure

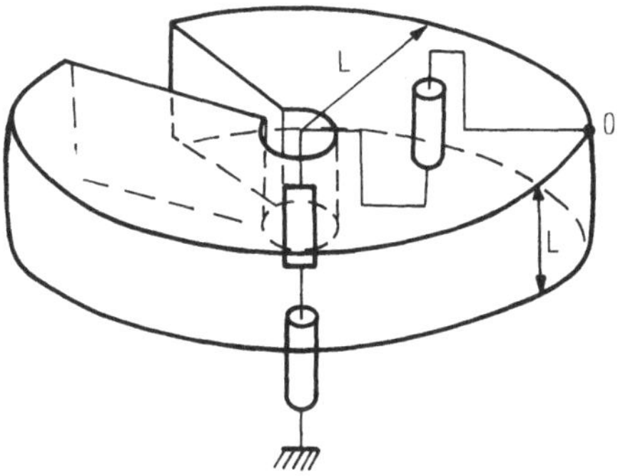

Fig. 1.43. Workspace of RPR - structure

Workspace is a cylinder of radius 2L and height L

$$V \approx 4\pi L^3 \approx 13L^3$$

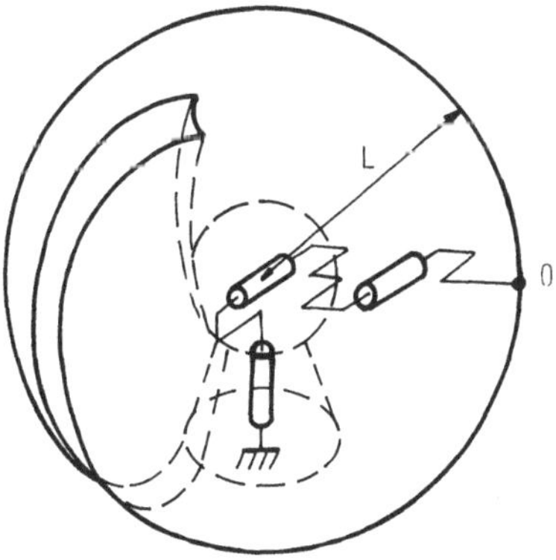

Fig. 1.44. Workspace of RRR - structure

Workspace is a sphere of radius 2L

$$V \approx \frac{32}{3}\pi L^3 \approx 34L^3$$

This comparison demonstrates the evident superiority of the RRP and RRR structures, possessing a workspace approximatelly 30 times greater than the PPP - structure. The RPP and RPR - structures, with workspaces approximatelly 10 times greater, thus offer medium sized workspaces.

References

[1] Vukobratović M., Potkonjak V., Applied Dynamics and CAD of Manipulation Robots, Monograph, Springer-Verlag, 1985.

[2] Robototehnika, edited by E.P.Popov and E.I. Jurievitch, Mashinostroenie, Moscow, 1984.

[3]*) Gorla B., Renaud M., Modeles des Robots Manipulateurs, Application a leur Commande, Cepadues-Editions, Toulouse, 1984.

[4] Stepanenko Yu., Dynamics of Spatial Mechanisms (in Russian), Mathematical Institute, Belgrade, 1974.

[5] Introduction to Robotics, edited by Vukobratović M., Springer-Verlag, 1988.

[6] Vukobratović M., Stokić D., Applied Control of Manipulation Robots: Analysis, Synthesis and Exercises, Springer-Verlag, 1989.

*) Text in 1.3.1-1.3.12 including figures was taken over from ref. [3].

Chapter 2
Computer Forming of Mathematical Model of Manipulation Robots Dynamics

2.1 General About Computer-Oriented Procedures for Forming of Mathematical Models of Robot Dynamics

Active spatial mechanisms are multivariable mechanical systems of exceptionally complex and variable structure. Their main features include, a large number of links (segments) and degrees of freedom of their relative motions as well as frequent alterations in their kinematic chain configurations.

Additionally, contemporary robotic mechanisms are required to move at higher speeds whilst still retaining so far attained degree of precision. At present the translational composite speeds of the manipulator tip, for medium range robots, exceeds 3 m/s while angular velocities of some degrees of freedom are in excess of 8 rad/s.

Numerous applications of manipulation robots may include, manipulation of various loads, assisting metal machining during a manufacturing process, spray-painting, electric-arc and spot welding, automatic assembly etc. Such applications imply that the kinematic configuration of a robot varies during a working cycle. These kinematic configurations are clearly recognizable within a working cycle and can range from manipulator motion towards an object, object manipulation according to specified requirements, such as positioning onto a mechaning tool or a conveyer within a system, to the assembly insertion process. During all phases of manipulating systems work the dynamical parameters change (payload, inertia moments, etc.) as well as boundary conditions (manipulating mechanism movement in free work space and appearance of dynamical reactions under constrained robot gripper movement in mechanical assembly and metal machining). It therefore implies that different mathematical models are required for the dynamics description of each particular phase during the manipulation of a robot, so that an accurate representation of the relevant working regime can be obtained. Furthermore, during the process of robot design, it is imperative that various mechanical configurations are analysed and that mathematical models of high fidelity for their study of dynamics are available, in order to enable a choice of appropriate manipulation mechanisms based on adopted criteria and given constraints.

Manual generation of mathematical models of complex spatial mechanisms leads to unavoidable errors. Transfering this tedious task to a computer oriented procedures for forming robot motion differential equations is a significant contribution to the efficient and adequate study of dynamics of robotic manipulators, their control system laws synthesis and corresponding microcomputer implementation.

In reaching the solution of the problem of forming equations of motion, differentiation of any analytic expression should be avoided, since numerical differentiation on a digital computer is considered quite unfavourable. Thus in the process of mathematical modelling of active spatial mechanisms, the ultimate aim becomes obtaining a sufficiently general algorithm for a digital computer. Such an algorithm would be required to [2, 3, 4]:

- assemble the mechanism in accordance to the adopted mechanical (kinematic) scheme,
- calculate positions, velocities and accelerations of the spatial mechanism link,
- form the differential equations of motion,
- integrate the formed equations in compliance with imposed specific conditions and on the basis of a iterative procedure or calculate the driving forces (torques) required to execute the desired motion.

The first operation of such an automatic procedure is therefore assembling the mechanism. In that case any alterations in the structure of the mechanism are easily modelled by a corresponding change of parameters contained in the input information. The remaining part of the procedure runs automatically until the constructing mechanism motions, or the calculation of their driving forces are finished. The automatic assembly of mechanisms is a feature of great importance in solving the delicate task of selecting appropriate kinematic schemes of the robotic mechanism, which, if done manually, would require a separate mathematical model, in analytical form, for every possible configuration of the mechanism. Consequently, the concept of automatic generation of mathematical models of robotic mechanism motions would be considered as an important step towards systematic evaluation of robot mechanism and the synthesis of the control algorithms in various tasks of applied and particularly industrial robotics.

The last fifteen years or so, saw the advent of powerfull methods for mathematical modelling of robotic mechanism. Systematization of methods can be achieved by adopting various criteria. However, it is customary that the adopted criterion is founded on the laws of mechanics according to which the method was formulated. According to this criterion, three basic groups of methods are distinguished. Namely:

1. Methods founded upon the second order Lagrange's equations,
2. Methods based on Newton-Euler's dynamic equations (general theorems of mechanics),
3. Methods based on Appel's equations.

The way of method program implementation can be adopted as another criterion for systematization. According to this criterion, basically two groups of computer-oriented methods can be identified:

1. Numeric methods,
2. Symbolic methods.

In the case of numeric methods each variable participating in forming the model is treated like a symbol to which only one numeric value is adjoined. On the contrary, with symbolic methods the variable is treated as a set of symbols, describing its dependence on the parameters and states (positions and velocities) of the system. Due to its convenience, only the method based on the general theorems of mechanics will be presented here, which belongs to the group of numeric-iterative methods. For studying robot dynamics and the evaluation of its dynamic properties, such method is completely satisfactory. Symbolic methods, i.e. the methods generating the mathematical models of robots in symbolic form are user-oriented methods, serving for the synthesis of the dynamic control laws and current microcomputer implementation. For the precise derivation of the mathematical model the definition of basic notions relating to the physical model of the robotic manipulator mechanism will be defined.

Basic assumptions relating to the physical model of robotic manipulators can be reduced to the following:

1. Mechanism links are modelled by rigid bodies,
2. The kinematic chain is neither branched nor closed,

3. There exists no kinematic coupling between degrees of freedom (no mechanical coupling exists between individual degrees of freedom of the robot mechanism).

Assumptions that were given above, are generally justifiable within the class of manipulation robots whereas they do not hold for the class of locomotion robots, with active multi-purpose platforms and complex mechanical systems. Therefore, in the text to follow the term robot will refer to a manipulation robot (robotic manipulator).

Let us consider the model of the robot mechanism shown in Fig. 2.1. The model consists of n rigid bodies which represent the links of the mechanism chain. These links are interconnected by prismatic (sliding) or revolute joints. For a precise derivation of differential equations of the robot dynamics, let us introduce a number of basic definitions that have already been briefly mentioned in Sections 1.2 and 1.3 of Chapter 1.

Configuration

Configuration of the mechanism is represented by an arranged n-tuple $(J_1,\ldots,J_n)$, such that for each $i \in N=\{1,\ldots,n\}$, $J_i \in \{R, P\}$, where R denotes a rotational (revolute) and P a prismatic (sliding) joint.

To illustrate this, the configuration RPP-RRR stands for a mechanism with 6 joints, where the second and third joints are sliding, and the remaining ones are rotational.

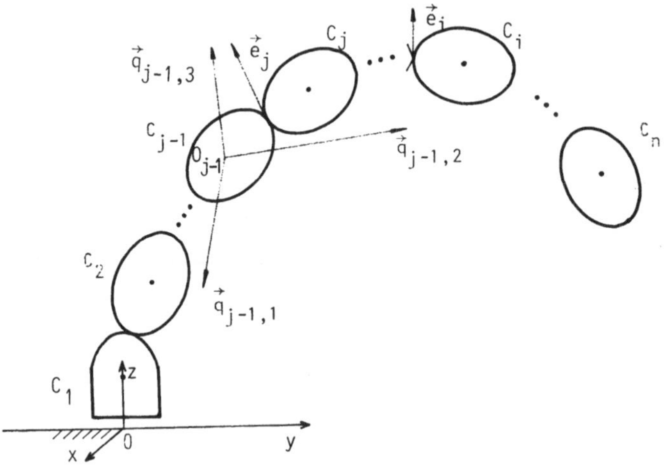

Fig. 2.1. Model of robot mechanism with n links and n joints

Link

A link is defined by an arranged set of parameters $C_i(K_i, \mathcal{D}_i)$, where K_i represents a set of kinematic, and $\mathcal{D}_i$ a set of dynamic parameters, and $i \in N$ - mechanism link index.

Sets K_i and $\mathcal{D}_i$ may be defined in various ways. In the basic Newton--Euler method, as well as in the method which will be developed in this section, these sets are assumed to have the form

$$K_i = (Q_i, \tilde{R}_i, \tilde{E}_i), \qquad \mathcal{D}_i = (m_i, \underline{\underline{J}}_i)$$

with $Q_i = (\vec{q}_{i1}, \vec{q}_{i2}, \vec{q}_{i3})$ - the internal (local) ortonormal coordinate system attached to the link i;

$\tilde{R}_i = \{\tilde{r}_{ik}^{*)}\}$ - the set of the distance vectors from points Z_{ik} to the origin of coordinate system Q_i, where the points Z_{ik} represent centers of joint between the i-th and k-th mechanism links, $k \in \{1,\ldots,k_i\}$;

$\tilde{E}_i = \{\tilde{e}_{ik}\}$ - the set of the unit vectors of joint axes by which the i-th link C_i is connected to the remaining links C_k in points Z_{ik};

m_i - the mass of the i-th link;

$\underline{\underline{J}}_i$ - inertia tensor of the i-th link defined with respect to the local system Q_i.

We commonly accept that the axes of coordinate system Q_i are aligned along the principal axes of inertia, so the inertia tensor $\underline{\underline{J}}_i$ reduces to three moments of inertia $\underline{\underline{J}}_i = (J_{i1}, J_{i2}, J_{i3})$. In addition, the origin of coordinate system Q_i is accepted to coincide with the centre of mass of link C_i. Vectors whose projections are given with respect to the link coordinate system Q_i are marked with ~. An example of a robot link is shown in Fig. 2.2. In the proceeding text some definitions from Section 1.3 of this book are given in their complete form.

---

*) ~ denotes the vector projections with respect to the link coordinate system Q_i.

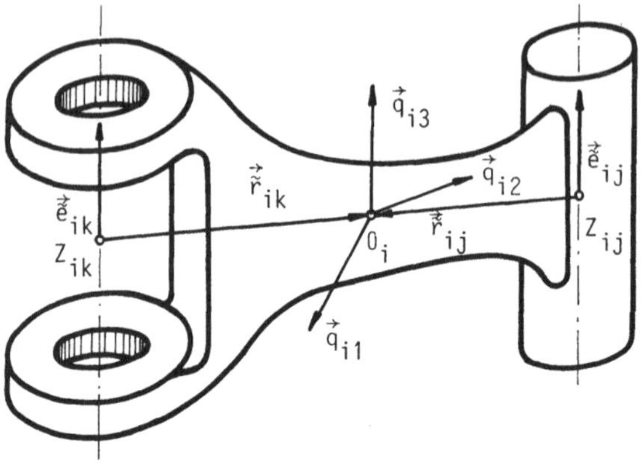

Fig. 2.2. Link C_i with two joints (Z_{ik} and Z_{ij}) with the centre of mass in point O_i

Kinematic pair

A kinematic pair P_{ik} represents a set of 2 adjacent links $\{C_i, C_k\}$ interconnected by a joint at point Z_{ik}.

The notion of class and subclass of a kinematic pair is introduce depending on the type of joint connection. A j-th class kinematic pair (j=1,...,5) is defined as a set of 2 adjacent links interconnected by a joint with s = 6-j degrees of freedom. A kinematic pair of j-th class and ℓ-th subclass is defined as a pair having r rotational and t linear d.o.f. in point Z_{ik}, where

$$r = \begin{cases} m-\ell+1 & \ell \leq m+1 \\ 0 & \ell > m+1 \end{cases}, \qquad t = s - r$$

m denotes the maximum possible number of rotational degrees of freedom in the j-th class. For example, classes 1, 2 and 3 permit 3 rotations (m=3), class 4 - two, and class 5 only one rotation (see Fig. 1.16).

Kinematic chain

A kinematic chain Λ_n is a set of n interconnected kinematic pairs, $\Lambda_n = \{P_{ik}\}$, $i \in N$, $k \in N$.

According to the structure of connections, chains are classified into simple, complex, open and closed.

A chain in which no link C_i enters more than two kinematic pairs is said to be a simple kinematic chain. On the other hand, a complex kinematic chain contains at least one link C_i, $i \in N$ which enters into more than two kinematic pairs.

An open kinematic chain possesses at least one link which belongs to one kinematic pair only. If each link C_i, $\forall i \in N$, enters into at least two kinematic pairs, the chain is said to be closed.

Joint coordinates

Scalar quantities which determine, in a unique manner, the relative position of the links of kinematic pair $P_{ik} = \{C_i, C_k\}$ to each other are manipulator joint coordinates q_{ik}^{ℓ}. The superscript $\ell \in \{1,\ldots,s\}$ where $s=6-j$ is the number of degrees of freedom, and j is the class of pair P_{ik}.

The definition of joint coordinates may be simplified if we consider fifth-class pairs only, i.e. pairs with a single degree of freedom of relative motion. This assumption does not reduce the generality of consideration, since a pair of any class may be represented by superposition of virtual fifth-class pairs. It is therefore particularly significant to introduce, for such pairs, a unique manner of defining the joint coordinates. Fig. 2.3 shows a revolute*) kinematic pair whose joint angle (coordinate) is denoted by q^i. For the revolute kinematic pair the corresponding joint coordinate q_{ik}^{ℓ} is defined as the angle of rotation at the joint about axis $\vec{e}_i$ and is seen as the angle between the projections of vectors $-\vec{r}_{i-1,i}$ and $\vec{r}_{ii}$ onto the plane perpendicular to $\vec{e}_i$. Fig. 2.4 shows the relevant prismatic (sliding) pair. The center of joint Z_i lies on the axis of joint $\vec{e}_i$ and is determined by vector $\vec{r}_{i-1,i}$ because this vector starts from Z_i and extends to the c.g. of link C_{i-1}. Point Z_i' at the axis of joint C_i is adopted in such a way that when $q^i=0$ it coincides with point Z_i. For $q^i \neq 0$ the distance between Z_i and Z_i' is q^i. Vector $\vec{r}_{ii}$ represents the radius vector between the point at link C_{i-1} and the centre of mass of link C_i, and is therefore dependent upon the coordinate q^i: $\vec{r}_{ii} = \vec{r}_{ii}^{\,o} + q^i\vec{e}_i$, where $\vec{r}_{ii}^{\,o}$ is radius vector between Z_i and O_i with $q^i=0$. It is understood that all values are referenced to the local coordinate frame.

---

*) In most cases in the textbook, the terms revolute and prismatic (sliding) for the corresponding types of kinematic pairs and joints of mechanism are used, while the degrees of freedom (joint coordinates) are described by the terms rotational and translational (linear).

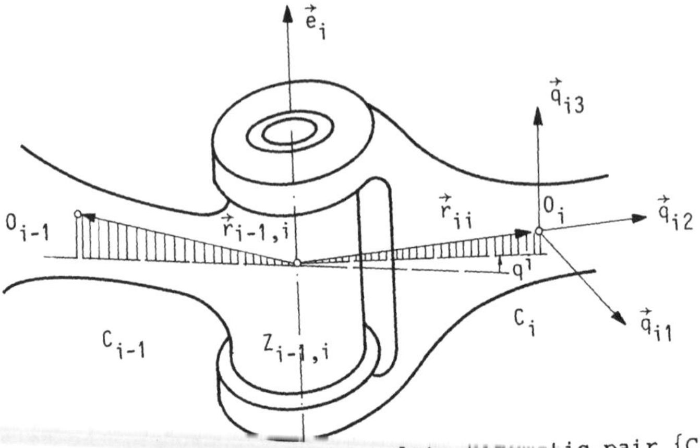

Fig. 2.3. Joint coordinate of a revolute kinematic pair $\{C_{i-1}, C_i\}$

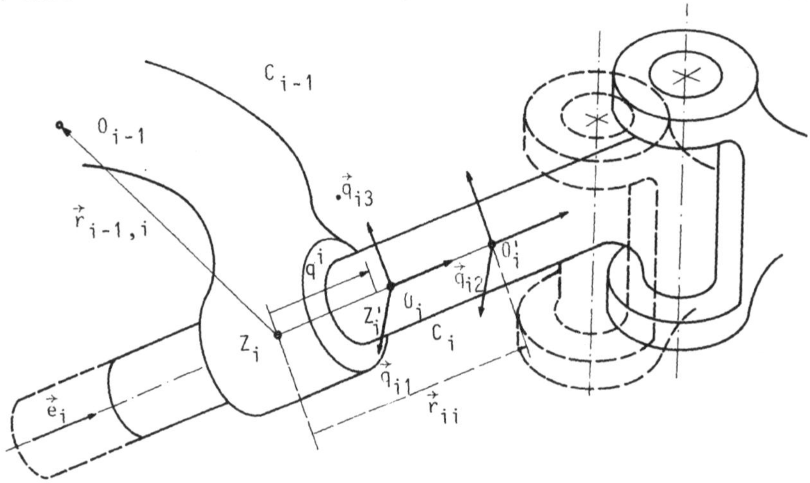

Fig. 2.4. Joint coordinate of a sliding kinematic pair $\{C_{i-1}, C_i\}$

External coordinates

Scalar quantities x_{ek}, $k \in \{1,\ldots,m\}$, which determine the position and (partially or completely) orientation of the n-th link of chain Λ_n with respect to a reference coordinate system are said to be the external coordinates of mechanism.

The specification of external coordinates, and of their number $m \leq 6$, depends on the type of manipulation task and will be briefly considered in Section 2.5.

Active mechanism

An active mechanism represents a system comprising: a) a mechanical part that can be modelled by an appropriate kinematic chain, and b) a set of driving units (actuators) which produce driving torques (forces).

To form the mathematical model of an active mechanism dynamics, it is necessary to:

1) identify the parameters of kinematic chain and actuators,
2) form the dynamic model of mechanism, and
3) form the models of actuators.

The dynamic model of an open active mechanism represents a set of n non-linear differential equations which describe system motion in the space of joint coordinates under the action of driving forces.

In the following text we will present one of the recursive numerical methods for forming dynamic models of open active mechanisms with the aid of a digital computer. This method consists of following stages: determination of the "home" position, determination of the actual position of links with respect to reference frame, a kinematic stage, and a dynamic stage. We will use the following notation for describing these stages. N and I denote the sets of indices

$$N = \{i:\ i \in (1,\dots,n)\}, \qquad I = \{j:\ j \in (1,\dots,i)\}$$

where n is the number of kinematic pairs. ξ_i is a symbol which, depending on the type of kinematic pair i, has the value 0 or 1:

$$\xi_i = \begin{cases} 0 & \text{- the i-th pair is the revolute} \\ 1 & \text{- the i-th pair is the sliding one} \end{cases}$$

Symbol $\bar{\xi}_i = 1-\xi_i$ will also be used, as well as the notation introduced by the preceding definitions.

Stage 1: Determination of "home" position - mechanism assembly

It is assumed that the active mechanism under consideration may be described by a simple, open kinematic chain Λ_n consisting of sets of kinematic pairs $\{C_{i-1}, C_i\}$, $i \in N$. The kinematic pair $\{C_o, C_1\}$ represents in fact the first link of the mechanism which is connected to fundament C_o by joint Z_{1o}. It is necessary to determine the position of all links with respect to the reference frame, under the condition that all joint coordinates equal zero, $q^i=0$, $i \in N$.

Let us assume all kinematic parameters $K_j^o = (Q_j^o, R_j^o, E_j^o)$ for $j \in (1,\ldots$ $\ldots, i-1)$ to be known, where the upperscript o means that $q^j=0$, $\forall j \in N$. The task of determining the "home" position now reduces to determining $K_i^o = (Q_i^o, R_i^o, E_i^o)$. Sets R_i and E_i for the case of a simple kinematic chain have the following form:

$$R_i = [\vec{r}_{ii}, \vec{r}_{i,i+1}] \quad \text{and} \quad E_i = [\vec{e}_i, \vec{e}_{i+1}],$$

where $\vec{e}_i$ and $\vec{e}_{i+1}$ stand for vectors $\vec{e}_{ii}$ and $\vec{e}_{i,i+1}$.

On the basis of the assumption that K_{i-1}^o is known i.e. that the mechanism is assembled up to the (i-1)th link, it follows that $\vec{e}_i^{\,o}$ and $\vec{r}_{i-1,i}^{\,o}$ are also known. Further, let us note the set of 3 orthogonal unit vectors in point $Z_{i-1,i}$: $\vec{e}_i^{\,o}$, $\vec{a}_i^{\,o}$ and $\vec{e}_i^{\,o} \times \vec{a}_i^{\,o}$, where $\vec{a}_i^{\,o} = -\text{ort}^{*)}(\vec{e}_i^{\,o} \times (\vec{r}_{i-1,i}^{\,o} \times \vec{e}_i^{\,o}))$, whose components are known with respect to the reference coordinate system (Fig. 2.5). On the other hand, let us note the set of vectors $\vec{\tilde{e}}_i$, $\vec{\tilde{a}}_i$ and $\vec{\tilde{e}}_i \times \vec{\tilde{a}}_i$, where $\vec{\tilde{a}}_i = -\text{ort}(\vec{\tilde{e}}_i \times (\vec{\tilde{r}}_{ii} \times \vec{\tilde{e}}_i))$, whose components are known with respect to the local coordinate system Q_i. Under the condition $q^i = 0$, these two sets of vectors coincide. Since $Q_i^o = [\vec{q}_{i1}^{\,o}\ \vec{q}_{i2}^{\,o}\ \vec{q}_{i3}^{\,o}]$ represents transformation matrix of the C_i-th link of local coordinate system into the reference system, it follows that

$$\vec{e}_i^{\,o} = Q_i^o \vec{\tilde{e}}_i, \quad \vec{a}_i^{\,o} = Q_i^o \vec{\tilde{a}}_i, \quad \vec{e}_i^{\,o} \times \vec{a}_i^{\,o} = Q_i^o(\vec{\tilde{e}}_i \times \vec{\tilde{a}}_i) \tag{2.1}$$

and this completely determines the matrix Q_i^o:

$$Q_i^o = [\vec{e}_i^{\,o}\ \vec{a}_i^{\,o}\ \vec{e}_i^{\,o} \times \vec{a}_i^{\,o}][\vec{\tilde{e}}_i\ \vec{\tilde{a}}_i\ \vec{\tilde{e}}_i \times \vec{\tilde{a}}_i]^T \tag{2.2}$$

Let us note that matrix transformation has been used instead of its inverse, due to the orthogonality of vectors $\vec{\tilde{e}}_i$, $\vec{\tilde{a}}_i$ and $\vec{\tilde{e}}_i \times \vec{\tilde{a}}_i$.

It is now necessary to determine the remaining elements of set K_i^o, i.e. sets R_i^o and E_i^o. Since the transformation matrix Q_i^o has been determined, the following relations evidently hold

---

*) Ort (•) denotes the unit vector of (•).

$$\vec{r}^{\,o}_{ii} = Q^o_i \vec{\bar{r}}_{ii}, \quad \vec{r}^{\,o}_{i,i+1} = Q^o_i \vec{\bar{r}}_{i,i+1}, \quad \vec{e}^{\,o}_{i+1} = Q^o_i \vec{\bar{e}}_{i+1} \tag{2.3}$$

Recursiveness required for calculating Q^o_i $\forall i \in N$, for all vectors contained in R^o_i and E^o_i, has thus been established.

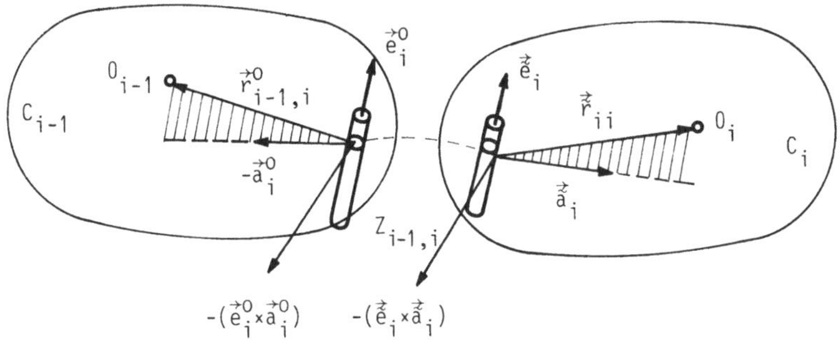

Fig. 2.5. Relative position of link C_i with respect to C_{i-1} with $q^i=0$

Appendix 1 contains derivations of the transformation matrices together with comments on specifities of robotic mechanisms.

Stage 2: Mechanism position

This stage consists of determining the elements of the set of kinematic variables $K_i = (Q_i, R_i, E_i)$, where the mechanism position is determined by joint coordinates q^i, $i \in N$.

Applying the theorem of finite rotations (Rodrigues' formula) for the case of revolute joints ($\bar{\xi}_i=1$) or the theorem of the translation of rigid body ($\xi_i=1$), one obtains

$$\vec{q}_{ij} = \begin{bmatrix} \vec{e}_i \times (\vec{q}^{\,o}_{ij} \times \vec{e}_i)\bar{\xi}_i \\ (\vec{e}_i \times \vec{q}^{\,o}_{ij})\bar{\xi}_i \\ (\vec{e}_i \cdot \vec{q}^{\,o}_{ij})\vec{e}_i \bar{\xi}_i + \vec{q}^{\,o}_{ij}\xi_i \end{bmatrix}^T \begin{bmatrix} \cos q^i \\ \sin q^i \\ 1 \end{bmatrix}, \quad j=1,2,3. \tag{2.4}$$

We have thus determined the matrix $Q_i = [\vec{q}_{i1}, \vec{q}_{i2}, \vec{q}_{i3}]$ $\forall i \in N$ which represents the transformation matrix from the i-th link coordinate system O_i into the reference system. Now, we directly obtain

$$R_i = \{\vec{r}_{ii}, \vec{r}_{i,i+1}\} = \{Q_i \vec{\bar{r}}_{ii} + q^i \vec{e}_i \xi_i, \; Q_i \vec{\bar{r}}_{i,i+1}\} \tag{2.5}$$

$$E_i = \{\vec{e}_i, \vec{e}_{i+1}\} = \{Q_i\tilde{\vec{e}}_i, Q_i\tilde{\vec{e}}_{i+1}\} \tag{2.5}$$

where it is evident that $\vec{r}_{ij} = \vec{r}_{i-1,j} - \vec{r}_{i-1,i} + \vec{r}_{ii}$, $j\in I$, $i\in N$. Determination of $\vec{e}_{i+1}$ provides the recursiveness of calculating the transformation matrix Q_i according to (2.4), and thus the elements of sets (2.5). Application of Denavit-Hartenberg parameters is usual for the forming of robotic mechanisms mathematical models. While in kinematic models these parameters are more advantageous than Rodrigues' formula of finite rotations, for the forming of dynamic models this advantage is, in the general case, on the side of the Rodrigues' formula. In Appendix 2 the Denavit-Hartenberg approach to forming the kinematic robot model is given, as well as the relation between this approach and the one based on Rodrigues' formula.

Stage 3: Mechanism kinematics

This stage consists of forming the set of kinematic quantities $\mathcal{L}_i = \{\Omega_i, W_i\}$, where

$$\Omega_i = \{\vec{\omega}_i\}, \qquad W_i = \{\vec{\varepsilon}_i, \vec{w}_i\}, \text{ with}$$

$\vec{\omega}_i$ - angular velocity of the i-th link,

$\vec{\varepsilon}_i$ - angular acceleration of the i-th link, and

$\vec{w}_i$ - linear acceleration of the i-th link.

These kinematic quantities are functions of the velocities and accelerations of joint coordinates $\dot{q}^i$ and $\ddot{q}^i$ and variables formed in the preceding stage: $K_i = (Q_i, R_i, E_i)$. All relations considered in this stage are recursive. Applying the basic theorems about rigid body kinematics, one obtains [2, 3, 4]:

$$\begin{aligned}
\vec{\omega}_i &= \vec{\omega}_{i-1} + \dot{q}^i\vec{e}_i\bar{\xi}_i \\
\vec{\varepsilon}_i &= \vec{\varepsilon}_{i-1} + \dot{q}^i\vec{\omega}_{i-1}\times\vec{e}_i\bar{\xi}_i + \ddot{q}^i\vec{e}_i\bar{\xi}_i \\
\vec{w}_i &= \vec{w}_{i-1} - \vec{\varepsilon}_{i-1}\times\vec{r}_{i-1,i} - \vec{\omega}_{i-1}\times(\vec{\omega}_{i-1}\times\vec{r}_{i-1,i}) + \\
&+ \vec{\varepsilon}_i\times\vec{r}_{ii} + \vec{\omega}_i\times(\vec{\omega}_i\times\vec{r}_{ii}) + (2\dot{q}^i\vec{\omega}_{i-1}\times\vec{e}_i + \ddot{q}^i\vec{e}_i)\xi_i
\end{aligned} \tag{2.6}$$

The basic relations of rigid body kinematics are given in Appendix 3 of this book.

It follows that angular and linear accelerations are functions of the second derivatives of joint coordinates $\ddot{q}^j$, $j \in I$. However, to form the dynamic model matrices, expressions $\vec{\varepsilon}_i$ and $\vec{w}_i$ should be rearranged so that accelerations $\ddot{q}^j$, $j \in N$, figure explicitly [6]:

$$\vec{\varepsilon}_i = [\vec{\alpha}_{i1} \cdots \vec{\alpha}_{ii} \; 0 \cdots 0]\ddot{q} + \vec{\theta}_i + \vec{\varepsilon}_o$$
$$\vec{w}_i = [\vec{\beta}_{i1} \cdots \vec{\beta}_{ii} \; 0 \cdots 0]\ddot{q} + \vec{\eta}_i + \vec{w}_o \qquad (2.7)$$

where $\ddot{q} = [\ddot{q}^1 \cdots \ddot{q}^n]^T$ and $\vec{\varepsilon}_o$, $\vec{w}_o$ - angular and linear accelerations of the fundamental (basic) mechanism link. The set of kinematic quantities W_i thus reduces to

$$W_i = \{\vec{\alpha}_{ij}, \vec{\beta}_{ij}, \vec{\theta}_i, \vec{\eta}_i, j \in I\}.$$

On the basis of (2.6) and (2.7) the following relations are obtained:

$$\sum_{j=1}^{i} \vec{\alpha}_{ij}\ddot{q}^j + \vec{\theta}_i + \vec{\varepsilon}_o = \sum_{j=1}^{i-1} \vec{\alpha}_{i-1}\ddot{q}^j + (\ddot{q}^i\vec{e}_i + \dot{q}^i\vec{\omega}_{i-1} \times \vec{e}_i)\bar{\xi}_i + \vec{\theta}_{i-1} + \vec{\varepsilon}_o$$

$$\sum_{j=1}^{i} \vec{\beta}_{ij}\ddot{q}^j + \vec{\eta}_i + \vec{w}_o = \sum_{j=1}^{i-1} \vec{\beta}_{i-1,j}\ddot{q}^j + \vec{\eta}_{i-1} + \vec{w}_o - (\sum_{j=1}^{i-1} \vec{\alpha}_{i-1,j}\ddot{q}^j + \vec{\theta}_{i-1} + \vec{\varepsilon}_o) \times$$
$$\times \vec{r}_{i-1,i} - \vec{\omega}_{i-1} \times (\vec{\omega}_{i-1} \times \vec{r}_{i-1,i}) + (\sum_{j=1}^{i-1} \vec{\alpha}_{i-1}\ddot{q}^j + \vec{\theta}_{i-1} + \vec{\varepsilon}_o) \times$$
$$\times \vec{r}_{ii} + (\ddot{q}^i\vec{e}_i + \dot{q}^i\vec{\omega}_{i-1} \times \vec{e}_i) \times \vec{r}_{ii}\bar{\xi}_i + \vec{\omega}_i \times (\vec{\omega}_i \times \vec{r}_{ii}) +$$
$$+ (\ddot{q}^i\vec{e}_i + 2\dot{q}^i\vec{\omega}_{i-1} \times \vec{e}_i)\xi_i$$

Equating coefficients of equal j, one obtains:

$$\vec{\alpha}_{ij} = \vec{\alpha}_{i-1,j}, \qquad j \neq i$$
$$\vec{\alpha}_{ij} = \vec{e}_j\bar{\xi}_j, \qquad j \in I, \; i \in N$$
$$\vec{\theta}_i = \vec{\theta}_{i-1} + \dot{q}^i(\vec{\omega}_{i-1} \times \vec{e}_i)\bar{\xi}_i, \quad \vec{\theta}_o = 0, \qquad i \in N$$
$$\vec{\beta}_{ij} = \vec{\beta}_{i-1,j} + \vec{\alpha}_{i-1,j} \times (\vec{r}_{ii} - \vec{r}_{i-1,i}), \qquad j \in \{1,2,\ldots,i-1\}, \; i \in N \qquad (2.8)$$
$$\vec{\beta}_{ii} = (\vec{e}_i \times \vec{r}_{ii})\bar{\xi}_i + \vec{e}_i\xi_i$$

$$\vec{\eta}_i = \vec{\eta}_{i-1}+\vec{\theta}_{i-1}\times(\vec{r}_{ii}-\vec{r}_{i-1,i})+\dot{q}^i(\vec{\omega}_{i-1}\times\vec{e}_i)\times\vec{r}_{ii}\bar{\xi}_i+\vec{\gamma}_{ii}-\vec{\gamma}_{i-1,i}+$$

$$+\ 2\dot{q}^i(\vec{\omega}_{i-1}\times\vec{e}_i)\xi_i-\vec{\varepsilon}_o\times(\vec{r}_{i-1,i}-\vec{r}_{ii})$$

$$\vec{\eta}_o=0, \quad \vec{\gamma}_{ij} = \vec{\omega}_i\times(\vec{\omega}_i\times\vec{r}_{ij}, \quad j\in I, \quad i\in N$$

In this manner the recursive relations determining the required vector coefficients $\vec{\alpha}_{ij}$, $\vec{\beta}_{ij}$, $\vec{\theta}_i$ and $\vec{\eta}_i$, have been obtained. For example, vector coefficient $\vec{\eta}_i$ for the i-th kinematic pair is described using the vector coefficient of the (i-1)-th pair. If $\vec{\eta}_{i-1}$ is then described using η_{i-2} and so on the following can be deduced:

$$\vec{\eta}_i = \vec{\eta}_i(\vec{\varepsilon}_o=0)+\vec{\varepsilon}_o\times\sum_{k=1}^{i}(\vec{r}_{kk}-\vec{r}_{k-1,k}) = \vec{\eta}_i(\vec{\varepsilon}_o=0)+\vec{\varepsilon}_o\times\vec{r}_{i,1}$$

where: $\vec{\eta}_i(\varepsilon_o=0)$ - vector coefficient $\vec{\eta}_i$ for $\vec{\varepsilon}_o=0$,

$\vec{r}_{i,1}$ - radius vector from the first joint to the centre of mass of the i-th link.

Using these coefficients, the required accelerations (linear and angular) as functions of relative (generalized) accelerations, can be obtained.

Stage 4: Mechanism dynamics [2, 3, 4]

This stage includes evaluation of dynamic quantities $\Delta_i=\{F_i, M_i\}$, $i\in N$, where

$F_i = \{\vec{F}_i\}$, $M_i = \{\vec{M}_i\}$, with

$\vec{F}_i$ - inertial force at the centre of mass of the i-th link, and

$\vec{M}_i$ - moment of the inertial force of the i-th link.

The inertial force may be calculated using the second Newton's law

$$\vec{F}_i = -m_i\vec{w}_i = [\vec{a}_{i1} \cdots \vec{a}_{ii}\ 0 \cdots 0]\ddot{q} + \vec{a}_i^o - m_i\vec{w}_o \tag{2.9}$$

where m_i - is the mass of the i-th link. Comparing this with (2.7), we obtain

$$\vec{a}_{ij} = -m_i\vec{\beta}_{ij}, \quad \vec{a}_i^o = -m_i\vec{\eta}_i \tag{2.10}$$

The moments of inertial forces are determined from Eulers' dynamic equations (see Appendix 4):

$$\tilde{M}_i^1 = -J_{i1}\tilde{\varepsilon}_i^1 + (J_{i2}-J_{i3})\tilde{\omega}_i^2\tilde{\omega}_i^3$$

$$\tilde{M}_i^2 = -J_{i2}\tilde{\varepsilon}_i^2 + (J_{i3}-J_{i1})\tilde{\omega}_i^3\tilde{\omega}_i^1 \tag{2.11}$$

$$\tilde{M}_i^3 = -J_{i3}\tilde{\varepsilon}_i^3 + (J_{i1}-J_{i2})\tilde{\omega}_i^1\tilde{\omega}_i^2$$

where ~ indicates that vectors $\vec{\tilde{M}}$, $\vec{\tilde{\varepsilon}}$ and $\vec{\tilde{\omega}}$ are taken as projections on the local coordinate system. The upper index indicates the number of the projection, i.e.,

$$\tilde{\varepsilon}_i^j = \vec{\varepsilon}_i \cdot \vec{q}_{ij} \quad \text{or} \quad \tilde{M}_i^j = \vec{M}_i \cdot \vec{q}_{ij} \qquad i \in I, \quad j=1,2,3$$

For transforming to the fixed coordinate system transformation matrix Q_i, for the i-th segment is used:

$$\vec{M}_i = Q_i \vec{\tilde{M}}_i \tag{2.12}$$

Multiplying left and right hand sides of expression (2.11) by Q_i and replacing acceleration projections $\tilde{\varepsilon}_i^j$ by $\tilde{\varepsilon}_i^j = \vec{\varepsilon}_i \cdot \vec{q}_{ij}$, one obtains the projections of the moments of inertial forces onto the axes of the absolute system:

$$\begin{aligned} M_i^j = -[& (Q_i^{j1}J_{i1}q_{i1}^1 + Q_i^{j2}J_{i2}q_{i2}^1 + Q_i^{j3}J_{i3}q_{i3}^1)\varepsilon_i^1 + \\ & + (Q_i^{j1}J_{i1}q_{i1}^2 + Q_i^{j2}J_{i2}q_{i2}^2 + Q_i^{j3}J_{i3}q_{i3}^2)\varepsilon_i^2 + \\ & + (Q_i^{j1}J_{i1}q_{i1}^3 + Q_i^{j2}J_{i2}q_{i2}^3 + Q_i^{j3}J_{i3}q_{i3}^3)\varepsilon_i^3] + \lambda_i^j, \qquad j=1,2,3 \end{aligned} \tag{2.13}$$

where

$$\vec{\lambda}_i = \begin{bmatrix} \lambda_i^1 \\ \lambda_i^2 \\ \lambda_i^3 \end{bmatrix} = Q_i \begin{bmatrix} (\vec{\omega}_i \cdot \vec{q}_{i2})(\vec{\omega}_i \cdot \vec{q}_{i3}) \cdot (J_{i2}-J_{i3}) \\ (\vec{\omega}_i \cdot \vec{q}_{i3})(\vec{\omega}_i \cdot \vec{q}_{i1}) \cdot (J_{i3}-J_{i1}) \\ (\vec{\omega}_i \cdot \vec{q}_{i1})(\vec{\omega}_i \cdot \vec{q}_{i2}) \cdot (J_{i1}-J_{i2}) \end{bmatrix} \tag{2.14}$$

Since $Q_i^{jk} = q_{ik}^j$, the expression for the moment of inertial force of the i-th link, can be written as:

$$\vec{M}_i = -T_i\vec{\varepsilon}_i + \vec{\lambda}_i \tag{2.15}$$

where T_i is the 3×3 matrix containing the elements:

$$T_i^{jk} = \sum_{\ell=1}^{3} Q_i^{j\ell} J_{i\ell} q_{i\ell}^k = \sum_{\ell=1}^{3} q_{i\ell}^j q_{i\ell}^k J_{i\ell} \tag{2.16}$$

Substituting the expression for angular accelerations (2.7) into (2.15) follows that,

$$\vec{M}_i = \sum_{j=1}^{i} \vec{b}_{ij}\ddot{q}^j + \vec{b}_i^o - T_i\vec{\varepsilon}_o \tag{2.17}$$

where $\vec{b}_{ij} = -T_i\vec{\alpha}_{ij}$; $\vec{b}_i^o = -T_i\vec{\theta}_i + \vec{\lambda}_i$.

In cases when the mechanism link can be considered as a cane (see Appendix 5) three main moments of inertia are replaced by a set $J_i = \{J_{is}, J_{iN}\}$. Then, the moment of inertial forces can be described directly in the fixed coordinate system (where projections of angular velocities onto the fixed axes do not figure, see Eq. (2.21)). We, now, introduce the equivalent angular acceleration $\vec{\tau}_i$ such that the moment of inertial forces thereby produced is equal to that resulting from the action of angular velocity $\vec{\omega}_i$ [2]:

$$\vec{\tau}_i = (\vec{\omega}_i \cdot \vec{s}_i)(\vec{s}_i \times \vec{\omega}_i), \quad i \in I, \tag{2.18}$$

where $\vec{s}_i$ is the unit vector along the cane axis which can be determined using: $\vec{s}_i = \vec{r}_{ii}/|\vec{r}_{ii}|$. In compliance with (2.13) and $\tilde{\varepsilon}_i^j = \vec{\varepsilon}_i \vec{q}_{ij}$ the inertial moment becomes [2]:

$$\vec{M}_i = -J_i(\vec{\varepsilon}_i + \vec{\tau}_i), \quad i \in I \tag{2.19}$$

Acceleration $\vec{\varepsilon}$ is seperated into components normal to the cane axis (N) and those parallel to the cane axis (s)

$$\vec{\varepsilon}_{iN} = (\vec{s}_i \times \vec{\varepsilon}_i) \times \vec{s}_i; \quad \vec{\varepsilon}_{is} = (\vec{\varepsilon}_i \cdot \vec{s}_i)\vec{s}_i \tag{2.20}$$

Because $\vec{\tau}_i$ is normal to the cane axis the following holds:

$$\vec{M}_i = -J_{iN}(\vec{\varepsilon}_{iN} + \vec{\tau}_i) - J_{is}\vec{\varepsilon}_{is}, \quad i \in I \tag{2.21}$$

Using relations (2.8) giving the coefficients of angular accelerations, equation (2.21) takes the form of a summation i.e:

$$\vec{M}_i = \sum_{j=1}^{i} \vec{C}_{ij}\ddot{q}^j + \vec{C}_i^o \tag{2.22}$$

where,

$$\begin{aligned}\vec{C}_{ij} &= -J_{iN}(\vec{s}_i \times \vec{\alpha}_{ij}) \times \vec{s}_i - J_{is}(\vec{\alpha}_{ij} \cdot \vec{s}_i)\vec{s}_i \\ \vec{C}_i^o &= -J_{iN}[(\vec{s}_i \times (\vec{\theta}_i + \vec{\varepsilon}_o)) \times \vec{s}_i + \vec{\tau}_i] - J_{is}[((\vec{\theta}_i + \vec{\varepsilon}_o) \cdot \vec{s}_i)\vec{s}_i]\end{aligned} \tag{2.23}$$

and $\vec{\tau}_i$ is given by expression (2.18).

Apart from the inertial forces and moments, external forces and moments $\vec{G}_i$ and $\vec{M}_i^G$ also act upon the links so that the total forces and moments can be expressed in the form

$$\vec{F}_i^u = \vec{F}_i + \vec{G}_i; \quad \vec{M}_i^u = \vec{M}_i + \vec{M}_i^G \tag{2.24}$$

Substituting expressions (2.9) and (2.17) into (2.24) one obtains,

$$\begin{aligned}\vec{F}_j^u &= \sum_{k=1}^{j} \vec{a}_{jk}\ddot{q}^k + \vec{a}_j^o - m_j\vec{w}_o + \vec{G}_j \\ \vec{M}_j^u &= \sum_{k=1}^{j} \vec{b}_{jk}\ddot{q}^k + \vec{b}_j^o - T_j\vec{\varepsilon}_o + \vec{M}_j^G\end{aligned} \tag{2.25}$$

Let the kinematic chain be fictively disconnected at the i-th joint and consider the equilibrium of the mechanism free end (Fig. 2.6). The action of the "rejected" mechanism part is substituted by a force $\vec{R}_i$ and moment $\vec{M}_i^*$. This force and moment will be termed as the total reaction force at the i-th joint. In determining the overall reaction the external forces $\vec{F}_j^u$ and moments $\vec{M}_j^u$ (j=i,i+1,...,n) are reduced to the centre of the i-th joint. Reactions of drives of all subsequent joints need not be taken into account since each drive acts upon two adjacent links with forces and moments of equal magnitude but opposite direction which are thus annulled in the summation process.

The overall reactions can therefore be written as [2]:

$$\vec{R}_i = -\sum_{j=i}^{n} \vec{F}_j^u = -\sum_{j=i}^{n} \left(\sum_{k=1}^{j} \vec{a}_{jk}\ddot{q}^k + \vec{a}_j^o - m_j\vec{w}_o + \vec{G}_j \right) \tag{2.26}$$

$$\vec{M}_i^* = -\sum_{j=i}^{n} (\vec{M}_j^u+\vec{r}_{ji}\times\vec{F}_j^u) = -\sum_{j=i}^{n} [\sum_{k=1}^{j} (\vec{b}_{jk}+\vec{r}_{ji}\times\vec{a}_{jk})\ddot{q}^k +$$

$$+ \vec{r}_{ji}\times\vec{a}_j^o-m_j\vec{r}_{ji}\times\vec{w}_o+\vec{b}_j^o-T_j\vec{\varepsilon}_o+\vec{M}_j^G+\vec{r}_{ji}\times\vec{G}_j] \tag{2.26}$$

where n - number of links in the chain.

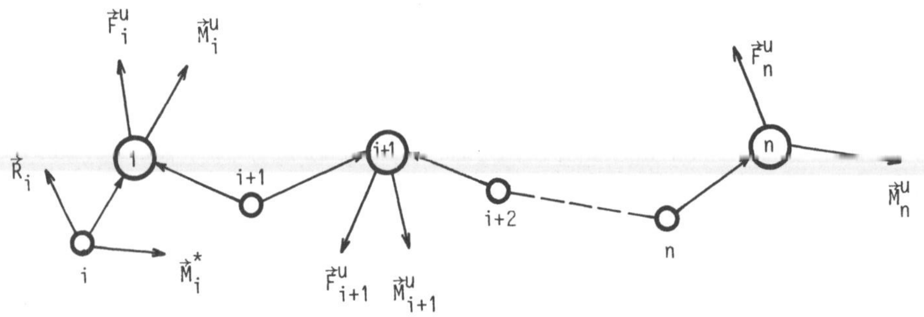

Fig. 2.6. Equilibrium of the free chain

To obtain the reactions, it is appropriate to establish the recursive relations between the reactions in two adjacent joints. From (Fig. 2.7a) which illustrates a single mechanism member, it is evident that [2]:

$$\vec{R}_i = \vec{R}_{i+1}-\vec{F}_i^u$$

$$\vec{M}_i^* = \vec{M}_{i+1}^*+(\vec{r}_{ii}-\vec{r}_{i+1,i})\times\vec{R}_{i+1}+\vec{r}_{ii}\times\vec{F}_i^u-\vec{M}_i^u \tag{2.27}$$

Following the chain length from the last to the first member according to (2.27) all the reactions can be determined.

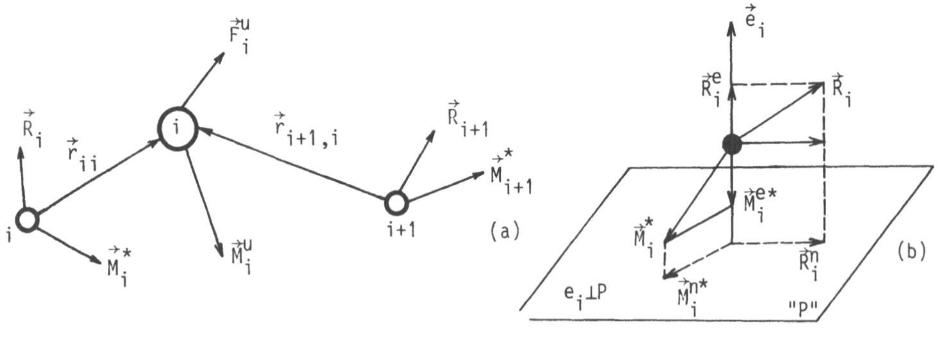

Fig. 2.7. Reaction at the i-th member

Disassemble the overall reaction force $\vec{R}_i$ and moment $\vec{M}_i^*$ at the i-th joint, to their components parallel and perpendicular to vector $\vec{e}_i$ (Fig. 2.7b). Because the perpendicular components can not cause any movement in the mechanism, they are only acting as load on the joint, and are necessary when considering the forces of friction. One of the parallel components (moment for sliding kinematic pairs and forces for revolute pairs) also contributes to the force of friction present at the i-th joint. The parallel component of $\vec{M}_i^*$ with the revolute joint, i.e. the parallel component of $\vec{R}_i$ with the sliding kinematic pair must be compensated by the corresponding drives in the joints to keep the mechanism in equilibrium. If the joint is revolute:

$$P_i^M = \vec{M}_i^* \cdot \vec{e}_i \tag{2.28}$$

and for the sliding kinematic pair:

$$P_i^F = \vec{R}_i \cdot \vec{e}_i \tag{2.29}$$

Conditions (2.28), (2.29) and (2.26) enable the determination of the required forces and moments at each joint of the robot mechanism, so that in the case of a revolute kinematic pair ($\xi_i=0$) one obtains:

$$P_i^M = -\vec{e}_i \cdot \sum_{k=1}^{n} \sum_{j=\max(i,k)}^{n} (\vec{b}_{jk} + \vec{r}_{ji} \times \vec{a}_{jk}) \ddot{q}^k -$$

$$- \vec{e}_i \cdot \sum_{j=i}^{n} (\vec{r}_{ji} \times (\vec{a}_j^o + \vec{G}_j - m_j \vec{w}_o) + \vec{b}_j^o - T_j \vec{\varepsilon}_o)$$

whereas for the linear drive ($\xi_i=1$)

$$P_i^F = -\vec{e}_i \cdot \sum_{k=1}^{n} \sum_{j=\max(i,k)}^{n} \vec{a}_{jk} \ddot{q}^k - \vec{e}_i \cdot \sum_{k=1}^{n} \sum_{j=1}^{n} (\vec{a}_j^o + \vec{G}_j - m_j \vec{w}_o)$$

These expressions can be written in matrix form as

$$P = H(q, \theta)\ddot{q} + h(q, \dot{q}, \theta) \tag{2.30}$$

where $P = [P_1 \cdots P_n]^T$ - vector*) of driving moments P^M (forces P^F), $q = [q^1 \cdots q^n]^T$ - joint coordinates, vector $\theta = [\theta^1 \cdots \theta^n]^T$ - geometric and dynamic parameters vector. According to the expressions given above the

*) *It is clear that the "drive" vector has the direction of the rotational (or translational) axis and P^M or P^F represent the projections of this vector onto the axis oriented like $\vec{e}$.*

elements of matrix (2.30) take the form

$$[H_{ik}] = [-\vec{e}_i \cdot \sum_{j=\max(i,k)}^{n} ((\vec{b}_{jk}+\vec{r}_{ji}\times\vec{a}_{jk})\bar{\xi}_i+\vec{a}_{jk}\xi_i)]$$

$$[h_i] = [-\vec{e}_i \sum_{j=i}^{n} ((\vec{r}_{ji}\times(\vec{a}_j^{\,o}+\vec{G}_j-m_j\vec{w}_o)+(\vec{b}_j^{\,o}-T_j\vec{\varepsilon}_o))\bar{\xi}_i+(\vec{a}_j^{\,o}-m_j\vec{w}_o+\vec{G}_j)\xi_i)] \tag{2.31}$$

2.2 Complete Mathematical Models of Manipulation Robots

Complete mathematical models of robots consist of the mechanism model S^M and the model of the actuators S^i which drive the mechanical degrees of freedom of the mechanism.

In general, robotic actuators can be described using linear (D.C. motors) or nonlinear time varying models (hydraulic, brushless D.C. servomotors,` AC servomotors) which are described in Appendix 6 of this textbook. Here, the actuators are presented by linear model with constant coefficients, which is the case of D.C. permanent magnet actuators often applied to various types of robotic mechanisms:

$$S^i: \quad \dot{x}^i = A^i x^i + f^i M_i^* + b^i N(u^i), \quad x^i(t_o) = x_o^i \tag{2.32}$$

where: x^i - state vector of the i-th actuator

u^i - control input to the i-th actuator

M_i^* - moment (load) acting on the i-th actuator*)

A^i - matrix of subsystem S^i

b^i - input distribution vector

f^i - load distribution vector

$N(u^i)$ - nonlinearity of the amplitude saturation type

$$N(u^i) = \begin{cases} -u_m^i & \text{for} \quad u^i < -u_m^i \\ u^i & \text{for} \quad -u_m^i \le u^i \le u_m^i \\ u_m^i & \text{for} \quad u^i > u_m^i \end{cases} \tag{2.33}$$

---

*) *For a revolute joint driven by a D.C. motor or wane - hydraulic actuator* $P_i^M = M_i^*$

The mathematical model of the complete system S can be described by the dynamic model of the robot mechanism S^M (2.30) and a set of actuator models S^i (2.32) for $\forall i \in I$.

Looking at the case where for each subsystem state vector S^i, x^i, two coordinates coincide with q^i (angle or linear displacement) and $\dot{q}^i$ (angular or linear velocity) of the i-th degree of freedom.

The state vector of the complete system is given as $x = (x^{1T}, x^{2T}, \ldots$ $\ldots, x^{nT})^T$, while the order of the complete system $N = \sum_{i=1}^{n} n_i$, where, n_i - the order of the subsystem (actuator). The input to the system is $u = (u^1, u^2 \cdots u^n)^T$.

Because one coordinate in each state vector coincides with $\dot{q}^i$, vector P (2.30) can be expressed as a function of the state vector x:

$$S^M: \quad P = (I_n - HTF)^{-1}[HT(Ax+BN(u))+h] \tag{2.34}$$

where $T = \mathrm{diag}(T^i)$, $\ddot{q}^i = T^i \dot{x}^i$, $A = \mathrm{diag}(A^i)$, $B = \mathrm{diag}(b^i)$, $F = \mathrm{diag}(f^i)$ and $N(u) = (N(u^1), N(u^2), \ldots, N(u^n))^T$. Substituting (2.34) into (2.32) and reducing to a unique model, one can obtain the model of system S, in other words mathematical manipulation robot model,

$$S: \quad \dot{x} = \hat{A}(x) + \hat{B}(x)N(u) \tag{2.35}$$

where $\hat{A}(x) = Ax + F(I_n - HTF)^{-1}[HTAx+h]$, $\hat{B} = B + F(I_n - HTF)^{-1}HTB$.

Using the procedure for computational assembly of mathematical models of active mechanisms, described in the preceding section and on the basis of adopted actuator models (2.32), the complete mathematical model of manipulation robot dynamics S (2.35) can be elaborated by a digital computer and used both in the performance analysis of a robotic system and in control synthesis of various manipulation tasks. Worth mentioning is that the relationship between q^i and $\dot{q}^i$ from the mechanical part of system S^M and state vector coordinates x^i need not be linear nor such that two coordinates of state vector x^i coincide with q^i and $\dot{q}^i$, as was assumed. In this case the relationship between q^i, $\dot{q}^i$ and x^i is generally given by a nonlinear function

$$q^i = g_1^i(x^i), \quad \dot{q}^i = g_2^i(x^i), \quad \forall i \in I \tag{2.36}$$

If as in the previous case the state vector is described in the form

$x = (x^{1T}, x^{2T}, \ldots, x^{nT})^T$, the mechanical part of system S can be expressed as a function of the state vector x

$$S^M: \quad P = (I_n - HG(x)F)^{-1}[HG(x)(Ax+BN(u))+h] \qquad (2.37)$$

where the symbols are as previously defined in (2.32), $G(x)=\mathrm{diag}(G^i(x^i))$, $G^i(x^i)=\partial g_2^i(x^i)/\partial x^i$. Substituting (2.37) into (2.32) and unifying these models, the total system model is obtained in the form analogous to (2.35), with the matrices of the system given as $\hat{A}=Ax+F(I_n-HG(x)F)^{-1}\cdot$ $\cdot[HG(x)Ax+h]$, $\hat{B}=B+F(I_n-HG(x)F)^{-1}HG(x)B$.

Friction effects

In concluding the section about mathematical modelling the effects of friction forces on the robotic system will be considered, since for example the actuator models contain viscous friction elements (e.g. coefficient F_v in A.6.1). Any viscous friction components in the mechanical part of the system can be included by the introduction of a viscous friction term $M_{vf}=F_{vi}\dot{q}^i$ in every joint, where F_{vi} is the coefficient of friction. In this manner the model of robot mechanism dynamics (2.30) takes the form of $P = H(q, \theta)\ddot{q}+h(q, \dot{q}, \theta)+F_v\dot{q}$ where $F_v=$ $= \mathrm{diag}[F_{vi}]$.

The problem of dry friction is more complex in nature and can be considered in two ways. The first approach is derived from the energy analysis i.e. friction component is determined with the energy losses it produces. Namely, one has to take into account the transmission losses using reducer efficiency coefficient, i.e. its mechanical efficiency η. Due to reducer losses, the necessary motor power (output mechanical power) is $P_i\dot{q}^i/\eta_i$. Due to the same reasons the available driving torque P_i^m is obtained as $P_i^m=P_i/N_{vi}\eta_i$, where N_{vi} is gearing ratio. Mechanical efficiency is a function of velocity and can be obtained from catalogue data sheets. This methodology suffers from a disadvantage that the values of friction moments thereby obtained are not sufficiently accurate especially the friction moment at start up. The second approach, applied to all complex spatial systems, involves exact evaluation of dry friction components on the basis of various theoretically valid assumptions which are difficult to satisfy in practice. Coherent with the primary reason for neglecting these effects (the trend towards direct drive motors) mentioned in the Preface, these forces are not included in the dynamic model of the robot mechanism.

2.3 Influence of Mechanical Vibrations on Dynamic Behaviour of Manipulation Robots

The working conditions that industrial robots are generally operating in, often include mechanical vibrations resulting from machine impacts, aperiodic or periodic motions which are usually unbalanced, etc. These vibrations induce time varying forces which are commonly stochastic in nature. An example of such an effect, in particular a sliding motion registered on a track, is shown as a record in Fig. 2.7 whereas Fig. 2.8 shows the record of the vertical acceleration component. It is evident that in the absence of impulse loading, the random acceleration signal can be approximated by "white noise". In order to obtain the differential equations of motion it is assumed that the kinematic chain is rigidly fixed onto the vibrating platform and that the accelerations of the base member equal to the corresponding accelerations of the platform. With the aim of using a model of manipulation robot dynamics which includes the initial acceleration vector, the mathematical model in (2.30) and (2.31) will be transformed into a form which explicitly defines the initial acceleration vectors. Expression for η_i (2.8) for cases when $\vec{\varepsilon}_o = 0$ differs from that when $\vec{\varepsilon}_o \neq 0$ by $\vec{\varepsilon}_o \times \sum_{k=1}^{i} (\vec{r}_{kk} - \vec{r}_{k-1,k})$. Similarly there exists a discrepancy for values of $\vec{a}_i^o$ (2.10). Namely, $\vec{a}_i^o$ when $\vec{\varepsilon}_o \neq 0$ differs from $\vec{a}_i^o(\vec{\varepsilon}_o = 0)$ by $-m_i \vec{\varepsilon}_o \times \sum_{k=1}^{i} (\vec{r}_{kk} - \vec{r}_{k-1,k})$. Bearing this in mind and substituting $-m_i \vec{\varepsilon}_o \times \sum_{k=1}^{i} (\vec{r}_{kk} - \vec{r}_{k-1,k})$ into (2.31) one obtains

$$
\begin{aligned}
h(q, \dot{q}, \theta) = [h_i] = [-\vec{e}_i \sum_{j=i}^{n} \{(\vec{r}_{ji} \times [-m_j \vec{\varepsilon}_o \times \sum_{k=1}^{i} (\vec{r}_{kk} - \vec{r}_{k-1,k}) - \\
- m_j \vec{w}_o + \vec{G}_j] + (\vec{b}_j^o - T_j \vec{\varepsilon}_o)) \bar{\xi}_i + [-m_i \vec{\varepsilon}_o \times \sum_{k=1}^{i} (\vec{r}_{kk} - \vec{r}_{k-1,k}) - \\
- m_j \vec{w}_o + G_j] \xi_i \}] \qquad (2.38)
\end{aligned}
$$

In equation (2.38) for the revolute pair $\xi_i = 0$, $(\bar{\xi}_i = 1)$ vectors $\vec{e}_i \sum_{j=i}^{n} \vec{r}_{ji} \times m_j \vec{w}_o$ and $\vec{e}_i \sum_{j=i}^{n} \vec{r}_{ji} \times [m_j \vec{\varepsilon}_o \times \sum_{k=1}^{i} (\vec{r}_{kk} - \vec{r}_{k-1,k})]$ can be transformed, using the cross products rule as follows

$$
\vec{e}_i \sum_{j=i}^{n} \vec{r}_{ji} \times m_j \vec{w}_o = \vec{e}_i \sum_{j=i}^{n} (m_j \vec{r}_{ji} \times \vec{w}_o) = \vec{w}_o \cdot (\vec{e}_i \times \sum_{j=i}^{n} m_j \vec{r}_{ji}) \qquad (2.39)
$$

$$
\vec{e}_i \sum_{j=i}^{n} \vec{r}_{ji} \times [m_j \vec{\varepsilon}_o \times \sum_{k=1}^{i} (\vec{r}_{kk} - \vec{r}_{k-1,k})] = \sum_{j=i}^{n} m_j \vec{\varepsilon}_o \times \sum_{k=1}^{i} (\vec{r}_{kk} -
$$

$$-\vec{r}_{k-1,k})\cdot(\vec{e}_i\times\vec{r}_{ji})=\vec{\varepsilon}_o\sum_{j=i}^{n} m_j\sum_{k=1}^{i}(\vec{r}_{kk}-\vec{r}_{k-1,k})\times(\vec{e}_i\times\vec{r}_{ji}) \quad (2.40)$$

Substituting (2.39) and (2.40) into expression (2.38) and using (2.30) the following matrix equation is obtained

$$H(q)\ddot{q} + h_1(q)w_o + h_2(q)\varepsilon_o + \zeta(q, \dot{q}) = P \quad (2.41)$$

where, $H_{ik} = -\vec{e}_i \sum_{j=\max(i,k)}^{n} (\vec{b}_{jk}+\vec{r}_{ji}\times\vec{a}_{jk})$

$$h_{1i} = \vec{e}_i\times\sum_{j=i}^{n} m_j\vec{r}_{ji}$$

$$h_{2i} = \vec{e}_i\sum_{j=i}^{n} T_j + \sum_{j=i}^{n} m_j\vec{r}_{j,1}\times(\vec{e}_i\times\vec{r}_{ji}) \quad (2.42)$$

$$\zeta_i = -\vec{e}_i\sum_{j=i}^{n} [\vec{r}_{ji}\times(\vec{a}_j^{\,o}(\vec{\varepsilon}_o=0) + \vec{G}_j+\vec{b}_j^{\,o}]$$

Similarly for the sliding pair $\xi_i=1$, $(\bar{\xi}_i=0)$ vector $\vec{e}_i\sum_{j=i}^{n} m_j\vec{\varepsilon}_o\times\sum_{k=1}^{i}(\vec{r}_{kk}-\vec{r}_{k-1,k})$ can be transformed into

$$\vec{e}_i\sum_{j=i}^{n} m_j\vec{\varepsilon}_o\times\sum_{k=1}^{i}(\vec{r}_{kk}-\vec{r}_{k-1,k}) = \vec{\varepsilon}_o\cdot(\sum_{j=i}^{n} m_j\sum_{k=1}^{i}(\vec{r}_{kk}-\vec{r}_{k-1,k})\times\vec{e}_k) \quad (2.43)$$

Substituting (2.43) into (2.38) and using (2.31) h_1, h_2 and ζ can therefore be described as

$$H_{ik} = -\vec{e}_i\sum_{j=k}^{n}\vec{a}_{jk}$$

$$h_{1i} = \vec{e}_i\sum_{j=i}^{n} m_j$$

$$h_{2i} = \sum_{j=i}^{n} m_j\sum_{i=1}^{j}(\vec{r}_{ii}-\vec{r}_{i-1,i})\times\vec{e}_i = \sum_{j=i}^{n} m_j(\vec{r}_{j,1}\times\vec{e}_i) \quad (2.44)$$

$$\zeta_i = -\vec{e}_i\sum_{j=i}^{n}(\vec{a}_j^{\,o}(\vec{\varepsilon}_o=0)+\vec{G}_j)$$

The inclusion of dynamic models of actuators by which the joints of the mechanism are powered, gives the complete dynamic model of the robot, which represent its dynamics within the considered working regime.

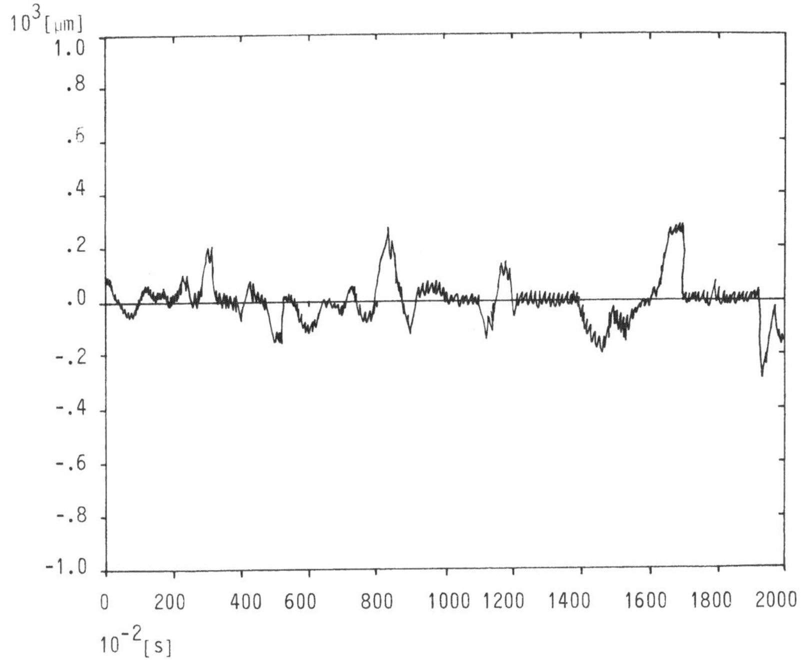

Fig. 2.7. Displacement in vertical direction

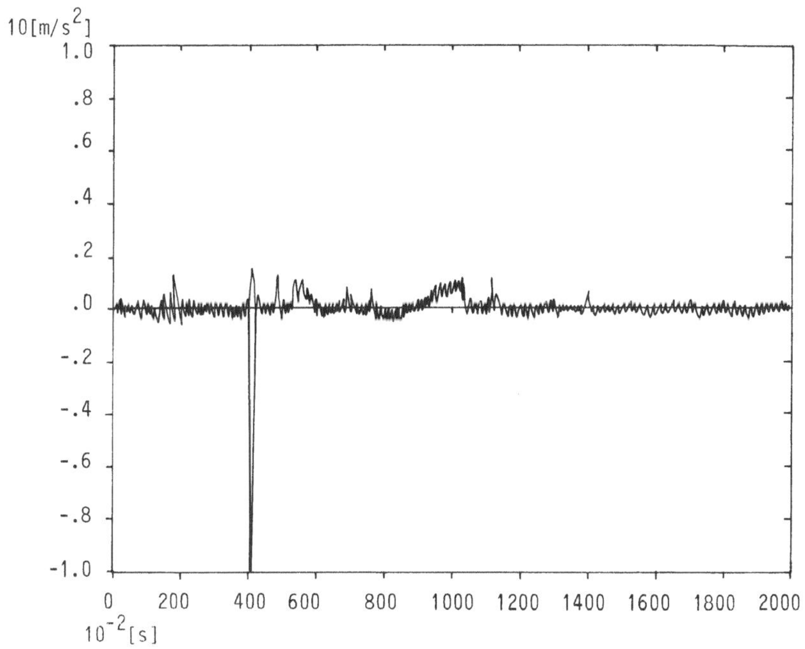

Fig. 2.8. Acceleration in vertical direction

E X A M P L E

On the basis of a mathematical model of the form given by (2.41), (2.42), (2.44) and with the use of programme VIBRO and main programme in Appendix 9, a simulation of the manipulation robot dynamics at the fundament under mechanical vibrations is presented. Geometric and dynamic parameters of the robot UMS-2 (Fig. 2.9) are given in tables 2.1 and 2.2, and the actuator parameters in table 2.3.

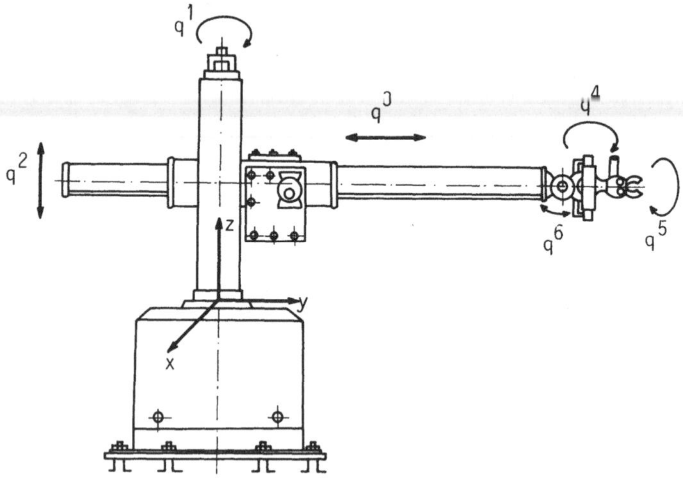

Fig. 2.9. Manipulation robot UMS-2

The motion of joint coordinates was considered, the initial and terminal positions of which are given in table 2.4.

Motion between the initial and final positions was performed on a rectilinear segment with the adopted parabolic velocity profile. Variations of joints coordinates and velocities can be represented in the following form

$$q(t) = q^o + \lambda(t)(q^F - q^o), \qquad \dot{q}(t) = \dot{\lambda}(t)(q^F - q^o), \qquad 0 \le t \le T \tag{2.45}$$

where q^o - initial position

q^F - terminal position

Joint	Joint type*)	Unit vectors of joint axes - $\vec{e}_i$	Vectors $\vec{r}_{ii}$[m]	Vectors $\vec{r}_{i,i+1}$[m]
1	R	0., 0., 1.	0., 0., 0.188	0., 0., -0.192
2	P	0., 0., 1.	0., 0., 0.01	0., 0., -0.01
3	P	0., 1., 0.	0., 0.01, 0.	0., -0.44, 0.
4	R	0., 1., 0.	0., 0.025, 0.	0., -0.025, 0.
5	R	1., 0., 0.	0., 0.025, 0.	0., -0.025, 0.
6	R	0., 0., 1.	0., 0.025, 0.	0., -0.025, 0.

Table 2.1. Geometric parameters of robot

Link	Mass m_i[kg]	Moments of inertia [kgm^2]		
		J_{xi}	J_{yi}	J_{zi}
1	10.	0.	0.	0.0294
2	7.	0.055	0.	0.055
3	4.15	J_S = 0.318, J_N = 0.318		
4	0.5	0.00015	0.00010	0.00015
5	0.5	0.00015	0.00010	0.00015
6	0.5	0.00015	0.00010	0.00015

Table 2.2. Dynamical parameters of robot

Actuator	$F_v\left[\frac{Nm}{rad/s}\right]$	$C_M\left[\frac{Nm}{A}\right]$	$C_E\left[\frac{V}{rad/s}\right]$	$r_R[\Omega]$	J_M[kgm^2]	L_R[H]	N_v[-]	N_m[-]
1	0.00580	1.50000	1.43000	1.60000	0.00003	0.00230	31.2	31.2
2	40.30000	125.40000	120.30000	1.60000	0.00003	0.00230	2616	2616
3	14.50000	75.50000	72.20000	1.60000	0.00003	0.00230	1570	1570
4	0.00750	1.36000	0.32500	3.32000	0.00001	0.00245	25	10
5	0.00750	1.36000	0.32500	3.32000	0.00001	0.00245	25	10
6	0.00750	1.36000	0.32500	3.32000	0.00001	0.00245	25	10

Table 2.3. Actuator parameters

Parameters of electromechanical actuators are presented in Appendix 6.

---

*) R - *revolute*, P - *prismatic*

$\lambda(t) \in [0,1]$ - scalar parameter specifying velocity distribution along the trajectory

$$\lambda(t) = \frac{6}{T^2}\left(\frac{1}{2} - \frac{1}{3}\frac{t}{T}\right)t^2$$

T - time duration of motion

Degrees of freedom	q^1 [rad]	q^2 [m]	q^3 [m]	q^4 [rad]	q^5 [rad]	q^6 [rad]
Initial position	0.2	0.	0.03	0.3	0.5	0.2
Terminal position	0.6	0.10	0.12	0.8	1.0	0.7

Table 2.4. Initial and terminal positions

The motion was simulated under the conditions of mechanical vibrations which are represented by the vectors $\vec{w}_o$ - translational and $\vec{\varepsilon}_o$ - angular accelerations. It is assumed that translational acceleration in the vertical sense w_{oz} and angular acceleration about this direction i.e. ε_{oz}, are the only accelerations acting on the manipulator. They are both expressed as simple sinusoidal functions:

$$w_{oz} = -e_\ell \Omega^2 \sin\Omega t, \qquad \varepsilon_{oz} = -e_r \Omega^2 \sin\Omega t \tag{2.46}$$

where e_ℓ - linear amplitude of platform motion

e_r - angular amplitude

Ω - angular frequency.

This form of vibrations corresponds to vibrations resulting from centrifugal forces generated in machines with unbalanced rotors.

Fig. 2.10 shows the deviation of joint coordinate q^2 from the nominal position for the case of excitation in the form (2.46) and adopted values:

$$e_\ell = 0.001[m], \qquad e_r = 0.001[rad], \qquad \Omega = 100[s^{-1}]$$

Only the deviation of the linear coordinate q^2, which is codirectional with translational acceleration w_o is shown. Dotted line denotes the actual and solid line the nominal position at time instant t. Using the well known relationship between the external and joint coordinates (Section 2.5), the deviation of the manipulator tip from the nominal trajectory in the external coordinate space, can be determined. Satisfactory trajectory tracking can be achieved with the introduction of subsystem (mechanical degrees of freedom of the robot) feedback loops together with load feedback loops.

With the aim of providing a global insight into the effects of feedback, on the characteristics of robot behaviour in the presence of vibrations Fig. 2.11 shows the deviation of q^2 from the nominal trajectory.

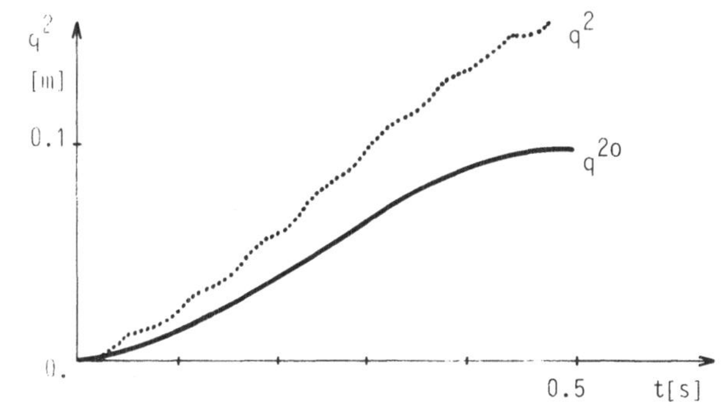

Fig. 2.10. Deviation of position q^2 from nominal value

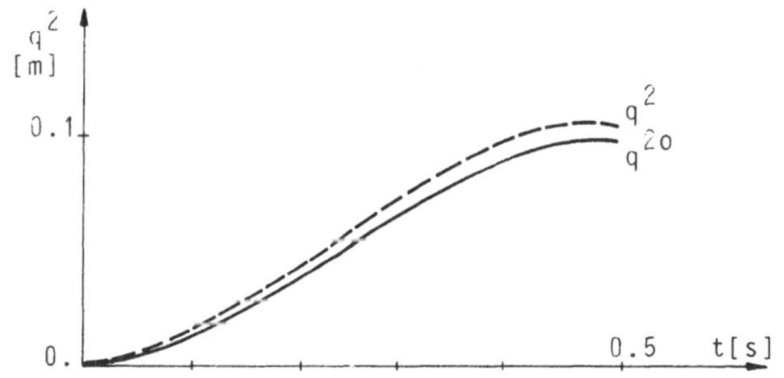

Fig. 2.11. Deviation of trajectory q^2 in the presence of feedback loops

The problem of maintaining high robot performance in conditions of active mechanical vibrations will not be considered in this textbook. Reference [13] deals with this problem in more detail.

2.4 Dynamics of Manipulation Robots with Gripper Constrained Motion

As mentioned in Ch. 1 the kinematic structure of the robot changes during the execution of a task i.e. from open it is transferred to closed kinematic chains. We shall consider closed kinematic chains whose dynamics are derived from those of an open kinematic chain by imposing some constraints on the motion of the last manipulator link. Problems of this type are often encountered in practical tasks, such as: grinding, polishing, engraving, writing, fitting, bilateral manipulation etc. The effects produced by impacts on contact with the working surface will not be considered.

*Mathematical model of the manipulator with constraints on gripper motion**)

In order to form the mathematical model which describes the dynamics of the closed configuration manipulator, let us first consider an open configuration manipulation mechanism which is subjected to an external force $\vec{F}_A$ and moment $\vec{M}_A$ (Fig. 2.11).

We assume that the mechanism consists of n segments each, with a single d.o.f. and that the mechanism position is described by n-dimensional vector of generalized coordinates q. Using general theorems of mechanics, analogously to section 2.2, the dynamic model of such a mechanism is obtained,

$$H\ddot{q} + h = P + D_1F_A + D_2M_A \qquad (2.47)$$

where: H - inertial matrix, dimension (n×n), $\ddot{q}$ - vector of generalized accelerations of degrees of freedom, order (n×1), h - vector of gravitational, centrifugal and Coriolis forces, order (n×1), P - driving torques (forces) vector at mechanism joints, order (n×1), D_1 and D_2 are adjoint matrices of force F_A and moment M_A respectively, (dimensions (n×3)).

It can be shown that matrices D_1 and D_2 are of the form:

---

*) In writing this section references [11, 12] were used.

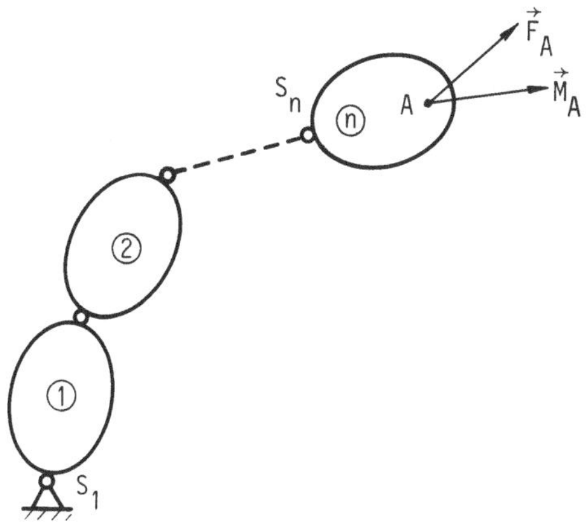

Fig. 2.11. Open kinematic chain subjected to action of external force $\vec{F}_A$ and moment $\vec{M}_A$

$$D_{1\ (n\times 3)} = \begin{bmatrix} d_{11}^T \\ \vdots \\ d_{1n}^T \end{bmatrix}, \qquad d_{1i} = \begin{cases} \vec{e}_i \times \vec{r}_{iA}, & \bar{\xi}_i = 1^{*)} \\ \vec{e}_i, & \bar{\xi}_i = 0 \end{cases} \tag{2.48}$$

$$D_{2\ (n\times 3)} = \begin{bmatrix} d_{21}^T \\ \vdots \\ d_{2n}^T \end{bmatrix}, \qquad d_{2i} = \begin{cases} \vec{e}_i, & \bar{\xi}_i = 1 \\ 0, & \bar{\xi}_i = 0 \end{cases} \tag{2.49}$$

In (2.48) $\vec{r}_{iA}$ represents the distance vector from axis of the i-th mechanism joint to the point at which the force acts. Using the notation from preceding text, $\vec{r}_{iA}$ can be given as follows

$$\vec{r}_{iA} = \overrightarrow{S_i A} = \sum_{k=i}^{n-1} (\vec{r}_{kk} - \vec{r}_{k,k+1}) + \vec{r}_{nn} + \vec{p} \tag{2.50}$$

where $\vec{p}$ - distance vector from centre of mass of the last link to the point at which the force acts.

Dynamic model (2.47) can be expressed in more suitable form as follows:

$$H\ddot{q} + h = P + DR_A \tag{2.51}$$

where

---

*) $\bar{\xi}_i$ - defines type of joint, $\bar{\xi}_i=1$ (revolute) $\bar{\xi}_i=0$ (prismatic).

$$D_{(n\times 6)} = [D_{1\,(n\times 3)} \mid D_{2\,(n\times 3)}]$$
$$R_{A\,(6\times 1)} = \left[\begin{array}{c} F_{A\,(3\times 1)} \\ \hline M_{A\,(3\times 1)} \end{array}\right] \tag{2.52}$$

With the aim of defining the robot functional motion we introduce a generalized vector of position X_g. For a robot with six d.o.f. X_g has the following form:

$$X_g = [x_A\ y_A\ z_A\ \theta\ \varphi\ \psi]^T \tag{2.53}$$

where Descartes' coordinates (x_A, y_A, z_A) and Euler's angles (θ, φ, ψ) define the position and gripper orientation, respectively (Fig. 2.12).

Let us now consider a manipulation task where the gripper cannot move freely, but is subjected to constraints. In this manner one obtains a closed chain. The imposed constraints reduce the number of d.o.f. Let n_r be the reduced number of d.o.f. If n is the number of d.o.f., then $n_r \leq n$, and equivalence is satisfied when there are no constraints. We shall introduce n_r free and independent parameters $(u_1, \ldots, u_{nr})$, in order to define the constrained position of the gripper. The reduced vector of position X_r thus obtained,

$$X_r = [u_1\ \cdots\ u_{nr}]^T \tag{2.54}$$

defines the relative gripper position with respect to constraints. Constrained motion can be expressed in second order Jacobian form

$$\ddot{X}_g = J_r \ddot{X}_r + A_r \tag{2.55}$$

where J_r - reduced Jacobian, dimension $(n\times n_r)$ and A_r - adjoined reduced vector, of order $(n\times 1)$.

It is well known*) that for motion without constraints

*) If the functional relationship, between the external and joint coordinates of the manipulator tip is used: $X_g = f(q)$, second derivative gives: $\ddot{X}_g = \frac{\partial^2 f}{\partial q^2}\dot{q}^2 + \frac{\partial f}{\partial q}\ddot{q}$, which with $\frac{\partial f}{\partial q} = J$ and $A = \frac{\partial^2 f}{\partial f^2}\dot{q}^2$ becomes $\ddot{X}_g = J\ddot{q} + A$ where J-$(n\times n)$ Jacobian matrix, A - adjoined $(n\times 1)$ vector and $\ddot{X}_g$ - acceleration vector in external coordinates. Section 2.5 deals with this in more detail.

$$\ddot{X}_g = J\ddot{q} + A \tag{2.56}$$

From the equivalence of expressions (2.55) and (2.56) one obtains

$$\ddot{q} = J^{-1}J_rX_r + J^{-1}(A_r-A) \tag{2.57}$$

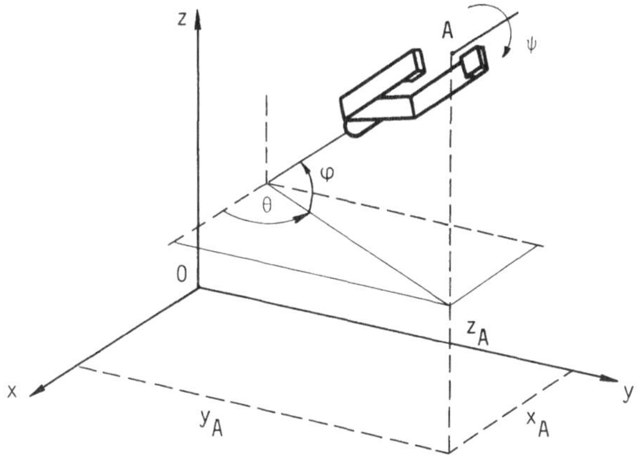

Fig. 2.12. External coordinates for defining motions

Prior to substituting (2.57) in equation (2.51) it must be pointed out that reaction forces R_A resulting from the imposed constraints appear and are included in the dynamic model (2.51). Depending on the type of constraint and the manipulator configuration, there exist $6-(n-n_r)$ scalar conditions which must be satisfied by the six-component reaction force R_A. These conditions can be written in matrix form

$$ER_A = 0 \tag{2.58}$$

where E - matrix, dimension $(6-n+n_r)\times 6$.

Equations (2.51) and (2.57) together with condition (2.58) define the closed configuration dynamic model.

During the calculation of the nominal drive it is assumed that the forces required for the execution of the manipulation task are known quantities. Therefore F_A and M_A, in other words R_A are known and satisfy condition (2.58). The required drives P are calculated using matrix equations (2.51). However, if unknown motion and reactions need to be calculated, (2.57) is substituted into (2.51) giving

$$HJ^{-1}J_r\ddot{X}_r + h - DR_A = P - HJ^{-1}(A_r - A) \tag{2.59}$$

Combining (2.58) and (2.59) allows the following matrix equation to be written,

$$\left[\begin{array}{c|c} HJ^{-1}J_r & -D \\ \hline 0 & E \end{array}\right]\left[\begin{array}{c} \ddot{X}_r \\ \hline R_A \end{array}\right] + \left[\begin{array}{c} h \\ \hline 0 \end{array}\right] = \left[\begin{array}{c} P \\ \hline 0 \end{array}\right] + \left[\begin{array}{c} -HJ^{-1}(A_r - A) \\ \hline 0 \end{array}\right] \tag{2.60}$$

the dimensions of which are

$$\left[\begin{array}{c|c} (n \times n_r) & (n \times 6) \\ \hline ((6-n+n_r) \times n_r & ((6-n+n_r) \times 6 \end{array}\right]\left[\begin{array}{c} (n_r \times 1) \\ \hline (6 \times 1) \end{array}\right] + \left[\begin{array}{c} (n \times 1) \\ \hline ((6-n+n_r) \times 1) \end{array}\right] =$$

$$= \left[\begin{array}{c} (n \times 1) \\ \hline ((6-n+n_r) \times 1) \end{array}\right] + \left[\begin{array}{c} (n \times 1) \\ \hline ((6-n+n_r) \times 1) \end{array}\right]$$

Equation (2.60) represents a system of n_r+6 equations which can be solved for $\ddot{X}_r$ and R_A. Unknown motion q can be obtained using expression (2.57). Dimensionality of the system of equations (2.60) can be reduced by introducing the reduced reaction vector R_{Ar}, whose dimension is $(n-n_r)\times 1$.

This is easily done because the six-component reaction force has $n-n_r$ independent components. R_A can now be expressed as follows,

$$R_A = GR_{Ar} \tag{2.61}$$

where G - $(6\times(n-n_r))$ dimensional matrix.

Substituting (2.61) into (2.59) we obtain the matrix equation

$$HJ^{-1}J_r\ddot{X}_r + h - DGR_{Ar} = P - HJ^{-1}(A_r - A) \tag{2.62}$$

in other words

$$[HJ^{-1}J_r \mid -DG]\left[\begin{array}{c} \ddot{X}_r \\ \hline R_{Ar} \end{array}\right] = P - h - HJ^{-1}(A_r - A) \tag{2.63}$$

whose dimensions are

$$[(n\times n_r) \mid (n\times(n-n_r))] \left[\frac{(n_r\times 1)}{((n-n_r)\times 1)} \right] = (n\times 1)$$

Thus, (2.63) represents a system of n equations which can be solved for n unknowns ($\ddot{X}_r$, R_{Ar}).

These theoretical results will be applied to one of the possible constraint types. This will be a constraint of the surface type because of its practical significance.

Gripper moving along a surface

We consider again a manipulator with the gripper which can not move freely but its point A is forced to move along a given surface (Fig. 2.13)

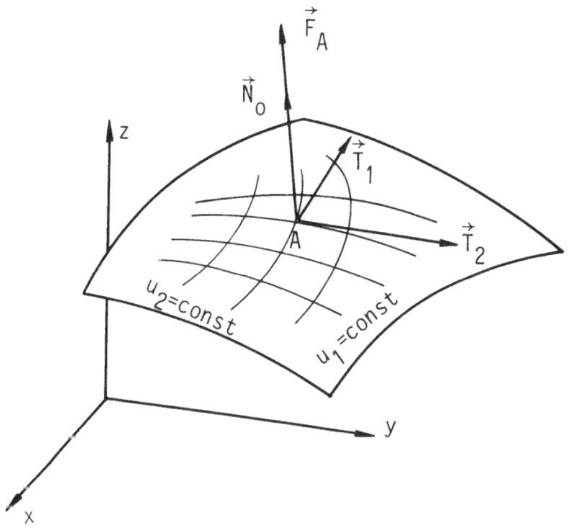

Fig. 2.13. Point A at a mobile surface

Let us define the relative position of the gripper point A with respect to a surface by means of two parameters u_1 and u_2. This leads to the parametric form of the moving surface (nonstationary constraint):

$$x = f_x(u_1, u_2, t)$$

$$y = f_y(u_1, u_2, t) \qquad (2.64)$$

$$z = f_z(u_1, u_2, t)$$

Now n_r=n-1 and the reduced position vector for a manipulator with six d.o.f. will be:

$$X_r = [u_1\ u_2\ \theta\ \varphi\ \psi]^T, \quad \text{for } n=6 \tag{2.65}$$

i.e., $u_3 = \Theta$, $u_4 = \varphi$, $u_5 = \psi$.

The first and the second derivative of (2.64) give:

$$\begin{aligned}
\dot{x} &= \frac{\partial f_x}{\partial u_1}\dot{u}_1 + \frac{\partial f_x}{\partial u_2}\dot{u}_2 + \frac{\partial f_x}{\partial t} \\
\dot{y} &= \frac{\partial f_y}{\partial u_1}\dot{u}_1 + \frac{\partial f_y}{\partial u_2}\dot{u}_2 + \frac{\partial f_y}{\partial t} \\
\dot{z} &= \frac{\partial f_z}{\partial u_1}\dot{u}_1 + \frac{\partial f_z}{\partial u_2}\dot{u}_2 + \frac{\partial f_z}{\partial t}
\end{aligned} \tag{2.66}$$

and

$$\ddot{x} = \underbrace{\frac{\partial^2 f_x}{\partial u_1^2}\dot{u}_1^2 + 2\frac{\partial^2 f_x}{\partial u_1 \partial u_2}\dot{u}_1\dot{u}_2 + \frac{\partial^2 f_x}{\partial u_2^2}\dot{u}_2^2 + 2\frac{\partial^2 f_x}{\partial u_1 \partial t}\dot{u}_1 + 2\frac{\partial^2 f_x}{\partial u_2 \partial t}\dot{u}_2 + \frac{\partial^2 f_x}{\partial t^2}}_{\alpha_x} + \frac{\partial f_x}{\partial u_1}\ddot{u}_1 + \frac{\partial f_x}{\partial u_2}\ddot{u}_2$$

$$\ddot{y} = \underbrace{\frac{\partial^2 f_y}{\partial u_1^2}\dot{u}_1^2 + 2\frac{\partial^2 f_y}{\partial u_1 \partial u_2}\dot{u}_1\dot{u}_2 + \frac{\partial^2 f_y}{\partial u_2^2}\dot{u}_2^2 + 2\frac{\partial^2 f_y}{\partial u_1 \partial t}\dot{u}_1 + 2\frac{\partial^2 f_y}{\partial u_2 \partial t}\dot{u}_2 + \frac{\partial^2 f_y}{\partial t^2}}_{\alpha_y} + \frac{\partial f_y}{\partial u_1}\ddot{u}_1 + \frac{\partial f_y}{\partial u_2}\ddot{u}_2$$

$$\ddot{z} = \underbrace{\frac{\partial^2 f_z}{\partial u_1^2}\dot{u}_1^2 + 2\frac{\partial^2 f_z}{\partial u_1 \partial u_2}\dot{u}_1\dot{u}_2 + \frac{\partial^2 f_z}{\partial u_2^2}\dot{u}_2^2 + 2\frac{\partial^2 f_z}{\partial u_1 \partial t}\dot{u}_1 + 2\frac{\partial^2 f_z}{\partial u_2 \partial t}\dot{u}_2 + \frac{\partial^2 f_z}{\partial t^2}}_{\alpha_z} + \frac{\partial f_z}{\partial u_1}\ddot{u}_1 + \frac{\partial f_z}{\partial u_2}\ddot{u}_2 \tag{2.67}$$

The set of equations (2.67) can be written in the compact matrix form as follows:

$$\begin{bmatrix} \ddot{x}_A \\ \ddot{y}_A \\ \ddot{z}_A \end{bmatrix} = \begin{bmatrix} \frac{\partial f_x}{\partial u_1} & \frac{\partial f_x}{\partial u_2} \\ \frac{\partial f_y}{\partial u_1} & \frac{\partial f_y}{\partial u_2} \\ \frac{\partial f_z}{\partial u_1} & \frac{\partial f_z}{\partial u_2} \end{bmatrix} \begin{bmatrix} \ddot{u}_1 \\ \ddot{u}_2 \end{bmatrix} + \begin{bmatrix} \alpha_x \\ \alpha_y \\ \alpha_z \end{bmatrix} \tag{2.68}$$

Now, the Jacobian form (2.56) can be obtained. The reduced Jacobian and the reduced adjoint matrix respectively are:

$$J_r = \left[\begin{array}{cc|c} \frac{\partial f_x}{\partial u_1} & \frac{\partial f_x}{\partial u_2} & \\ \frac{\partial f_y}{\partial u_1} & \frac{\partial f_y}{\partial u_2} & O_{(3\times(n-3))} \\ \frac{\partial f_z}{\partial u_1} & \frac{\partial f_z}{\partial u_2} & \\ \hline \multicolumn{2}{c|}{O_{((n-3)\times 2)}} & I_{((n-3)\times(n-3))} \end{array}\right], \qquad A_r = \left[\begin{array}{c} \alpha_x \\ \alpha_y \\ \alpha_z \\ \hline O_{((n-3)\times 1)} \end{array}\right] \tag{2.69}$$

where α_x, α_y, α_z are as defined in eq. (2.67) and I is a unit matrix of the corresponding dimension.

Let us consider the reactions. It is clear that there exists a reaction force $\vec{F}_A$ perpendicular to the surface and the reaction moment equals zero ($\vec{M}_A=0$) (Fig. 2.13). If we define two tangents

$$\vec{T}_1 = \left\{\frac{\partial f_x}{\partial u_1}, \frac{\partial f_y}{\partial u_1}, \frac{\partial f_z}{\partial u_1}\right\}$$

$$\vec{T}_2 = \left\{\frac{\partial f_x}{\partial u_2}, \frac{\partial f_y}{\partial u_2}, \frac{\partial f_z}{\partial u_2}\right\} \tag{2.70}$$

having the unit vectors

$$\vec{T}_{o1} = \vec{T}_1/|\vec{T}_1|, \qquad \vec{T}_{o2} = \vec{T}_2/|\vec{T}_2| \tag{2.71}$$

then it holds that $\vec{F}_A \perp \vec{T}_{o1}$ and $\vec{F}_A \perp \vec{T}_{o2}$ or

$$\vec{T}_{o1} \cdot \vec{F}_A = 0, \qquad \vec{T}_{o2} \cdot \vec{F}_A = 0 \tag{2.72}$$

The conditions (2.72) together with the condition $\vec{M}_A = 0$ can be written in the form (2.58) where:

$$E = \left[\begin{array}{c|c} T_{o1}^T & O_{(1\times 3)} \\ \hline T_{o2}^T & O_{(1\times 3)} \\ \hline O_{(3\times 3)} & I_{(3\times 3)} \end{array}\right]_{(5\times 6)} \tag{2.73}$$

Now, all the elements of the dynamic model (2.60) are determined and the model can be solved.

The concept of independent reaction components can also be applied. Then the unit vector $\vec{N}_o$ perpendicular to the surface can be obtained as:

$$\vec{N}_o = \vec{T}_{o1} \times \vec{T}_{o2} \tag{2.74}$$

Now it holds that $\vec{F}_A || \vec{N}_o$ and, accordingly,

$$\vec{F}_A = S\, \vec{N}_o \tag{2.75}$$

where

$$S = |\vec{F}_A| \tag{2.76}$$

is the independent component.

Now, the reduced reaction vector is:

$$R_{Ar} = [S]_{(1\times 1)} \tag{2.77}$$

since $n - n_r = 1$. The matrix G which transforms the reaction components (eq. (2.61)) is:

$$G = \left[N_o^T \,\vdots\, O_{(1\times 3)}\right]^T \tag{2.78}$$

and has the dimensions (6×1).

Now, all the elements of the dynamic model (2.63) are determined and the model can be solved.

Let us introduce the friction force:

$$\vec{F}_f = -\mu S \vec{v}_{Aro} \tag{2.79}$$

where μ is friction coefficient and $\vec{v}_{Aro}$ is the unit vector:

$$\vec{v}_{Aro} = \vec{v}_{Ar}/|\vec{v}_{Ar}| \tag{2.80}$$

and $\vec{v}_{Ar}$ is the relative velocity of gripper point A with respect to the surface. This relative velocity can be expressed as the difference:

$$\vec{v}_{Ar} = \vec{v}_A - \vec{v}_s \tag{2.81}$$

where $\vec{v}_A$ is the velocity of point A and $\vec{v}_s$ is the velocity of the corresponding surface point. Since $\vec{v}_A$ can be obtained from (2.66) and $\vec{v}_s = \{\frac{\partial f_x}{\partial t}, \frac{\partial f_y}{\partial t}, \frac{\partial f_z}{\partial t}\}$, one obtains:

$$v_{Ar} = \begin{bmatrix} \frac{\partial f_x}{\partial u_1} & \frac{\partial f_x}{\partial u_2} \\ \frac{\partial f_y}{\partial u_1} & \frac{\partial f_y}{\partial u_2} \\ \frac{\partial f_z}{\partial u_1} & \frac{\partial f_z}{\partial u_2} \end{bmatrix} \begin{bmatrix} \dot{u}_1 \\ \dot{u}_2 \end{bmatrix} \tag{2.82}$$

The friction force produces an additional component of generalized forces and thus the model (2.62) becomes:

$$HJ^{-1}J_r X_r + h - D(G+G')S = P - HJ^{-1}(A_r - A) \tag{2.83}$$

where

$$G' = [-\mu v_{Aro}^T \mid O_{(1\times3)}]^T \tag{2.84}$$

Finally the dynamic model (2.63) becomes:

$$[HJ^{-1}J_r \,\vdots\, -D(G+G')]\begin{bmatrix}\ddot{X}_r \\ \hdashline S\end{bmatrix} = P-h-HJ^{-1}(A_r-A) \tag{2.85}$$

and it can be solved for n unknowns ($\ddot{X}_r$, S). For a manipulator having six d.o.f., this model is solved and the simulation results are given at the end of this section.

Cases of restricted gripper motion

A number of characteristic examples of constrained gripper motion which are often met in practice will be given. Let us consider a task of writing on a given surface (Fig. 2.14).

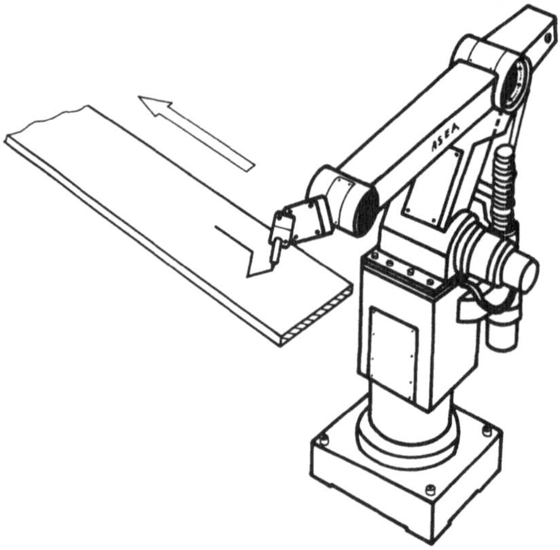

Fig. 2.14. Task of writing

If a manipulator having five degrees of freedom is to be used, four independent parameters have to be defined. In this case they are, $u_1(t)$ and $u_2(t)$ - defining position and angles $\theta(t)$ and $\varphi(t)$ - defining the orientation of the "pen" (Fig. 2.15). Thus the reduced vector of position $X_r(t)$ is given and if the value of the force to be attained while writing is known, nominal dynamics can be solved.

Another example of constrained gripper motion to be considered is in the task of grinding (Fig. 2.16).

Case (a) considers a flat surface rotating as shown in the figure.

However, in practice the surface is not ideally flat and the axis of rotation is not perpendicular to the surface. This further implies that the motion of the surface is not a simple rotation. For this reason, reaction force $\vec{F}_A$ and friction force $\vec{F}_f$ are not constant and therefore lead to vibrations of the workpiece and the gripper.

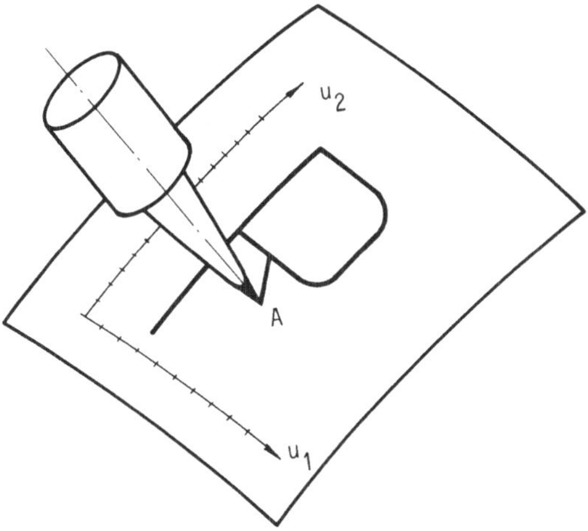

Fig. 2.15. Writing on a surface

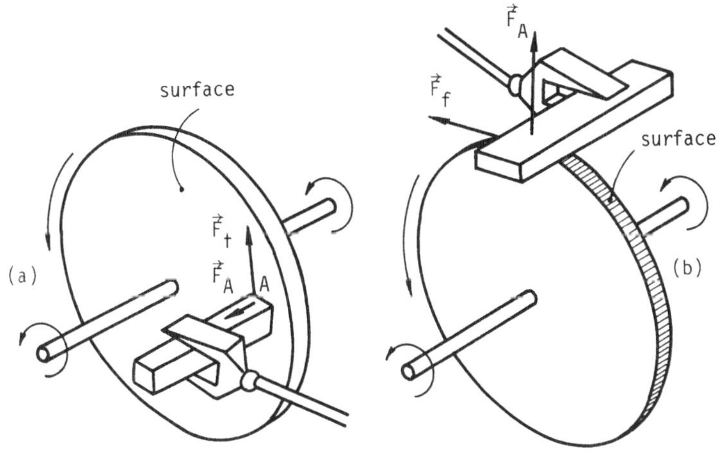

Fig. 2.16. Task of grinding

Case shown in Fig. 2.16(b) considers a cylindrical surface. The rotation is not ideal because the cylinder is not ideal, furthermore the axis of rotation does not coincide with the centre of mass axis. Consequently the reaction force varies and produces vibrations of the workpiece. An industrial robot performing a grinding task is shown in Fig. 2.17.

Let another example of constrained gripper motion be considered; this is the process of assembling machine parts, which is a very characteristic and probably the most delicate manipulation task in industrial applications.

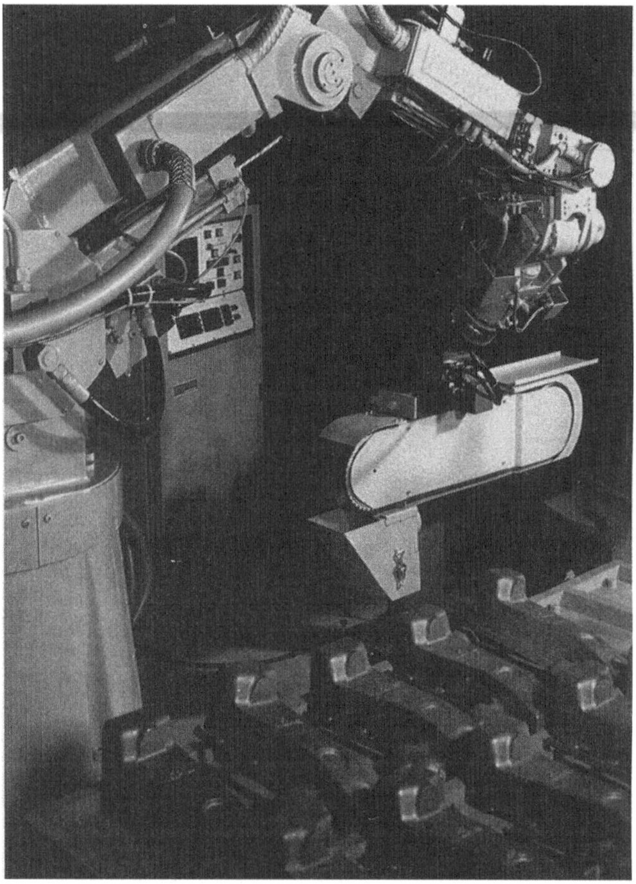

Fig. 2.17. Manipulation task of grinding

In Fig. 2.18, a manipulation task of joining machine parts by inserting a cylindrical shaft is given. With insertion, two problems are identified: inserting a cylindrical object (Fig. 2.19) and inserting a rectangular object (Fig. 2.20). In the case of a cylindrical object being inserted $n_r=2$ and the reduced vector of position $X_r = [u_1\ u_2]^T$ (Fig. 2.19). The reaction force and moment satisfy the condition $\vec{h}\cdot\vec{F}_A = 0$, $\vec{h}\cdot\vec{M}_A = 0$ which can be written in the form (2.58).

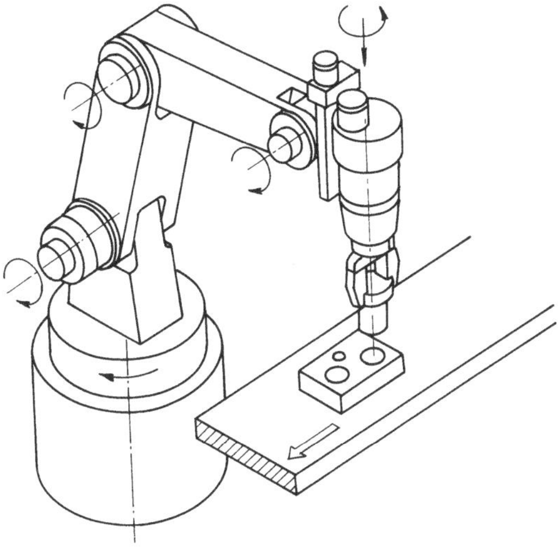

Fig. 2.18. A manipulation task of joining

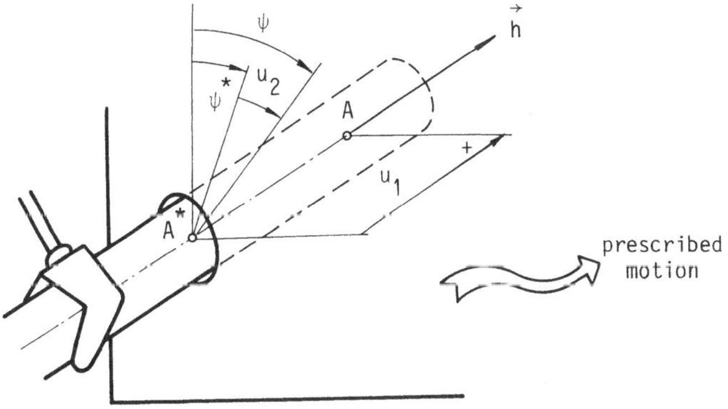

$u_1 = \overline{A^*A}$ - relative translational motion

$u_2 = \psi - \psi^*$ - relative rotation about $\vec{h}$

Fig. 2.19. Insertion of a cylindrical object

For the case of a rectangular object being inserted $n_r=1$ and $X_r = [u_1]$ (Fig. 2.20), and the force $\vec{F}_A$ satisfies condition $\vec{h}\cdot\vec{F}_A=0$.

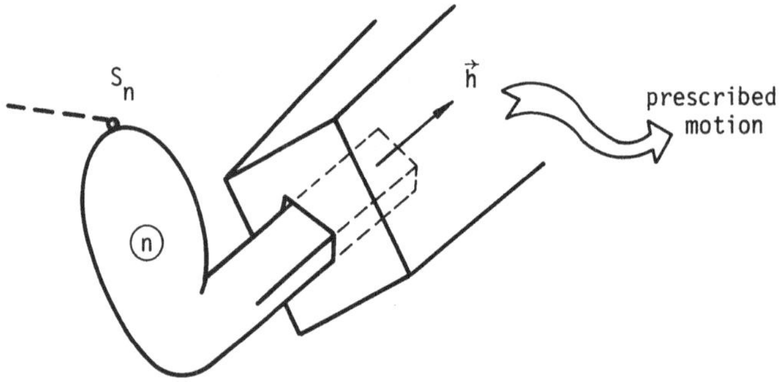

Fig. 2.20. Insertion of a rectangular object

The task of automatic fitting (assembling), in other words its dynamics and control synthesis, will be dealt with in the next volume of this textbook series.

Following the presented considerations, simulation result for constraints of the surface type are given.

E X A M P L E

As an example of non-stationary constraint of the surface type a manipulation robot performing the task of grinding will be considered. In the context of this manipulation task the dynamics of open as well as closed kinematic chains will be encountered. During the approach of the gripper (cutting-tool) towards the workpiece, its motion is unconstrained, whereas at the instant that contact occurs, the chain closes and constrained motion is observed. The nominal dynamics of free and constrained motions, as well as the external force resulting from resistance to cutting are calculated.

Let us consider a manipulation robot performing the task of grinding a workpiece which is moving at a constant speed of 0.1 m/s (Fig. 2.21). During the time T_1 the manipulator moves freely i.e. the gripper (cutting tool) is approaching the workpiece. We assume that contact between the gripper is free of impact and that the end of period T_1 is also the beginning of period T_2 (point A_1 in Fig. 2.21). Motion between points A_1 and A_2 is constrained and the manipulator is considered as a closed chain. Motion of the manipulation robot relative to the surface, is defined using the parameters u_1 and u_2, as shown in the preceding text. During the grinding period parameter u_1 is constant whereas parameter u_2 is of the form

$$u_2(t) = \frac{L}{T} t = 0.2t \tag{2.86}$$

where L[m] - length (L = 0.4), t[s] - total grinding time (T = 2).

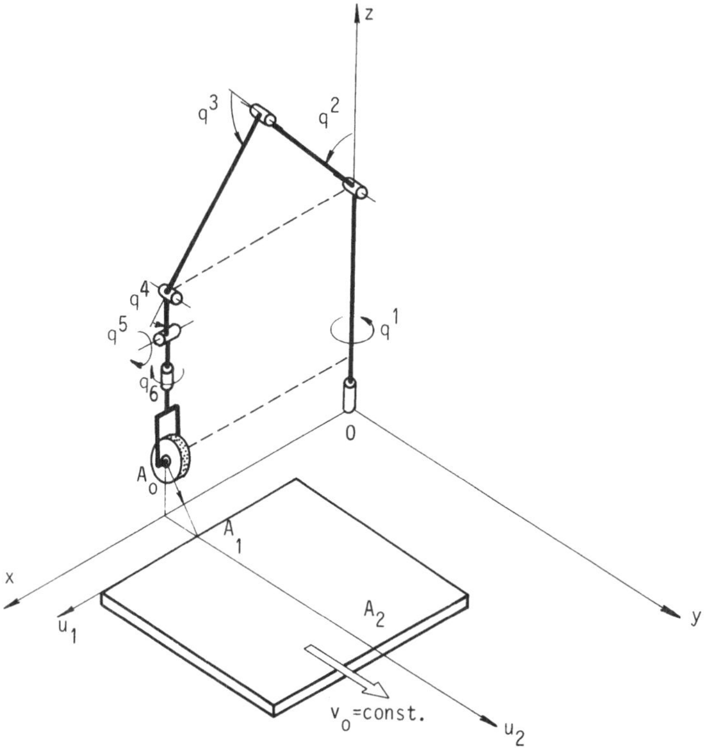

Fig. 2.21. Manipulation robot performing the task of grinding

Using expressions (2.64)-(2.69) one obtains the reduced Jacobian J_r and the adjoined vector A_r of the form:

$$J_r = \left[\begin{array}{cc|c} \frac{\partial f_x}{\partial u_1} & \frac{\partial f_x}{\partial u2} & \\ \frac{\partial f_y}{\partial u_1} & \frac{\partial f_y}{\partial u_2} & O_{(3\times(n-3))} \\ \frac{\partial f_z}{\partial u_1} & \frac{\partial f_z}{\partial u_2} & \\ \hline O_{((n-3)\times 2)} & & I_{((n-3)\times(n-3))} \end{array}\right] = \left[\begin{array}{cc|ccc} 1 & 0 & 0 & 0 & 0 \\ 0 & 1 & 0 & 0 & 0 \\ 0 & 0 & 0 & 0 & 0 \\ \hline 0 & 0 & 1 & 0 & 0 \\ 0 & 0 & 0 & 1 & 0 \\ 0 & 0 & 0 & 0 & 1 \end{array}\right] \tag{2.87}$$

$$A_r = \begin{bmatrix} \alpha_x \\ \alpha_y \\ \alpha_z \\ \hline O_{((n-3)\times 1)} \end{bmatrix} = \begin{bmatrix} 0 \\ 0 \\ 0 \\ \hline 0 \\ 0 \\ 0 \end{bmatrix} \tag{2.88}$$

Tangents $\vec{T}_1$ and $\vec{T}_2$, defined by expression (2.70) are of the form:

$$\vec{T}_1 = \left\{\frac{\partial f_x}{\partial u_1}, \frac{\partial f_y}{\partial u_1}, \frac{\partial f_z}{\partial u_1}\right\} = \{1, 0, 0\}$$

$$\vec{T}_2 = \left\{\frac{\partial f_x}{\partial u_2}, \frac{\partial f_y}{\partial u_2}, \frac{\partial f_z}{\partial u_2}\right\} = \{0, 1, 0\} \tag{2.89}$$

and the unit vectors (2.71) are

$$\vec{T}_{o1} = \vec{T}_1/|\vec{T}_1| = \{1, 0, 0\}$$

$$\vec{T}_{o2} = \vec{T}_2/|\vec{T}_2| = \{0, 1, 0\} \tag{2.90}$$

From the theory of metal cutting it is evident that during flat grinding two mutually perpendicular reaction components exist, i.e. radial F_r and tangential F_t. Furthermore it is known, from the relevant literature, that the ratio F_r/F_t is within the limits of 1.5-3. In this example the ratio adopted is $F_r/F_t=2$.

The reaction force F_A (force resisting to cutting F_r) satisfies condition (2.72), and matrix E defined by expression (2.73) is of the form,

$$E = \begin{bmatrix} T_{o1}^T & O_{(1\times 3)} \\ \hline T_{o2}^T & O_{(1\times 3)} \\ \hline O_{(3\times 3)} & I_{(3\times 3)} \end{bmatrix} = \left[\begin{array}{ccc|ccc} 1 & 0 & 0 & 0 & 0 & 0 \\ \hline 0 & 1 & 0 & 0 & 0 & 0 \\ \hline 0 & 0 & 0 & 1 & 0 & 0 \\ 0 & 0 & 0 & 0 & 1 & 0 \\ 0 & 0 & 0 & 0 & 0 & 1 \end{array}\right] \tag{2.91}$$

Thus conditions (2.58) are obtained

$$ER_A = \begin{bmatrix} 1 & 0 & 0 & 0 & 0 & 0 \\ 0 & 1 & 0 & 0 & 0 & 0 \\ 0 & 0 & 0 & 1 & 0 & 0 \\ 0 & 0 & 0 & 0 & 1 & 0 \\ 0 & 0 & 0 & 0 & 0 & 1 \end{bmatrix} \begin{bmatrix} F_{Ax} \\ F_{Ay} \\ F_{Az} \\ M_{Ax} \\ M_{Ay} \\ M_{Az} \end{bmatrix} = \begin{bmatrix} F_{Ax} \\ F_{Ay} \\ M_{Ax} \\ M_{Ay} \\ M_{Az} \end{bmatrix} = 0 \qquad (2.92)$$

where: F_{Ax}, F_{Ay}, F_{Az} - projections of reaction forces

M_{Ax}, M_{Ay}, M_{Az} - projections of moments of reaction forces onto Cartesian axis x, y, z.

It is clear that the two force components and the three moment components are identically equal zero. The third force component F_{Az} is non--zero and is in fact equal to radial component F_r.

The unit vector $\vec{N}_o$, defined by expression (2.74) is of the form:

$$\vec{N}_o = \vec{T}_{o1} \times \vec{T}_{o2} = \{1,\ 0,\ 0\} \times \{0,\ 1,\ 0\} = \{0,\ 0,\ 1\} \qquad (2.93)$$

and matrix G (2.78) is defined as:

$$G = \left[N_o^T \,\vdots\, 0_{(1\times 3)} \right]^T = [0 \quad 0 \quad 1 \,\vdots\, 0 \quad 0 \quad 0]^T \qquad (2.94)$$

In the case under consideration the friction force equals the tangential component F_t because

$$v_{Ar} = \begin{bmatrix} \frac{\partial f_x}{\partial u_1} & \frac{\partial f_x}{\partial u_2} \\ \frac{\partial f_y}{\partial u_1} & \frac{\partial f_y}{\partial u_2} \\ \frac{\partial f_z}{\partial u_1} & \frac{\partial f_z}{\partial u_2} \end{bmatrix} \begin{bmatrix} \dot{u}_1 \\ \dot{u}_2 \end{bmatrix} = \begin{bmatrix} 1 & 0 \\ 0 & 1 \\ 0 & 0 \end{bmatrix} \begin{bmatrix} 0 \\ \dot{u}_2 \end{bmatrix} = [0 \quad \dot{u}_2 \quad 0]^T \qquad (2.95)$$

$$\vec{v}_{Aro} = \vec{v}_{Ar} / |\vec{v}_{Ar}| = [0 \quad \dot{u}_2 \quad 0]^T / \dot{u}_2 = [0,\ 1,\ 0]^T \qquad (2.96)$$

Finally, the friction force $\vec{F}_f$ (2.79) is of the form:

$$\vec{F}_f = -\mu S\vec{v}_{Aro} = -\mu S[0,\ 1,\ 0]^T = [0,\ \mu S,\ 0]^T = [0,\ -F_t,\ 0]^T \qquad (2.97)$$

Matrix G' (2.84) is thus defined as:

$$G' = \left[-\mu v^T_{Aro} \mid 0_{(1\times 3)}\right]^T = \left[0 \quad -\mu \quad 0 \mid 0 \quad 0 \quad 0\right]^T \qquad (2.98)$$

All the components of the dynamic model (2.85) are thus defined and it can be solved for $\ddot{X}_r$ and S.

Solving the second relation of expression (A.6.1) for $\ddot{q}$, given in Appendix 6 one obtains,

$$\ddot{q} = -\mathrm{diag}(\frac{F^i_v}{J^i_R})\dot{q} + \mathrm{diag}(\frac{C^i_M N^i_m}{J^i_R})i_R - \mathrm{diag}(\frac{1}{J^i_R})M^* \qquad (2.99)$$

It we substitute relation $i^i_R = (u^i - C_E\dot{q}^i)/r_R$ from (A.6.5) into (2.99), it follows that:

$$\ddot{q} = -\mathrm{diag}(\frac{F^i_v}{J^i_R})\dot{q} + \mathrm{diag}(\frac{C^i_M N^i_m}{J^i_R r^i_R})\left[u - \mathrm{diag}(C^i_E N^i_v)\dot{q}\right] - M^*\mathrm{diag}(\frac{1}{J^i_R}) =$$

$$= -\left[\mathrm{diag}(\frac{F^i_v}{J^i_R}) + \mathrm{diag}(\frac{C^i_M C^i_E N^i_v N^i_m}{J^i_R r^i_R})\right]\dot{q} + \mathrm{diag}(\frac{C^i_M N^i_m}{J^i_R r^i_R})u - \mathrm{diag}(\frac{1}{J^i_R})M^* \qquad (2.100)$$

From (2.100) the value for voltage u is obtained

$$u = \mathrm{diag}(\frac{J^i_R r^i_R}{C^i_M N^i_m})\ddot{q} + \left[\mathrm{diag}(\frac{r^i_R F^i_v}{C^i_M N^i_m}) + \mathrm{diag}(C^i_E N^i_v)\right]\dot{q} + \mathrm{diag}(\frac{r^i_R}{C^i_M N^i_m})M^* \qquad (2.101)$$

where $u = [u_1,\dots,u_n]^T$, $\ddot{q} = [\ddot{q}^1,\dots,\ddot{q}^n]^T$, $\dot{q} = [\dot{q}^1,\dots,\dot{q}^n]^T$, $M^* = [M^*_1,\dots,M^*_n]^T$, $\mathrm{diag}(\lambda^i) = \mathrm{diag}(\lambda^1, \lambda^2,\dots,\lambda^m)$, λ^i - corresponding parameter in brackets.

Replacing M^* in (2.101) by the expression

$$H\ddot{q} + h - D(G+G')S = P, \quad (M^* = P = P^M) \qquad (2.102)$$

which is obtained using (2.57) and (2.83), one obtains:

$$u = \mathrm{diag}(\frac{J^i_R r^i_R}{C^i_M N^i_m})\ddot{q} + \left[\mathrm{diag}(\frac{r^i_R F^i_v}{C^i_M N^i_m}) + \mathrm{diag}(C^i_E N^i_v)\right]\dot{q} +$$

$$+ \mathrm{diag}(\frac{r^i_R}{C^i_M N^i_m})[H\ddot{q} + h - D(G+G')S] \qquad (2.103)$$

Expression (2.103) can be written more suitably as:

$$\mathrm{diag}\left(\frac{J_R^i r_R^i}{C_M^i N_m^i}\right)\ddot{q} + \mathrm{diag}\left(\frac{r_R^i}{C_M^i N_m^i}\right)H\ddot{q} = u - \left[\mathrm{diag}\left(\frac{r_R^i F_v^i}{C_M^i N_m^i}\right) + \mathrm{diag}(C_E^i N_v^i)\right]\dot{q} -$$

$$- \mathrm{diag}\left(\frac{r_R^i}{C_M^i N_m^i}\right)h + \mathrm{diag}\left(\frac{r_R^i}{C_M^i N_m^i}\right)D(G+G')S \qquad (2.104)$$

Multiplying (2.104) by $\mathrm{diag}(C_M^i N_m^i)$ and $\mathrm{diag}(r_R^i)^{-1}$ gives:

$$\mathrm{diag}(J_R^i)\ddot{q}+H\ddot{q}=\mathrm{diag}\left(\frac{C_M^i N_m^i}{r_R^i}\right)u-\left(\mathrm{diag}F_v^i+\mathrm{diag}\left(\frac{C_E^i C_M^i N_v^i N_m^i}{r_R^i}\right)\right)\dot{q}-h+D(G+G')S \quad (2.105)$$

or

$$[(\mathrm{diag}(J_R^i) + H]\ddot{q}=\mathrm{diag}\left(\frac{C_M^i N_m^i}{r_R^i}\right)u+D(G+G')S -$$

$$- \left[\mathrm{diag}F_v^i+\mathrm{diag}\left(\frac{C_E^i C_M^i N_v^i N_m^i}{r_R^i}\right)\right]\dot{q}-h \qquad (2.106)$$

Solving equation (2.106) for $\ddot{q}$ the joint accelerations vector is obtained of the form:

$$\ddot{q} =[H+\mathrm{diag}(J_R^i)]^{-1}\left\{\mathrm{diag}\left(\frac{C_M^i N_m^i}{r_R^i}\right)u+D(G+G')S -\right.$$

$$\left.- \left[\mathrm{diag}F_v^i+\mathrm{diag}\left(\frac{C_E^i C_M^i N_v^i N_m^i}{r_R^i}\right)\right]\dot{q}-h\right\} \qquad (2.107)$$

From the equivalence of relations (2.57) and (2.107) it follows:

$$J^{-1}J_r\ddot{X}_r+J^{-1}(A_r-A)=[H+\mathrm{diag}(J_R^i)]^{-1}\left\{\mathrm{diag}\left(\frac{C_M^i N_m^i}{r_R^i}\right)u+D(G+G')S -\right.$$

$$\left.-\left[\mathrm{diag}F_v^i+\mathrm{diag}\left(\frac{C_E^i C_M^i N_v^i N_m^i}{r_R^i}\right)\right]\dot{q}-h\right\}$$

or

$$[J^{-1}J_r\ddot{X}_r+J^{-1}(A_r-A)][H+\mathrm{diag}(J_R^i)] =\mathrm{diag}\left(\frac{C_M^i N_m^i}{r_R^i}\right)u+D(G+G')S -$$

$$-\left[\mathrm{diag}F_v^i+\mathrm{diag}\left(\frac{C_E^i C_M^i N_v^i N_m^i}{r_R^i}\right)\right]\dot{q}-h \qquad (2.108)$$

Introducing the degree of efficiency of the reducer $\eta = \frac{N_m}{N_v}$, (2.108) becomes:

$$[J^{-1}J_r\ddot{X}_r+J^{-1}(A_r-A)][H+\mathrm{diag}J_R^i]=\mathrm{diag}\left(\frac{C_M^i N_v^i \eta^i}{r_R^i}\right)u+D(G+G')S -$$

$$-\left[\mathrm{diag}F_v^i+\mathrm{diag}\left(\frac{C_E^i C_M^i (N_v^i)^2 \eta^i}{r_R^i}\right)\right]\dot{q}-h \qquad (2.109)$$

Denoting:

$$H^* = H + \mathrm{diag}(J_R^i), \quad U_{c_i} = \left[\mathrm{diag}\left(\frac{C_M^i N_v^i \eta^i}{r_R^i}\right)u\right]$$

$$U_i^* = \{[(\mathrm{diag}F_V^i + \mathrm{diag}(\frac{C_E^i C_M^i (N_V^i)^2}{r_R^i})]\dot{q} + h\}_i$$

expression (2.109) can more suitably be written as:

$$H^*J^{-1}J_r\ddot{X}_r - D(G+G')S = U_c - U^* - H^*J^{-1}(A_r - A) \tag{2.110}$$

or

$$[H^*J^{-1}J_r \mid -D(G+G')]\begin{bmatrix}\ddot{X}_r \\ S\end{bmatrix} = U_c - U^* - H^*J^{-1}(A_r - A) \tag{2.111}$$

Geometric and dynamic parameters, as well as actuator parameters are given in Tables 2.5, 2.6 and 2.7 respectively.

The joint coordinates of the manipulator at the initial point A_o, are shown in Table 2.8 and Table 2.9 shows the initial and terminal points on a nominal trajectory and the time duration of the movement.

Joint	Joint type	Unit vector of joint axis $\vec{e}_i$ **)	Vectors $\vec{r}_{ii}$ **) [m]	Vectors $\vec{r}_{i,i+1}$ **) [m]
1	R*)	0., 0., 1.	0., 0., 0.4	0., 0., -0.4
2	R	1., 0., 0.	0., 0.4, 0.	0., -0.4, 0.
3	R	1., 0., 0.	0., 0., -0.4	0., 0., 0.4
4	R	1., 0., 0.	0., 0.075, 0.	0., -0.075, 0.
5	R	0., 0., 1.	0., 0.075, 0.	0., -0.075, 0.
6	R	0., 1., 0.	0., 0.15, 0.	0., -0.15, 0.

Table 2.5. Geometric parameters of the manipulator

Link	Mass m_i [kg]	Moments of inertia [kgm^2] J_{xi}	J_{yi}	J_{zi}
1	0.	0.	0.	0.2
2	5.	0.25	0.01	0.25
3	5.	0.25	0.25	0.01
4	1.	0.002	0.002	0.002
5	1.	0.002	0.002	0.002
6	2.	0.01	0.002	0.01

Table 2.6. Dynamic parameters

*) R - revolute (rotational)

**) All vectors are in joint coordinate system

Actuator	C_M $\left[\frac{Nm}{A}\right]$	C_E $\left[\frac{V}{rad/s}\right]$	r_R [Ω]	F_v $\left[\frac{Nm}{rad/s}\right]$	N_v [-]	N_m [-]	η [-]	J_M [kgm^2]
1	1.5	1.43	1.6	0.0058	31.17	24.94	0.8	0.00003
2	22.32	27.90	1.8	3.15	150	120	0.8	0.00079
3	14.88	18.6	1.8	1.4	100	80	0.8	0.00079
4	3.52	4.4	0.85	0.24	100	80	0.8	0.00001
5	3.52	4.4	0.85	0.24	100	80	0.8	0.00001
6	3.52	4.4	0.85	0.24	100	80	0.8	0.00001

Table 2.7. Actuator parameters

Point	Joint coordinates					
	q^1	q^2	q^3	q^4	q^5	q^6
A_o	-1.57080	-0.52360	-2.09439	-0.52360	0	0

Table 2.8. Joint coordinates

Point on a trajectory	x [m]	y [m]	z [m]	q^4 [rad]	q^5 [rad]	q^6 [rad]	Movement	Duration of movement [s]
A_o	0.8	0.	0.2	-0.52360	0.	0.	$A_o \rightarrow A_1$	1
A_1	0.8	0.15	0.	-0.52360	0.	0.		
A_2	0.8	0.75	0.	-0.52360	0.	0.	$A_1 \rightarrow A_2$	1

Table 2.9. Initial and terminal points on the nominal trajectory*)

Adopting a trapezoidal velocity profile with the acceleration and retardation time of 0.2 [s] and using expressions (2.30) and (2.51) for unconstrained and constrained motion respectively, the nominal drive moments at mechanism joints are calculated. During the calculation of the drive between points A_1 and A_2, the nominal value of force F_A=30[N] was adopted, as a value to be reached during grinding and it is assumed to act upon the centre of mass of the last link. The adopted value for the friction coefficient, obtained from the literature was 0.3.

*) x, y, z - coordinates of manipulator tip with respect to the fixed coordinate system.

Results of nominal dynamics for robot joints 1 and 3 in the form of the records normalized on their indicated maximal values are shown in Figs. 2.22. and 2.23, respectively.

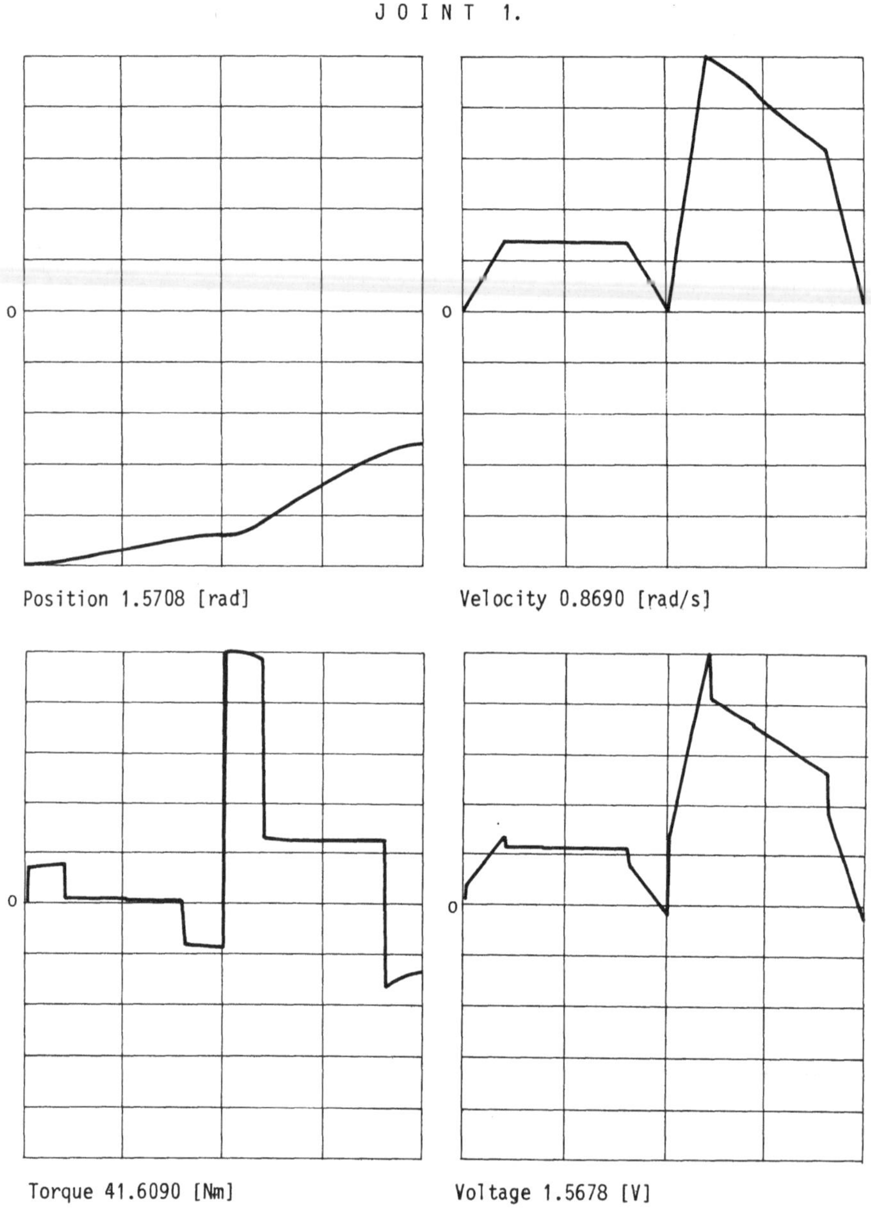

Fig. 2.22. Nominal values: position (q), velocity ($\dot{q}$), torque (P) and voltage (u)

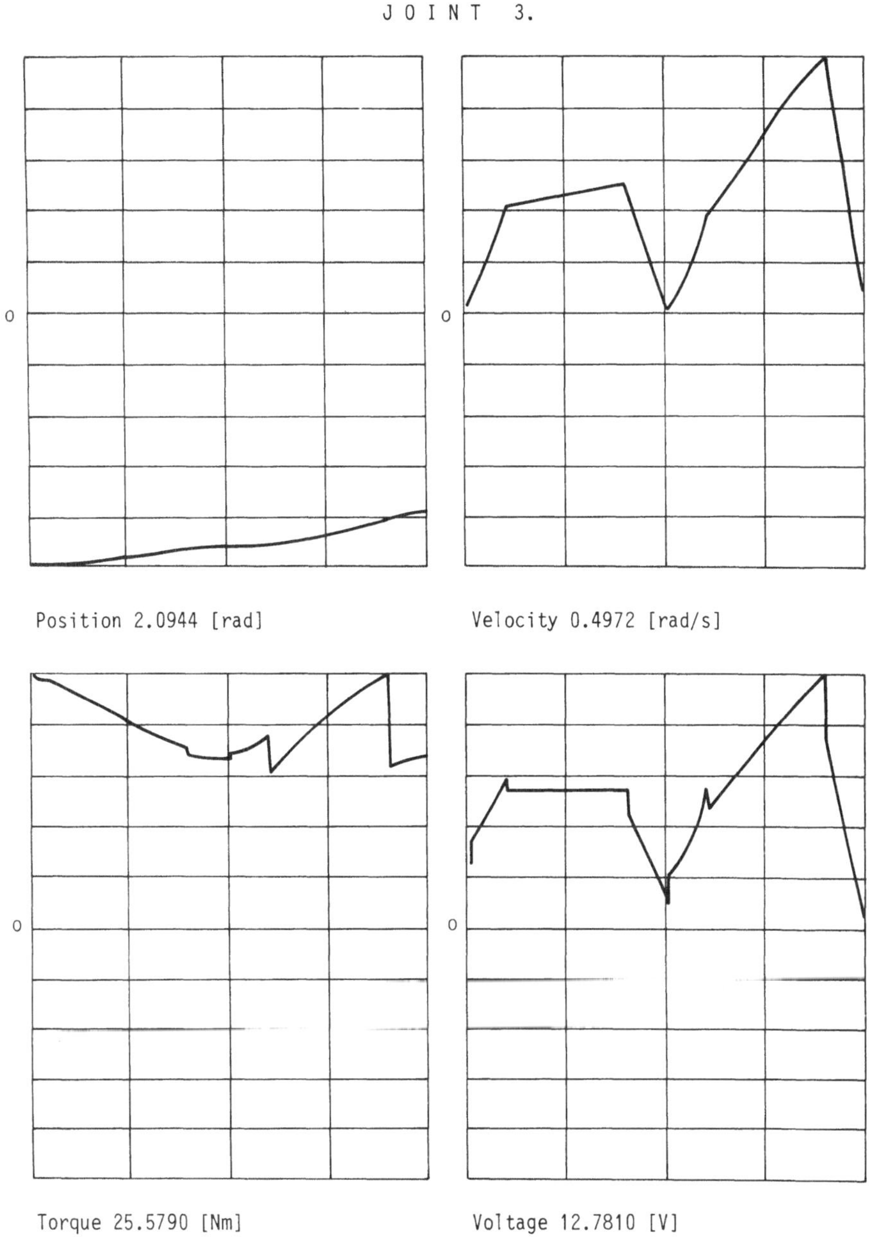

Fig. 2.23. Nominal values: position, velocity, torque and voltage

2.5 Dynamic Analysis of Manipulation Robots

Mathematical models in the form (2.30) allow us to deal with two fundamental problems of dynamics, the direct and the inverse problem solution. The direct problem involves the calculation of driving forces and torques required to realize the prescribed motion. The inverse problem, on the other hand, involves determination of the motion laws of joint coordinates when driving forces or torques are known.

Practical tasks in robotics place emphasize on the problem of the direct solution i.e. driving forces are calculated for the prescribed motion (known joints trajectories).

However, in practice a manipulation task is practically never given in joint coordinates of the mechanism, but is instead described by the so called "external coordinates" X, e.g.: the law of manipulator tip motion and gripper orientation in external space. In order to assign properly a manipulation task a unique relationship should exist between the "external coordinates" (vector X) and the generalized (joint) coordinates (vector q).

For the case when the unique relationship is nonexistent e.g. when there is excess number of degrees of freedom (system redundancy), a specific problem is presented in the synthesis of trajectories and will not be considered in this textbook.

Let us designate by η the function which transforms the generalized (internal) coordinates q into external ones X:

$$X = \eta(q) \tag{2.112}$$

where q and X are n-dimensional vectors.

The function η is one-place and can always be determined (not explicitly but as a computational algorithm). The problem lies in the dificulty of calculating q from such a system of equation (2.112) resulting from the impossibility to express q either explicitly, or even approximately, numerically because of the complexity of the system which has to be solved.

In generating mathematical models it is necessary that q and $\dot{q}$ are known in each time, but if we want to determine the drives P, we have to know $\ddot{q}$, too.

Let us explain the procedure for generating the mentioned model on the basis of a prescribed profile (trajectories) of external coordinates X and known state q and $\dot{q}$.

By double differentiation of (2.112):

$$\dot{X} = \frac{\partial \eta}{\partial q} \dot{q} \tag{2.113}$$

$$\ddot{X} = \frac{\partial \eta}{\partial q} \ddot{q} + \frac{\partial^2 \eta}{\partial q^2} \dot{q}^2 \tag{2.114}$$

and denoting

$$X^a \triangleq \ddot{X}, \qquad J = \frac{\partial \eta}{\partial q}, \qquad A = \frac{\partial^2 \eta}{\partial q^2} \dot{q}^2$$

equation (2.114) becomes

$$X^a = J\ddot{q} + A \tag{2.115}$$

where J, $(n\times n)$ Jacobian matrix, n is the number of degrees of freedom of the manipulation mechanism, and A has the dimensions $(n\times 1)$. Matrices A and J are functions of state q, $\dot{q}$. It is therefore necessary that X^a is given in succesive time instances. X^a represents the acceleration profile of manipulator mechanism tip. It is worth reminding ourselves that the acceleration profile X^a can be simply obtained from the velocity profile which is often given on the basis of the requirements of the manipulation task. Let us now mention that the acceleration in external coordinates of the manipulation mechanism is described by the linear acceleration vector at the gripper centre of mass or at its tip ($\vec{w}$) and by the angular acceleration vector of the gripper (its last link) $\vec{\varepsilon}$, $X^a = \begin{bmatrix} x \\ \varepsilon \end{bmatrix}$. If J is not a singular matrix, the required generalized acceleration $\ddot{q}$ will be

$$\ddot{q} = J^{-1}(X^a - A) \tag{2.116}$$

On the basis of the calculated $\ddot{q}$ and the mathematical model (2.30) the driving forces and torques required to execute the prescribed motion can be evaluated. The given procedure is shown by a block-diagram of Fig. 2.24 and is suitable for computer implementation.

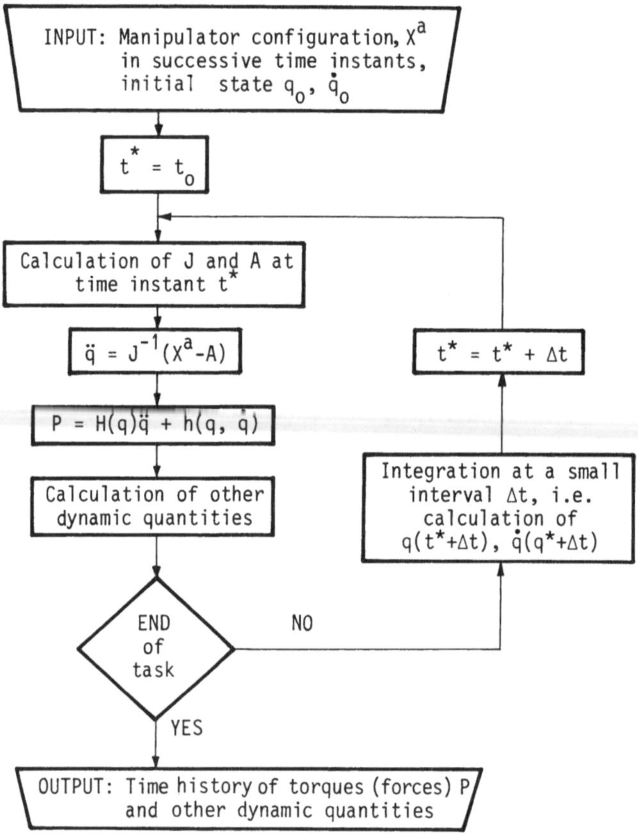

Fig. 2.24. Block diagram of the algorithm

In practical applications of industrial robotics, prescribing the trajectories or velocity profiles of manipulator mechanism tip is common practice so that the synthesis of robot trajectories in the task of moving the manipulator tip along a prescribed trajectory and transfer of a workpiece of a given orientation along its prescribed trajectory will be analysed briefly.

Task of transfering the manipulator tip along a prescribed trajectory

Let us consider the simplest manipulation task, that of transfering the manipulator tip with the gripper and a workpiece, along a prescribed trajectory in space. In principle, this task can be reduced to specifying the initial and terminal positions of the manipulator tip without prescribing any particular trajectory along which the manipulator tip should move. In this case, control could be synthesized by minimizing some of the criteria. However, due to restrictions in the working space within which the robot operates the choice of alternative trajectories is very limited such that optimization is almost impossible. Therefore it appears that the synthesis of robot trajectories, based on functional requirements imposed by restrictions on the workspace is a better alternative. Here, as in other sections of the textbook, we confine ourselves to non-redundant manipulation mechanisms i.e. mechanisms that are assumed to comprise of rigid links and have six degrees of freedom.

Because any point in the work-space without obstacles (defined by kinematic limitations of the manipulator considered) can be reached by a manipulator having only three degrees of freedom, the minimal (basic) manipulator configuration is sufficient for the realization of such a task. For this reason we confine ourselves to observing only three d.o.f. and consider the remaining three d.o.f. (gripper) fixed (i.e. q^4, q^5, q^6 = const).

We consider the problem of synthesizing the nominal trajectory $x^{oi}(t)$, and programmed inputs $u^{oi}(t)$, for the given manipulator tip motion (its prescribed trajectory).

In this case, there is a unique correspondence between angles of the three d.o.f. and the tip coordinates (point D in Fig. 2.25) in the orthogonal coordinate frame.

The relationship between the coordinates of point D(x, y, z) and angular coordinates can be described in the following manner:

$$\vec{X}_D = (x, y, z)^T = f(q) = f(q^1, q^2, q^3) \tag{2.117}$$

In determining the trajectory of manipulator angles when trajectory of point D in space is prescribed, we shall consider small increments of manipulator tip motion along the prescribed trajectory $\Delta\vec{X}_D = (\Delta x, \Delta y, \Delta z)^T$. Under the assumption of small increments, the following relationship between angles and orthogonal coordinates holds:

$$\left(\frac{\partial f_i}{\partial q^1}\right)\Delta q^1 + \left(\frac{\partial f_i}{\partial q^2}\right)\Delta q^2 + \left(\frac{\partial f_i}{\partial q^3}\right)\Delta q^3 = \Delta X_{Di}, \quad i=1,2,3 \tag{2.118}$$

It follows that:

$$A\Delta q = \Delta\vec{X}_D, \qquad \Delta q = (\Delta q^1, \Delta q^2, \Delta q^3)^T \tag{2.119}$$

where the elements of matrix A are:

$$a_{ij} = \frac{\partial f_i}{\partial q^j}, \qquad i,j=1,2,3$$

and are evaluated for the previous point on the manipulator trajectory. Being:

$$\Delta q = q(t_\ell)-q(t_{\ell-1}), \qquad \Delta\vec{X}_D = \vec{X}_D(t_\ell)-\vec{X}_D(t_{\ell-1})$$

the elements of matrix A are calculated for $q(t_{\ell-1})$ and the angles at the next point on the trajectory of tip D are evaluated according to (2.119).

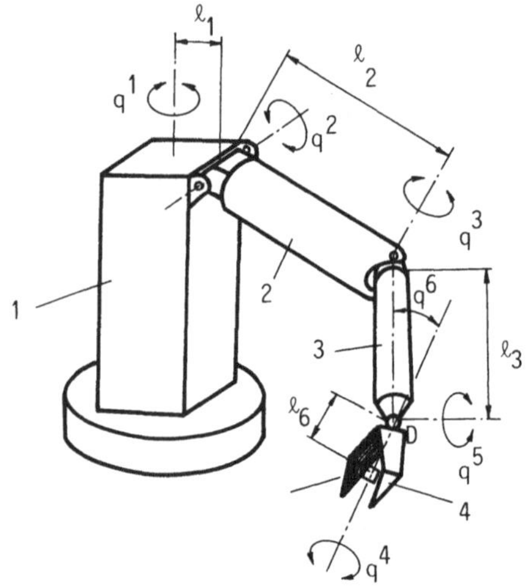

Fig. 2.25. UMS-1 manipulator (six d.o.f)

In principle, the algorithm for synthesizing the prescribed kinematics of the minimal configuration is stated as follows. The trajectory $\vec{X}^o_D(t)$ along which the point D should be moved is prescribed with a given velocity distribution $\dot{\vec{X}}^o_D(t)$, $\forall t\in T$, starting from the initial position of the manipulator (with angles $q^o(0)$ and coordinates $\vec{X}^o_D(0)$). For the point $q^o(0)$ the matrix A is calculated from (2.119) and the desired $\dot{\vec{X}}_D(0)$. Thus,

$$\dot{q}^o(0) = A^{-1}(0)\dot{\vec{X}}^o_D(0) \qquad \text{for} \qquad \Delta t\to 0 \tag{2.120}$$

assuming that the matrix A is nonsingular.

Let us consider sufficiently short time intervals $\Delta t = t_\ell - t_{\ell-1}$, during which the values of matrix $A(t_\ell)$ elements don't changed significantly. For sufficiently small Δt we may assume that:

$$q^o(t_\ell) = q^o(t_{\ell-1}) + \dot{q}^o(t_{\ell-1})\Delta t \tag{2.121}$$

Thus, one calculates the angles for the next point on the desired trajectory of the manipulator tip $\vec{X}_D^o(t)$. Using the new $q^o(t_\ell)$, the matrix $A(t_\ell)$ is calculated and then the necessary angular velocity $\dot{q}^o(t_\ell)$ is calculated from (2.120). Thus, one obtains the nominal trajectories for all coordinates of the state vector $x^o(t)$, corresponding to the minimal configuration (three angles and three angular velocities are calculated).

When the nominal trajectories $x^o(t)$, $\forall t \in T$, are calculated on the basis of the dynamic robot model (2.30) the nominal driving torques $P^o(t)$, $\forall t \in T$ can be calculated.

The motion of the manipulator UMS-1 for a specific control task is presented in Fig. 2.26. The manipulator tip should be moved from the point A, defined by $x^o(0) = (q^1(0), \dot{q}^1(0), \ldots, \dot{q}^3(0))^T = (0.1, 0, -0.8, 0, 0.1, 0)^T$, to the point B, defined by $x^o(\tau) = (-0.4, 0, -0.9, 0, 1.9, 0)^T$ in the time interval $\tau_s = 1.8$ s (this means that the manipulator has to reach the terminal position at a precisely defined time instant, $T_s = \{t: t=\tau=\tau_s\}$). The manipulator tip should move along a straight line between the points A and B. The tip acceleration should be constant and change its sign once during the movement. Namely, the manipulator tip acceleration should be:

$$a_{tip} = \begin{cases} a_{max} & \text{if} \quad ||\vec{X}_D^o(t) - \vec{X}_D^o(0)|| \le 0.5 \text{ dist}, \\ -a_{max} & \text{if} \quad ||\vec{X}_D^o(t) - \vec{X}_D^o(0)|| > 0.5 \text{ dist}, \end{cases} \tag{2.122}$$

where $a_{max} = \frac{4dist}{\tau^2}$, $dist = ||\vec{X}_D^o(\tau) - \vec{X}_D^o(0)||$.

The results of the nominal dynamics synthesis for this particular motion are presented in Fig. 2.27 - 2.28. In Fig. 2.26 the motion is shown in three - quarter projection in space (the gripper is assumed to be fixed with respect to the third member). In Fig. 2.27 the minimal trajectories are presented for all three manipulator angles and in Fig. 2.28 the corresponding driving torques are presented.

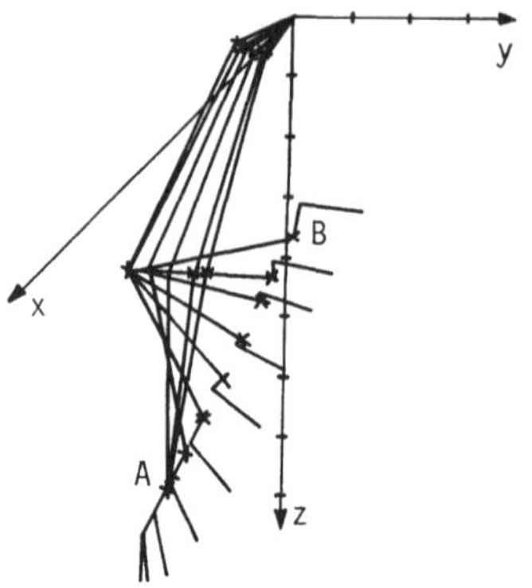

Fig. 2.26. Transfer of manipulator UMS-1 tip along a straight line (with fixed gripper)

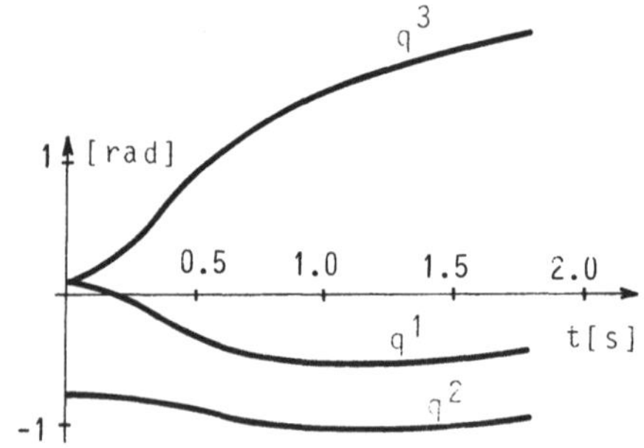

Fig. 2.27. Nominal trajectories of minimal configuration

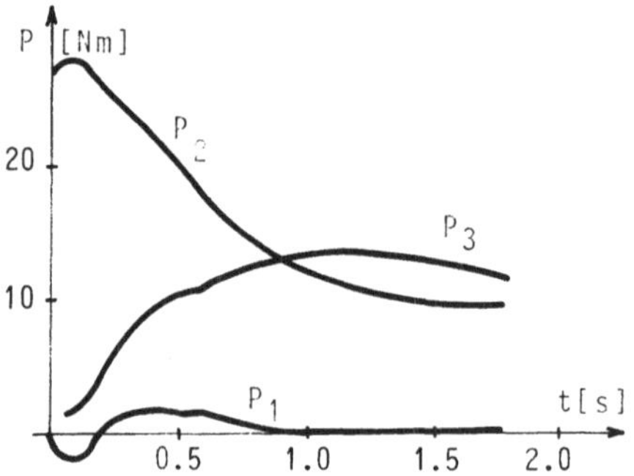

Fig. 2.28. Nominal driving torques of minimal configuration

Task of transferring the workpiece of a desired orientation along a prescribed trajectory

The task considered in this paragraph is an extension of the task discussed in the preceding section. Here, in addition to transferring the manipulator tip along a prescribed trajectory in space, it is necessary to achieve a particular orientation of the workpiece during the transfer, i.e., to ensure that some axis of the working object has a defined orientation with respect to the axes of the absolute coordinate system. In order to realize this task, the three d.o.f of the minimal configuration are not sufficient. Manipulators with $n \leq 6$ d.o.f are therefore considered (Fig. 2.25).

We now consider a problem of nominal trajectory $x^0(t)$ synthesis i.e., desired motion of the object with a prescribed orientation with respect to the axes of the absolute system. The determination of trajectories of the six angles and six angular velocities when the trajectory and the orientation of the workpiece are defined in space, is not an easy task. However, if at the level of nominal dynamics the system is decoupled into two functional subsystems so that one subsystem consists of the first three d.o.f with their actuators (minimal manipulator configuration) and the other subsystem consists of three d.o.f of the gripper with their actuators, then the problem of nominal trajectory synthesis $q^o(t)$ and $\dot{q}^o(t)$, $\forall t \in T$ is considerably simplified. In this case, the vector is divided into the state vectors of the two subsystems $x_o^o = (x^{1T}, x^{2T}, x^{3T})^T$ and $x_x^o = (x^{4T}, x^{5T}, x^{6T})^T$.

Since the manipulator tip trajectory is defined, three angles and three angular velocities of the first subsystem are uniquely defined. The angles and angular velocities can be calculated as in the preceding section, since the task for the first subsystem is reduced to that of finding out the manipulator tip motion along a prescribed trajectory in space.

When the nominal trajectories $x_o^o(t)$ for the positioning subsystem x_o^o are calculated, the nominal trajectories for the gripper subsystem have to be synthesized. This subsystem should provide the desired orientation of the workpiece during its transfer along a prescribed trajectory in space. The desired orientation of the object can be obtained by prescribing the trajectory of the object tip with respect to the trajectory of the manipulator tip (gripper joint).

The object tip coordinates in the Cartesian coordinate system are functions of all six manipulator angles:

$$(X_{D_i}) = f_{p_i}(q^1, q^2, \ldots, q^6), \quad i=1,2,3 \tag{2.123}$$

where $(X_{D_i})^T = (x_p, y_p, z_p)^T$ are the coordinates of the object tip. Since (2.118 - 2.121) define the trajectories of the first three d.o.f q^{io}, i=1,2,3 one obtains:

$$(X_{D_i}) = f_{p_i}(q^{1o}, q^{2o}, q^{3o}, q^4, q^5, q^6), \quad i=1,2,3 \tag{2.124}$$

i.e. the coordinates of the object tip are functions of the gripper angles. Thus, there exists a unique relationship between the coordinates of the object tip (or the gripper) and the three angles of the gripper.

On the basis of (2.123) the following holds:

$$[A_{p1} \mid A_{p2}]\begin{bmatrix} \Delta q_o^o \\ \hline \Delta q_x^o \end{bmatrix} = \Delta \vec{X}_p$$

where the elements of matrices A_{p1} and A_{p2} (3×3) are given by:

$$a_{ij}^{p1} = \frac{\partial f_{pi}}{\partial q^j}; \qquad a_{ij}^{p2} = \frac{\partial f_{pi}}{\partial q^{j+3}}, \qquad i,j=1,2,3$$

Since the orientation of the object is defined, the object tip trajectory $\vec{X}_D(t)$ is defined as well as the trajectory of the object tip velocity $\dot{\vec{X}}_p(t)$. As with the calculation of the minimal configuration trajectories we shall observe small increments of the object tip movement along the trajectory $\vec{X}_D^o(t)$, during which the matrices A_{p1} and A_{p2} do not change significantly. Let us assume the same time intervals $\Delta t_\ell = t_\ell - t_{\ell-1}$ that we assumed in the synthesis of the minimal configuration trajectories. When the positions of the manipulator and the gripper at the instant t_ℓ are known, the matrices $A_{p1}(t_\ell)$, $A_{p2}(t_\ell)$ can be calculated and since $\dot{q}_o^o(t_\ell)$ are already known, we obtain:

$$\dot{q}_x^o(t_\ell) = A_{p2}(t_\ell)^{-1}(\dot{\vec{X}}_D^o(t_\ell) - A_{p1}(t_\ell)\dot{q}_o^o(t_\ell)) \tag{2.125}$$

with the assumption that the matrix A_{p2} is nonsingular. One thus calculates the gripper angular velocities and the gripper angles trajectories. If the matrix A_{p2} is singular, the singular points should be carefully studied. One thus calculates the nominal angles $q^o(t)$ and nominal velocities $\dot{q}^o(t)$ for both functional subsystems $\tilde{S}_o$ and $\tilde{S}_x$. If a distribution (profile) of the manipulator tip velocity is defined, then the accelerations $\ddot{q}^o(t)$ for the six d.o.f are defined. Now, the driving torques $P^o(t)$, $\forall t \in T$ can be calculated according to the model of the mechanical part of the system S^M (2.30).

Figs. (2.29 - 2.30) present the results of the nominal trajectories and programmed inputs synthesis for a particular task for the manipulator UMS-1 (Fig. 2.25). The working object should move from the position A to the position B (Fig. 2.29e). The object position at the point A is defined by the manipulator coordinates. The gripper joint $\vec{X}_A(0)=(0.425, 0.167, 0.571)^T$, while the gripper center of gravity coordinates are $\vec{X}_{DA}(0)=(0.455, 0.203, 0.630)^T$. The position of the object at the point B is given by the manipulator tip coordinates $\vec{X}_B^o(\tau)=(0.368, 0.365,$

$0.355)^T$, while the coordinates of the gripper c.o.g. are $\vec{X}^o_{DB}(\tau)=(0.391, 0.410, 0.410)^T$ [m]. The manipulator angles corresponding to position A are $q^o(0)=(0, 1.1, 0.5, 0, 0, 0)^T$ [rad], while the manipulator angles for the point B are $q^o(\tau)=(0.08, 0.94, 1.25, 0., 0.78, -1.50)^T$ [rad]. The movement should be performed in $\tau=\tau_s=0.9s$. The working object should be moved in such way that the tip of manipulator should move along a straight line between points $A(\vec{X}^o_A(0))$ and $B(\vec{X}^o_B(\tau))$ and the gripper c.o.g. should also move along a straight line between its two terminal positions $A(\vec{X}^o_{DA}(0))$ and $B(\vec{X}^o_{DB}(\tau))$. A further requirement is for the acceleration of manipulator tip to be constant and that it changes its sign once, i.e., that the manipulator tip has the acceleration given by (2.122), which in this case was limited at $a_{max}=1.48$ m/s^2 (dist=0.3 m and τ = 0.9s). Because the conditions for object orientation during transfer of object are so defined that two d.o.f. are sufficinet for them to be satisfied, matrix A_{p2} and Eq. (2.125) are singular. For this reason one d.o.f. is fixed (in this case $q^4(t)=0$) and the conditions for gripper orientations are satisfied via the remaining two d.o.f. of the orientation subsystem.

The angular velocities of the free (not fixed) d.o.f $\dot{q}^5(t)$ and $\dot{q}^6(t)$ are calculated on the basis of (2.125), where matrices $A_{p2}(t)$ and $A_{p1}(t)$ have dimensions (2×2) and (2×3), respectively.

Fig. 2.29a shows the distance covered by the manipulator tip from the starting point A as a function of time. Fig. 2.29b shows the desired manipulator tip velocity during transfer of the workpiece, (according to Eq. (2.122)). The manipulator tip coordinates in an orthogonal absolute system are shown in Fig. 2.29c. It is required that the velocities of angles at starting and terminal positions are zero, $\dot{q}^o(0)=0$, $\dot{q}^o(\tau)=0$. The differences between the coordinates of the gripper centre and tip coordinates of the minimal (basic) configuration are shown in 2.29d, which describes the change of gripper orientation during object transfer. Fig. 2.30a shows the nominal trajectories of the positioning subsystem angles for the given task. Fig. 2.30b shows the nominal trajectories of the gripper angles (orientation subsystem). With these nominal trajectories, and using the given mathematical model of the mechanism (2.30), the nominal driving torques for all angular coordinates are calculated and shown in Figs. 2.30c and 2.30d.

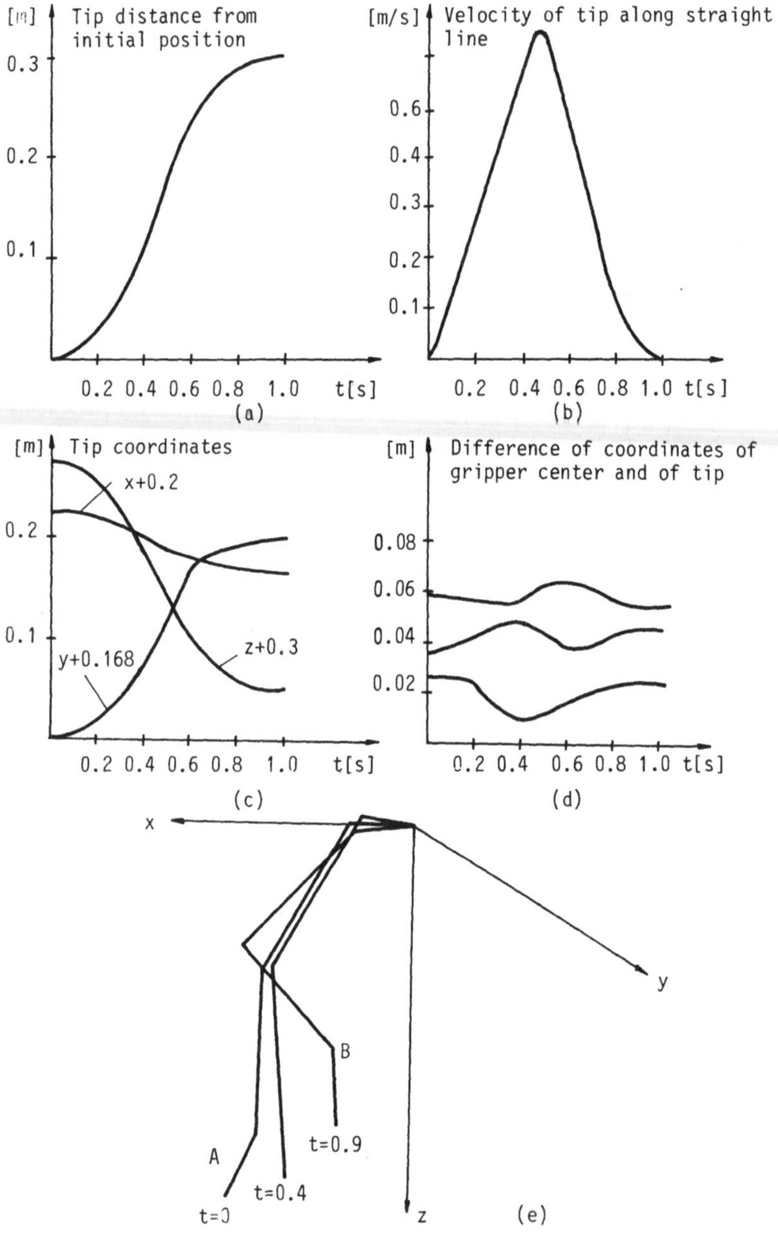

Fig. 2.29. Trajectories of tip and gripper centre of gravity (UMS-1)

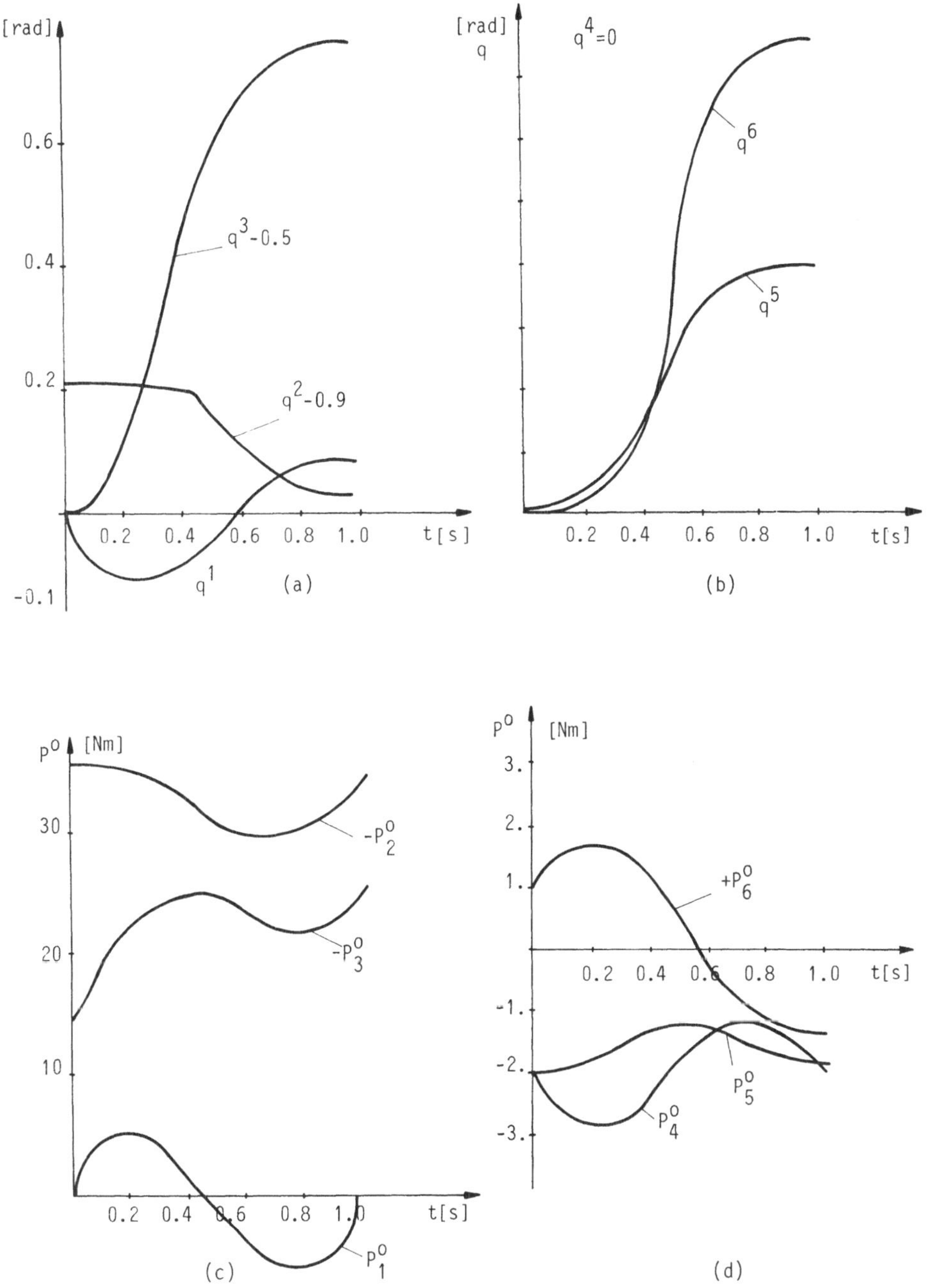

Fig. 2.30. Trajectories of angles and driving torques (UMS-1)

Calculation of other dynamic characteristics

In the previous paragraphs it was shown how driving forces and torques P(t) which produce the prescribed motion of a manipulator are calculated. As the output we also obtained the trajectory in the state space i.e. q(t), $\dot{q}(t)$. This calculation is the basis of the algorithm for dynamic analysis of manipulator motion. But, we are often also interested in some other dynamic characteristics, some of which will be discussed here.

Diagrams of torque vs. r.p.m.

One interesting characteristic which can be obtained as the output of the algorithm shown in previous text, is the diagram of torque vs. motor speed of rotation. Since the angular speed of a motor is usually expressed in terms of revolutions per minute (r.p.m.), we talk about torque-r.p.m. diagram. Because we are considering manipulator mechanisms driven by electric D.C. motors, this torque - r.p.m. diagram can be computed for each joint i.e. each motor (actuator).

Let us first consider the D.C. actuators. The dynamic analysis algorithm, given in the preceding text, computes the torque P_i and the relative (generalized) velocity $\dot{q}^i$ for each joint and each time instant. The r.p.m. for the i-th joint is $n_i = \frac{60}{2\pi}\dot{q}^i$ ($\dot{q}^i$ is expressed in rad/s), so each time instant gives one point of the P_i-n_i diagram. This diagram is valid for the shaft of the joint considered. But we are usually interested in the diagram of the motor itself. Then the reducer in the joint must be taken into account. Let us consider a reducer with the speed reduction ratio equal to N_{Vi}. Then the motor r.p.m. is:

$$n_i^m = N_{Vi} n_i = N_{Vi} \frac{60}{2\pi} \dot{q}^i \tag{2.126}$$

If the reducer has the mechanical efficiency $\eta(N, n)$, then the torque multiplier ratio is $N_V \cdot \eta(N, n)$ and so the motor torque is:

$$P_i^m = \frac{P_i}{N_{Vi} \cdot \eta_i} \tag{2.127}$$

Hence the motor diagram P_i^m - n_i^m is obtained by calculating (2.126) and (2.127) in each time instant t_o, t_1, t_2,... One example of a torque - r.p.m. diagram is shown in Fig. 2.31. (Only a qualitative presentation of the diagram is given).

Such diagrams are very useful during the synthesis and choice of D.C. servosystems. The producer gives the P^m_{max}-n^m motor characteristic in the catalogue, where P^m_{max} is the maximal motor torque at motor r.p.m.=n^m. By comparing the necessary characteristic, obtained by means of the algorithm described, with the one from catalogue, one can conclude whether the chosen motor suits its application. The usage of these characteristics will be shown in the text to follow.

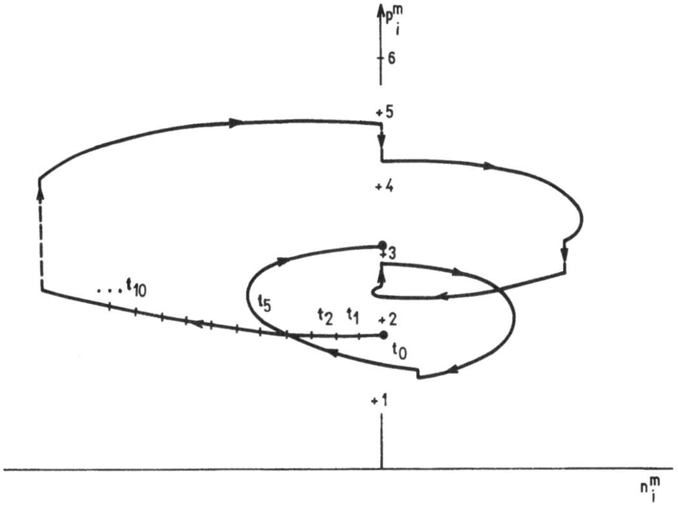

Fig. 2.31. A torque - r.p.m. diagram

Calculation of the power needed and the energy consumed

The next interesting characteristics (e.g. for the choice of actuators) are the power requirements in each joint. For i-th joint the power needed at some time instant is obtained as $Q_i = P_i \cdot \dot{q}^i$. But the power produced by the motor has to be larger because of the power loss in the reducer. So the necessary motor power in manipulator joint is $Q^m_i = P_i \cdot \dot{q}^i / \eta_i$, where η_i is the mechanical efficiency of the reducer. It should be pointed out that this is the output mechanical power of the motor.

The energy consumption may also be easily computed. Let us consider the i-th joint and its corresponding actuator. Let $E_i^{(k)}$ be the energy consumed in the first k time steps, and let ΔE_i^k be the energy consumed in the k-th time step (time interval Δt_k). The total energy consumed by the actuator of j-th joint is being evaluated during the time-iterative procedure of dynamic analysis, as summation of the energy at each step Δt_k:

$$E_i^{(k)} = E_i^{(k-1)} + \Delta E_i^k \tag{2.128}$$

To calculate ΔE_i^k we adopt the medium driving torque value on the interval:

$$P_{imed}^k = \frac{1}{2}(P_i^{k-1} + P_i^k), \tag{2.129}$$

where the upper index indicates the k-th time instant. Now,

$$\Delta E_i^k = P_{imed}^k \cdot \Delta q^{ik} \tag{2.130}$$

where $\Delta q^{ik} = q^{ik} - q^{i(k-1)}$.

It should be pointed out that this discussion on energy has dealt with the mechanical power and mechanical energy only. If we want to calculate the energy which has to be taken from an energy source (e.g. from an electric battery), then we should take care of the energy lost in actuators. In this paragraph we give only some ideas of such energy consumption calculation. If a manipulator is driven by D.C. electromotors then the energy loss follows from resistance and friction effects. Let us consider one joint and the corresponding motor. The power required from a source can be computed as $Q=u \cdot i_r$ where u is control voltage and i_r is rotor current. Now, in a time step Δt_k the energy increment is:

$$\Delta E_i^k = u_i^k i_{ri}^k \Delta t_k = Q_i \Delta t_k$$

where the lower index represents the number of joint and the upper index indicates the k-th time instant.

It is clear that the energy consumed by the whole manipulator is the sum of energies consumed by the actuators.

Tests of dynamic characteristics

Knowledge of dynamic characteristics is very useful in the design process and application of robotic systems. For these reasons, an algorithm for testing these characteristics will be shown. It is an automatic procedure which calculates the relevant dynamic characteristics, on the basis of chosen values of manipulator parameters and further tests whether the results satisfy the given conditions. In this manner the user has acquired information whether the chosen robot parameters are adequate. Otherwise, a correction is made. It should be mentioned that, in principle, each dynamic characteristic which is calculated can be tested. Here, we mention only the most relevant tests.

Tests of a D.C. electromotor

Suppose that we have chosen a D.C. electromotor as the actuator for the i-th joint of the manipulator. All the calculations refer to some defined manipulation task with a prescribed execution time.

Test 1 [11]. The P^m_{max}-n^m characteristic of the chosen motor can be found in the catalogue. It is the diagram of maximal torque as a function of motor r.p.m. If the diagram is not given directly, it can be constructed from the data given in the catalog. In manipulator systems, we often use permanent magnet D.C. motors. Once again, worth mentioning is that for payload not exceeding 100 kg, most commonly used actuators today are D.C. electromotors. For such motors the P^m_{max}-n^m characteristic has a polygonal form (straight lines in Fig. 2.32). In the catalog we find sometimes the value of maximal motor torque (corresponding to the point A in Fig. 2.32) and the maximal angular speed (point B in Fig. 2.32). These two values define the maximal characteristic (this is a straight line). Let the value of maximal torque for $n^m \to 0$ (point A) be marked by P^m_M and the value of angular speed for $P^m \to 0$ (point B) be marked by n^m_M. The torque P^m_M is often called stall torque, and the rotation speed n^m_M is called no-load speed. When this speed is expressed in terms of r.p.m. it is marked by n^m_M and if expressed in terms of rad/s then we mark it by ω^m_M. It should be said that there is sometimes a difference between the real value of maximal torque (P^m_{Mr}) and its theoretical value (P^m_M) (the real value of P^m_{Mr} is less then P^m_M). In such a case the maximal characteristic P^m_{max}-n^m has an upper bound P^m_{Mr} (point C in Fig. 2.32). This characteristic defines the feasible domain. The real P^m-n^m characteristics must be wholy within this domain. In each iteration a new point of the diagram is obtained and the algorithm checks whether it is within the permissible domain. If it is, a new iteration starts, and if not, the algorithm signals that there is a violation of the constraint.

The constraint considered follows from the mathematical model of D.C. actuator. Complete actuator models are given in Appendix 6 and we here give only a simplified derivation of torque-speed constraint. According to the second Kirchoff's law:

$$u = r_R i_R + C_E \dot{q}^m \tag{2.131}$$

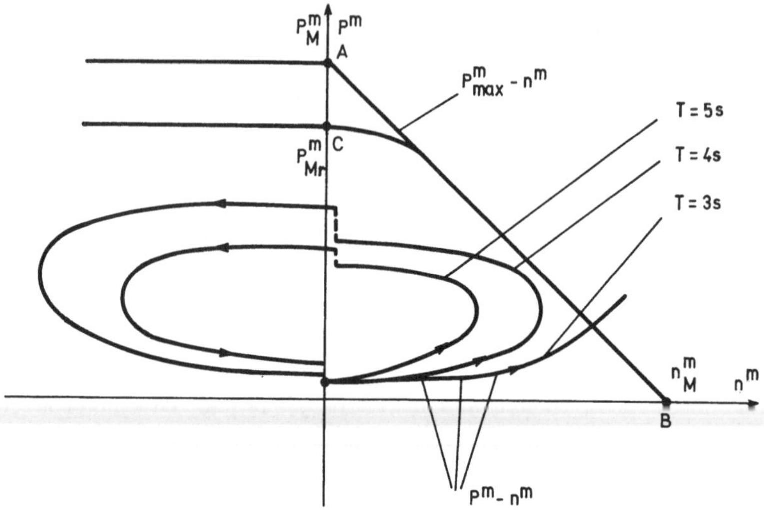

Fig. 2.32. P^m - n^m diagrams

where u is input voltage, r_R is rotor resistance, i_R is rotor current, C_E is the constant of electromotor force, and $\dot{q}^m$ is angular speed (rad/s). If we neglect rotor acceleration effects and friction term, the motor output torque is:

$$P^m = C_M i_R \tag{2.132}$$

where C_M is the torque constant. Combining (2.131) and (2.132) one obtains:

$$P^m = \frac{C_M}{r_R} u - \frac{C_M C_E}{r_R} \dot{q}^m \tag{2.133}$$

If angular speed is expressed in terms of r.p.m. ($n^m = \frac{60}{2\pi} \dot{q}^m$), then:

$$P^m = \frac{C_M}{r_R} u - \frac{C_M C_E 2\pi}{r_R 60} n^m \tag{2.134}$$

Let us introduce u_{max} as the constraint on maximal input voltage. Then, from (2.133) and (2.134) it follows:

$$\frac{P_{max}^m}{P_M^m} + \frac{n^m}{n_M^m} = 1 \tag{2.135}$$

where $P_M^m = \frac{u_{max} C_M}{r_R}$ is stall torque and $n_M^m = \frac{60}{2\pi} \cdot \frac{u_{max}}{C_E}$ is no-load speed.

This constraint of maximal input voltage can be represented by a straight line in P^m-n^m plane ((1.) in Fig. 2.33). We use this constraint in quadrants I and III of the P^m-n^m plane. For quadrants II and IV we introduce the constraint of maximal rotor current in order to keep this current smaller than the stall current value: $|i| \leq i_M = \frac{u_{max}}{r_R}$. This constraint is represented by a horizontal line (2.) in P^m-n^m plane (Fig. 2.33). Finally, we introduce the constraint of maximal allowed speed n_{max} (i.e. $n_{max} = n^m_M$, line (3.). Fig. 2.33).

If viscous friction is not neglected then no-load speed becomes $\omega^m_M = u_{max}/(C_E + r_R F_V/C_M)$ where F_v is the viscous friction coefficient. This modified constraint is represented by a dotted line in Fig. 2.33.

One example of the constraint is shown in Fig. 2.32. Straight lines represent the constraint P^m_{max}-n^m. It can be concluded that the P^m-n^m plots spread when the working speed increases, i.e., the execution time T decreases. For T=5s and T=4s the diagrams are wholly within the permissible domain. This means that the chosen actuator can produce manipulator work at this speed. For T=3s the diagram extends beyond the permissible domain i.e. the constraint is violated and the motor cannot produce manipulator work at that speed.

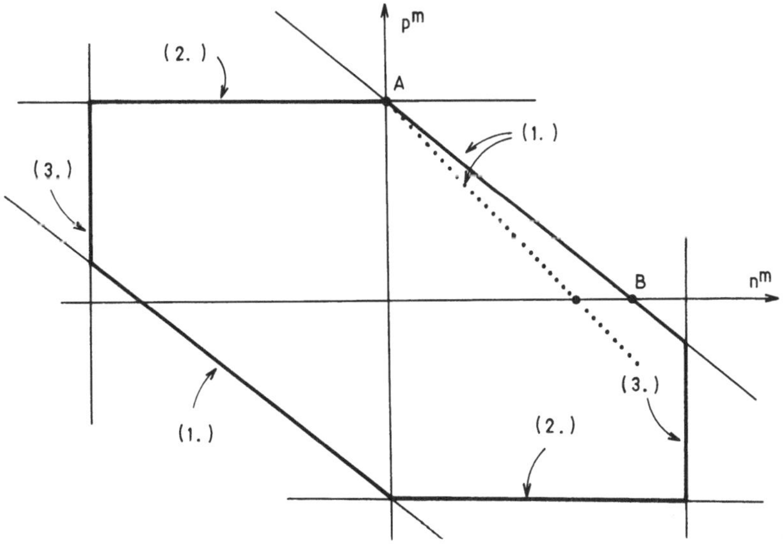

Fig. 2.33. Constraints of D.C. motor

<u>Test 2 [11].</u> Another test can be made from the standpoint of necessary motor power. The algorithm computes the power needed in each joint i.e. $Q_i^m = P_i^m \dot{q}^{mi}/\eta_i$. By comparing this function (for the joint considered) with the maximal power which can be produced by the chosen motor we conclude whether the motor has been chosen correctly.

Here we give only a short presentation of power test procedure. Besides the power $Q_i^m = P_i^m \dot{q}^{mi}$ which is a product of torque and angular speed, we introduce a notion of "dynamic power" (or acceleration power) DQ as a product of torque and acceleration i.e. $DQ_i^m = P_i^m \ddot{q}^{mi}$.

Now, a very useful characteristic is a diagram connecting power and dynamic power. Each time instant gives one point having the coordinates Q^m and DQ^m. In this way, at the end of manipulation task we obtain Q^m-DQ^m characteristic (Fig. 2.34). This characteristic has to be within the feasible domain which is defined by a straight line connecting the maximal values Q_M^m and DQ_M^m (points B and A respectively, in Fig. 2.34). If the diagram violates this constraint then the test is negative. Q_M^m is the maximal motor power and can be expressed in the form $Q_M^m = P_M^m \omega_M^m / 4 = (P_M^m)^2 r_R / 4C_M C_E$ where P_M^m is the stall torque of motor, ω_M^m is no-load speed, r_R is rotor resistance, and finally C_M and C_E are constants of torque and electromotor force. Maximal value of DQ_M^m can be obtained by $DQ_M^m = (Q_M^m/T_{em})$. T_{em} is called electromechanical constant and has the form $T_{em} = J_R r_R / C_M C_E$, where J_R is the rotor moment of inertia. It should be said that the real value of maximal dynamic power DQ_{Mr}^m can be less than theoretical value DQ_M^m.

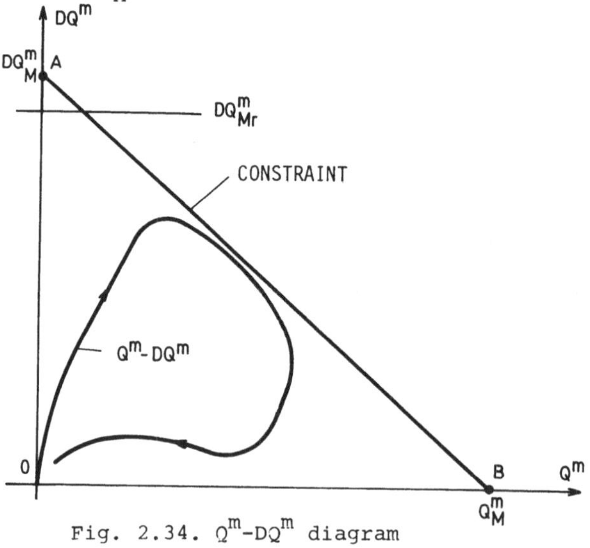

Fig. 2.34. Q^m-DQ^m diagram

Test 3 [11]. Finally, we introduce the test of motor heating. For this test the method of equivalent torque is used. We compute this equivalent motor torque by

$$P^m_{eq} = \sqrt{\frac{\sum_k (P^{mk})^2 \Delta t_k}{\sum_k \Delta t_k}} \qquad (2.136)$$

where P^{mk} is motor torque at the k-th time instant and Δt_k is time increment. The summation is performed over the whole manipulation task or over a finite time interval corresponding to a part of manipulation task. The equivalent torque is now tested against nominal motor torque P^m_{nom}. If $P^m_{eq} \leq P^m_{nom}$ the test is positive. If $P^m_{eq} > P^m_{nom}$ then there is overheating of motor and the test is negative.

Testing of a hydraulic actuator can be done in a way similar to the test of a D.C. electromotor. With the D.C. motor these maximal capabilities were defined by a linear maximal characteristic P^m_{max}-n^m (Figs. 2.32 and 2.33). With the hydraulic actuator this appears to be rather different. Maximal torque (or force) is not coupled strongly with actuator speed. Theoretically, the torque does not depend on motor speed. For instance, with the rotational hydraulic actuator the torque can be expressed as $P^m = \Delta p \frac{V'}{2\pi}$, where Δp is difference of pressure and V' is the unit volume (per one full revolution). The maximal torque P^m_{max} follows directly from the maximal pressure differential and has a constant value with respect to motor speed. The motor speed can be defined by $n^m = \frac{V_s}{V'}\, 60$ where V_s is the motor flow. Thus the maximal speed n^m_{max} is determined by maximum flow.

Choice of optimal design parameters

Mathematical models of robot mechanism dynamics can provide information which could aid the appropriate dimensioning as well as the optimal choice of parameters of a manipulation robot.

With this in mind it is necessary to select the optimality criteria. From the efficiency standpoint, with industrial manipulation robots two aspects are distinguished, namely, the working speed and the energy consumption. For this reason, during the process of evaluation and optimization of a manipulator mechanism, three criteria are defined:

(a) working speed/velocity criterion (time criterion)

(b) energy consumption criterion

(c) combined criterion (combination of (a) and (b)).

We will explain briefly the essence of these criteria. An example of the velocity criterion will be given.

(a) *Velocity criterion.* Let T be the time necessary for the execution of a given manipulation task. Optimization according to this criterion involves the choice of a manipulator configuration which allows the maximum working speed i.e. shortest time T.

(b) *Energy criterion.* Let E be the total energy used by the manipulator in the execution of a task. Optimization according to the energy criterion involves the selection of a configuration which would yield the minimal consumption of energy.

(c) *Combined criterion.* It presents the combination of the two criteria mentioned above such that it takes into consideration the working speed as well as energy consumed.

All of them can be used as criteria for minimization when choosing the optimal values of parameters in the design process. However, they can also be of use in the evaluation and comparison of various manipulation robots which the market offers. Criteria (a) and (b) can be used in the optimization procedures whereas criterion (c) is used in the evaluation and comparison of robotic systems.

Limitations. The choice of optimal parameters is performed by the minimization of the adopted criteria. To this end it is necessary to introduce limitations. If a reachability limit is introduced (robot kinematic ability constraint), the following limitations are of practical significance:

(i) *Limitations on the drive.* This limit is set by a requirement that the drive units in the joints are able to produce the driving forces (torques) necessary for performing a set manipulation task.

(ii) *Strength limitations.* The stresses in links of the manipulator should not exceed allowable values.

(iii) *Stiffness limitations.* The position and orientation errors arising from elastic deformations of manipulator links should not exceed allowable values.

EXAMPLE [6]

The task of transfer in such way a payload of 5kg mass along the trajectory ABCA (Fig. 2.35), that during this motion the gripper maintains its initial orientation in space is given to a manipulation robot UMS-1 (Fig. 2.25). Let the manipulator move along the trajectory segments AB, BC and CA in equal times and during these motions the velocity profile be triangular (2.36). Let T be the total time for the execution of this task.

Robot links are represented by circular tubes having a constant relationship between the inner and outer radii, i.e., $\psi = \frac{r}{R} = 0.75$. It is of course possible to adopt a link of a different cross section.

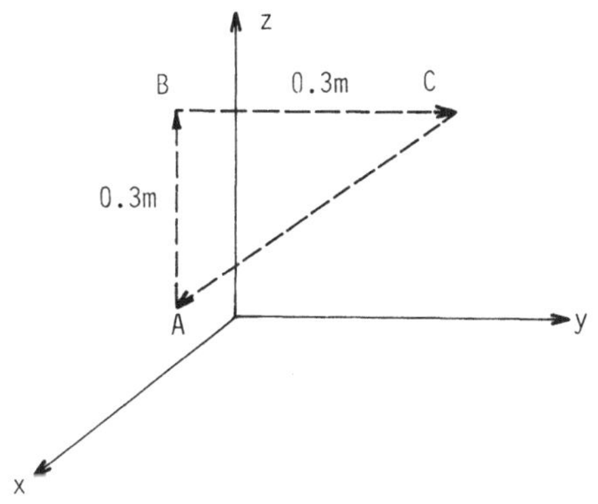

Fig. 2.35. Payload transfer trajectory

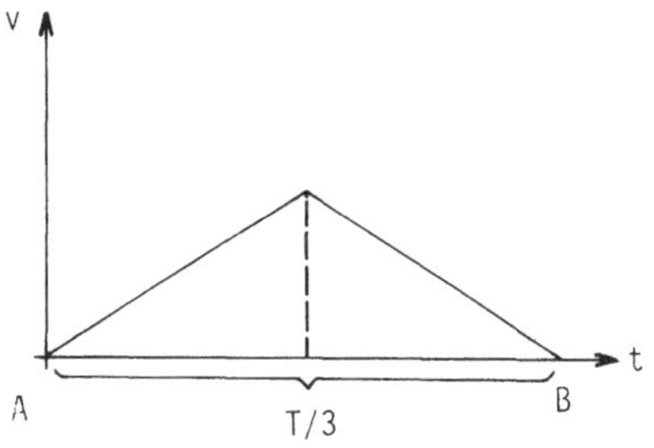

Fig. 2.36. Triangular velocity profile

As material of which the link is made, an alloy AℓMg3 was chosen with the following characteristics: density ρ=2700 kg/m^3, permitted bending stress $\sigma_d = \frac{1}{k}$ 30 daN/mm^2, torsion stress $\tau_d = \frac{1}{k}$ 25 daN/mm^2, Young's modulus $E_j = 7.848 \cdot 10^{10}$ N/m^2 and the safety coefficient k=5.

As driving motors for the manipulator joints, D.C. permanent magnet INDOX motors were chosen, Frame 23, type 2315-p20-0, produced by INDIANA GENERAL. The reduction ratio in revolute joints is 100.

Optimization according to the velocity criterion was adopted, based on the following procedure. The optimization process is performed as follows: for a selected value of R a series of simulations are performed with a successive reduction of task duration time T. This procedure is repeated until limitation (II) is violated, i.e., when the selected motors are unable to realise robot motion at that speed. R is then reduced and the procedure repeated. Thus the curve $T_{min}(R)$ i.e., minimal execution time for various values of R, is obtained. Let us first consider only limitations (I)-(II). The procedure of reducing R is repeated until the stress limitation (I) is violated i.e. stress in the links exceeds allowable limits. If a further reduction of R is desired, time T would have to be increased. Consequently, the minimum time (maximum velocity) appears in both limitations (I) and (II). Fig. 2.37 illustrates the results (for the anthropomorphic manipulator of Fig. 2.25), i.e., curve $T_{min}(R)$ and limitations (I) and (II). The minimum appears at the point M_1 designated by a circle. Let us now introduce limitations on stiffness (III). We impose the condition that the manipulator tip linear deviation due to segment elasticity should be less than 1 mm. In this case we consider only the quasi-static deflection due to nominal dynamics. By introducing limitation (III) the permissible domain is narrowed and the minimum point moves into position M_2, designated by a square in Fig. 2.37. The coordinates $(R^T_{opt}, T^{abs}_{min})$ correspond to the point M_2. The dotted line in Fig. 2.37 represents the corresponding energy consumptions. The abrupt decline of energy consumption to the left of M_1 is due to the abrupt increase of working time T i.e., the velocity decline. Besides the dependences in Fig. 2.37 which enable the optimal selection of parameter values, the dependence $T_{min}(R)$ can be automatically generated, for different manipulation robot structures. On the basis of these dependences the adequate kinematic scheme of the manipulation robot can be chosen when a specific task is performed. The dependences of energy consumption on the velocity i.e., time T for same robot structures, can also be obtained. Furthermore, curves describing the dependence of consumed energy E on the radius R, or on some other characteristic parameter of the link cross-section, can be obtained. It is of interest to study the dependence of stress in cylindrical links on their radius R, if such links are being used. Let $\sigma_{max}(R, T)$ be the maximum stress appearing in the link during task execution in time T, where R is the outer radius. The family of curves $\sigma_{max}(R, T)$, for various times T can show violations of limitations on allowable stress σ_d. Other functional relationships are possible and are computer generated on the basis of complete dynamic models of robotic mechanism. These are used directly in the design and performance evaluation of robotic systems. Ref. [6] gives a broader insight into

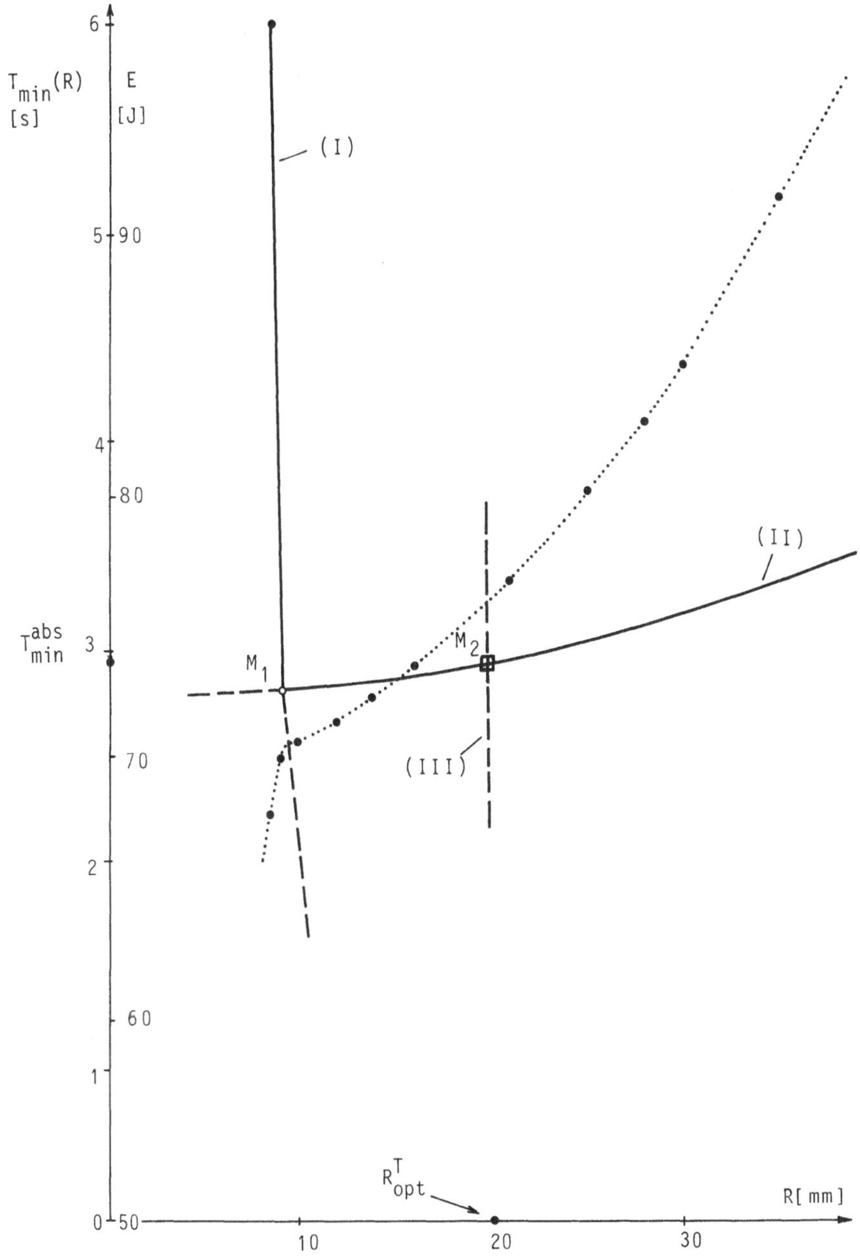

Fig. 2.37. Dependence $T_{min}(R)$ for the UMS-1 manipulator

the applicability of mathematical models of robot dynamics in the tasks of contemporary design of robotic mechanisms.

2.6 Dynamics of Flexible Manipulation Robots

Previous considerations of manipulation robot dynamics were based on the supposition about rigid links and joints of the mechanism, which represent, to a certain extent, an idealization of actual robotic structures.

Elastic deformations of the robot mechanism cause deviations of position and orientation of its point from the foreseen ones. Their intensity depends on external forces and the system parameters, determining the properties of elasticity, e.g. compliance of the mechanical structure, and are defined by its form, dimensions and materials data. Motion of elastic systems is accompanied by perturbational kineto-elastodynamic effects, the character of which is oscillatory. Hence, the knowledge of the dynamic behaviour of manipulation robots as systems of real-elastic bodies becomes very important from the aspect of mechanical structure design and control synthesis.

The structural compliance of the links and joints are influencing on the compliance of the robot mechanical structure as a whole. The compliance of the links is demonstrated by deviation of the internal coordinates from the nominal, due to elastic deformations during transfer of forces or torques in the corresponding actuators, reducers, transmissions, etc. The results of experimental investigations proove that with majority of contemporary industrial robots the compliance of the mechanical structure in the allowed load range is relatively weakly expressed. The dominant influence onto somewhat larger deflexion of the manipulator tip position in some cases, is that of joints compliance, more precisely, compressibility of the hydraulic actuator, or reducer elasticity (Harmonic Drive), or deflexion of the transmision with relatively far dislocated drives of joints, etc. The contribution of joint compliance hereby does not surpass 20% of the total compliance of the mechanical structure. The allowed payload mass of industrial robots most often amounts to 5% of their mechanical part mass, the eigenfrequences of which are relatively high and quite out of the band - width of the control system. Based on this, it can be concluded, that the rigid bodies model represent a justified and sufficiently exact approximation of ideal (rigid) robotic systems.

However, their performances and dimensions are opposite to some economic indices, which are suggesting the development of leightweight robotic structures, either by diminishing their rigidity by applying conventional structural materials, or by applying new, composite materials, with which today already practically the same structural properties along with 2-3 times smaller robot link mass can be achieved. Diminishing the link masses of the robot enables increassing its working speeds, diminishing energy consumption, increasing the payload, installation of smaller and cheaper actuators, etc. Thus, diminishing the cross-section dimensions of the robot links (when conventional materials are used) leads to smaller rigidity, i.e. increasing the compliance of the mechanical structure, so that in this case the elastodynamic effects cannot be neglected. Knowing their intensities and frequence characteristics is indispensable in order to realize in scope of modern design and control theory, the appropriate solutions, aimed at removing the unfavourable oscillatory effects and realizing the required accuracy and reliability of system. This, on one side, has been influencing a powerful growth of interest in the course of the last years for the development of manipulation robots mathematical models, containing elastic effects of the mechanical structure. On the other hand, strong support to the study of these problems is derived from new application fields of robots, using mechanical structures of relatively large span. The limited actuator abilities, significantly larger dimensions and specific demands, being posed to these systems, do not allow increasing the rigidity of their structures, as in the case of industrial robots, so that structural compliance becomes their significant mark. At the beginning, large manipulator arms, were developed for cosmic research. Later it showed, that such systems could find a very attractive application in civil works, shipbuilding, forestry, etc. Mathematical models of robotic mechanism dynamics, based on physical models of elastic bodies, are much more complex, than the earlier presented models of rigid body dynamics. Hence in the following text some basic aspects of elastic robots modelling will be given and in more detail will be presented an approximate method, which can be used in the design procedure and the choice of robot parameters aimed at obtaining the necessary stiffness of the structure (for given constraint of elastic deviations) with relatively rigid robots. We shall pay our attention to the derivation of linearized model of system. The reasons for this lie in the facts that the linear problems in theory of oscillations are much better explored and that many standard techniques for solving linear problems are available.

Fundamentals of elastic body motion

During deformation all points of the body move relative to their undeformed position, changing by that the relative positions, i.e. mutual distances and angles. According to the basic postulate of deformable bodies kinematics, motion of elastic bodies consists of these components (Fig. 2.38): translation, rotation and pure deformation. The first two correspond to ideally rigid body motion while the third one encloses the relative displacements (extensions and shears) in the near vicinity of the point considered. The deviation vector is now depending not only on time, but also on the position of the considered point on the body $\vec{u}_i = \vec{u}_i(x, y, z, t)$ in the adopted coordinate system, so that the system possesses an infinite number of degrees of freedom.

As deformation of real bodies is accompanied by very complex processes, very difficult to be described mathematically in exact way, we are compelled to introduce certain suppositions and thus partly idealize the problem. Most often considering the deformations is based on the model of the ideally - elastic Hooke's body, with which the deformation process is reversible and for which the linear dependence between stress and deformation holds. Second supposition refers to the intensity of elastic deformations. It is usually considered, that the local elastic displacement are of such order of magnitude, that the analysis of deformations can be carried out with sufficient exactness based on the linear elasticity theory. In other words, it is supposed, that the intensities of the deformation vectors are very small, compared with the body dimensions and that their increments with respect to coordinates ($\partial u_i/\partial x, \ldots$, etc.) are much less than one.

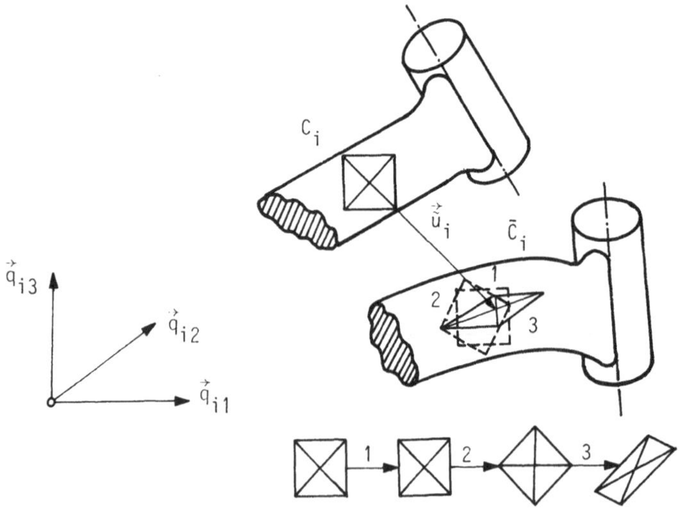

Fig. 2.38. Components of elastic displacement: 1 - translation, 2 - rotation, 3 - pure deformation

The dynamic conditions of body equilibrium in the mechanics of continuum are being established on a body segment of infinitesimally small dimensions and then these relations are widened to the whole considered domain. In this way, dynamic equation are obtained in the form of partial or integro-differential equations, which define, together with the initial and boundary conditions, a very complex boundary - value problem. This problem might be solved for limited number of elementary and simple technical systems. Hence we are compelled to use approximate methods with more complex systems, like manipulation robots. Solving of the dynamics problem is now transferred to the discrete analysis domain, i.e. to considering equivalent systems with a finite number of degrees of freedom, the motion of which is described by set of ordinary differential equations.

The basic idea of discretization consists of representing the elastic deformations vector in the form of a function of space variables and a finite number of time-dependent parameters, $\vec{\tilde{u}}_i(\tilde{x}_i, \tilde{y}_i, \tilde{z}_i, t) = \vec{\tilde{u}}_i(\tilde{x}_i, \tilde{y}_i, \tilde{z}_i, u_i^1, \ldots, u_i^{N_i})$, which according to the supposition of small deformations, can be expanded into a power series with respect to small parameters u_i^α, $(\alpha = 1, \ldots, N_i)$. Taking the first-order members only, we obtain:

$$\vec{\tilde{u}}_i(\tilde{x}_i, \tilde{y}_i, \tilde{z}_i, t) = \sum_{\alpha=1}^{N_i} \vec{\tilde{f}}_i^\alpha(\tilde{x}_i, \tilde{y}_i, \tilde{z}_i)\, u_i^\alpha(t) \tag{2.137}$$

where $\vec{\tilde{f}}_i^\alpha$ are interpolation functions, u_i^α are generalized elasticity coordinates, $\tilde{x}_i$, $\tilde{y}_i$, $\tilde{z}_i$ are coordinates of the position vector of the considered point before deformation with respect to the local coordinate system.

Depending on the manner of defining and choice of interpolation functions, three discretizing methods can be distinguished, being used most frequently in the structural analyzis. In the lumped mass method the system is substituted by a finite number of concentrated masses in chosen points, interconnected by springs ($\tilde{f}_i^\alpha$ are considered as unit impuls functions, while u_i^α represent the intensities of mass displacements in the local axes directions). In the method of finite elements the interpolation functions are defined within a subdomain of finite dimension bodies (finite elements), while u_i^α represent displacements of the points of their connections (nodes). If $\tilde{f}_i^\alpha$ are chosen in such way, that they describe qualitatively the oscillation forms (the modes) of the elastic body, and that u_i^α represent their amplitudes, we are coming to the assumed-modes method. The first two methods (lumped mass method and finite elements method) are methods of physical discretization while the third one uses mathematical discretization.

Let it be mentioned, that for the vectors of elastic rotations $\vec{\varphi}_i(x, y, z, t)$ too, affected by the deformations due to their corelation with the displacement vector in the linear elasticity theory $\vec{\varphi}_i = \frac{1}{2}\, \mathrm{rot}\vec{\tilde{u}}_i$, holds a formula, analogous, to expression (2.137).

Dynamics of robots with compliant joints

Let us first suppose that the mechanism links are rigid, i.e. the mechanical structure compliance is due only to the elasticity of the driving-transmission system. For description of the compliance of this mechanism part, we will use the lumped parameter model. The actuators, reducers and transmissions are represented by rigid bodies interconnected by torsional, or linear springs. For the sake of problem simplification, one spring with equivalent rigidity can be introduced (Fig. 2.39).

Let us denote by Δq^i the small angular (or linear) deflections of the equivalent spring. Due to fact that they correspond to the deviations of internal coordinates, and if we neglect the mass of the spring, it follows:

$$q^i = q^{io} + \Delta q^i, \qquad P_i^M = P_i^{Mo} = K_{i-1,i}\Delta q^i \tag{2.138}$$

where q^i is the internal coordinate of compliant joint, q^{io} is the internal coordinate of "rigid" joint, P_i^M is the load on output shaft of compliant joint, P_i^{Mo} is the load of "rigid" joint, $K_{i-1,i}$ is the spring stiffness.

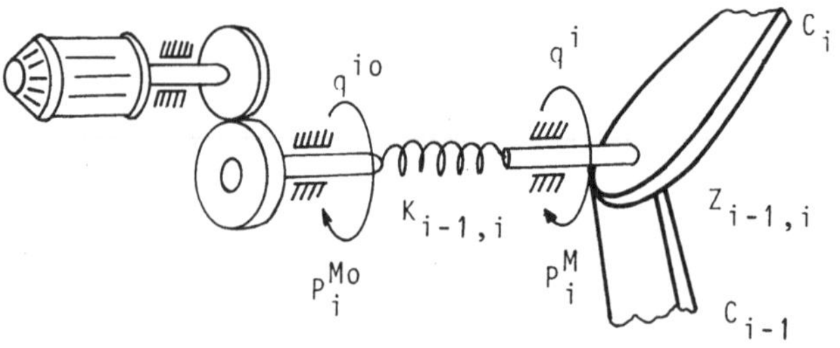

Fig. 2.39. Model of joint compliance

As during mechanism motion, according to the assumption of rigid links, no changes happen except the internal coordinates values and their derivatives, the dynamic model of a manipulation robot with compliant joints can be directly given as:

$$P = H(q, \theta)\ddot{q} + h(q, \dot{q}, \theta)$$
$$P = K_z \Delta q \qquad (2.139)$$

where $\Delta q = [\Delta q^1, \ldots, \Delta q^n]$ is the vector of the internal coordinates elastic deviations, $K_z = \mathrm{diag}[K_{i-1,i}]$ is a n×n diagonal stiffness matrix H and column-matrix h defined in (2.31).

Since the linear oscillations problems are better investigated, it is more purposeful to use the model linearized with respect to small Δq, instead of the non-linear model (2.139). Its derivation has been presented in detail in Chapter 3.

Dynamic model of robot with elastic links

Now, let us consider only the effects of manipulator links elasticity upon the mechanism dynamics, assuming that the manipulator joints are ideally rigid [28].

Let us define the vector of elastic displacements of an arbitrary point of the i-th link, as relative displacement with respect to the position of the "rigid skeleton" C_i, taken by this link in the kinematic chain as ideally rigid, whereby the previous links have already been deformed (Fig. 2.40). Such approach enables us, on one side, to enclose the coupling effects, and on the other, to derive the linearized dynamic model of oscillations. Namely, the vectors of relative elastic displacements can be treated as small values, while the total displacement (e.g. of the manipulator tip), relative to the positive of the equivalent rigid robot with system of larger span can be significantly expressed. By definition, the vector of elastic angular displacements $\vec{\tilde{\varphi}}_i(\tilde{x}_i, \tilde{y}_i, \tilde{z}_i, t) = \vec{\varphi}_i(\vec{\tilde{r}}_i, t)$ defines the relative rotations with respect to the local tri-hedron of axes, connected to the rigid skeleton ($\vec{\tilde{r}}_i$ is the position of the arbitrary point before deformation - see Fig. 2.40). From the aspect of defining total displacements and orientation with respect to referent coordinate system, we are interested in the linear and angular displacements of the i-th link tip of joint Z_i (Fig. 2.40):

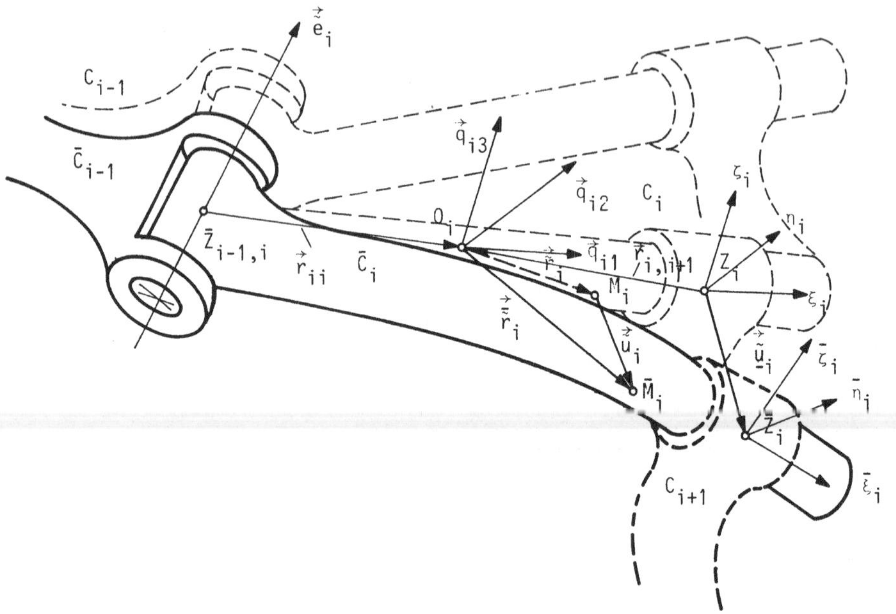

Fig. 2.40. Local elastic displacements

$$\underline{\vec{\tilde{u}}}_i = \vec{\tilde{u}}_i(-\vec{\bar{r}}_{i+1,i}, t) \quad \text{and} \quad \underline{\vec{\tilde{\varphi}}}_i = \vec{\tilde{\varphi}}_i(-\vec{\bar{r}}_{i+1,i}, t),$$

because they are transferred to the rest of the chain, causing its linear and angular displacements. Let us define the matrix of elastic rotations of the i-th link tip, $\underline{E}_i$, as a transformation matrix from the tri-hedron $\bar{Z}_i\bar{\xi}_i\bar{\eta}_i\bar{\zeta}_i$ into the corresponding tri-hedron $Z_i\xi_i\eta_i\zeta_i$, connected to the rigid skeleton, with its axes parallel to the axes of the internal coordinate system in O_i. For small rotations holds:

$$\underline{E}_i = I + \begin{bmatrix} 0 & -\underline{\tilde{\varphi}}_{i,3} & \underline{\tilde{\varphi}}_{i,2} \\ \underline{\tilde{\varphi}}_{i,3} & 0 & -\underline{\tilde{\varphi}}_{i,1} \\ \underline{\tilde{\varphi}}_{i,2} & \underline{\tilde{\varphi}}_{i,1} & 0 \end{bmatrix} = I + \underline{\underline{E}}_i \qquad (2.140)$$

where I is the unit diagonal matrix, while the elements of the skew-symetric matrix $\underline{\underline{E}}_i$ are the projections of the rotation vector of joint $\bar{Z}_i$ onto the axes of the local tri-hedron $\underline{\tilde{\varphi}}_{i,j}$ (j=1,2,3). Since deflexion of the i-th robot link does not lead to a relative change of orientation of the tri-hedron in $\bar{Z}_i$ and the local coordinate system of the (i+1)-st link, which is the same as the orientation of the local tri-hedrons of these two links in the equivalent rigid mechanism, by successive transformations we obtain the transformation matrix of the elastic mechanism:

$$\bar{Q}_i = Q_1\underline{E}_1 Q_{12}\underline{E}_2 Q_{23} \cdots Q_{i-2,i-1}\underline{E}_{i-1} Q_i = Q_i + \Delta Q_i$$
$$\Delta Q_i = \sum_{j=1}^{i-1} Q_j \underline{E}_j Q_{ji} \tag{2.141}$$

where Q_i is transformation matrix of the equivalent rigid system, Q_{ji} is the transformation matrix from the i-th to the j-th internal coordinate system, $Q_{ji}=Q_j^T Q_i$, ΔQ_i is the increment due to elastic rotation. The upper and lower dashed symbols denote variables describing motion of deflected mechanism w.r. to reference frame and variables representing elastic displacements and rotation of the link end points, respectively.

Using notations from Fig. 2.40, the position vector of arbitrary point of the i-th link of elastic mechanism can be represented as the sum $\bar{\vec{r}}_i=\vec{r}_i+\vec{u}_i$, while analogously to the previous, in the reference coordinate system it can be represented as [28]:

$$\bar{\vec{r}}_{M_i} = \vec{r}_{M_i} + \Delta\vec{r}_{M_i}$$
$$\vec{r}_{M_i} = \vec{r}_{i,1} + \vec{r}_i = \sum_{k=1}^{i-1} Q_k(\tilde{\vec{r}}_{kk}-\tilde{\vec{r}}_{k,k+1}) + Q_i(\tilde{\vec{r}}_{ii}+\tilde{\vec{r}}_i) \tag{2.142}$$
$$\Delta\vec{r}_{M_i} = \Delta\vec{r}_{i,1} + \vec{u}_i = \sum_{k=1}^{i-1} [\Delta Q_k(\tilde{\vec{r}}_{kk}-\tilde{\vec{r}}_{k,k+1})+Q_k\underline{\tilde{\vec{u}}}_k] + Q_i\tilde{\vec{u}}_i,$$

where $\vec{r}_{M_i}$ is the position vector of the corresponding point of the equivalent rigid system with respect to the reference system, while $\Delta\vec{r}_{M_i}$ is the vector of its elastic displacement. Let it be noted, that it is obtained by superposition of displacements due to relative deflections and rotation of all links in the chain part from the basis up to the considered point. By using this concept, all of the other kinematic vectors can also be expressed in the form of a sum of nominal and additive components due to elasticity. Hereby we will assume that the angular velocity and acceleration vectors define the i-th local tri-hedron rotation, so that their nominal components are the same as in (2.6). As with elastic links we must define linear acceleration of some arbitrary point, for the nominal component we can use the expression for the local tri-hedron center, adding the position vector $\vec{r}_i$ everywhere to the vector $\vec{r}_{ii}$. Applying the theorem of relative motion and taking also the components of the relative elastic displacement for the additive components, we obtain:

$$\Delta\vec{\omega}_i = \Delta\vec{\omega}_{i-1} + \dot{q}^i\Delta\vec{e}_i\bar{\xi}_i + \vec{\omega}_i^{e\ell}$$

$$\Delta\vec{v}_i = \Delta\vec{v}_{o_i} + \Delta\vec{\omega}_i \times \vec{r}_i + \vec{\omega}_i \times \vec{u}_i + \vec{v}_i^{e\ell}$$

$$\Delta\vec{v}_{o_i} = \Delta\vec{v}_{o_{i-1}} + \Delta\vec{\omega}_i \times \vec{r}_{ii} - \Delta\omega_{i-1} \times \vec{r}_{i-1,i} + \vec{\omega}_{i-1} \times \underline{\vec{u}}_{i-1} + $$
$$+ \xi_i \dot{q}^i \Delta\vec{e}_i + \underline{\vec{v}}_{i-1}^{e\ell}$$

$$\Delta\vec{\varepsilon}_i = \Delta\vec{\varepsilon}_{i-1} + \ddot{q}^i \Delta\vec{e}_i \bar{\xi}_i + \dot{q}^i (\vec{\omega}_{i-1} \times \Delta\vec{e}_i + \Delta\vec{\omega}_{i-1} \times \vec{e}_i) \bar{\xi}_i +$$
$$+ \underline{\vec{\varepsilon}}_{i-1}^{e\ell} + \vec{\omega}_{i-1} \times \underline{\omega}_{i-1}^{e\ell}$$

$$\Delta\vec{w}_i = \Delta\vec{w}_{o_i} + \Delta\vec{\varepsilon}_i \times \vec{r}_i + \vec{\omega}_i \times (\Delta\vec{\omega}_i \times \vec{r}_i) + \Delta\vec{\omega}_i \times (\vec{\omega}_i \times \vec{r}_i) +$$
$$+ \vec{\omega}_i \times (\vec{\omega}_i \times \vec{u}_i) + 2\vec{\omega}_i \times \vec{v}_i^{e\ell} + \vec{w}_i^{e\ell} + \vec{\varepsilon}_i \times \vec{u}_i$$

$$\Delta\vec{w}_{o_i} = \Delta\vec{w}_{o_{i-1}} - \Delta\vec{\varepsilon}_{i-1} \times \vec{r}_{i-1,i} + \vec{\varepsilon}_{i-1} \times \underline{\vec{u}}_{i-1} - \vec{\omega}_{i-1} \times (\Delta\vec{\omega}_{i-1} \times \vec{r}_{i-1,i}) -$$
$$- \Delta\omega_{i-1} \times (\vec{\omega}_{i-1} \times \vec{r}_{i-1,i}) + \vec{\omega}_{i-1} \times (\vec{\omega}_{i-1} \times \vec{u}_{i-1}) + \underline{w}_{i-1}^{e\ell} +$$
$$+ 2\vec{\omega}_{i-1} \times \underline{v}_{i-1}^{e\ell} + \xi_i [\ddot{q}^i \Delta\vec{e}_i + 2\dot{q}^i (\Delta\vec{\omega}_i \times \vec{e}_i + \vec{\omega}_i \times \Delta\vec{e}_i)] +$$
$$+ \Delta\vec{\varepsilon}_i \times \vec{r}_{ii} + \vec{\omega}_i \times (\Delta\vec{\omega}_i \times \vec{r}_{ii}) + \Delta\vec{\omega}_i \times (\vec{\omega}_i \times \vec{r}_{ii}) \qquad (2.143)$$

where $\Delta\vec{e}_i = \Delta Q_i \vec{\tilde{e}}_i$, $\vec{u}_i = Q_i \vec{\tilde{u}}_i$, $\vec{\omega}_i^{e\ell}$, $\vec{\varepsilon}_i^{e\ell}$, $\vec{v}_i^{e\ell}$ and $\vec{w}_i^{e\ell}$ are the vectors of angular and linear velocities and accelerations of the relative elastic displacements and rotations, expressed in the reference coordinate system:

$$\vec{v}_i^{e\ell} = Q_i \sum_{\alpha=1}^{N_i} \vec{\tilde{f}}_i^{\alpha} \dot{u}_i^{\alpha}$$
$$\qquad (2.144)$$
$$\vec{\omega}_i^{e\ell} = Q_i \sum_{\alpha=1}^{N_i} \vec{\tilde{\varphi}}_i^{\alpha} \dot{u}_i^{\alpha}$$

Analogous relations hold for the acceleration vectors, which are linear forms of the generalized elastic accelerations. The vectors of interpolating functions $\vec{\tilde{\varphi}}_i^{\alpha}$, due to the connection between the rotation angles and linear elastic displacements are:

$$\vec{\tilde{\varphi}}_i^{\alpha}(\tilde{x}_i, \tilde{y}_i, \tilde{z}_i) = \frac{1}{2} \underline{\underline{R}} \vec{\tilde{f}}_i^{\alpha}(\tilde{x}_i, \tilde{y}_i, \tilde{z}_i)$$

$$\underline{\underline{R}} = \begin{bmatrix} 0 & -\partial/\partial\tilde{z} & \partial/\partial\tilde{y} \\ \partial/\partial\tilde{z} & 0 & -\partial/\partial\tilde{x} \\ -\partial/\partial\tilde{y} & \partial/\partial\tilde{x} & 0 \end{bmatrix} \tag{2.145}$$

whereby $\underline{\underline{R}}$ is the skew-symmetric matrix-operator of partial differentiation. Using the coordinates of vector $\vec{\tilde{\varphi}}_i^\alpha$, the component $\underline{\underline{E}}_i$ of the transformation matrix of elastic rotations $\underline{E}_i$ (2.140) can be expressed as explicit form of generalized elastic coordinates u_i^α:

$$\underline{E}_i = \sum_{\alpha=1}^{N_i} \underline{\underline{E}}_i^\alpha u_i^\alpha$$

where elements of the matrix $\underline{\underline{E}}_i^\alpha$ are formed by the coordinates of vector $\vec{\tilde{\varphi}}_i^\alpha$ according to (2.140). By that, based on (2.141) the increment of the transformation matrix ΔQ_i is explicitly defined.

For the sake of deriving the dynamic equations of the flexible robotic mechanisms, let us, as in the case of rigid body dynamics (Fig. 2.6), disconnect fictitiously the robot kinematic chain in the i-th joint. Let us substitute the action of the rejected part of mechanism, including the i-th drive, by reaction forces and moments, which naturally, will not be the same as in "rigid" dynamics. Let us set the conditions of dynamic equilibrium of all forces, acting on the free end, taking care now about the internal forces equilibrium at elastic deformation. These conditions are expressed by the general dynamic equation, which in the form of Lagrange-D'Alembert's principle, applied on the considered mechanism part reads:

$$\sum_{j=i}^{n} \int_{V_j} (\vec{f}_j - \vec{\bar{w}}_j) \cdot \delta \vec{\bar{r}}_{M_j} \, dm - \int_{V_j} \delta\varepsilon_j^T \sigma_j \, dv = -(\vec{R}_i^{e\ell} \cdot \delta \vec{\bar{r}}_{Z_{i-1,i}} + \vec{M}_i^{e\ell *} \cdot \delta \vec{\bar{\gamma}}_{Z_{i-1,i}}) \tag{2.146}$$

where $\vec{f}_j$ is the external force per unit mass, $\delta\vec{\bar{r}}_{M_j}$ is the virtuel displacement of the arbitrary point of the j-th link ($j>i$) in the deformed kinematic chain, $\vec{\bar{w}}_j = \vec{w}_j + \Delta\vec{w}_j$ is the vector of its acceleration given by (2.6) and (2.143), ε_j and σ_j are the strain and stress vectors, $\vec{R}_i^{e\ell}$ and $\vec{M}_i^{e\ell *}$ are the total force and moment in the joint $Z_{i-1,i}$ of the flexible mechanism, $\delta\vec{\bar{r}}_{Z_{i-1,i}}$ and $\delta\vec{\bar{\gamma}}_{Z_{i-1,i}}$ are the vectors of the virtuel displacements and rotations of the joint $Z_{i-1,i}$. Dynamic equations of motion as function of generalized coordinates and their derivatives are derived by means of the virtuel work method. First, let us impose the increment δq^i to the internal coordinates only, which causes virtuel displacements [28]:

$$\delta\vec{\bar{r}}^{q^i}_{M_j} = (\vec{r}^{q^i}_{M_j}+\Delta\vec{r}^{q^i}_{M_j})\delta q^i$$

$$\vec{r}^{q^i}_{M_j} = \bar{\xi}_i\vec{e}_i\times(\vec{r}_{j,i}+\vec{r}_j) + \xi_i\vec{e}_i$$

$$\Delta\vec{r}^{q^i}_{M_j} = \bar{\xi}_i\Delta\vec{e}_i\times(\vec{r}_{j,i}+\vec{r}_j)+\bar{\xi}_i\vec{e}_i\times(\Delta\vec{r}_{j,i}+\vec{u}_j)+\xi_i\Delta\vec{e}_i \tag{2.147}$$

$$\delta\vec{\bar{r}}^{q^i}_{Z_{i,i-1}} = \xi_i\vec{\bar{e}}_i\delta q^i$$

$$\delta\vec{\bar{\gamma}}^{q^i}_{Z_{i,i-1}} = \bar{\xi}_i\vec{\bar{e}}_i\delta q^i$$

where $\delta\vec{\bar{r}}^{q^i}_{M_j}$ is the vector of virtuel displacement of some arbitrary point of the j-th link due to increment $\delta q^i(j=i,\dots,n)$, $\vec{r}^{q^i}_{M_j}$ and $\Delta\vec{r}^{q^i}_{M_j}$ are the vector influence coefficients of displacement (nominal and additional component, respectively), expressing the influence of δq^i onto $\delta\vec{\bar{r}}_{M_j}$; by $\vec{r}_{j,i}$ and $\Delta\vec{r}_{j,i}$ are denoted the vectors of the nominal distance of the origin of the j-th internal coordinate system from joint $Z_{i,i-1}$ and its additional component (2.142), respectively.

The vectors $\delta\vec{\bar{r}}^{q^i}_{Z_{i,i-1}}$ and $\delta\vec{\bar{\gamma}}^{q^i}_{Z_{i,i-1}}$ have not been divided into additional and nominal components because on the corresponding virtuel displacements, under the supposition of ideal joints, work is done by the driving force and torque only:

$$\vec{\bar{M}}^{e\ell*}_i\cdot\delta\vec{\bar{\gamma}}^{q^i}_{Z_{i,i-1}} = P^M_i\delta q^i, \qquad \vec{\bar{R}}^{e\ell}_i\cdot\delta\vec{\bar{r}}_{Z_{i,i-1}} = P^F_i\delta q^i$$

Now, let us impose to the i-th link the virtuel elastic displacements $\delta\vec{u}_i$. Hereby work will be done by the internal elastic forces of the i-th link only and the external forces acting on this and the other links of the considered chain part. The work of external forces, acting on the other links, is done along virtuel displacements being transfered over the end of the i-th link. If we suppose that the joints are ideally rigid, which is justified due to the fact that the links are often reinforced and that now we are not considering the compliance of joints, the work of the reaction forces will be zero. The vectors of displacements will be:

$$\delta\vec{r}_{M_j}^{\,u_i} = \sum_{\alpha=1}^{N_1} (\vec{r}_{M_j}^{\,u_i} + \Delta\vec{r}_{M_j}^{\,u_i})\,\delta u_i^{\alpha}$$

where the influence coefficients of virtuel displacement are:

$$\vec{r}_{M_j}^{\,u_i^{\alpha}} = \begin{cases} \vec{f}_i^{\alpha} = Q_i\vec{\tilde{f}}^{\alpha} & \text{for} \quad j=i \\ \underline{\vec{f}}_i^{\alpha} + \underline{\vec{\varphi}}_i^{\alpha}\times(\vec{r}_{j,i+1}+\vec{r}_j) & \text{for} \quad j>i \end{cases} \tag{2.147a}$$

$$\Delta\vec{r}_{M_j}^{\,u_i^{\alpha}} = \begin{cases} \Delta\vec{f}_i^{\alpha} = \Delta Q_i\vec{f}_i^{\alpha} & \text{for} \quad j=i \\ \Delta\underline{\vec{f}}_i^{\alpha} + \Delta\underline{\vec{\varphi}}_i^{\alpha}\times(\vec{r}_{j,i+1}+\vec{r}_j)+\underline{\vec{\varphi}}_j^{\alpha}\times(\Delta\vec{r}_{j,i+1}+\vec{u}_j) & \text{for} \quad j>i \end{cases}$$

$\vec{r}_{M_j}^{\,u_i^{\alpha}}$ is "nominal component" which describes displacements in the kinematic chain just due to virtuel elastic deformations of the i-th link, while the additional term $\Delta\vec{r}_{M_j}^{\,u_i^{\alpha}}$ takes into account that the rest of the mechanism has already been deformed, too.

Work of the internal forces will be determined by means of the stress--strain displacement relations of the linear elasticity theory:

$$\begin{aligned} \sigma_i &= D_i\varepsilon_i \\ \varepsilon_i &= \mathcal{L}\cdot\tilde{u}_i = \mathcal{L}\tilde{\mathbf{F}}_i u_i \end{aligned} \tag{2.148}$$

where $\sigma_i = [\sigma_{\tilde{x}\tilde{x}}\ \sigma_{\tilde{y}\tilde{y}}\ \sigma_{\tilde{z}\tilde{z}}\ \sigma_{\tilde{x}\tilde{y}}\ \sigma_{\tilde{y}\tilde{z}}\ \sigma_{\tilde{x}\tilde{z}}]_i^T$,

$\varepsilon_i = [\varepsilon_{\tilde{x}\tilde{x}}\ \varepsilon_{\tilde{y}\tilde{y}}\ \varepsilon_{\tilde{z}\tilde{z}}\ 2\varepsilon_{\tilde{x}\tilde{y}}\ 2\varepsilon_{\tilde{y}\tilde{z}}\ 2\varepsilon_{\tilde{x}\tilde{z}}]_i^T$.

D_i is the symmetric elasticity matrix of the material, $\mathcal{L}$ is the differentiation matrix-operator, $\tilde{\mathbf{F}}_i = [\tilde{f}_i^1,\dots,\tilde{f}_i^{\alpha},\dots,\tilde{f}_i^{N_i}]$ is a $(3\times N_i)$ matrix of interpolation functions, $u_i = [u_i^1,\dots,u_i^{\alpha},\dots,u_i^{N_i}]^T$ is the vector of generalized elastic coordinates of the i-th link. Matrices $\mathcal{L}_i$ and D_i (for homogeneous isotropic elastic medium) have the form:

$$\mathcal{L} = \begin{bmatrix} \partial/\partial\tilde{x} & 0 & 0 \\ 0 & \partial/\partial\tilde{y} & 0 \\ 0 & 0 & \partial/\partial\tilde{z} \\ \partial/\partial\tilde{y} & \partial/\partial\tilde{x} & 0 \\ 0 & \partial/\partial\tilde{z} & \partial/\partial\tilde{y} \\ \partial/\partial\tilde{z} & 0 & \partial/\partial\tilde{x} \end{bmatrix}$$

$$D_i = \frac{E_i}{2(1+\nu)} \begin{bmatrix} \frac{2(1-\nu)}{1-2\nu} & \frac{2\nu}{1-2\nu} & \frac{2\nu}{1-2\nu} & 0 & 0 & 0 \\ & \frac{2(1-\nu)}{1-2\nu} & \frac{2\nu}{1-2\nu} & 0 & 0 & 0 \\ & & \frac{2(1-\nu)}{1-2\nu} & 0 & 0 & 0 \\ & \text{symmetrical} & & 1 & 0 & 0 \\ & & & & 1 & 0 \\ & & & & & 1 \end{bmatrix}$$

where E is Young's modulus and ν is Poisson's ratio. Based on (2.148) follows:

$$\int_{V_i} \delta\varepsilon^T \sigma dV = \delta u^{iT} K_i u^i$$
$$K_i = \int_{V_i} (\mathcal{L}\tilde{\mathbf{F}}_i)^T D_i \mathcal{L}\tilde{\mathbf{F}}_i dV \tag{2.149}$$

where K_i is the structural stiffness matrix of the i-th link.

In the finite elements method several specific procedures were developed for calculating K_i, because solving the integral (2.149) with system of higher order can be purposeless. Choice and definition of the interpolating functions can be found in numerous references.

Substituting relations (2.146-2.149) into (2.145), equalizing all members with the virtual increments of generalized coordinates, neglecting members of second and higher order with respect to small u_i^α and their derivatives, by explicit separating second derivatives $\ddot{q}^i$, as well as the elastic coordinates u_i^α and their derivatives (due to linear dependence of all values) and by repeating the described procedure for i= 1,...,n, we derive the complete model of the manipulation robot with elastic links dynamics, in the form [28]:

$$H^{qq}(q,\ \theta)\ddot{q} + h^{q}(q,\ \dot{q},\ \theta) + H^{qu}(q,\ \theta)\ddot{u} + C^{qu}(q,\ \dot{q},\ \theta)\dot{u} + $$
$$+ K^{qu}(q,\ \dot{q},\ \ddot{q},\ \theta)u = P^{q}$$

$$H^{uu}(q,\ \theta)\ddot{u} + (C^{uu}(q,\ \dot{q},\ \theta) + D^{uu}(t))\dot{u} + (K^{uu}(q,\ \dot{q},\ \ddot{q},\ \theta) +$$
$$+ K_s^{uu}(\theta))u + H^{uq}(q,\ \theta)\ddot{q} + h^{u}(q,\ \dot{q},\ \theta) = P^{u} \qquad (2.150)$$

where the first equation is describing the motion of the manipulation system, and the second one its structural dynamics. As we have expressed, in deriving the model, all values in the form of a sum of nominal and additional components, matrix H^{qq} and vector h^{q} have the same structure like the corresponding matrices of the equivalent rigid system (2.31) (we have introduced additional indexes in order to make a difference as compared with the additional effects and designated their order). The remaining matrices of the system express the dynamic effects of elasticity. Their elements depend on system parameters, internal coordinates and their derivatives as well as on so-called generalized inertial and stiffness parameters. These parameters, obtained on the basis of (2.146) by solving of volume integrals, qualitatively represents the effects of deformation motion upon the system dynamics. Hereby $u = [u^1,...,u^n]$ is the $(N\times 1)$ vector of the mechanism elastic coordinates $(N = \sum_{i=1}^{n} N_i)$, sets of $(n\times N)$ matrices $\{H^{qu},\ C^{qn},\ K^{qu}\}$ and $(N\times N)$ matrices $\{H^{uu},\ C^{uu},\ K^{uu}\}$ are expressing the inertial generalized gyroscopic effects and the effects of geometric rigidity, D^{uu} is a $(N\times N)$ matrix of structural damping and K_s^{uu}=block diag $[K_i]$ is a block matrix of structural stiffness. The $(N\times n)$ matrix H^{uq} and the $(N\times 1)$ vector h^{u} express the influence of inertial, gyroscopic, gravitational and external forces (more precisely of their nominal components, while the additional ones are enclosed in the matrix of geometrical stiffness) onto the mechanism oscillatory motion. The vectors P^{q} and P^{u} of dimensions $(n\times 1)$ and $(N\times 1)$, respectively, represent the vectors of generalized active forces in the mechanism joints and the micro-elastic control forces, acting onto the mechanism links (forces of active damping - active stiffness forces, etc.).

A detailed definition of the matrices and vectors of the complete general dynamics model (2.150) would require a lot of space, ranging out of this book's contents. Forming of some of the system matrices in the lumped-mass approach will be demonstrated for the approximate model case.

Equations (2.150) together with the relation:

$$P^q = K_z \Delta q$$

describing the elasticity of joints, define the complete model of the robot mechanism dynamics with compliant mechanical structure.

Endpoint compliance

For the designers of manipulation robots mechanical structure, apart from the frequency characteristics and damping capacity, the most interesting in any way is the knowledge of the endpoint compliance - the compliance at the mechanism gripper, under the action of forces, attacking most frequently this mechanism link (Fig. 2.11).

Practically, the problem of compliance calculation in the endpoint reduces to definition of the relations between the vector of position and orientation deviations of the gripper tip and the external forces and moments in the reference coordinate system, which can be represented by the relation:

$$\Delta x_A = C_A F_A, \tag{2.151}$$

where in the general case $\Delta x_A = [\Delta x, \Delta y, \Delta z, \Delta \theta, \Delta \varphi, \Delta \psi]^T$ is the deviation vector of the external coordinates (Fig. 2.12), $F_A = [F_{Ax}, F_{Ay}, F_{Az}, M_{Ax}, M_{Ay}, M_{Az}]^T$ is the vector of external forces and moments, while C_A is the matrix of manipulator endpoint compliance. For determining of this matrix, let us first define the increment Δx_A, as a sum of deflexions due to compliance of joints and the mechanical structure, structure. The vector Δx_A after neglecting all quadratic and higher order terms with respect to Δq^i and u_i^α, can be expressed as function of the generalized coordinates:

$$\Delta x_A = J_A^q(q, \theta)\Delta q + J_A^u(q, \theta)u, \tag{2.152}$$

where J_A^q and J_A^u are Jacobian matrices (6×n) and (6×N), respectively, expressing the influence of the deviations of internal coordinates and elastic links deformations onto the displacement of the gripper point A. These matrices are structured:

$$J_A^q = \begin{bmatrix} J_{A(3\times n)}^{q^v} \\ \hline J_{A(3\times n)}^{q^\omega} \end{bmatrix}, \quad J_A^u = \begin{bmatrix} J_{A(3\times n)}^{u^v} \\ \hline J_{A(3\times n)}^{u^\omega} \end{bmatrix}$$

where:

$$J_A^{q^v} = [\vec{r}_A^{q^1} \ \dots \ \vec{r}_A^{q^i} \ \dots \ \vec{r}_A^{q^n}]_{(3\times n)}$$

$$J_A^{q^\omega} = [\vec{\gamma}_A^{q^1} \ \dots \ \vec{\gamma}_A^{q^i} \ \dots \ \vec{\gamma}_A^{q^n}]_{(3\times n)}$$

$$J_A^{u^v} = [\underbrace{\vec{r}_A^{u_1^1} \ \dots \ \vec{r}_A^{u_1^{N_1}}}_{N_1} \ \dots \ \underbrace{\vec{r}_A^{u_i^1} \ \dots \ \vec{r}_A^{u_i^{N_i}}}_{N_i} \ \dots \ \underbrace{\vec{r}_A^{u_n^1} \ \dots \ \vec{r}_A^{u_n^{N_n}}}_{N_n}]_{(3\times N)}$$

$$J_A^{u^\omega} = [\underbrace{\vec{\gamma}_A^{u_1^1} \ \dots \ \vec{\gamma}_A^{u_1^{N_1}}}_{N_1} \ \dots \ \underbrace{\vec{\gamma}_A^{u_i^1} \ \dots \ \vec{\gamma}_A^{u_i^{N_i}}}_{N_i} \ \dots \ \underbrace{\vec{\gamma}_A^{u_n^1} \ \dots \ \vec{\gamma}_A^{u_n^{N_n}}}_{N_n}]_{(3\times N)} \tag{2.153}$$

where the vector displacement coefficients $\vec{r}_A^{q_i}$ and $\vec{r}_A^{u_i^\alpha}$ $(A\equiv A_n)$ are given by (2.147–2.147a), while the influence coefficients of rotation are:

$$\vec{\gamma}_A^{q^i} = \bar{\xi}_i \vec{e}_i, \quad \vec{\gamma}_A^{u_i^\alpha} = \begin{cases} \vec{\varphi}_i^\alpha & \text{for} \quad i<n \\ \vec{\varphi}_n^\alpha(\tilde{r}_n) & \text{for} \quad i=n \end{cases} \tag{2.154}$$

By applying the method of virtuel work we determine the relation between joints compliance and elastic displacements of links, due to external forces at point A. After neglecting the additional terms of the Jacobian matrices due to interconnected deflexions of the mechanism links and joints, we obtain:

$$K_z \Delta q = J_A^{q^T} \cdot F_A, \quad K_S \cdot u = J_S^{u^T} \cdot F_A \tag{2.155}$$

where J_A^q and J_A^u are "nominal" components of the Jacobian, defined by (2.153-2.154). By calculating Δq and u from the last equation, under the supposition that the stiffness matrices are regular (all mechanism joints are active and their rigidity is not equal zero, i.e. the vector of elastic displacements u represents the independent relative elastic displacements), and substituting in (2.152) we obtain the compliance matrix:

$$C_A = J_A^q K_z^{-1} J_A^{q^T} + J_A^u K_S^{-1} J_A^{u^T} \tag{2.156}$$

It should be stressed out, that in deriving the compliance matrix, we have neglected the quadratic members in the expression for displacement (2.152) and the additional members of the Jacobian in (2.155), so that it practically expresses the displacement, obtained by superposition of all displacements of the manipulator end due to the individual compliances of joints and links, whereby the rest of the chain is rigid. Of course this represents an approximation, because all mechanism parts are deformed simultaneously. It can be said, that the displacements, calculated in above way, represent the dominant part of real displacements and that they can be used with satisfactory accuracy in the process of mechanical structure design. However, when the displacements are calculated to be used for their proper compensations by means of the control system, particularly with robot mechanisms of large span (the Jacobian matrices elements of which attain bigger values) the additional terms should be taken into account, too. By this the expression for the compliance matrix becomes more complex and its definition demands deeper study, passing out of the scope of such conceived textbook. Volume 8 of the monograph series [28] is devoted to flexible manipulators.

APPROXIMATE METHOD FOR DYNAMIC ANALYSIS OF FLEXIBLE MANIPULATION ROBOTS)*

Differing from the cases of strongly elastic link of the manipulation mechanisms of large spans, where it is necessary to calculate, using the corresponding techniques, the elastic deformations and oscillations of the mechanical structure for their compensation by means of control signals, in this paragraph an approximate method is presented, having another purpose.

---

*) In writing this section references [6, 19] were used.

Namely, with "normally-rigid" manipulation mechanisms it is necessary to check their rigidity or to "design" them, so that some characteristic mechanism points (e.g. joints and manipulator tip) do not override values given in advance. Let it be reminded of the fact, that to-day manipulation robots are applied in very fast and precise industrial tasks. Fast and precise motion understands very short settling time of the manipulation robot tip in some points of its workspace. For achieving such robot performance it is necessary to ensure also sufficient rigidity of the mechanical structure, so that the accuracy of the control system can match the supposition of a rigid robot mechanism. In the oposite case, oscillations will occure in the points of robot motion terminations, reflecting itself on prolongued settling time. For preventing unsuitable dynamic behaviour of the manipulation robot during its operation the rigidity test in the robot design process must be carried out. The engineering method presented here [6, 19] is approximate and conservative. In broader scope it has been given in Vol. 1 of the monograph series [6].

One of the basic suppositions, on which the approximate model of flexible robot dynamics is based, concerns the intensity and character of the kineto-elastodynamic effects. Namely, we will consider that they do not cause essentially the deviations from the nominal motion of the robotic manipulator. In that way, practically all additional terms from the first equation of system (2.150) can be rejected and only the nominal dynamic model left. By neglecting the coupling elastic effects, being transferred from one link to the other along the chain and damping effects, the model of structural dynamics will reduce to [6]:

$$D\ddot{u} + u = c \tag{2.157}$$

where $D = K_S^{uu^{-1}} \cdot H^{uu}$, $c = -K_S^{uu^{-1}} \cdot h^u$.

Differing from the previous general considerations, these assumptions enable us to define the elastic displacements with respect to the position of the rigid manipulation system and to obtain them directly by solving the model (2.157).

Basic assumptions

The basic idea of the approach to the elastic manipulator dynamics is to consider the manipulator as an open chain of elastic canes and to consider weights, inertial forces of nominal motion, nominal driving forces and torques and nominal reactions in joints as known external forces and moments. These values are computed in the block of nominal dynamics (Appendix 9).

Let us introduce the values $\vec{u}_i$, i=1,...,n and $\vec{\varphi}_i$, i=1,...,n, which will distinguish elastic manipulator deviation from the nominal motion of the rigid mechanical structure of robot, $\vec{u}_i$ is the linear deviation vector of joint S_{i+1} from its nominal motion (Fig. 2.41), $\vec{\varphi}_i$ is the rotation vector, i.e., the angular deviation of the point S_{i+1} relative to the nominal orientation (Fig. 2.42). Under the assumption that the deviations are small, the angles can be treated as vectors. Thus, $\vec{u}_i$, $\vec{\varphi}_i$, i=1,...,n represent the characteristic values of the elastic oscillations, which can be called micro-motion. Total motion is regarded as superposition of nominal and micro motion. Let us introduce matrix notation: u_i is a (3×1) matrix corresponding to the vector $\vec{u}_i$, u - a (3n×1) matrix

$$u = \begin{bmatrix} u_1 \\ \vdots \\ u_n \end{bmatrix} \tag{2.158}$$

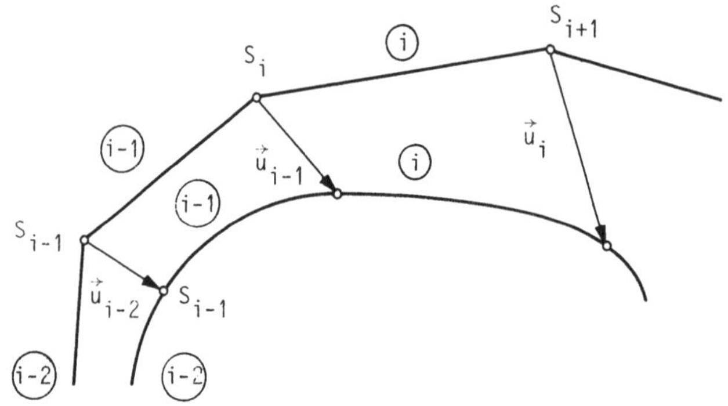

Fig. 2.41. Linear elastic deviation-deflection

The task is now to present the computer oriented method for the formation of the mathematical model of micro-motion for known nominal motion. Thus, a procedure will be derived by which, for known characteristics and values of nominal dynamics in some time instant t^*, the matrices D(3n×3n) and c(3n×1) can be calculated so that:

$$D\ddot{u} + u = c \tag{2.159}$$

and by means of which the model of micro-motion for time instant t^*

would be formed. The elastic oscillations are not considered as damped because we are especially interested in the maximal elastic deviation. Damping forms may easily be included in the calculation.

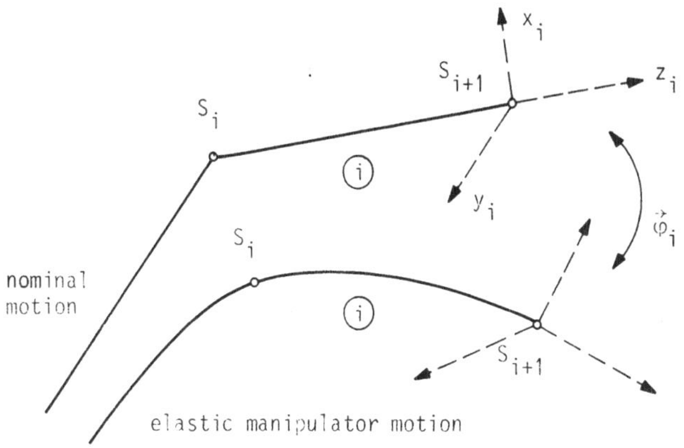

Fig. 2.42. Angular elastic deviation-shape (orientation change in space)

One of the assumptions which is adopted is that micro-motion due to link elasticity does not influence the generalized (relative) coordinates in the mechanism joints, the time history of which remains the same as in the nominal motion. Thus, the deviations are due to elastic link deformations only. Such an assumption permits the incorporation of the block of micro-dynamics into existing algorithm for the simulation of nominal motion and nominal dynamics (Appendix 9). This incorporation can be performed according to the block-scheme in Fig. 2.43.

In the input of such a supplemented algorithm, in addition to the former input values, the initial deviation values $u(t_o)$ and $\dot{u}(t_o)$ are also included.

In the output, besides the nominal dynamic values, the time histories of the linear deviation u and the angular displacement φ, as well as the other micro-dynamical values, are obtained.

It still remains to derive the matrices D and c determining the mathematical model of micro-dynamics.

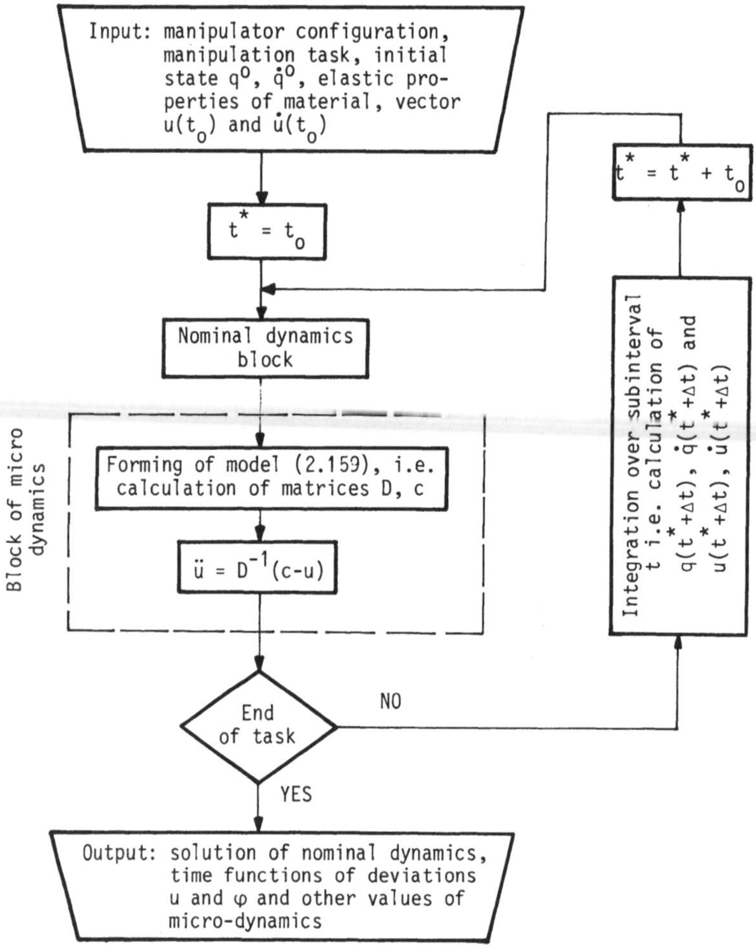

Fig. 2.43. Block-scheme of the simulation algorithm with the micro-dynamics block implemented

Deriving the model of elastic oscillations

Let us introduce two basic assumptions [6]:

(1) Small elastic deflections, i.e. small deviations from nominal motion,

(2) Superposition of the nominal and micro motion will be considered, assuming that the micro motion, i.e., elastic oscillations, do not influence the nominal dynamics.

These assumptions enable to write the mathematical model of micro-dynamics in the form (2.159), and then incorporate it as a separate block into the algorithm for the simulation of nominal dynamics, as shown in Fig. 2.43.

Kinematic and dynamic connections

The deflection $\vec{u}_i$ consists of three components (Fig. 2.44):

$$\vec{u}_i = \vec{u}_{i-1} + \vec{u}_i^{e\ell} + \vec{\varphi}_{i-1} \times \vec{\ell}_i \tag{2.160}$$

where $\vec{u}_i^{e\ell}$ represents the elastic deflection of link i due to its elastic deformation under the action of forces and moments. $\vec{\varphi}_{i-1} \times \vec{\ell}_i$ represents the component of deflection $\vec{u}_i$ due to the tilt of link i-1.

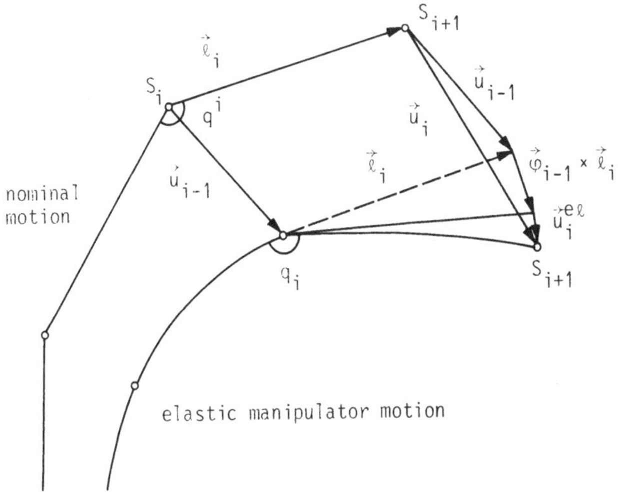

Fig. 2.44. Components of link deflection

The recurrent formula for deflection (2.160) in matrix notation is:

$$u_i = u_{i-1} + u_i^{e\ell} - \underline{\underline{\ell}}_i \varphi_{i-1} \tag{2.161}$$

where $\underline{\underline{\ell}}_i$ is the 3×3 matrix

$$\underline{\underline{\ell}}_i = \begin{bmatrix} 0 & -\ell_{i_z} & \ell_{i_y} \\ \ell_{i_z} & 0 & -\ell_{i_x} \\ -\ell_{i_y} & \ell_{i_x} & 0 \end{bmatrix} \tag{2.162}$$

corresponding to the vector $\vec{\ell}_i = \{\ell_{i_x}, \ell_{i_y}, \ell_{i_z}\}$.

For the angular deviation,

$$\vec{\varphi}_i = \vec{\varphi}_{i-1} + \vec{\varphi}_i^{e\ell} \tag{2.163}$$

where $\vec{\varphi}_i^{e\ell}$ is the angular elastic deviation of link i due to its deformation. In matrix notation:

$$\varphi_i = \varphi_{i-1} + \varphi_i^{e\ell} \tag{2.164}$$

Let us now find expressions for the elastic deflection $\vec{u}_i^{e\ell}$ and tilt $\vec{\varphi}_i^{e\ell}$. Let us consider the link i (Fig. 2.45). The mass of each cane will be considered as two concentrated masses in points S_i and S_{i+1}. Let us denote the mass at the point S_i by $\mu_i^d = \frac{m_i}{2}$ and in S_{i+1} by $\mu_i^{up} = \frac{m_i}{2}$. Such an approximatiom permits the simple inclusion of the motor masses in the joints by adding the motor mass in the joint S_i to the mass μ_i^d or μ_{i-1}^{up}. It is clear that inclusion of the motor masses reduces the error which appears due to mass division.

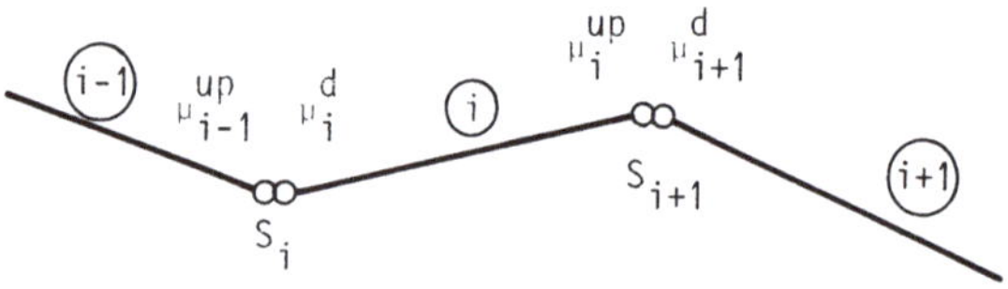

Fig. 2.45. Concentrated masses in joints

The link i will be considered to have its lower end S_i fixed and the upper end S_{i+1} will be considered free, replacing the action of the next link by reactions. Thus, the following forces and moments act on the free end (Fig. 2.46):

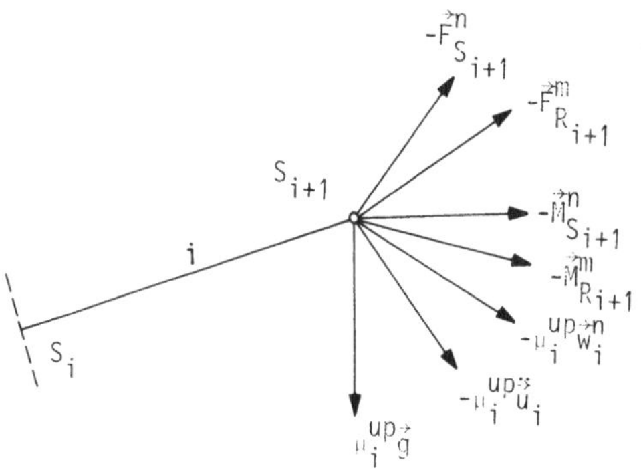

Fig. 2.46. Forces and moments at the "free" end of a link

- $\vec{F}^n_{S_{i+1}}$ - nominal force in the joint S_{i+1}

  This force can be determined by [6]:

  $$\vec{F}^n_{S_i} = \vec{F}^n_{R_i} + \xi_i \vec{P}^n_i$$

  where $\vec{P}_i$, $i=1,\ldots,n$ are driving forces and torques in the joints, and $\vec{F}_{R_i}$, $i=1,\ldots,n$ are reaction forces in the joints;

- $\vec{F}^m_{R_{i+1}}$ is the micro-reaction force in joint S_{i+1} due to micro-motion.

  Nominal reaction force $\vec{F}^n_{R_{i+1}}$ is perpendicular to the translation axis if S_{i+1} is a prismatic joint. Since the generalized coordinate does not depend on micro-motion, the micro-reaction $\vec{F}^m_{R_{i+1}}$ is not, in general, perpendicular to the translation axis;

- $\vec{M}^n_{S_{i+1}}$ is the nominal moment in S_{i+1}

  This moment can be determined by:

  $$\vec{M}^n_{S_i} = \vec{M}^n_{R_i} + \bar{\xi}_i \vec{P}^n_i$$

  where $\vec{M}_{R_i}$, $i=1,\ldots,n$ are reaction moments in the joints;

- $\vec{M}^m_{R_{i+1}}$ is the micro-moment of the reaction in joint S_{i+1}; as in the case of the micro-reaction force, $\vec{M}^m_{R_{i+1}}$, in general, is not perpendicular to the rotation axis (if the joint S_{i+1} is revolute), which is the case for the nominal reaction moment $\vec{M}^n_{R_{i+1}}$;

- $\mu_i^{up}\vec{g}$ is the gravity force ($\vec{g} = \{0, 0, -9.81\}$);

- $\mu_i^{up}\vec{w}_i^n$ is the nominal inertial force ($\vec{w}_i^n$ is the nominal acceleration of the point S_{i+1});

- $\mu_i^{up}\ddot{\vec{u}}_i$ is the micro-inertial force.

Let us assume that the reactions $\vec{F}^n_{S_{i+1}}$, $\vec{M}^n_{S_{i+1}}$, $\vec{F}^m_{R_{i+1}}$ and $\vec{M}^m_{R_{i+1}}$ in the joint S_{i+1} act on the next link i.e. the link i+1. So, in this joint, $-\vec{F}^n_{S_{i+1}}$, $-\vec{M}^n_{S_{i+1}}$, $-\vec{F}^m_{R_{i+1}}$ and $-\vec{M}^m_{R_{i+1}}$ act on the link i.

The elastic deflection $u_i^{e\ell}$ can be written in matrix form:

$$u_i^{e\ell} = \alpha_i(-F^n_{S_{i+1}} - F^m_{R_{i+1}} + \mu_i^{up}g - \mu_i^{up}w_i^n - \mu_i^{up}\ddot{u}_i) + \\ + \beta_i(-M^n_{S_{i+1}} - M^m_{R_{i+1}}) \tag{2.165}$$

and the elastic tilt $\varphi_i^{e\ell}$:

$$\varphi_i^{e\ell} = \gamma_i(-F^n_{S_{i+1}} - F^m_{R_{i+1}} + \mu_i^{up}g - \mu_i^{up}w_i^n - \mu_i^{up}\ddot{u}_i) + \\ + \delta_i(-M^n_{S_{i+1}} - M^m_{R_{i+1}}) \tag{2.166}$$

α_i, β_i, γ_i, δ_i are matrix influence coefficients (3×3).

In matrix notation μ_i^{up} and μ_i^d are diagonal 3×3 matrices, with the masses along the diagonal:

$$\mu_i^{up} = \begin{bmatrix} \mu_i^{up} & & \\ & \mu_i^{up} & \\ & & \mu_i^{up} \end{bmatrix}, \quad \mu_i^d = \begin{bmatrix} \mu_i^d & & \\ & \mu_i^d & \\ & & \mu_i^d \end{bmatrix}$$

Let us now consider the isolated link i, with the forces and moments acting on it (Fig. 2.47).

Let us apply to the link D'Alambert's principle of the equilibrium of forces:

$$F^n_{S_i} + F^m_{R_i} + G_i + F^n_{I_i} - \mu^d_i \ddot{u}_{i-1} - \mu^{up}_i \ddot{u}_i - F^n_{S_{i+1}} - F^m_{R_{i+1}} = 0 \quad (2.167)$$

where G_i is the total gravity force and $F^n_{I_i}$ is the total inertial force of nominal motion for the whole link. Dynamic equilibrium of the nominal motion yields:

$$F^n_{S_i} + G_i + F^n_{I_i} - F^n_{S_{i+1}} = 0 \quad (2.168)$$

so eq. (2.167) becomes:

$$F^m_{R_i} = \mu^d_i \ddot{u}_{i-1} + \mu^{up}_i \ddot{u}_i + F^m_{R_{i+1}} \quad (2.169)$$

a recurrent formula for the reaction micro-forces. The boundary condition for the recursion is $F^m_{R_{n+1}} = 0$.

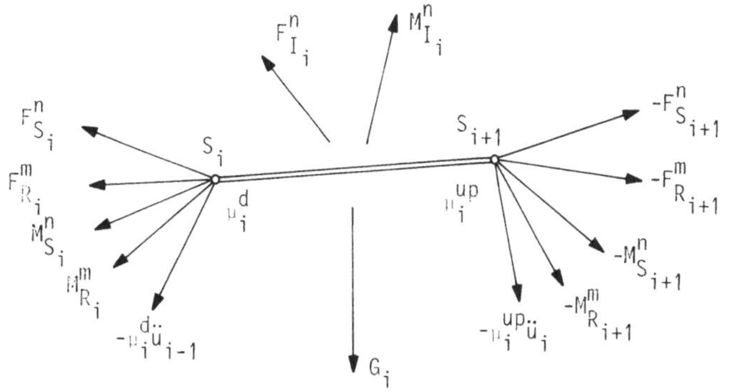

Fig. 2.47. Forces and moments acting on the i-th link

Further, let us apply to the i-th link D'Alambert's principle of equilibrium of moments for the point S_i [6]:

$$M^n_{S_i} + M^m_{R_i} + M_{G_i} + M^n_{IS_i} - \underline{\underline{\ell}}_i \mu^{up}_i \ddot{u}_i - \underline{\underline{\ell}}_i F^n_{S_{i+1}} - \underline{\underline{\ell}}_i F^m_{R_{i+1}} -$$

$$- M^n_{S_{i+1}} - M^m_{R_{i+1}} = 0 \quad (2.170)$$

where M_{G_i} is the moment of gravitational forces and $M^n_{IS_i}$ is the nominal moment of the inertial forces relative to the point S_i.

Dynamic equilibrium of nominal motion gives:

$$M^n_{S_i} + M_{G_i} + M^n_{IS_i} - \underline{\underline{\ell}}_i F^n_{S_{i+1}} - M^n_{S_{i+1}} = 0 \qquad (2.171)$$

so (2.170) becomes:

$$M^m_{R_i} = \underline{\underline{\ell}}_i F^m_{R_{i+1}} + \underline{\underline{\ell}}_i \mu_i^{up} \ddot{u}_i + M^m_{R_{i+1}} \qquad (2.172)$$

a recursive formula for micro-moments of reactions. The boundary condition is $M^m_{R_{n+1}} = 0$.

The equations (2.161), (2.164), (2.165), (2.166) and (2.169), (2.172), $i=1,\ldots,n$ determine the mathematical model of micro-dynamics and permit the calculation of $\ddot{u}_1,\ldots,\ddot{u}_n$.

In further text this set of equations will be transformed into the matrix form (2.157). The formalism of block-matrices will be used [6].

Let us introduce the $3n\times 1$ block-vectors in the following way. Let a_i, $i=1,\ldots,n$ be a set of 3×1 vectors. Then introduce the block vector $a = [a_1^T \ldots a_n^T]^T$. In that way we introduce the block-vectors F^n_S, F^m_R, w^n, M^n_S, M^m_R, φ, u, g. Furtheron, let us introduce the block-diagonal matrices $3n\times 3n$. Let $b_1,\ldots,b_n$ be a set of (3×3) matrices. Then the block-diagonal matrix is $b = \mathrm{diag}[b_1 \ldots b_n]$. Let us introduce the following block-diagonal matrices: α, β, γ, δ, μ^d, μ^{up}, $\underline{\underline{\ell}}$. Then, the $3n\times 3n$ block-matrices for index shifting will be used [6]:

$$\sigma^{(\ell)} = \begin{bmatrix} 0 & & 0 & 0 \\ I & \cdots & \vdots & 0 \\ & \ddots & \vdots & \\ 0 & & I & 0 \end{bmatrix}, \qquad \sigma^{(r)} = \begin{bmatrix} 0 & I & & \\ & \vdots & \ddots & 0 \\ 0 & \vdots \; 0 & & \ddots \\ 0 & \cdots & \cdots & I \\ 0 & 0 & & 0 \end{bmatrix} \qquad (2.173)$$

and summations block-matrices $3n\times 3n$:

$$\Sigma_{(dt)} = \begin{bmatrix} I & & \\ \vdots & \ddots & 0 \\ I & \cdots & I \end{bmatrix}, \qquad \Sigma_{(upt)} = \begin{bmatrix} I & \cdots & I \\ & \ddots & \vdots \\ 0 & & I \end{bmatrix} \qquad (2.174)$$

where I is the 3×3 unit matrix.

Starting from expressions (2.161), (2.164)-(2.166), (2.169), (2.172) and introducing the block matrices, the mathematical model of micro-dynamics is obtained in matrix form (2.157), where [6]:

$$D = \sum_{(dt)} (f + h\sigma^{(r)} \sum_{(upt)} \underline{\underline{\ell}} [\mu^{up} + \sigma^{(r)} \sum_{(upt)} (\mu^{d}{}_{\sigma}{}^{(\ell)} + \mu^{up})] \tag{2.175}$$

$$f = \alpha - \underline{\underline{\ell}}\sigma^{(\ell)} \sum_{(dt)} \gamma, \quad h = \beta - \underline{\underline{\ell}}\sigma^{(\ell)} \sum_{(dt)} \delta \tag{2.176}$$

$$c = \sum_{(dt)} [f(-\sigma^{(r)} F_S^n + \mu^{up}(g - w^n)) - h\sigma^{(r)} M_S^n] \tag{2.177}$$

Matrices D and c determine the mathematical model of micro-motion. For a known nominal dynamics these matrices are calculated from expression (2.175)-(2.177). After calculating $\ddot{u}$ from (2.157), other microdynamic values φ, F_R^m, M_R^m, u are calculated from the following expressions [6]:

$$F_R^m = \sum_{(upt)} (\mu^{d}{}_{\sigma}{}^{(\ell)} + \mu^{up}) \ddot{u} \tag{2.178}$$

$$M_R^m = \sum_{(upt)} (\underline{\underline{\ell}}\sigma^{(r)} F_R^m + \underline{\underline{\ell}}\mu^{up}\ddot{u}) = \sum_{(upt)} \underline{\underline{\ell}}(\sigma^{(r)} F_R^m + \mu^{up}\ddot{u}) \tag{2.179}$$

$$\varphi = \sum_{(dt)} [\gamma(-\sigma^{(r)} F_S^n - \sigma^{(r)} F_R^m + \mu^{up} g - \mu^{up} w^n - \mu^{up}\ddot{u}) + \delta(-\sigma^{(r)} M_S^n - \sigma^{(r)} M_R^n)] \tag{2.180}$$

$$u = \sum_{(dt)} [\alpha(-\sigma^{(r)} F_S^n - \sigma^{(r)} F_R^m + \mu^{up} g - \mu^{up} w^n - \mu^{up}\ddot{u}) + \beta(-\sigma^{(r)} M_S^n - \sigma^{(r)} M_R^m) - \underline{\underline{\ell}}\sigma^{(\ell)}\varphi] \tag{2.181}$$

Influence coefficients

Each link of the mechanism is considered to be fixed at its lower end S_i. The reduced mass of the cane μ_i^{up} is concentrated at its free end S_{i+1} where the forces and moments act (Fig. 2.46). Let us consider the cane AB with its fixed end at the point A and with reduced mass μ^B concentrated in point B (Fig. 2.48). System $O_s x_s y_s z_s$ is the coordinate

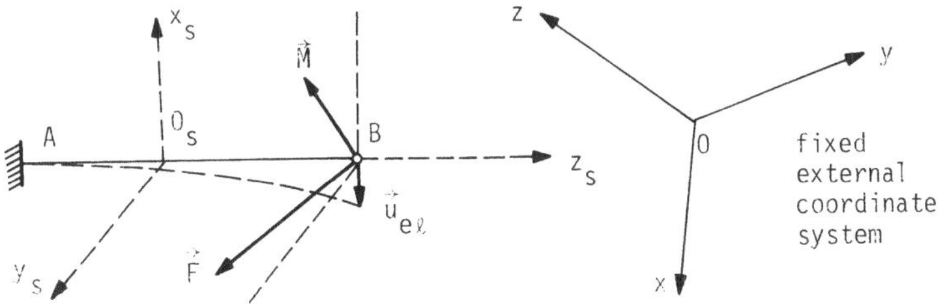

Fig. 2.48. Deformations of a fixed-end cane

system of the cane and Oxyz is the absolute, or external coordinate system. Suppose that at the point B there is an arbitrary force $\vec{F}$ and moment $\vec{M}$. Let the cane be deformed elastically so that at the point B there is a linear deflection $\vec{u}^{e\ell}$ and angular deviation $\vec{\varphi}^{e\ell}$. For $\vec{u}^{e\ell}$ in the cane coordinate system the equation of elastic displacements can be written as [6]:

$$u^{e\ell}_{x_s} = \alpha_{x_s}(F_{x_s} - \mu^B w^B_{x_s}) - \beta_{x_s} M_{y_s}, \qquad u^{e\ell}_{y_s} = \alpha_{y_s}(F_{y_s} - \mu^B w^B_{y_s}) - \beta_{y_s} M_{x_s},$$

$$u^{e\ell}_{z_s} = \alpha_{z_s}(F_{z_s} - \mu^B w^B_{z_s}) + 0 \cdot M_{z_s} \tag{2.182}$$

where $u^{e\ell}_{\varkappa_s}$ is the projection (component) of the linear deviation onto the x_s - axis, F_{x_s} and M_{x_s} - projections onto the same axis of the force and moment respectively, $w^B_{x_s}$ - the projection of the acceleration of point B; likewise for the other axes, y_s and z_s.

α_{x_s} is the influence coefficient for the bending deflection along the x_s - axis, under the action of the force at the point B. Similarly, α_{y_s} is defined in terms of the y_s - axis.

α_{z_s} is the influence coefficient for extension along the z_s - axis under the action of the force at B.

β_{x_s} is the influence coefficient for bending deflection along the x_s - axis due to the moment acting at B; likewise for the y_s - axis.

All the coefficients α and β are for linear deviation.

The equations (2.182) can be combined:

$$\vec{\tilde{u}}^{e\ell} = \tilde{\alpha}(\vec{\tilde{F}} - \mu^B \vec{\tilde{w}}^B) + \tilde{\beta}\vec{\tilde{M}} \tag{2.183}$$

where:

$$\tilde{\alpha} = \begin{bmatrix} \alpha_{x_s} & 0 & 0 \\ 0 & \alpha_{y_s} & 0 \\ 0 & 0 & \alpha_{z_s} \end{bmatrix}, \qquad \tilde{\beta} = \begin{bmatrix} 0 & \beta_{x_s} & 0 \\ -\beta_{y_s} & 0 & 0 \\ 0 & 0 & 0 \end{bmatrix} \tag{2.184}$$

and the tilde over vector means that the vector is expressed by three projections onto the axes of the cane coordinate system $O_s x_s y_s z_s$.

Let Q be the transformation matrix from the local to external coordinate system. Then:

$$\vec{F} = Q\tilde{\vec{F}}, \qquad \tilde{\vec{F}} = Q^{-1}\vec{F} \tag{2.185}$$

and likewise for other vectors. Now eq. (2.183) can be written in terms of the external system:

$$\vec{u}^{e\ell} = \alpha(\vec{F}-\mu^B\vec{w}^B) + \beta\vec{M} \tag{2.186}$$

where: $\alpha = Q\tilde{\alpha}Q^{-1}$, $\beta = Q\tilde{\beta}Q^{-1}$ (2.187)

By a procedure like that for linear displacement, the equation for angular displacement $\vec{\varphi}^{e\ell}$ is found to be [6]:

$$\begin{aligned}
\varphi^{e\ell}_{x_s} &= -\gamma_{x_s}(F_{y_s} - \mu^B w^B_{y_s}) + \delta_{x_s} M_{x_s} \\
\varphi^{e\ell}_{y_s} &= \gamma_{y_s}(F_{x_s} - \mu^B w^B_{x_s}) + \delta_{y_s} M_{y_s} \\
\varphi^{e\ell}_{z_s} &= 0\cdot(F_{z_s} - \mu^B w^B_{z_s}) + \delta_{z_s} M_{z_s}
\end{aligned} \tag{2.188}$$

where γ_{x_s} is the influence coefficient for the bending angle around the x_s - axis due to the force acting at B. Likewise for the y_s - axis.

δ_{x_s} is the influence coefficient for the bending angle around the x_s - axis due to the moment acting at B. Likewise for the y_s - axis.

δ_{z_s} is the influence coefficient for the torsion around the z_s - axis due to the moment acting at B.

Let us combine the relation (2.188):

$$\tilde{\vec{\varphi}}^{e\ell} = \tilde{\gamma}(\tilde{\vec{F}}-\mu^B\tilde{\vec{w}}^B) + \tilde{\delta}\vec{M} \tag{2.189}$$

where:

$$\tilde{\gamma} = \begin{bmatrix} 0 & -\gamma_{x_s} & 0 \\ \gamma_{y_s} & 0 & 0 \\ 0 & 0 & 0 \end{bmatrix}, \qquad \tilde{\delta} = \begin{bmatrix} \delta_{x_s} & 0 & 0 \\ 0 & \delta_{y_s} & 0 \\ 0 & 0 & \delta_{z_s} \end{bmatrix} \tag{2.190}$$

Introducing the transformation matrix, (2.189) becomes:

$$\vec{\varphi}^{e\ell} = \gamma(\vec{F} - \mu^B \vec{w}^B) + \delta\vec{M} \tag{2.191}$$

where:

$$\gamma = Q\tilde{\gamma}Q^{-1}, \qquad \delta = Q\tilde{\delta}Q^{-1} \tag{2.192}$$

The coefficients α_{x_s}, α_{y_s}, α_{z_s}, β_{x_s}, β_{y_s}, γ_{x_s}, γ_{y_s}, δ_{x_s}, δ_{y_s}, δ_{z_s} hold for the cane coordinate system and can be found from tables, for a given form of cross section. Let us mention that eqs. (2.186) and (2.191) also hold when the end point A of the cane is moving, in which case, the cane coordinate system is moving, too. In this case, $\vec{w}^B$ is the absolute acceleration of the point B with respect to the external (fixed) system, $\vec{u}_i^{e\ell}$ and $\vec{\varphi}_i^{e\ell}$ represent the cane elastic deflection and tilt, and the transformation matrix Q and the matrix coefficients α, β, γ, δ are functions of time.

Let us now apply these considerations to the active articulated mechanism. The link i can be regarded as having a fixed - end at a moving point S_i (Fig. 2.46, 2.49). Taking into account the assumption that displacements from nominal motion are small, the matrix influence coefficients will be calculated for nominal motion. The body - fixed system

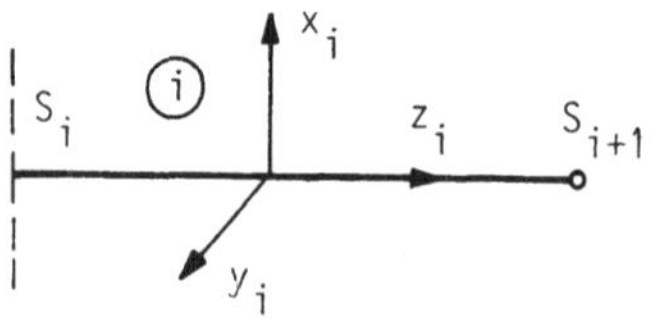

Fig. 2.49. Link as a fixed - end cane

of i-th link, i.e. $O_i x_i y_i z_i$ represents the cane coordinate system. Thus, the influence coefficients $\alpha_{x_s}, \ldots, \delta_{z_s}$ are taken from tables for the adopted cross - section and they represent input data. The algorithm gives the matrices $\tilde{\alpha}_i$, $\tilde{\beta}_i$, $\tilde{\gamma}_i$, $\tilde{\delta}_i$ according to (2.184), (2.190) and then the matrix coefficients α_i, β_i, γ_i, δ_i according to:

$$\alpha_i = Q_i \tilde{\alpha}_i Q_i^{-1}, \qquad \beta_i = Q_i \tilde{\beta}_i Q_i^{-1}$$
$$\gamma_i = Q_i \tilde{\gamma}_i Q_i^{-1}, \qquad \delta_i = Q_i \tilde{\delta}_i Q_i^{-1} \tag{2.193}$$

The most common links of manipulation robots are cylindrical or rectangular tubes (Figs. 2.50, 2.51).

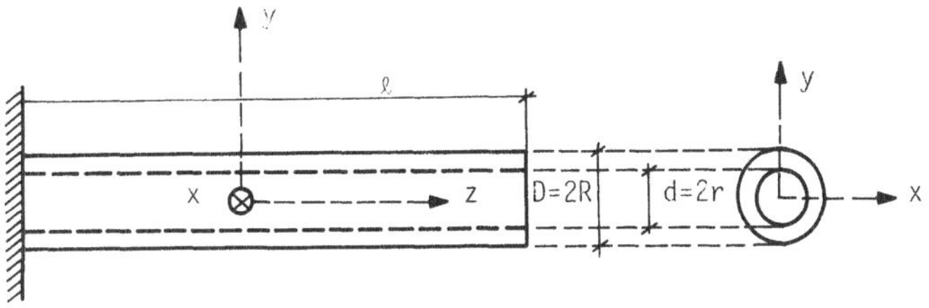

Fig. 2.50. Cylindrical tube link

Let us first consider a cane in the form of a cylindrical tube with one fixed end (Fig. 2.50). The moments of inertia for such a cross-section are:

$$I_x = I_y = \frac{R^4\pi}{4}(1-\psi^4) = \frac{D^4\pi}{64}(1-\psi^4) \tag{2.194}$$

where $\psi = r/R$.

If at the free end of the cane a force is acting, the influence coefficients α_x, α_y of the deflextion due to bending and the influence coefficient α_z due to extension are:

$$\alpha_x = \frac{\ell^3}{3EI_y}, \qquad \alpha_y = \frac{\ell^3}{3EI_x}, \qquad \alpha_z = \frac{\ell}{EA} \tag{2.195}$$

where E is Young's modulus for the adopted material and A is the area of cross-section:

$$A = R^2\pi(1-\psi^2) = \frac{D^2\pi}{4}(1-\psi^2)$$

The influence coefficients γ_x, γ_y of the tilt due to bending are:

$$\gamma_x = \frac{\ell^2}{2EI_x}, \quad \gamma_y = \frac{\ell^2}{2EI_y}, \quad \gamma_z = 0 \tag{2.196}$$

If at the end of the cane a moment is acting, the influence coefficients of the deflection due to bending are:

$$\beta_x = \frac{\ell^2}{2EI_y}, \quad \beta_y = \frac{\ell^2}{2EI_x}, \quad \beta_z = 0 \tag{2.197}$$

and the influence coefficients δ_x, δ_y of the tilt due to bending and δ_z due to torsion are, respectively:

$$\delta_x = \frac{\ell}{EI_x}, \quad \delta_y = \frac{\ell}{EI_y}, \quad \delta_z = \frac{32\ell}{(1-\psi^4)GD^4\pi} \tag{2.198}$$

where G is the shear modulus:

$$G = \frac{E}{2(1+\nu)} \tag{2.199}$$

and ν is Poisson's coefficients.

Let us now consider a cane in the form of a rectangular tube (Fig. 2.51):

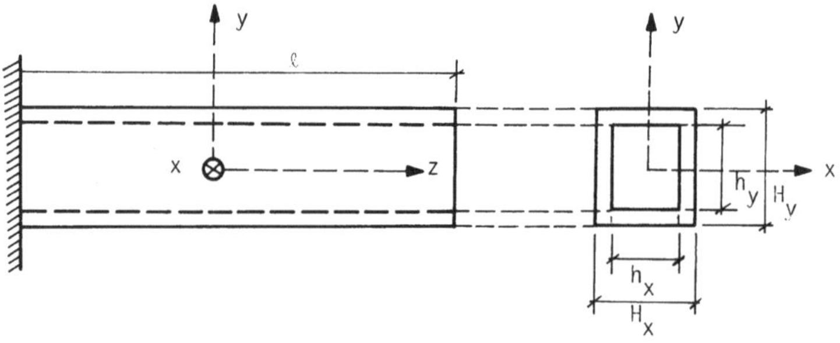

Fig. 2.51. Link in the form of a rectangular tube

The moments of inertia of this cross-section are:

$$I_x = \frac{H_x H_y^3 - h_x h_y^3}{12}, \quad I_y = \frac{H_x^3 H_y - h_x^3 h_y}{12} \tag{2.200}$$

For the moments of inertia so calculated, the expressions for the influence coefficients (2.195 - 2.198) are valid. The difference is that for the influence coefficients α_z due to extension in expression

(2.195) the area A is calculated as $A = H_x H_y - h_x h_y$. There is also a difference in the case of torsion (2.198), where

$$\delta_z = \frac{\ell}{GJ}, \quad J = \frac{1}{12}(H_x^3 H_y + H_x H_y^3 - h_x^3 h_y - h_x h_y^3). \tag{2.201}$$

Application of the method

The micro-dynamic analysis gives time functions of the deflection (u), of the tilt (φ) and the "micro-reactions" (F_R^m, M_R^m), i.e. $\vec{u}_i$, $\vec{\varphi}_i$, $\vec{F}_{R_i}^m$, $\vec{M}_{R_i}^i$, $i=1,\ldots,n$ for a series of time-instants.

By projecting the micro-reactions in joint S_i

$$\xi_i \vec{F}_{R_i}^m + \bar{\xi}_i \vec{M}_{R_i}^m \tag{2.202}$$

onto the axis $\vec{e}_i$ (the revolute or prismatic joint axis), the additional load torque due to elastic displacements is:

$$\Delta P_i = \vec{e}_i(\xi_i \vec{F}_{R_i}^m + \bar{\xi}_i \vec{M}_{R_i}^m) \tag{2.203}$$

If one considers a manipulator with six links and six d.o.f., it is most interesting to know the deflection ($\vec{u}_6$) and the tilt ($\vec{\varphi}_6$) of the tip. One can thus determine the magnitude of the positioning and orientation error (relative to nominal motion) due to links elasticity. This is particularly useful in the "dynamic method" for evaluation and systematic choice of robotic mechanisms.

Finally, it is necessary to underline once again that the task of the described method is the approximate calculation of robot arm rigidity at given manipulator tip deflection. This conservative but simple method gives the results which, for the conventional three-link manipulator mechanism do not exceed by 20% the deflections calculated based on exact but strongly complex and time consuming procedures.

2.7 Dynamics of Cooperative Manipulation Robots

Introduction

In some industrial applications of robotic systems it is sometimes more convenient or even necessary to use more than one robot to accomplish a given task. An example of this situation is when a single manipulator cannot handle the object either because it is beyond manipulator's load capacity, or when the geometrical properties of the object make it difficult to manipulate (long bar). The cooperative work of manipulators offers some possibilities to solve more sophisticated problems than the insufficient load capacity of a single manipulator. The idea is to use manipulators with complementary features and to make the best use of good properties of both manipulators. For example, we can combine a heavy-load manipulator of lower precision with another, more sensored and more precise, but lower capacity manipulator. In the task of assembling two parts, the first manipulator would carry the heavier part and bring it in the most appropriate position for the other manipulator to carry out the assembling. When tracking of trajectories is considered, the first manipulator would realize some simple tracking algorithm, while taking over the most part of the load, and the other manipulator would be less loaded, but would be engaged in correcting the motion so that good and precise tracking is achieved. In this chapter the cooperation of two manipulators, handling an object in such a way that there is no relative motion among the grippers of manipulators and the object, will be considered. The goal is to carry out the kinematic and dynamic analysis of the cooperative work and to derive the dynamic model, which can serve, for example, for further investigation of control laws.

Before proceeding, some constraints will be posed on the physical properties of manipulators and the object, for the sake of simplicity of the dynamic model of cooperative work. It is supposed that manipulators are represented as open kinematic chains of rigid links, connected via ideal kinematic pairs of the fifth class (Fig. 1.16), and that each manipulator has six degrees of freedom. It is supposed, too, that the object is a rigid body and that the manipulators and the object are not subject to external constraints of motion. Only the trajecto-

---

*) This section is written by M. Djurović.

ries, wherein singular positions of both manipulators are absent, are considered. The manipulators will be denoted as M_1 and M_2.

Parameters of cooperative work

Besides the parameters of manipulators involved in cooperation, it is necessary to define the parameters that characterize the object, its relative position to the grippers of manipulators, and the relative position of the manipulators themselves. For this purpose, three local reference systems are introduced, T_1, which is bound to the last segment of manipulator M_1, T_2, which is bound to the last segment of manipulator M_2 and T_o, which is bound to the object, with the origin in the center of mass of the object, and with axis along the main axis of inertia. The parameters are (Figs. 2.52 and 2.53):

$\vec{r}_i^L$ - relative position of the system T_o with respect to the system $T_i (i=1, 2)$ given in the system T_i.

R_i^o - transformation matrix that maps vectors from system T_o to vectors in system $T_i (i=1, 2)$.

$\vec{\rho}$ - relative position of the base of M_2 with respect to the base coordinate system of M_1.

R_1^2 - transformation matrix that maps vectors from the base coordinate system of M_2 to vectors in the base coordinate system of M_1.

m_o - mass of the object.

J_{xx}, J_{yy}, J_{zz} - moments of inertia of the object in the system T_o.

It should be emphasized that this set of parameters is not the minimal one, since each transformation matrix is a function of the three corresponding independent parameters (Euler angles, for example), but is adopted here for convenience.

Kinematic relations

Since the last segments of manipulators are firmly connected via the manipulated object, the motions of manipulators are not independent. This implies that the number of degrees of freedom of the overall system is decreased. In this paragraph the set of holonomic constraints posed on the motion of the manipulators will be determined.

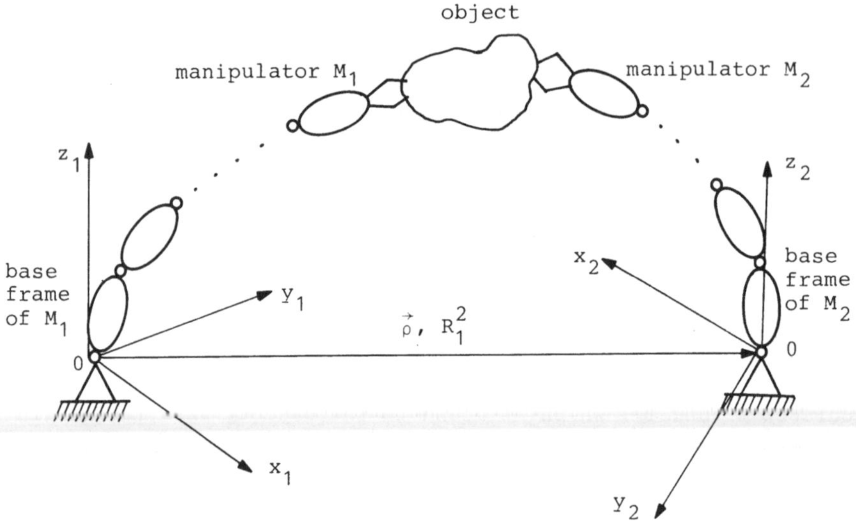

Fig. 2.52. Relative position of manipulators

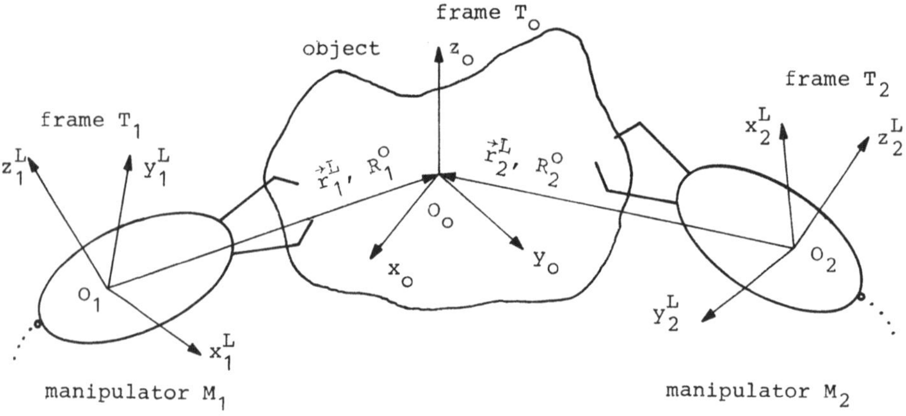

Fig. 2.53. Relative position of grippers and the object

In order to obtain the most convenient form of kinematic relations among internal coordinates of manipulators, we shall consider these relations in terms of internal velocities, because they will get a linear form. The treatise of constraints will be carried out in local coordinate systems, T_1, T_2 and T_o, because they are mutually motionless, and we can expect that in this way the constraints will get the simplest form.

Coordinates of an arbitrary vector $\vec{a}_1^L$, in system T_1, can be expressed via coordinates of the same vector in system T_2, $\vec{a}_2^L$:

$$\vec{a}_1^L = R_1^o R_2^{oT} \vec{a}_2^L$$

Superscript $()^L$ is used to emphasize that the value is related to the local coordinate system.

If by $\vec{v}_i^L$ and $\vec{\omega}_i^L$ we denote the linear and angular velocity, respectively, of the local coordinate frame T_i, expressed in T_i, we get the following relation:

$$\vec{\omega}_1^L = R_1^o R_2^{oT} \vec{\omega}_2^L,$$

since mutually motionless coordinate systems, T_1 and T_2, have the same angular velocity, and:

$$\vec{v}_1^L = R_1^o R_2^{oT} [\vec{v}_2^L + \vec{\ell}^L \times \vec{\omega}_2^L]$$

where

$$\vec{\ell}^L = R_2^o R_1^{oT} \vec{r}_1^L - \vec{r}_2^L$$

is the relative position of the origin of T_2 with respect to T_1. If we substitute the cross product $\vec{\ell}^L \times \vec{\omega}_2^L$ with corresponding matrix product, $\ell_M \vec{\omega}_2^L$, where ℓ_M is:

$$\ell_M = \begin{bmatrix} 0 & -\ell_z^L & \ell_y^L \\ \ell_z^L & 0 & -\ell_x^L \\ -\ell_y^L & \ell_x^L & 0 \end{bmatrix}$$

we obtain (O_3 is null matrix of order three):

$$\begin{bmatrix} \vec{v}_1^L \\ \vec{\omega}_1^L \end{bmatrix} = \begin{bmatrix} R_1^o R_2^{oT} & R_1^o R_2^{oT} \ell_M \\ O_3 & R_1 R_2^{oT} \end{bmatrix} \begin{bmatrix} \vec{v}_2^L \\ \vec{\omega}_2^L \end{bmatrix} \tag{2.204}$$

Let us consider the Jacobians J_i^L, (i=1, 2), defined by relation:

$$\begin{bmatrix} \vec{v}_i^L \\ \vec{\omega}_i^L \end{bmatrix} = J_i^L \dot{q}_i, \quad i = 1, 2 \tag{2.205}$$

where q_i is the vector of internal coordinates of manipulator M_i. These Jacobians act as operators that map vectors from the space of internal velocities to velocities in local coordinate systems, and should not be confused with Jacobians that map to velocities in the external absolute coordinate system (usually bound to the base of the corresponding manipulator). From (2.204) and (2.205) the relation between internal velocities of manipulators is obtained [31]:

$$J_1^L \dot{q}_1 = Q_1^2 J_2^L \dot{q}_2 \tag{2.206}$$

where Q_1^2 denotes the matrix:

$$Q_1^2 = \begin{bmatrix} R_1^O R_2^{OT} & R_1 R_2^{OT} \ell_M \\ O_3 & R_1 R_2^{OT} \end{bmatrix}$$

If the time derivation operator is applied to equation (2.205), the following relation between internal accelerations of the manipulator M_i and accelerations of M_i's tip with respect to local coordinate frame T_i is obtained:

$$\begin{bmatrix} \vec{w}_i^L \\ \vec{\varepsilon}_i^L \end{bmatrix} = J_i^L \ddot{q}_i + L_i(q_i, \dot{q}_i), \qquad i = 1, 2 \tag{2.207}$$

where $L_i(q_i, \dot{q}_i)$ denotes $\dot{J}_i \dot{q}_i$.

The relation between internal accelerations is obtained by applying time derivation operator to (2.206).

$$J_1^L \ddot{q}_1 + L_1(q_1, \dot{q}_1) = Q_1^2 (J_2^L \ddot{q}_2 + L_2(q_2, \dot{q}_2)) \tag{2.208}$$

Let us define Γ_1 and Γ_2 by relations [31]:

$$\Gamma_1(q_1, q_2) = (J_2(q_2))^{-1} (Q_1^2)^{-1} J_1(q_1)$$

$$\Gamma_2(q_1, \dot{q}_1, q_2, \dot{q}_2) = J_2^{-1}((Q_1^2)^{-1} L_1 - L_2)$$

Now relations between internal velocities and accelerations of manipulators can be expressed in the following form:

$$\dot{q}_2 = \Gamma_1 \dot{q}_1 \tag{2.209}$$

$$\ddot{q}_2 = \Gamma_1 \ddot{q}_1 + \Gamma_2 \tag{2.210}$$

Observe that Γ_1 acts like an operator that maps internal velocities of M_1 to internal velocities of M_2. Since there are exactly six independent holonomic constraints among internal velocities of manipulators, expressed in differential form via equation (2.209), it follows that there is a loss of six degrees of freedom, and that the overall system has: (DOF of M_1)+(DOF of M_2)-(N^o of constraints)=six DOF. We could some to the same conclusion by observing the fact that when the motion of one manipulator is given, the motion of the tip of other manipulator is defined and inverse kinematics uniquely determines the motion of other manipulator in its internal coordinates. The set of six independent variables has to be chosen, which will serve as generalized coordinates, to describe the dynamic behaviour of the system. We adopt that set to be the set of internal coordinates of M_1, and the rest of equations in this chapter will be reduced and mapped to the joint space of M_1. In other words, when internal coordinates (or velocities and accelerations) of M_1 are known, internal coordinates of M_2 can be determined via constraint relations. It should be noted that this choice in no way emphasizes M_1 as a "leader" or "master" manipulator, and has a motive in a convenience of representation of system motion.

Object motion

The motion of the object under the influence of forces and torques applied by manipulators will now be treated. We adopt the following notation (see Fig. 2.54):

$\vec{F}_i^L$, $\vec{M}_i^L$ - Force and torque applied upon the object by manipulator M_i, given in system T_i, i = 1, 2, and reduced to the origin of T_i;

$\vec{w}_o^L$, $\vec{\varepsilon}_o^L$, $\vec{\omega}_o^L$ - linear and angular acceleration, and angular velocity, respectively, of the object in system T_o;

J_o - Tensor of inertia of the object in system T_o;

$R_B^1(q_1)$ - Transformation matrix between system T_1 and the absolute reference system bound to the base of M_1;

$\vec{g}$ - Vector of the gravitational acceleration given in the absolute system.

The equation of the force balance acting upon the object, in the system T_1 is [31]:

$$\vec{F}_1^L + R_1^o R_2^{oT} \vec{F}_2^L = m_o R_1^o \vec{w}_o^L - m_o R_B^{1T}(q_1)\vec{g}, \text{ where}$$

$$R_1^o \vec{w}_o^L = \vec{w}_1^L + \vec{\varepsilon}_1^L \times \vec{r}_1^L + \vec{\omega}_1^L \times (\vec{\omega}_1^L \times \vec{r}_1^L), \text{ and finally,}$$

$$\vec{F}_1^L + R_1^o R_2^{oT} \vec{F}_2^L = m_o[\vec{w}_1^L + \vec{\varepsilon}_1^L \times \vec{r}_1^L + \vec{\omega}_1^L \times (\vec{\omega}_1^L \times \vec{r}_1^L)] - m_o R_B^{1T}(q_1)\vec{g} \qquad (2.211)$$

The equation of the torque balance, acting upon the object, with respect to the origin of the system T_o, in the system T_1, is:

$$\vec{M}_1^L - \vec{r}_1^L \times \vec{F}_1^L + R_1^o R_2^{oT} (\vec{M}_2^L - \vec{r}_2^L \times \vec{F}_2^L) = R_1^o (J_o \vec{\varepsilon}_o^L - (J_o \vec{\omega}_o^L) \times \vec{\omega}_o^L) \tag{2.212}$$

If $\vec{\varepsilon}_o^L$ and $\vec{\omega}_o^L$ are mapped in T_1, it follows:

$$\vec{M}_1^L - \vec{r}_1^L \times \vec{F}_1^L + R_1^o R_2^{oT} (\vec{M}_2^L - \vec{r}_2^L \times \vec{F}_2^L) = R_1^o J_o R_1^{oT} \vec{\varepsilon}_1^L - $$

$$- (R_1^o J_o R_1^{oT} \vec{\omega}_1^L) \times (R_1^{oT} \vec{\omega}_1^L) \tag{2.213}$$

Since matrix R_1^o is unitary, and matrix J_o diagonal, it can be easily shown that $R_1^o J_o R_1^{oT} = J_o$. If we use the following notation:

$$\begin{aligned} K_1(\dot{q}_1) &= \vec{\omega}_1^L \times (\vec{\omega}_1^L \times \vec{r}_1^L) \\ K_2(\dot{q}_1) &= (J_o \vec{\omega}_1^L) \times (R_1^{oT} \vec{\omega}_1^L) \end{aligned} \tag{2.214}$$

and, instead of vector cross products, introduce corresponding matrix products:

$$\vec{r}_i^L \times \vec{F}_i^L \Longleftrightarrow V_i \vec{F}_i^L \Rightarrow V_i = \begin{bmatrix} 0 & -r_{iz}^L & r_{iy}^L \\ r_{iz}^L & 0 & -r_{ix}^L \\ -r_{iy}^L & r_{ix}^L & 0 \end{bmatrix}$$

the equations of force (2.211) and torque (2.213) balance can be written in more compact form [31]:

$$\begin{aligned} \vec{F}_1^L + R_1^o R_2^{oT} \vec{F}_2^L &= m_o [\vec{w}_1^L - V_1 \vec{\varepsilon}_1^L + K_1] - m_o R_B^{1T}(q_1) \vec{g} \\ \vec{M}_1^L - V_1 \vec{F}_1^L + R_1^o R_2^{oT} (\vec{M}_2^L - V_2 \vec{F}_2^L) &= J_o \vec{\varepsilon}_1^L - K_2 \end{aligned} \tag{2.215}$$

Let us introduce the following notation:

$$A = \begin{bmatrix} I_3 & O_3 \\ -V_1 & I_3 \end{bmatrix}, \qquad B = \begin{bmatrix} R_1^o R_2^{oT} & O_3 \\ -R_1^o R_2^{oT} V_2 & R_1^o R_2^{oT} \end{bmatrix}$$

$$C^* = \begin{bmatrix} m_o I_3 & -m_o V_1 \\ O_3 & J_o \end{bmatrix}, \qquad C = C^* J_1^L(q_1)$$

$$D = C^{*}L_1(q_1, \dot{q}_1) - \begin{bmatrix} m_o R_B^{1T}(q_1)\vec{g} - m_o K_1 \\ K_2 \end{bmatrix},$$

$$F_i = \begin{bmatrix} \vec{F}_i^L \\ \vec{M}_i^L \end{bmatrix}, \qquad i = 1, 2$$

where I_3 denotes unit matrix of order three, and O_3 denotes null matrix of order three. Considering equation (2.207), the differential equations of the object motion (2.215) can now be unified in the matrix form [31]:

$$AF_1 + BF_2 = C\ddot{q}_1 + D \tag{2.216}$$

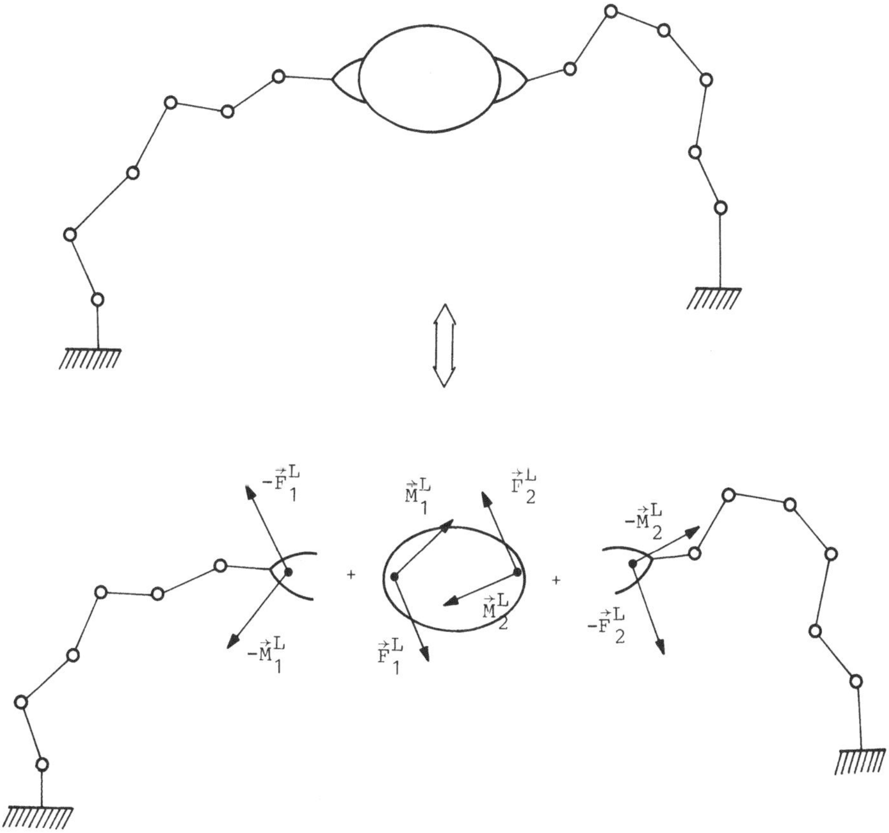

Fig. 2.54. Dynamics decomposition

Extension of the dynamic model of manipulator

In this paragraph we shall extend the dynamic model of manipulator (2.30, 2.31) to include the effects of external forces and torques exerted at the manipulator's tip. With this additional effect, the dynamic model of manipulator has the form:

$$H(q)\ddot{q} + h(q, \dot{q}) = P + P_e$$

where term P_e stems from the influence of external force/torque vector, as seen from the joint space of the manipulator. The problem is to determine the relation between this term and external force/torque vector. We shall do that using the principle of virtual work.

Let δr represent the infinitesimal displacement of the manipulator's tip in the effector coordinate frame, and F represent the force/torque vector, defined in the same frame, acting upon the manipulator's tip. If we neglect inertial effects and suppose that joints are frictionless, we get the following equation for virtual work δW:

$$\delta W = P_e^T \delta q - F^T \delta r = (P_e - J^{LT} F)^T \delta q$$

Since δq represent linearly independent admissible displacements, the term $(P_e - J^{LT}F)$ should be zero, for δW to vanish. Finally, we obtain:

$$P_e = J^{LT} F$$

and the dynamic model of the manipulator is:

$$H(q)\ddot{q} + h(q, \dot{q}) = P + J^{LT} F \qquad (2.217)$$

Dynamic model of cooperation

In order to obtain the dynamic model of cooperation, it is necessary to unify dynamic models of the manipulators and the model of interaction, represented by equation (2.216). It can be done by eliminating the forces of interaction F_i (i=1, 2) from the mentioned models.

The dynamic models of the manipulators M_1 and M_2 are (2.217) [31]:

$$H_1(q_1)\ddot{q}_1 + h_1(q_1, \dot{q}_1) = P_1 - J_1^{LT}(q_1)F_1 \tag{2.218}$$

$$H_2(q_2)\ddot{q}_2 + h_2(q_2, \dot{q}_2) = P_2 - J_2^{LT}(q_2)F_2 \tag{2.219}$$

Negative sign of the term that corresponds to external force/torque vector stems from the notation adopted before (see Fig. 2.54). If we calculate F_1 from (2.216) and replace it in (2.218) we get the following equation:

$$\begin{aligned} &H_1\ddot{q}_1 + h_1 = P_1 - J_1^{LT}A^{-1}[C\ddot{q}_1+D-BF_2], \text{ that is,} \\ &[H_1+J_1^{LT}A^{-1}C]\ddot{q}_1 + h_1 + J_1^{LT}A^{-1}D = P_1 + J_1^{LT}A^{-1}BF_2 \end{aligned} \tag{2.220}$$

If we calculate F_2 from equation (2.219) and replace $\ddot{q}_2$ using the equation (2.210), it follows:

$$F_2 = -(J_2^{LT})^{-1}[H_2[\Gamma_1\ddot{q}_1+\Gamma_2]+h_2-P_2] \tag{2.221}$$

By replacing F_2 from (2.221) in (2.220) and grouping the terms, the following equation is obtained:

$$\begin{aligned} &[H_1 + J_1^{LT}A^{-1}B(J_2^{LT})^{-1}H_2\Gamma_1 + J_1^{LT}A^{-1}C]\ddot{q}_1 + \\ &h_1 + J_1^{LT}A^{-1}B(J_2^{LT})^{-1}[H_2\Gamma_2+h_2] + J_1^{LT}A^{-1}D = \\ &P_1 + J_1^{LT}A^{-1}B(J_2^{LT})^{-1}P_2 \end{aligned} \tag{2.222}$$

For the sake of clarity, let us introduce the following notation:

$$\Phi(q_1, q_2) = J_1^{LT}A^{-1}B(J_2^{LT})^{-1}$$

$$X(q_1, q_2) = H_1 + \Phi H_2\Gamma_1 + J_1^{LT}A^{-1}C$$

$$\chi(q_1, q_2, \dot{q}_1, \dot{q}_2) = h_1 + \Phi[H_2\Gamma_2+h_2] + J_1^{LT}A^{-1}D$$

With this notation, equation (2.222) becomes:

$$X(q_1, q_2)\ddot{q}_1 + \chi(q_1, q_2, \dot{q}_1, \dot{q}_2) = P_1 + \Phi(q_1, q_2)P_2 \tag{2.223}$$

and that is the dynamic model of the cooperative work of manipulators.

Corollary

The procedure carried out in order to obtain the dynamic model of cooperative work of manipulators can be summarized in the following steps:

Step one: Determine the constraints posed on the internal coordinates of manipulators.

Step two: Chose the set of generalized coordinates as independent variables for representing the system motion.

Step three: Breakdown the problem of the system motion in the problem of motion of subsystems (manipulators and the object) under the influence of the rest of the system, represented by corresponding forces and torques of interaction.

Step four: Determine the dynamic model of the object motion under the influence of manipulators. The dynamic models of manipulators under the influence of external forces and torques are known and adopted in form (2.218) and (2.219).

Step five: Map all equations in the space of generalized coordinates and eliminate introduced forces and torques of interaction.

If we observe (2.223), a strong similarity between this model and the dynamic model of a single manipulator can be noticed. The term $X(q_1, q_2)\ddot{q}_1$ corresponds to the inertial forces, the term $\chi(q_1, q_2, \dot{q}_1, \dot{q}_2)$ corresponds to the sum of Coriolis, centrifugal and gravitational forces, P_1 is the vector of driving torques produced by actuators of M_1, and term $\Phi(q_1, q_2)P_2$ is the vector of driving torques produced by M_2 and mapped to the joint space of M_1. The matrix $\Phi(q_1, q_2)$ acts like an operator that maps forces from the joint space of M_2 to the forces in the joint space of M_1. The matrix $X(q_1, q_2)$ is an equivalent inertial matrix, and is obtained as a sum of corresponding inertial matrices of M_1, M_2 and the object, multiplied by necessary mapping operators. The similar conclusion can be made by observing the definition of $\chi(q_1, q_2, \dot{q}_1, \dot{q}_2)$. When we compare the system of one manipulator and the system of two cooperating manipulators, we note the common property considering number of DOF, but the difference arises when number of inputs is considered. The direct dynamic problem, it means determination of the system motion when input torques are known can be resolved for both manipulation systems. The inverse dynamic problem - determination of inputs required to achieve prescribed system motion cannot be resolved for the system of two cooperating manipulators due to the redundancy of driving system. Namely, the system has more actuators than the number of DOF. To determine the input torques, one must pose additional requirements on the system. This offers many possibilities, because one can freely choose what additional criterion to apply.

This chapter contains nine appendices, first five of which present concise explanations relevant to the basics of this textbook while the other four represent necessary support for detailed explanations relating to the presented computer methods.

Appendix 6 contains dynamic models of electric, hydraulic and pneumatic servo-actuators of various types and complexity. With the aim of clarifying and accepting the principle involved in the presented computer method of forming mathematical models of open configuration robotic systems, Appendix 7 is given.

Appendix 8 gives a mathematical model, in iterative form, of the widely used "ASEA" mechanism, which allows the modelling of local closed chains on the basis of already adopted procedure.

Appendix 9 contains the complete program for digital simulation of open chain configuration dynamics, based on the presented procedure using Newton-Euler's dynamic equations. This appendix contains the program for digital linearization, too.

Appendix 9 contains the program VIBRO which represents a programming support for solving the problem of accuracy, when dealing with robotic mechanisms subjected to mechanic vibrations of the fundament.

References

[1] Artobolevskii I.I., Theory of Mechanisms and Machines (in Russian), "NAUKA", Moscow, 1975.

[2] Stepanenko Yu., Dynamics of Spatial Mechanisms (in Russian), Mathematical Institute, Belgrade, 1974.

[3] Vukobratović M., Stepanenko Yu., "Mathematical Models of General Anthropomorphic Systems", Mathematical Biosciences, Vol. 17, 1973.

[4] Stepanenko Yu., Vukobratović M., "Dynamics of Articulated Open--Chain Active Mechanisms", Mathematical Biosciences, Vol. 28, No 1/2, 1976.

[5] Vukobratović M., Legged Locomotion Robots and Anthropomorphic Mechanisms, research monograph, "Mihailo Pupin" Institute, Belgrade, 1975, also published in Russian, "MIR", Moscow 1976, in Jappanese, Tokyo 1975, in Chinese, Peking, 1983.

[6] Vukobratović M., Potkonjak V., Scientific Fundamentals of Robotics 1: Dynamics of Manipulation Robots, Springer-Verlag, 1982, also published in Jappanese, Tokyo 1986.

[7] Vukobratović M., Potkonjak V., "Contribution to Automatic Forming of Active Chain Models via Lagrangian Form", Transactions of the ASME Journal of Applied Mechanics, No 1, 1979.

[8] Witenburg J., Dynamics of Systems of Rigid Bodies, B.G. Teubner, Stutgart, 1977.

[9] Popov E.P., Vereschagin A.F., Zenkevitch S.A., Manipulation Robots: Dynamics and Algorithms (in Russian), "NAUKA", Moscow, 1978.

[10] Vukobratović M., Kirćanski N., Scientific Fundamentlas of Robotics 4: Real-Time Dynamics of Manipulation Robots, Springer--Verlag, 1985

[11] Vukobratović M., Potkonjak V., Scientific Fundamentals of Robotics 6: Applied Dynamics and CAD of Manipulation Robots, Springer-Veralg, 1985.

[12] Vukobratović M., Vujić D., "Contribution to Solving Dynamic Robot Control in a Machining Process", Mechanism and Machine Theory, Vol. 22, No 5, 1987.

[13] Vukobratović M., Vujić D., "Nominal Tracking Simulation in Conditions of Mechanical Vibrations Impact on the Manipulation Robots", Mechanism and Machine Theory, Vol. 22, No 5, 1987.

[14] Vukobratović M., Katić D., Potkonjak V., "Computer-Assisted Choice of Electrohydraulic Servosystems for Manipulation Robots Using Complete Mathematical Models", Mechanism and Machine Theory, Vol. 22, No 5, 1987.

[15] Akselrod B.V., Vujić D., Vukobratović M., Gradeckii V.G., Tchernousko F.L., "Dynamics Modelling of Manipulators under

Fundament Vibrations", (in Russian), Journal of Rigid Body Mechanics, ANUSSR, No 2, Moscow, 1987.

[16] Vukobratović M., Kirćanski M., Scientific Fundamentals of Robotics 3: Kinematics and Trajectory Synthesis of Manipulation Robots, Springer-Verlag, 1986

[17] Paul R., Robots Manipulators: Manipulators: Mathematics, Programming and Control, The MIT Press, 1981.

[18] Truckenbrodt A., "Dynamics and Control Methods for Moving Flexible Structures and Their Application to Industrial Robots", Proc. of 5th World Congress on Theory of Machines and Mechanisms, publ. ASME, 1979.

[19] Vukobratović M., Potkonjak V., "Computer Method for Dynamic Modelling of a Manipulators with Elastic Properties", Journal of Mechanism and Machine Theory, Vol. 17, No 2, 1982.

[20] Sunada H.W., Dynamic Analysis of Flexible Spatial Mechanism and Robotic Manipulators, Ph. D. Theiss, University of California, Los Angeles, 1981.

[21] Akulenko L.D., Mihailov S.A. Tchernousko F.L., "Modelling of Dynamics with Elastic Segments", (in Russian), Journal of Rigid Body Mechanics ANUSSR, No 3, 1981.

[22] Vukobratović M., Cvetković V., "Computer-Oriented Algorithm Modelling of Active Spatial Mechanisms for Application in Robotics", Trans. on Systems, Man and Cybernetics, Vol. SMC-12, No. 6, 1982.

[23] Rivin E., "Analysis of Structural Compliance for Manipulators", Proc. of Internat. Conference on Robotics and Factories in the Future, North Carolina, USA, 1984.

[24] Love A.E.H. Treatise on the Mathematical Theory of Elasticity, 4th. Ed., Cambridge Univ. Press, 1927.

[25] Bathe K.J., Finite Elements Procedures in Engineering Analysis, Prentice Hall, 1982.

[26] Meirovitch L., Analytical Methods in Vibrations, The MacMillan Co., 1967.

[27] Wanner H.C., "Highflexible Manipulator Systems", (in German), Robotersysteme 2, 1986.

[28] Vukobratović M., Šurdilović D., Dynamics and Control of Flexible Manipulation Robots, Scientific Fundamentals of Robotics 8, Springer-Verlag, 1989.

[29] Zheng F.Y., Luh S.Y.J., "Joint Torques for Control of Two Coordinated Moving Robots", Proceedings of the IEEE Robotics and Automation Conference, San Francisco, CA, 1986.

[30] Lilov L., "Structure, Kinematics and Dynamics of Multibody Systems", (in Russian), Advances in Mechanics, No 1-2, 1983.

[31] Djurović M., Vukobratović M., "Contribution to Dynamic Modelling of Cooperative Manipulation", Mech. & Mach. Th. (to appear), 1989.

Chapter 3
Computer Method for Linearization and Parameter Sensitivity of Manipulation Robots Dynamic Models

3.1 Introduction

In Chapter 2 the basic method for automatic setting of mathematical models of open-chain manipulation robot dynamics, based on general theorems of mechanics was presented. By including dynamic models of actuator units (Appendix 6) in robot mechanism dynamics, a complete dynamic model of an open-chain active mechanism manipulation robot is formed. With the presented local closed chain of parallelogram type of a robotic mechanisms which are met in industrial application, practically all mechanical structures of contemporary robotic manipulators are represented. Thus the necessary foundation is established for controlling robotic manipulators based upon required information on their dynamics. Mathematical models of robotic manipulators dynamics are derived by applying the algorithm presented in Chapter 2 of this book.

The derived mathematical model of a robotic mechanism, as a complete mathematical model of robot dynamics (including the actuators' dynamics) enables the calculation of driving forces and torques as well as of the open-loop control signals. The open-loop control signals in the forthcoming, second book of this textbook series dedicated to control synthesis, are termed programme or nominal control. These nominal control signals are "sufficient" in cases of unperturbed (ideal) working conditions of the system. This is never the case with real technical systems, due to perturbations present of the initial conditions type or small parameter variations and inaccurate information on system parameters (in this case parameters of mechanism and actuator units). In order to perform precise tracking of desired (nominal) robot trajectories, it is required to synthesize dynamic control for the compensation of the aforementioned system perturbations.

Mathematical models of active mechanisms in perturbed regimes are also nonlinear. However, when only small perturbations are present, they can be represented by the corresponding linearized models. It will be seen in the book dealing with problems of control that the linearized model of dynamics will be used in the synthesis of optimal manipulation robot control.

Beside the linearized models of robotic mechanism dynamics, parameter sensitivity models are also of practical significance. It was mentioned already that parameter variations are inherent both to manipulator mechanism and to its actuator units so that one of the main reasons for introducing feedback loops in the control law synthesis was to account for the presence of the parameter variations. Mathematical models of system parameter sensitivity are providing insight into the rank of parameters, so the required tolerances of which can be determined. It is clear that parameters of a higher sensitivity rank, which means that they have more influence on system performance, will be awarded narrower tolerances and conversely, parameters having a low sensitivity rank will be awarded broad tolerances. All this could have some practical significance in parameter identification as well as in the production technology of components and parts of robotic mechanisms. It will also be seen in the forthcoming books of the series that a sensitivity function for such parameters will be used in the synthesis of adaptive control algorithms where significant changes may arise, e.g. in the mass of the working object (mass of the last manipulator segment with the working object).

Because any change in the mass of particular segment causes a corresponding variation of the inertia tensor, it is necessary to perform the sensitivity analysis of those parameters which have some bearing on control synthesis of manipulation robots.

3.2 Method of Computer Linearization of Dynamic Models Based on General Theorems of Mechanics

In Chapter 2 dynamic equations of motions for an open-chain spatial robotic mechanism of the form (2.3.30) have been derived:

$$S_M^o: \quad P^o = H(q^o)\ddot{q}^o + h(q^o, \dot{q}^o), \quad q^o(t_o), \ \dot{q}^o(t_o) \ \text{- given} \qquad (3.2.1)$$

S_M^o is the mathematical model of robot mechanism dynamics without actuator units. Superscript "o" denotes the nominal dynamic state, i.e., state of the system in the absence of perturbations, P^o is the vector of nominal driving torques (forces), q^o is the vector of nominal generalized coordinates, $H(q^o)$ is the inertial matrix of the mechanism dynamic system, skew-symmetric type, $h(q^o, \dot{q}^o)$ is a column matrix of centrifugal, Coriolis and gravitational forces.

Mathematical model of the deviation of mechanism dynamics (3.2.1) from the nominal trajectory q^o is given in the form [2, 5]:

$$S_M:\ \Delta P = H^o(t, \Delta q)\Delta\ddot{q} + h^o(t, \Delta q, \Delta\dot{q}),\ \Delta q(t_o),\ \Delta\dot{q}(t_o) - \text{given}, \quad (3.2.2)$$

where $\Delta q = q(t) - q^o(t)$ - deviation vector of the generalized coordinates, $\Delta P = P(t) - P^o(t)$ - deviation vector of driving torques (forces), H^o and h^o are previously devined matrices of the deviation model.

However, if only small perturbations are present (the case most frequently encountered in industrial manipulation) the non-linear model of deviations (3.2.2) may be substituted by its first order approximation:

$$S_{M\ell}:\ \delta P = H(q^o)\delta\ddot{q} + h_v(q^o, \dot{q}^o)\delta\dot{q} + h_p(q^o, \dot{q}^o, \ddot{q}^o)\delta q, \quad (3.2.3)$$

where δq - small deviation of the generalized coordinates vector from their nominal values, δP - deviation of the driving torques vector from $P^o(t)$, h_v and h_p are matrices of the linearized system:

$$h_v(q^o, \dot{q}^o) = \frac{\partial h(q^o, \dot{q}^o)}{\partial \dot{q}} \quad (3.2.4)$$

$$h_p(q^o, \dot{q}^o, \ddot{q}^o) = \frac{\partial H(q^o)}{\partial q}\ddot{q}^o + \frac{\partial h(q^o, \dot{q}^o)}{\partial q}. \quad (3.2.5)$$

The model (3.2.3) is correct for the case when $||\delta q(t_o)|| < \varepsilon_p$ and $||\delta\dot{q}(t_o)|| < \varepsilon_v$ where ε_p and ε_v are sufficiently small positive real number.

By solving the model dynamic equations (3.2.3) with respect to $\delta\ddot{q}$, which is enabled by the regularity of matrix $H(q^o)$ and introducing the state vector $\xi = [\delta q^T \mid \delta\dot{q}^T]^T$ of the mechanical part of the system, model (3.2.3) is obtained in the form:

$$S_{M\ell}:\ \dot{\xi} = \left[\begin{array}{c|c} 0 & I_n \\ \hline -H^{-1}h_p & -H^{-1}h_v \end{array}\right]\xi + \left[\begin{array}{c} 0 \\ \hline H^{-1} \end{array}\right]\delta P,\quad \xi(t_o) - \text{given}, \quad (3.2.6)$$

where I_n is unit $(n\times n)$ matrix.

As mentioned previously, model of actuator units dynamics (2.32):

$$S^i: \quad \dot{x}^i = A^i x^i + f^i P_i + b^i N(u^i), \quad x^i(t_o) = x_o^i, \qquad (3.2.7)$$

should be added to the mathematical model of the robotic mechanism, where x^i - subsystem (actuator unit) state vector, S^i; f^i and b^i are constant matrices, defined in Chapter 2, u^i - scalar input to the actuator unit, $N(u^i)$ - saturation type nonlinearity P_i - driving torque (force) of the i-th actuator.

Linearization of the model (3.2.7) yields:

$$S_\ell^i: \quad \delta\dot{x}^i = A^i \delta x^i + f^i \delta P_i + b^i N(t, \delta u^i), \quad \delta x^i(t_o) = \delta x_o^i, \qquad (3.2.8)$$

where δx^i, δP_i and δu^i are deviations of corresponding vectors.

Linearized model of the complete system is obtained by uniting the linearized model of mechanical part $S_{M\ell}$ (3.2.3) and model of actuators S_ℓ^i (3.2.8):

$$S_\ell: \quad \delta\dot{x} = \tilde{A}^o(x^o(t), u^o(t))\delta x + \tilde{B}^o(x^o(t))N(t, \delta u), \quad \delta x(t_o) = \delta x_o \qquad (3.2.9)$$

where $\delta x = [\delta x^{1T} \delta x^{2T} \dots \delta x^{nT}]^T$ - state vector, $\tilde{A}^o$ - $(N \times N)$ matrix, $\tilde{B}^o$ - $(N \times n)$ matrix, $\delta u = [\delta u^1 \dots \delta u^n]^T$ - input vector, n_i - order of subsystem (actuator) and $N = \sum_{i=1}^{n} n_i$ - order of complete system.

In order to automatize the process of linearizing the mechanism dynamic equations of motion or, in other words, to calculate the matrices of the linearized model (3.2.3): $h_p(q^o, \dot{q}^o)$ and $h_v(q^o, \dot{q}^o, \ddot{q}^o)$, it is required to develop a computer-orientated procedure for evaluating these matrices. The algorithm should not involve such operations which may generate cumulative errors (for instance - numerical differentiation). In order to form matrices h_p and h_v, it is necessary, to form matrices of partial derivatives in accordance to (3.2.4) and (3.2.5):

$$\frac{\partial H}{\partial q}, \quad \frac{\partial h(q^o, \dot{q}^o)}{\partial q}, \quad \frac{\partial h(q^o, \dot{q}^o)}{\partial \dot{q}}, \qquad (3.2.10)$$

where $q^o(t)$ is the prescribed continual nominal trajectory.

The algorithm forms matrices (3.2.10) simultaneously with matrices $H(q^o)$ and $h(q^o, \dot{q}^o)$ of the basic model (3.2.1).

In the automatic linearization procedure the following notation will be adopted: $J = \{j: j=1,\ldots,i\}$ is the set of indices:

$$\eta_{\ell i} = \sum_{k=1}^{i} \delta_{\ell k}, \qquad \delta_{\ell k} = \begin{cases} 1, & \text{if } i \geq \ell, \\ 0, & \text{otherwise,} \end{cases}$$

where $\delta_{\ell k}$ is Kronecker's symbol ($\delta_{\ell k} = 1$ for $\ell = k$, and 0 otherwise) ξ_i - a symbol that has value 1 if the i-th kinematic pair is translatory and 0 if the pair is revolute.

$\Delta_\ell A = \frac{\partial A}{\partial q}$, $\Delta_{\dot{\ell}} A = \frac{\partial A}{\partial \dot{q}}$ are partial derivatives of A with respect to q and $\dot{q}$, respectively,

The following theorems will also be used [2, 5]:

Theorem T.1.

The partial derivative of a noncomposite vector a_i *), $i \in I$, with respect to the generalized coordinate q^ℓ, $\ell \in I$, is given by:

$$\Delta_\ell \vec{a}_i = (\vec{e}_\ell \times \vec{a}_i)(1-\xi_\ell)\eta_{\ell i}, \tag{3.2.11}$$

where $\vec{e}_\ell$ denotes the unit vector of joint axis of ℓ-th kinematic pair and $\bar{\xi}_\ell = 1 - \xi_\ell$.

Proof: When the i-th kinematic pair is revolute ($\xi_\ell = 0$), $i \leq \ell$ and if $\eta_{\ell i} = 1$ (Fig. 3.1), the problem of determining:

$$\Delta_\ell \vec{a}_i = \lim_{\Delta q^\ell \to 0} \frac{\Delta \vec{a}_i}{\Delta q^\ell} \Bigg|_{\Delta q^j = 0,\ j \neq \ell}$$

reduces to the well-known theorem of elementary rotations from theoretical mechanics. According to this theorem we obtain $\lim_{\Delta q^\ell \to 0} \frac{\Delta \vec{a}_i}{\Delta q^\ell} = \vec{e}_\ell \times \vec{a}_i$. This prooves T.1. for $\xi_\ell = 0$ and $\ell \leq i$.

It is obvious from Fig. 3.1. that the value of $\Delta_\ell \vec{a}_i$ is 0 for $\ell > i$. In addition, for a translatory pair ℓ ($\xi_\ell = 1$), the value of $\vec{a}_i$ does not

---

*) Noncomposite vectors $\vec{a}_i$ are vectors that are assembled of vectors related to i-th member of a mechanism.

change, i.e., $\Delta_\ell \vec{a}_i = 0$, which completes the proof of T.1.

Theorem T.2.

The partial derivative of vector $\vec{r}_{ij}$, $i \in I$, $j \in J$, connecting the joint of j-th kinematic pair and the center of gravity of i-th mechanism link, with respect to the generalized coordinate q^ℓ, $\ell \in I$, is given by:

$$\Delta_\ell \vec{r}_{ij} = [(\vec{e}_\ell \times \vec{r}_{ij})\eta_{\ell,j-1} + (\vec{e}_\ell \times \vec{r}_{i\ell})(\eta_{\ell i} - \eta_{\ell,j-1})]\bar{\xi}_i +$$
$$+ \vec{e}_\ell(\eta_{\ell i} - \eta_{\ell,j-1})\xi_i, \tag{3.2.12}$$

where $\vec{e}_\ell$ is the unit vector of the ℓ-th link axis.

Proof: Using the relation $\vec{r}_{ij} = \vec{r}_{i\ell} + \vec{R}_{\ell j}$, where $\vec{R}_{\ell j}$ is the vector from the joint j to the joint ℓ and according to T.1, we obtain $\Delta_\ell \vec{r}_{ij}$ in form (3.2.12).

Theorem T.3.

The partial derivative of vector or scalar product of two vectors (which may be composite) $\vec{a}_i$ and $\vec{b}_i$ is given by [5]:

$$\Delta_\ell(\vec{a}_i \times \vec{b}_i) = (\Delta_\ell \vec{a}_i) \times \vec{b}_i + \vec{a}_i \times (\Delta_\ell \vec{b}_i). \tag{3.2.13}$$

Proof: Relation (3.2.13) is one of the basic formulae in vector analysis.

According to the method of general theorems (Newton-Euler method of forming dynamic equations of robotic mechanisms), joint angles have been adopted as generalized coordinates q^i, $i \in I$, of the mechanism. A Cartesian coordinate system $Q_i = (\vec{q}_{i1}, \vec{q}_{i2}, \vec{q}_{i3})$ is associated to each link of the mechanism, with the origin in its center of gravity (Fig. 3.2). Axes of the system lie along the main central inertial axes of the i-th link.

In the following text particular stages of the basic algorithm of forming dynamic equations (presented in Ch. 2), as well as the corresponding phases of the linearization algorithm will be described. A complete flow-chart of the primary algorithm as well as the linearization algorithm will be described. A complete flow-chart of the primary algorithm as well as the linearization algorithm is given in Fig. 3.3.

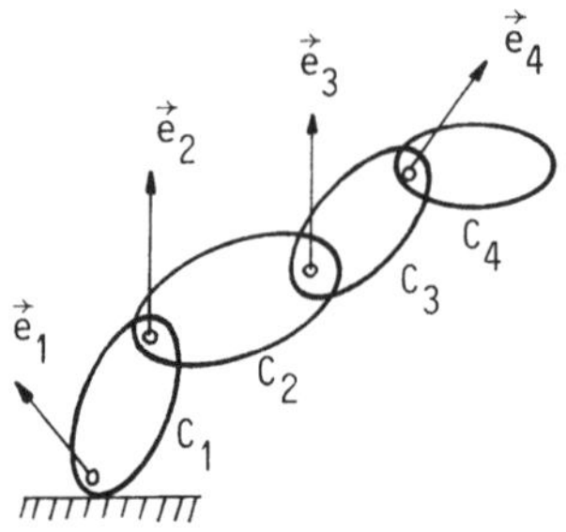

Fig. 3.1. Open kinematic chain

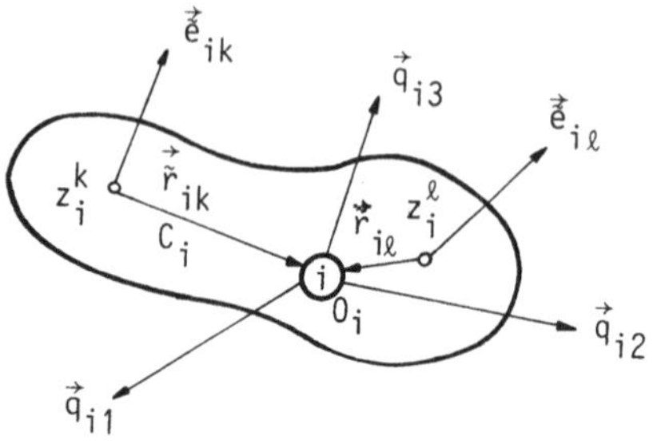

Fig. 3.2. Local coordinate system

As described in Chapter 2, the first stage of the procedure involves the "assembling" of links of i-th kinematic pair. It is assumed that the first link of the i-th pair is already "assembled" to the mechanism, i.e., that its initial position ($q^j=0$, $j\in J$) in the absolute coordinate system is determined. The assembling of the i-th pair is performed by making the vectors $\vec{e}_i$ and $\vec{\tilde{e}}_i$ coincide, under the condition $q^i=0$. From these conditions matrix $Q_i^o = [\vec{q}^{\,o}_{i1}\ \vec{q}^{\,o}_{i2}\ \vec{q}^{\,o}_{i3}]$ is determined, i.e. projections of vector $\vec{q}_{ij}$, $j=1,2,3$ onto the axis of the fixed coordinate system*).

According to the algorithm, forming of partial derivation is performed by using the Theorem T.1. Since vectors $\vec{q}^{\,o}_{ij}$, formed at the first stage of the algorithm are non-composite i.e. connected only to the i-th mechanism link, follows:

$$\Delta_\ell \vec{q}^{\,o}_{ij} = (\vec{e}_\ell \times \vec{q}^{\,o}_{ij})\bar{\xi}_\ell \eta_{\ell i}, \tag{3.2.14}$$

$$\Delta_\ell Q_i^o = [\Delta_\ell \vec{q}_{i1}\ \Delta_\ell \vec{q}^{\,o}_{i2}\ \Delta_\ell \vec{q}^{\,o}_{i3}]^T, \tag{3.2.15}$$

where $i, \ell\in I$, $j\in\{1, 2, 3\}$. According to the definition of symbol $\eta_{\ell i}$, relation (3.2.14) may be written in the form:

$$\Delta_\ell \vec{q}^{\,o}_{ij} = (\vec{e}_\ell \times \vec{q}^{\,o}_{ij})\bar{\xi}_\ell, \qquad \ell\in J, \tag{3.2.16}$$

---

*) $\vec{q}^{\,o}_{ij}$ denotes the vector $\vec{q}_{ij}$ in a stage of mechanism assembling ($q^i=0$)

while for all the remaining indices $\ell\in\{i+1,\ldots,n\}$ holds $\Delta_\ell \vec{q}_{ij}=0$, as it is clear from the flowchart in Fig. 3.3.

In the next stage of the primary algorithm the position of the i-th mechanism link is determined under condition $q^i=q^{io}$, where q^{io} is the prescribed value of the i-th generalized coordinate. The position of the i-th link is determined using projections of vectors $\vec{r}_{ii}$, $\vec{r}_{i,i+1}$, $\vec{r}_{ij}$, $i\in I$, $j\in J$ on the absolute coordinate system. These vectors, as well as the vector $\vec{e}_{i+1}$, are determined by applying the Rodrigue's finite rotations formula.

Because all the vectors (except $\vec{r}_{ij}$) formed in this stage are noncomposite, their partial derivatives are determined by T.1. The partial derivatives of $\Delta_\ell \vec{r}_{ij}$, $i\in I$, j, $\ell\in J$ are determined by Theorem T.2. All the operations are thus reduced to the calculation of vector products of the vectors formed in the primary algorithm (Fig. 3.3, block 2').

In the next step the angular and linear velocity, $\vec{\omega}_i$ and $\vec{v}_i$, and the angular and linear acceleration, $\vec{\varepsilon}_i$ and $\vec{w}_i$, of the i-th link are calculated. The superposition of the transfer and relative velocities and accelerations according to the known relations from rigid body dynamics, as was shown in Chapter 2, is applied. In order to prepare a computer-oriented algorithm convenient for the dynamic analysis of mechanism, expressions for $\vec{\varepsilon}_i$ and $\vec{w}_i$ are written in a slightly modified form using coefficients $\vec{\alpha}_{ij}$ and $\vec{\beta}_{ij}$, respectively. These coefficients are determined from recursive relations in Fig. 3.3, block 3.

Velocities and accelerations of the i-th link of the mechanism are composite vectors. For example, the angular velocity $\vec{\omega}_i$ may be expressed by the following sum:

$$\vec{\omega}_i = \sum_{k=1}^{i} \dot{q}^k \vec{e}_k \bar{\xi}_k, \qquad i\in I \tag{3.2.17}$$

Applying the Theorem T.1. we get

$$\Delta_\ell \vec{\omega}_i = \sum_{k=1}^{i} \dot{q}^k (\vec{e}_\ell \times \vec{e}_k) \bar{\xi}_k, \qquad \ell\in J \tag{3.2.18}$$

This relation may be expressed in the recurrent form:

$$\Delta_\ell \vec{\omega}_i = \Delta_\ell \vec{\omega}_{i-1} + \dot{q}^i (\vec{e}_\ell \times \vec{e}_i) \bar{\xi}_i \tag{3.2.19}$$

In the developed algorithm (Fig. 3.3, block 3), the relation (3.2.19) is used for the calculation of partial derivatives $\Delta_\ell \vec{\omega}_i$, $\ell \in J$, in order to minimize the number of numerical operations. By using the Theorems T.1, T.2, and T.3, recurrent relations with respect to the partial derivatives of coefficients $\vec{\alpha}_{ij}$ and $\vec{\beta}_{ij}$, $i \in I$, $j \in J$ have been formed. Partial derivatives of the coefficients $\vec{\alpha}_i^o$ and $\vec{\beta}_i^o$ with respect to generalized coordinates and velocities q^ℓ and $\dot{q}^\ell$ have also been formed in a similar way.

The next step of the primary algorithm is the calculation of the inertial force $\vec{F}_i$ and the moment $\vec{M}_i$ of the inertial force of the i-th link by using the values for angular and linear acceleration $\vec{\varepsilon}_i$ and $\vec{w}_i$, determined in the previous stage of the algorithm. In the algorithm shown in Fig. 3.3. all links can be described by the transversal (J_{Ni}) and longitudinal (J_{si}) moment of inertia, i.e., can be considered as canes. The algorithm is easily generalised for the case of a rigid body, as given in its general form in Chapter 2.

Similarly to the procedure from the preceding stage, the coefficients $\vec{a}_{ij}$ and $\vec{b}_{ij}$, $i \in I$, $j \in J$, are introduced, which are proportional to $\vec{\alpha}_{ij}$ and $\vec{\beta}_{ij}$ (Fig. 3.3, block 4).

The values $\Delta_\ell \vec{a}_{ij}$, $\Delta_\ell \vec{b}_{ij}$, $i \in I$; j, $\ell \in J$ have been determined using the Theorems T.1. and T.3. (Fig. 3.3, block 4').

All values relevant for the calculation of matrices of models (3.2.1) and (3.2.3) are formed during the process of "assembling" the mechanism links, with the index i changing from 1 to n. The matrices of the nonlinear model (3.2.1) are formed by using the general theorems which describe the translation of the center of mass of a rigid body and rotation about this point. Depending on the type of the i-th kinematic pair (sliding or revolute) the first or the second theorem is applied. In such way, all terms of the dynamic model matrices $H^{ik}(q)$ and $h^i(q, \dot{q})$ (Fig. 3.3, block 5) are determined.

For the case when the i-th pair is revolute, application of Theorems T.1. and T.3. gives:

$$\Delta_\ell H_{ik} = -(\vec{e}_\ell \times \vec{e}_i) \sum_{j=k}^{n} (\vec{b}_{jk} + \vec{r}_{ji} \times \vec{a}_{jk}) - \\ - \vec{e}_i \sum_{j=k}^{n} (\Delta_\ell \vec{b}_{jk} + \Delta_\ell \vec{r}_{ji} \times \vec{a}_{jk} + \vec{r}_{ji} \times \Delta_\ell \vec{a}_{jk}), \qquad (3.2.20)$$

INPUT PARAMETERS

↓

$i = 1$ ← (1)

↓

block 1, "assembling"

$$r_{Ni} = e_i\times(r_{i-1,i}\times e_i),\quad \tilde{r}_{Ni} = \tilde{e}_i\times(\tilde{r}_{ii}\times\tilde{e}_i),\quad a_i = -r_{Ni}/|r_{Ni}|,\quad \tilde{a}_i = \tilde{r}_{Ni}/|\tilde{r}_{Ni}|$$

$$b_i = e_i\times a_i,\quad \tilde{b}_i = \tilde{e}_i\times\tilde{a}_i,\quad Q_i^o = [q_{i1}^o\ q_{i2}^o\ q_{i3}^o] = [a_i\ e_i\ b_i][\tilde{a}_i\ \tilde{e}_i\ \tilde{b}_i]^T$$

↓

block 1'

$$\Delta_\ell q_{ij}^o = (e_\ell\times q_{ij}^o)(1-\xi_\ell)\eta_{\ell i},\quad \Delta_\ell Q_i^o = [\Delta_\ell q_{i1}^o\ \Delta_\ell q_{i2}^o\ \Delta_\ell q_{i3}^o]$$

↓

block 2, "positions"

$$q_{ij} = [q_{ij}^o\cos q^i+(1-\cos q^i)(e_i\cdot q_{ij}^o)e_i+e_i\times q_{ij}^o\sin q^i](1-\xi_i)+q_{ij}^o\xi_i,\qquad (j=1,2,3)$$

$$Q_i = [q_{i1}\ q_{i2}\ q_{i3}],\quad r_{ii} = Q_i\tilde{r}_{ii}+q^i e_i\xi_i,\quad r_{i,i+1} = Q_i\tilde{r}_{i,i+1},\quad e_{i+1} = Q_i\tilde{e}_{i+1}$$

$$r_{ij} = r_{i-1,j}-r_{i-1,i}+r_{ii},\qquad j=1,\dots,i-1$$

↓

block 2'

$$\Delta_\ell q_{ij} = (e_\ell\times q_{ij})(1-\xi_\ell)\eta_{\ell i},\quad \Delta_\ell Q_i = [\Delta_\ell q_{i1}\ \Delta_\ell q_{i2}\ \Delta_\ell q_{i3}],\quad \Delta_\ell r_{ii} =$$

$$= (e_\ell\times r_{ii})(1-\xi_\ell)\eta_{\ell i}+e_i\xi_i$$

$$\Delta_\ell r_{i,i+1} = (e_\ell\times r_{i,i+1})(1-\xi_\ell)\eta_{\ell i},\quad \Delta_\ell e_{i+1} = (e_\ell\times e_{i+1})(1-\xi_\ell)\eta_{\ell i}$$

$$\Delta_\ell r_{ij} = [(e_\ell\times r_{ij})\eta_{\ell,j-1}+(e_\ell\times r_{i\ell})(\eta_{\ell i}-\eta_{\ell,j-1})](1-\xi_\ell)+e_\ell(\eta_{\ell i}-\eta_{\ell,j-1})\xi_\ell$$

↓

block 3, "velocities and accelerations"

$$\omega_i = \omega_{i-1}+\dot{q}^i e_i(1-\xi_i),\quad v_i = v_{i-1}-\omega_{i-1}\times r_{i-1,i}+\omega_i\times r_{ii}+\dot{q}^i e_i\xi_i$$

$$\varepsilon_i = \sum_{j=1}^{i}\alpha_{ij}\ddot{q}^j+\alpha_i^o,\qquad w_i = \sum_{j=1}^{i}\beta_{ij}\ddot{q}^j+\beta_i^o$$

$$\alpha_{ii} = e_i(1-\xi_i),\quad \alpha_{ij} = \alpha_{i-1,j},\quad \alpha_i^o = \alpha_{i-1}^o+\dot{q}^i(\omega_{i-1}\times e_i)(1-\xi_i)$$

$$\beta_{ii} = (e_i\times r_{ii})(1-\xi_i)+e_i\xi_i,\quad \beta_{ij} = \beta_{i-1,j}+\alpha_{i-1,j}\times(r_{ii}-r_{i-1,i})$$

$$\beta_i^o = \beta_{i-1}^o+\alpha_{i-1}^o\times(r_{ii}-r_{i-1,i})+\dot{q}^i(\omega_{i-1}\times e_i)\times r_{ii}(1-\xi_i)+2\dot{q}^i(\omega_{i-1}\times e_i)\xi_i-\gamma_{i-1,i}+\gamma_{ii}$$

where $\gamma_{ij} = \omega_i\times(\omega_i\times r_{ij})$

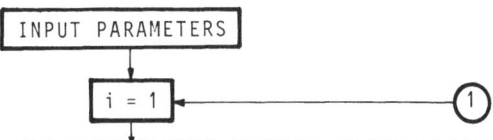

block 3'

$$\Delta_\ell\omega_i = \Delta_\ell\omega_{i-1}+\dot{q}^i(e_\ell\times e_i)(1-\xi_i),\ \Delta_\ell\alpha_{ij} = \Delta_\ell\alpha_{i-1,j},$$

$$\Delta_\ell\alpha_i^o = \Delta_\ell\alpha_{i-1}^o+\dot{q}^i(\Delta_\ell\omega_{i-1}\times e_i+\omega_{i-1}\times\Delta_\ell e_i)(1-\xi_i),\ \Delta_\ell\alpha_{ii} = \Delta_\ell e_i(1-\xi_i)$$

$$\Delta_{\dot\ell}\omega_i = \Delta_{\dot\ell}\omega_{i-1}+e_i(1-\xi_i)\delta_{\ell i},\ \Delta_{\dot\ell}\alpha_i^o = \Delta_{\dot\ell}\alpha_{i-1}^o+[(\omega_{i-1}\times e_i)\delta_{\ell i}+\dot{q}^i(\Delta_{\dot\ell}\omega_{i-1}\times e_i)](1-\xi_i)$$

$$\Delta_\ell\beta_{ij} = \Delta_\ell\beta_{i-1,j}+\Delta_\ell\alpha_{i-1,j}\times(r_{ii}-r_{i-1,i})+\alpha_{i-1,j}\times(e_\ell\times(r_{ii}-r_{i-1,i}))$$

$$\Delta_\ell\beta_{ii} = (\Delta_\ell e_i\times r_{ii}+e_i\times\Delta_\ell r_{ii})(1-\xi_i)+\Delta_\ell e_i\xi_i,\ \Delta_\ell\beta_i^o = \Delta_\ell\beta_{i-1}^o+\Delta_\ell\alpha_{i-1}^o\times(r_{ii}-r_{i-1,i})+$$

$$+\ \alpha_{i-1}^o\times(e_\ell\times(r_{ii}-r_{i-1,i}))+\dot{q}^i[(\Delta_\ell\omega_{i-1}\times e_i)\times r_{ii}+(\omega_{i-1}\times\Delta_\ell e_i)\times r_{ii}+(\omega_{i-1}\times e_i)\times$$

$$\times\ \Delta_\ell r_{ii}](1-\xi_i)+2\dot{q}^i(\Delta_\ell\omega_{i-1}\times e_i+\omega_{i-1}\times\Delta_\ell e_i)\xi_i-\Delta_\ell\gamma_{i-1,i}+\Delta_\ell\gamma_{ii}$$

where $\Delta_\ell\gamma_{ij} = \Delta_\ell\omega_i\times(\omega_i\times r_{ij})+\omega_i\times(\Delta_\ell\omega_i\times r_{ij})+\omega_i\times(\omega_i\times\Delta_\ell r_{ij})$

$$\Delta_{\dot\ell}\beta_i^o = \Delta_{\dot\ell}\beta_{i-1}^o+\Delta_{\dot\ell}\alpha_{i-1}^o\times(r_{ii}-r_{i-1,i})-\Delta_{\dot\ell}\omega_{i-1}\times(\omega_{i-1}\times r_{i-1,i})-\omega_{i-1}\times(\Delta_{\dot\ell}\omega_{i-1}\times r_{i-1,i})+$$

$$+\ \Delta_{\dot\ell}\omega_i\times(\omega_i\times r_{ii})+\omega_i\times(\Delta_{\dot\ell}\omega_i\times r_{ii})+[(\omega_{i-1}\times e_i)\times r_{ii}(1-\xi_i)+2(\omega_{i-1}\times e_i)\xi_i]\delta_{\ell i}\ +$$

$$+\ [\dot{q}^i(\Delta_{\dot\ell}\omega_{i-1}\times e_i)\times r_{ii}(1-\xi_i)+2\dot{q}^i(\Delta_{\dot\ell}\omega_{i-1}\times e_i)\xi_i]n_{\ell,i-1}$$

block 4, "inertial forces and moments"

$$F_i = -m_i w_i = \sum_{j=1}^{i} a_{ij}\ddot{q}^j + a_i^o,\quad a_{ij} = -m_i\beta_{ij},\quad a_i^o = -m_i\beta_i^o$$

$$M_i = \sum_{j=1}^{i} b_{ij}\ddot{q}^j + b_i^o,\quad b_{ij} = -J_{Ni}(s_i\times\alpha_{ij})\times s_i-J_{si}(\alpha_{ij}\cdot s_i)s_i$$

$$b_i^o = -J_{Ni}((s_i\times\alpha_i^o)\times s_i+\tau_i)-J_{si}(\alpha_i^o\cdot s_i)s_i,\ \tau_i = (\omega_i\cdot s_i)(s_i\times\omega_i),\ s_i = r_{ii}/|r_{ii}|$$

block 4'

$$\Delta_\ell a_{ij} = -m_i \Delta_\ell \beta_{ij}, \quad \Delta_\ell a_i^o = -m_i \Delta_\ell \beta_i^o, \quad \Delta_{\dot{\ell}} a_i^o = -m_i \Delta_{\dot{\ell}} \beta_i^o$$

$$\Delta_\ell b_{ij} = -J_{Ni}[(\Delta_\ell s_i \times \alpha_{ij}) \times s_i + (s_i \times \Delta \alpha_{ij}) \times s_i + (s_i \times \alpha_{ij}) \times \Delta_\ell s_i] -$$

$$-J_{si}[(\Delta_\ell \alpha_{ij} \cdot s_i) s_i + (\alpha_{ij} \cdot \Delta_\ell s_i) s_i + (\alpha_{ij} \cdot s_i) \Delta_\ell s_i \;, \quad \Delta_\ell s_i = e_\ell \times s_i$$

$$\Delta_\ell \tau_i = (\Delta_\ell \omega_i s_i + \omega_i \Delta_\ell s_i) \cdot (s_i \times \omega_i) + (\omega_i \cdot s_i) \cdot (\Delta_\ell s_i \times \omega_i + s_i \times \Delta_\ell \omega_i)$$

$$\Delta_\ell b_{io} = -J_{Ni}[(\Delta_\ell s_i \times \alpha_{io}) \times s_i + (s_i \times \Delta_\ell \alpha_{io}) \times s_i + (s_i \times \alpha_{io}) \times \Delta_\ell s_i + \Delta_\ell \tau_i] -$$

$$-J_{si}[(\Delta_\ell \alpha_{io} \cdot s_i) s_i + (\alpha_{io} \cdot \Delta_\ell s_i) s_i + (\alpha_{io} \cdot s_i) \Delta_\ell s_i]$$

$$\Delta_{\dot{\ell}} b_{io} = -J_{Ni}[(s_i \times \Delta_{\dot{\ell}} \alpha_{io}) \times s_i + \Delta_{\dot{\ell}} \tau_i] - J_{si}(\Delta_{\dot{\ell}} \alpha_i^o \cdot s_i) s_i$$

$$\Delta_{\dot{\ell}} \tau_i = (\Delta_{\dot{\ell}} \omega_i \cdot s_i)(s_i \times \omega_i) + (\omega_i \cdot s_i)(s_i \times \Delta_{\dot{\ell}} \omega_i)$$

$i = n$ — NO: $i+1$ → (1); YES ↓

block 5, "dynamic model matrices"

$$\xi_i = 0: \; H_{ik} = -e_i \cdot \sum_{j=k}^{n} (b_{jk} + r_{ji} \times a_{jk}), \quad h_i = -e_i \cdot \sum_{j=i}^{n} (r_{ji} \times (a_j^o + G_j) + b_j^o)$$

$$\xi_i = 1: \; H_{ik} = -e_i \cdot \sum_{j=k}^{n} a_{jk}, \quad h_i = -e_i \cdot \sum_{j=i}^{n} (a_j^o + G_j), \; G_j = (0, 0, m_j g)^T$$

block 5'

$$\xi_i = 0: \; \Delta_\ell H_{ik} = -\Delta_\ell e_i \cdot \sum_{j=k}^{n} (b_{jk} + r_{ji} \times a_{jk}) - e_i \cdot \sum_{j=k}^{n} (\Delta_\ell b_{jk} + \Delta_\ell r_{ji} \times a_{jk} + r_{ji} \times \Delta_\ell a_{jk})$$

$$\Delta_\ell h_i = -\Delta_\ell e_i \cdot \sum_{j=i}^{n} (r_{ji} \times (a_j^o + G_j) + b_j^o) - e_i \cdot \sum_{j=i}^{n} [\Delta_\ell r_{ji} \times (a_j^o + G_j) + r_{ji} \times \Delta_\ell a_j^o + \Delta_\ell b_j^o]$$

$$\Delta_{\dot{\ell}} h_i = -e_i \cdot \sum_{j=i}^{n} (r_{ji} \times \Delta_{\dot{\ell}} a_j^o + \Delta_{\dot{\ell}} b_j^o), \quad \xi_i = 1: \; \Delta_\ell H_{ik} = -\Delta_\ell e_i \cdot \sum_{j=k}^{n} a_{jk} - e_i \cdot \sum_{j=k}^{n} \Delta_\ell a_{jk}$$

$$\Delta_\ell h_i = -\Delta_\ell e_i \cdot \sum_{j=i}^{n} (a_j^o + G_j) - e_i \cdot \sum_{j=i}^{n} \Delta_\ell a_j^o, \quad \Delta_{\dot{\ell}} h_i = -e_i \cdot \sum_{j=i}^{n} \Delta_{\dot{\ell}} a_j^o$$

Fig. 3.3. Flow-chart diagram for nonlinear and linearized dynamic models generating where $\vec{b}_{ij}$, $\vec{r}_{ij}$, $\vec{a}_{ij}$ and $\vec{e}_j$ are formed in the primary algorithm, while $\Delta_\ell\vec{a}_{ij}$, $\Delta_\ell\vec{b}_{ij}$ and $\Delta_\ell\vec{r}_{ij}$ are formed in the expanded part of the algorithm.

A similar method is applied to get the partial derivatives of h^i, $i\in I$, so that the process of forming the matrices H and h, as well as their partial derivatives with respect to generalized coordinates and velocities, is unified and shown in the form of grouped blocks in the flow-chart*) (Fig. 3.3).

3.3 Sensitivity Analysis of Manipulation Robots Dynamic Models

As already established (2.30), the dynamic equations of motion of active, open-chain spatial mechanism can be presented in the following form:

$$S_M:\ H(q,\ \theta)\ddot{q}+h(q,\ \dot{q},\ \theta) = P, \quad q(t_o) = q_o,\ \dot{q}(t_o) = \dot{q}_o, \qquad (3.3.1)$$

where q(t) is generalized coordinates vector (angles and linear movements), θ is mechanism parameters vector, P = P(t) are driving forces (moments) vector, $H(q,\ \theta)$ is positive definite inertial matrix whose elements are continuously differentiable functions with respect to q and θ, $h(q,\ \dot{q},\ \theta)$ is column-matrix of centrifugal, Coriolis and gravitational forces.

In comparison with the previous definition of matrices of system S_M, model (3.3.1) incorporates the vector of parameters θ. Let us remind that matrices H and h are highly nonlinear functions of $\dot{q}$ and q and are linearly dependent on the vector of parameters θ. Variable parameters, whose variations affect the dynamic response of the complete system, are to be found in mathematical models of actuators (3.2.7), too. Considering that catalogue data of actuator parameters, used to form the coefficients of actuator mathematical models (3.2.7) differ from the corresponding data of the installed actuator units, it is necessary to analyse the problem of variation of these parameters. Sensitivity analysis based on parameters ranking allows to determine their admissible tolerances. In this section the variations of parameters of a robotic manipulator will be considered.

---

*) For simplicity in Fig. 3.3 symbols denoting vectors are excluded.

Parameter sensitivity of the mechanism driving forces

Parameters of a robot mechanism are of kinematic (geometrical) and dynamic character. Kinematic parameters are predominantly manipulator segment lengths which may be affected by manufacturing errors. From this point of view the effects of these errors (variations) on the accuracy of manipulation robot performance could be investigated. However, such effects will not be considered here because the mentioned problem can be practically eliminated by running a check on the dimensional accuracy of manufactured components. Therefore, we restrict ourselves to dynamic parameters of the robot mechanism, which are the mass and the inertia tensor of particular segment (link). These parameters are important not only for determining their deviation from nominal values, but also for their alteration (sometimes very significant) programmed or sporadic (depending on the type and application of the robot) and especially of the last segment (the gripper) which incorporates the workpiece (manipulation object).

We define the following sensitivity functions [1]:

$$\left.\frac{\partial P}{\partial \theta_i}\right|_{q=q^o(t)} \tag{3.3.2}$$

where $q^o(t)$ is the vector of internal coordinates (angles or translatory displacement) with nominal parameter values, θ_i is the vector of dynamic parameters (mass and moments of inertia) of segments.

Partial derivative of equation (3.3.1) with respect to parameter $\theta_{i\ell}$ ($\ell=1,\ldots,n_{pi}$), where n_{pi} is the number of dynamic parameters of the i-th segment, can be written as follows [1]:

$$\left.\frac{\partial P}{\partial \theta_{i\ell}}\right|_{q=q^o(t)} = H^i_\ell(q^o, \theta)\ddot{q}^o + h^i_\ell(q^o, \dot{q}^o, \theta), \tag{3.3.3}$$

where: $H^i_\ell(q^o, \theta) = \dfrac{\partial H(q^o, \theta)}{\partial \theta_{i\ell}}$, $h^i_\ell(q^o, \dot{q}^o, \theta) = \dfrac{\partial h(q^o, \dot{q}^o, \theta)}{\partial \theta_{i\ell}}$.

Matrices H^i_ℓ and h^i are efficiently calculated using any one from the available computational procedures for automatic forming of matrices H and h of the general dynamic model of open robotic mechanism (2.30) or (3.2.1). In our case the procedure used is based on fundamental theorems of mechanics, which was presented in detail in Chapter 2.

In order to envisage the effects of variations of the adopted dynamic parameters, mass and segments inertia tensor, let us remind briefly of the relations used in forming dynamic equations of motion of a robot mechanism link as derived in Chapter 2.

Dynamics of a mechanism is formed through a set of dynamic quantities:

$\vec{F}_i$ - inertial force acting at center of mass of i-th mechanism segment (link),

$\vec{M}_i$ - moment of inertial force of the i-th link.

Inertial force was represented using the second Newton's law by expressions (2.9) and (2.10), which were:

$$\vec{F}_i = -m_i\vec{w}_i = \sum_{j=1}^{i} \vec{a}_{ij}\ddot{q}^j + \vec{a}_i^o , \tag{3.3.4}$$

$$\vec{a}_{ij} = -m_i\vec{\beta}_{ij}, \qquad \vec{a}_i^o = -m_i\vec{\eta}_i. \tag{3.3.5}$$

Moments of inertial forces are determined from Euler's dynamic equations (2.11)-(2.17):

$$\vec{M}_i = \sum_{j=1}^{i} \vec{b}_{ij}\ddot{q}^j + \vec{b}_i^o, \tag{3.3.6}$$

$$\vec{b}_{ij} = -T_i\vec{\alpha}_{ij}, \qquad \vec{b}_i^o = -T_i\vec{\theta}_i + \vec{\lambda}_i, \tag{3.3.7}$$

where T_i - 3×3 matrix with elements T_i^{jk}:

$$T_i^{jk} = \sum_{\ell=1}^{3} Q_i^{j\ell} J_{i\ell} q_{i\ell}^k = \sum_{\ell=1}^{3} q_{i\ell}^j q_{i\ell}^k J_{i\ell}, \tag{3.3.8}$$

Q_i is the i-th segment transformation matrix (2.4) and its element $Q_i^{jk} = q_{ik}^j$ is the component of transformation matrix (Q_i) vector (column):

$$\vec{\lambda}_i = Q_i \begin{bmatrix} (\vec{\omega}_i\cdot\vec{q}_{i2})(\vec{\omega}_i\cdot\vec{q}_{i3})\cdot(J_{i2}-J_{i3}) \\ (\vec{\omega}_i\cdot\vec{q}_{i3})(\vec{\omega}_i\cdot\vec{q}_{i1})\cdot(J_{i3}-J_{i1}) \\ (\vec{\omega}_i\cdot\vec{q}_{i1})(\vec{\omega}_i\cdot\vec{q}_{i2})\cdot(J_{i1}-J_{i2}) \end{bmatrix}, \tag{3.3.9}$$

$\vec{\eta}_i$ and $\vec{\theta}_i$ from expression (3.3.5) and (3.3.7) are given by relation (2.8).

It should be mentioned that the case treated is for a robot mechanism whose first segment is firmly fixed to an immobile base ($\vec{\varepsilon}_o=0$, $\vec{w}_o=0$).

The effects of mass variations Δm_i on the mechanism dynamic are determined by altering the coefficients of expression (3.3.4). The influence of moment of inertia variations $\Delta J_{i\ell}$ ($\ell=1,2,3$) is determined by altering the dynamic coefficients of expression (3.3.6) and (3.3.7) i.e. matrix T_i and vector $\vec{\lambda}_i$. If we assume that:

$$m_i = m_{io} + \Delta m_i,$$
$$J_{i\ell} = J_{i\ell o} + \Delta J_{i\ell}, \qquad (\ell=1,2,3). \tag{3.3.10}$$

it follows that

$$\vec{a}_{ij} = \vec{a}_{ijo} - \Delta m_i \vec{\beta}_{ij},$$
$$\vec{a}_i^{\,o} = \vec{a}_{io}^{\,o} - \Delta m_i \vec{\eta}_i, \tag{3.3.11}$$

where $\vec{a}_{ijo} = -m_{io}\vec{\beta}_{ij}$ and $\vec{a}_{io}^{\,o} = -m_{io}\vec{\eta}_i$ are nominal values of $\vec{a}_{ij}$ and $\vec{a}_i^{\,o}$ respectively.

Analogous relations are obtained for vectors $\vec{b}_{ij}$ and $\vec{b}_i^{\,o}$. Matrix T_i can be represented in the form:

$$T_i = \sum_{\ell=1}^{3} [q_{i\ell}^1 \vec{q}_{i\ell} \vdots q_{i\ell}^2 \vec{q}_{i\ell} \vdots q_{i\ell}^3 \vec{q}_{i\ell}] J_{i\ell} = \sum_{\ell=1}^{3} Q_{i\ell} J_{i\ell}, \tag{3.3.12}$$

where $\vec{q}_{i\ell} = [q_{i\ell}^1 \; q_{i\ell}^2 \; q_{i\ell}^3]^T$.

Vector $\vec{\lambda}_i$ can be transformed similarly. To do that let us multiply as indicated in expression (3.3.9):

$$\vec{\lambda}_i = \begin{bmatrix} (q_{i3}^1 K_{i3} - q_{i2}^1 K_{i2}) J_{i1} + (q_{i1}^1 K_{i1} - q_{i3}^1 K_{i3}) J_{i2} + (q_{i2}^1 K_{i2} - q_{i1}^1 K_{i1}) J_{i3} \\ (q_{i3}^2 K_{i3} - q_{i2}^2 K_{i2}) J_{i1} + (q_{i1}^2 K_{i1} - q_{i3}^2 K_{i3}) J_{i2} + (q_{i2}^2 K_{i2} - q_{i1}^2 K_{i1}) J_{i3} \\ (q_{i3}^2 K_{i3} - q_{i2}^3 K_{i2}) J_{i1} + (q_{i1}^3 K_{i1} - q_{i3}^3 K_{i3}) J_{i2} + (q_{i2}^3 K_{i2} - q_{i1}^3 K_{i1}) J_{i3} \end{bmatrix},$$

where $K_{i\ell} = (\vec{\omega}_i \cdot \vec{q}_{i[\ell+1]})(\vec{\omega}_i \cdot \vec{q}_{i[\ell+2]})^{*)}$, $\ell=1,2,3$.

Introducing vector $\vec{\lambda}_{i\ell} = K_{i[\ell+1]}\vec{q}_{i[\ell+2]} - K_{i[\ell+1]}\vec{q}_{i[\ell+1]}$ one obtains:

$$\vec{\lambda}_i = \sum_{\ell=1}^{3} \vec{\lambda}_{i\ell} J_{i\ell}. \tag{3.3.13}$$

Substituting expressions (3.3.12) and (3.3.13) into (3.3.7):

$$\vec{b}_{ij} = -\sum_{\ell=1}^{3} Q_{i\ell} \vec{\alpha}_{ij} J_{i\ell}, \tag{3.3.14}$$

$$\vec{b}_i^{o} = \sum_{\ell=1}^{3} (Q_{i\ell}\vec{\theta}_i - \vec{\lambda}_{i\ell}) J_{i\ell}. \tag{3.3.15}$$

Taking into account moments of inertia variations in accordance to expression (3.3.10) one obtains [1]:

$$\begin{aligned} \vec{b}_{ij} &= \vec{b}_{ijo} - \sum_{\ell=1}^{3} Q_{i\ell} \vec{\alpha}_{ij} \Delta J_{i\ell} \\ \vec{b}_i^{o} &= \vec{b}_{io}^{o} - \sum_{\ell=1}^{3} (Q_{i\ell}\vec{\theta}_i - \vec{\lambda}_{i\ell}) \Delta J_{i\ell} \end{aligned} \tag{3.3.16}$$

where $\vec{b}_{ijo} = -\sum_{\ell=1}^{3} Q_{i\ell} \vec{\alpha}_{i\ell} J_{i\ell o}$ and $\vec{b}_{io}^{o} = -\sum_{\ell=1}^{3} (Q_{i\ell}\vec{\theta}_i - \vec{\lambda}_{i\ell}) J_{i\ell o}$ are nominal values of vectors $\vec{b}_{ij}$ and $\vec{b}_i^{o}$, respectively.

Having evaluated the components of vectors $\vec{a}_{ij}$, $\vec{a}_i^{o}$, $\vec{b}_{ij}$ and $\vec{b}_i^{o}$ and also their nominal values, from expression (3.3.4) inertial forces are obtained and expression (3.3.6) gives the moments of inertial forces acting upon the i-th link of the robot mechanism.

It is now possible to proceed to the forming of variations of elements in matrices H and h of dynamic equations (3.3.1) resulting from the variations of adopted dynamic parameters of the robot mechanism:

$$H_{ik} = -\vec{e}_i \cdot [\sum_{j=\max(i,k)}^{n} (\vec{b}_{jk} + \vec{r}_{ji} \times \vec{a}_{jk}) \bar{\xi}_i + \vec{a}_{jk} \xi_i], \tag{3.3.17}$$

---

*) $[\cdot]$ denotes the residue of the number in square brackets divided by 3: $[\ell] = \begin{cases} \ell & \ell \leq 3 \\ \ell-3 & \ell=4,5 \end{cases}$

$$h_i = [-\vec{e}_i \cdot \sum_{j=i}^{n} (\vec{r}_{ji} \times (\vec{a}_j^o + \vec{G}_j) + \vec{b}_j^o)\bar{\xi}_i + (\vec{a}_j^o + \vec{G}_j)\xi_i] \qquad (3.3.18)$$

All quantities that have appeared at various stages of forming the dynamic model (3.3.17) and (3.3.18) have been explained in detail in Chapter 2 and will thus not be repeated here. Elements of matrices H and h have been presented again for easier following of the text which will proceed to explain variations of these elements resulting from variations of dynamic parameters Δm_i and $\Delta J_{i\ell}$.

Components which characterize the influence of the i-th segment dynamic parameters on the dynamic behaviour of the whole system are easily recognisable on the right sides of equations (3.3.17) and (3.3.18). Substituting expressions (3.3.11) and (3.3.16) into (3.3.17) and (3.3.18) and segregating the nominal values in elements of matrices H_{jk} and h_j, one obtains [1]:

$$H_{jk} = H_{jko} + \sum_{\ell=1}^{3} \vec{e}_j \cdot Q_{i\ell}\vec{\alpha}_{ik}\bar{\xi}_j \eta_{ij}\Delta J_{i\ell} + \vec{e}_j[(\vec{r}_{ij} \times \vec{\beta}_{ik})\bar{\xi}_j + \vec{\beta}_{ik}\xi_j]\eta_{ij}\Delta m_i$$

$$h_j = h_{jo} + \{(\sum_{\ell=1}^{3} \vec{e}_j \cdot (Q_{i\ell}\vec{\theta}_i - \vec{\lambda}_{i\ell})\bar{\xi}_j \Delta J_{i\ell} + \vec{e}_j \cdot [(\vec{r}_{ij} \times (\vec{\eta}_i - \vec{g})\bar{\xi}_j +$$

$$+ (\vec{\eta}_i - \vec{g})\xi_j]\Delta m_i\}\eta_{ij}, \qquad (3.3.19)$$

where H_{jko} and h_{jo} are nominal values of elements of matrices H and h. Symbol $\eta_{ij}=1$ for $i \geq j$ and $\eta_{ij}=0$ otherwise.

Introduce the following notation:

$$H^i_{jk\ell} = \begin{cases} \vec{e}_j \cdot Q_{i\ell}\vec{\alpha}_{ik}\bar{\xi}_j \eta_{ij} & \ell=1,2,3 \\ \vec{e}_j \cdot [(\vec{r}_{ij} \times \vec{\beta}_{ik})\bar{\xi}_j + \vec{\beta}_{ik}\xi_j]\eta_{ij} & \ell=4 \end{cases}, \qquad (3.3.20)$$

$$h^i_{j\ell} = \begin{cases} \vec{e}_j(Q_{i\ell}\vec{\theta}_i - \vec{\lambda}_{i\ell})\bar{\xi}_j \eta_{ij} & \ell=1,2,3 \\ \vec{e}_j \cdot [(\vec{r}_{ij} \times (\vec{\eta}_i - \vec{g})\bar{\xi}_j + (\vec{\eta}_i - \vec{g})\xi_j]\eta_{ij} & \ell=4 \end{cases} \qquad (3.3.21)$$

Then, using expression (3.3.19) follows:

$$H_{jk} = H_{jko} + \sum_{\ell=1}^{n_{pi}} H^i_{jk\ell}\Delta\theta_{i\ell}, \quad h_j = h_{jo} + \sum_{\ell=1}^{n_{pi}} h^i_{j\ell}\Delta\theta_{i\ell}, \qquad (3.3.22)$$

where: $\Delta\theta_{i\ell} = \Delta J_{i\ell}$ $(\ell=1,2,3)$, $\Delta\theta_{i4} = \Delta m_i$, $n_{pi} = 4$.

Expressions (3.3.22) can be represented in matrix form:

$$H = H_o + \sum_{\ell=1}^{n_{pi}} H_\ell^i \Delta\theta_{i\ell},$$
$$h = h_o + \sum_{\ell=1}^{n_{pi}} h_\ell^i \Delta\theta_{i\ell}, \qquad (3.3.23)$$

where $H_o = H(q^o, \theta^o)$ and $h(q^o, \dot{q}^o, \theta^o)$ are matrices of model (3.3.1) with nominal values of parameters.

After substituting expression (3.3.23) into (3.3.1), one obtains that:

$$P = H(q^o, \theta^o)\ddot{q}^o + h(q^o, \dot{q}^o, \theta^o) + \sum_{\ell=1}^{n_{pj}} (H_\ell^i(q^o, \theta^o)\ddot{q}^o + h_\ell^i(q^o, \dot{q}^o, \theta^o))\Delta\theta_{i\ell}, \quad \text{or} \qquad (3.3.24)$$

$$\Delta P = \sum_{\ell=1}^{n_{pj}} S_\ell^i(q^o, \dot{q}^o, \ddot{q}^o, \theta^o)\Delta\theta_{i\ell}, \qquad (3.3.25)$$

where $S_\ell^i = H_\ell^i(q^o, \theta^o)\ddot{q}^o + h_\ell^i(q^o, \dot{q}^o, \theta^o)$.

From the last expression it follows that deviations of driving moments (forces) are linearly dependent on the limited parameter variations. Matrix S_ℓ^i is formed after linearizing matrices $H(q, \theta)$ and $h(q, \dot{q}, \theta)$ with respect to parameter θ.

Comparison of expressions (3.3.25) and (3.3.3) gives:

$$\left.\frac{\partial P}{\partial \theta_{i\ell}}\right|_{q=q^o(t)} = S_\ell^i(q^o, \dot{q}^o, \ddot{q}^o, \theta^o), \qquad (3.3.26)$$

for $\ell=1,\dots,n_{pi}$, $i=1,\dots,n$.

The method presented here holds for a general case when every mechanism segment is a rigid body of an arbitrary form. However, segments in the shape of a cane with diameter (or characteristic dimensions of the cross-section) much smaller than their length (see Appendix A.5) are often used. In simplified form these segments can be described by two moments of inertia, longitudinal (J_{is}) and transversal (J_{iN}). In this case vector coefficients of inertial forces can be represented in the form (see (2.23) for $\varepsilon_o=0$):

$$\vec{b}_{ij} = -J_{iN}(\vec{s}_i\times\vec{\alpha}_{ij})\times\vec{s}_i - J_{si}(\vec{\alpha}_{ij}\cdot\vec{s}_i)\vec{s}_i,$$
$$\vec{b}_i^o = -J_{iN}((\vec{s}_i\times\vec{\theta}_i)\times\vec{s}_i+\vec{\tau}_i) - J_{is}(\vec{\theta}_i\cdot\vec{s}_i)\vec{s}_i, \qquad (3.3.27)$$

and $\vec{\tau}_i$ is given by expression (2.18) i.e.:

$$\vec{\tau}_i = (\vec{\omega}_i\cdot\vec{s}_i)(\vec{s}_i\times\vec{\omega}_i), \qquad \vec{s}_i = \vec{r}_{ii}/|\vec{r}_{ii}|, \qquad i,j=1,\ldots,n$$

Assuming that:

$$m_i = m_{io}+\Delta m_i, \quad J_{iN} = J_{iNo}+\Delta J_{iN}, \quad J_{is} = J_{iso}+\Delta J_{is}, \qquad (3.3.28)$$

and substituting it into expression (3.3.27) we obtain,

$$\vec{b}_{ij} = \vec{b}_{ijo} - \Delta J_{iN}(\vec{s}_i\times\vec{\alpha}_{ij})\times\vec{s}_i - \Delta J_{is}(\vec{\alpha}_{ij}\cdot\vec{s}_i)\vec{s}_i,$$
$$\vec{b}_i^o = \vec{b}_{io}^o - \Delta J_{iN}((\vec{s}_i\times\vec{\theta}_i)\times\vec{s}_i+\vec{\tau}_i) - \Delta J_{is}(\vec{\theta}_i\cdot\vec{s}_i)\vec{s}_i, \qquad (3.3.29)$$

where $\vec{b}_{ijo}$ and $\vec{b}_{io}^o$ are nominal values of vectors $\vec{b}_{ij}$ and $\vec{b}_i^o$.

Denoting $\vec{b}_{ij1} = (\vec{s}_i\times\vec{\alpha}_{ij})\times\vec{s}_i$; $\vec{b}_{ij2} = (\vec{\alpha}_{ij}\cdot\vec{s}_i)\vec{s}_i$, $\vec{b}_{i1}^o = (\vec{s}_i\times\vec{\theta}_i)\times\vec{s}_i+\vec{\tau}_i$, $\vec{b}_{i2}^o = (\vec{\theta}_i\cdot\vec{s}_i)\vec{s}_i$, (3.3.19) acquire the form:

$$H_{jk} = H_{jko} + \vec{e}_j\cdot b_{ik1}\bar{\xi}_j\eta_{ij}\Delta J_{iN} + \vec{e}_j\cdot b_{ik2}\bar{\xi}_j\eta_{ij}\Delta J_{is} + $$
$$+ \vec{e}_j\cdot[(\vec{r}_{ij}\times\vec{\beta}_{ik})\bar{\xi}_j + \vec{\beta}_{ik}\xi_j]\eta_{ij}\Delta m_i,$$
$$(3.3.30)$$
$$h_j = h_{jo} + \vec{e}_j\cdot\vec{b}_{i1}^o\cdot\bar{\xi}_j\eta_{ij}\Delta J_{iN} + \vec{e}_j\cdot\vec{b}_{i2}^o\bar{\xi}_j\eta_{ij}\Delta J_{is} +$$
$$+ \vec{e}_j\cdot[(\vec{r}_{ij}\times(\vec{\eta}_i-\vec{g})\bar{\xi}_j + (\vec{\eta}_i-\vec{g})\xi_j]\eta_{ij}\Delta m_i.$$

The notations:

$$H_{jk\ell}^i = \begin{cases} \vec{e}_j\cdot\vec{b}_{ik\ell}\bar{\xi}_j\eta_{ij} & \ell=1,2 \\ \vec{e}_j\cdot[(\vec{r}_{ij}\times\vec{\beta}_{ik})\bar{\xi}_j+\vec{\beta}_{ik}\xi_j]\eta_{ij} & \ell=3 \end{cases},$$

$$h^j_{j\ell} = \begin{cases} \vec{e}_j \cdot \vec{b}^{\,o}_{i\ell} \bar{\xi}_j \eta_{ij} & \ell=1,2 \\ \vec{e}_j \cdot [(\vec{r}_{ij} \times (\vec{n}_i - \vec{g})\bar{\xi}_j + (\vec{n}_i - \vec{g})\xi_j]\eta_{ij}\Delta m_i & \ell=3 \end{cases} \tag{3.3.31}$$

give the elements of model matrices in the form (3.3.22) where $\eta_{pi}=3$. Using an analogous algorithm (expressions (3.3.23)-(3.3.26)) calculations for segments of an arbitrary form can be carried out.

The described algorithm is represented in block diagram form in Figs. 3.4. and 3.5.

Parameter sensitivity of the complete system

As was presented in Chapter 2 (2.32) and Appendix 6, mathematical models of actuators can be described by linear equations with time constant elements in canonical form

$$S^i: \quad \dot{x}^i = A^i(\theta)x^i + b^i(\theta)N(u^i) + f^i(\theta)P_i, \tag{3.3.32}$$

where x^i - state vector, $P_i = M_i^*$ - corresponding driving moment or force, u^i - i-th drive input, $N(u^i)$ - nonlinearity of the amplitude saturation type, θ - vector of parameters. In each subsystem S^i two components of vector x^i coincide with q^i and $\dot{q}^i$ - generalized coordinates in the dynamic model (3.3.1) of the mechanism S^M.

Actuators model (3.3.32) can be represented in matrix form for all d.o.f. of the mechanism:

$$S^A: \quad \dot{x} = A(\theta)x + F(\theta)P + B(\theta)N(u), \tag{3.3.33}$$

where $x = [x^{1T} \cdots x^{nT}]^T$, $N(u) = [N(u^1) \cdots N(u^n)]^T$, A, F and B are corresponding matrices of actuators system. Consider the complete manipulation system (2.35) consisting of the mechanical part S^M (3.3.1) and the appropriate system of actuators (3.3.33). Sensitivity matrix which describes the effect of parameter variations of the complete system on its dynamic behaviour, can be represented as $\left.\frac{\partial u}{\partial \theta_i}\right|_{q(t)=q^o(t)}$, (i=1,...,n) where u=u(t) - system control vector. In order to determine expression $\frac{\partial u}{\partial \theta_i}$ the model of actuators (3.3.33) should be used. Vector x in (3.3.33) is composed in the following form: $x = [q^T \,\vdots\, \dot{q}^T \,\vdots\, x_r^T]^T$. Let us introduce vector $x_q = [q^T \,\vdots\, \dot{q}^T]^T$ of mechanical part of the system

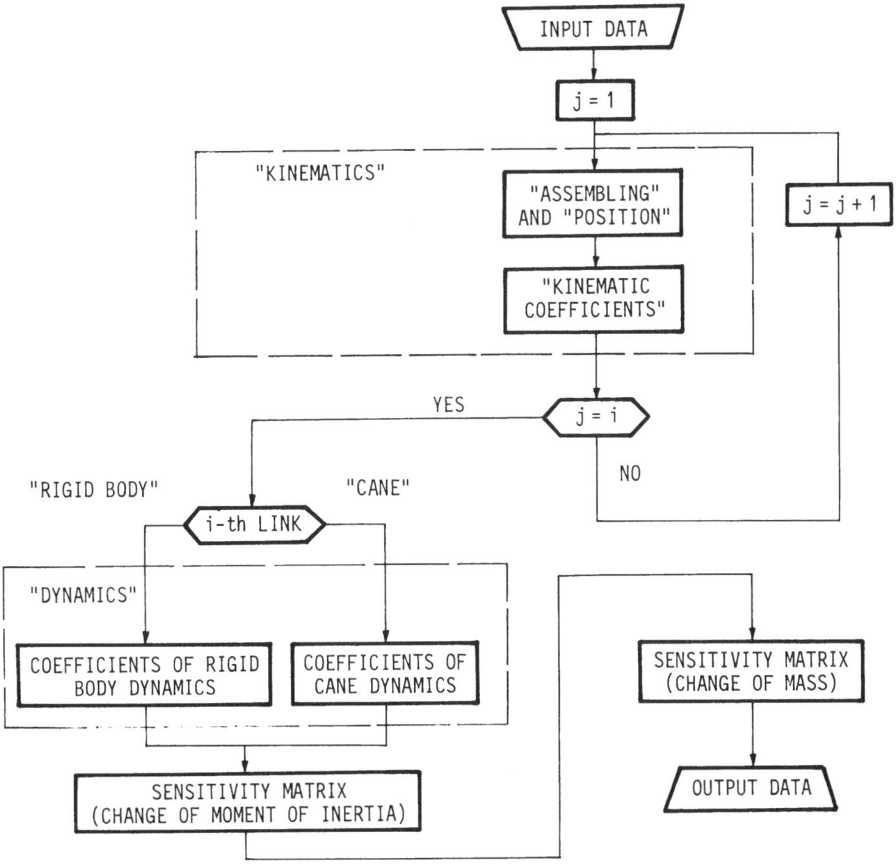

Fig. 3.4. Calculation of sensitivity matrix - global flow diagram

(S^M). In this case equation (3.3.33) can be represented in the form:

$$\begin{bmatrix} \dot{x}_q \\ \hline \dot{x}_r \end{bmatrix} = \left[\begin{array}{c|c} A_{q1} & A_{q2} \\ \hline A_{r1} & A_{r2} \end{array}\right] \begin{bmatrix} x_q \\ \hline x_r \end{bmatrix} + \begin{bmatrix} B_q \\ \hline B_r \end{bmatrix} u + \begin{bmatrix} F_q \\ \hline F_r \end{bmatrix} P. \quad (3.3.34)$$

It must be mentioned that matrix B_q is a zero matrix, $B_q = 0$, for a wide class of actuators (eq. electric d.c. servomotors, hydraulic drives). Differentiating (3.3.34) with respect to variables $\theta_{i\ell}$ ($\ell=1,\ldots$ $\ldots,n_{pi}$) under the condition:

$$\frac{\partial q}{\partial \theta_{i\ell}} = 0, \quad \text{i.e.} \quad \frac{\partial x_q}{\partial \theta_{i\ell}} = 0, \quad \text{we obtain:}$$

$\xi_j = \begin{cases} 1 & \text{j-th joint linear} \\ 0 & \text{j-th joint rotational} \end{cases}$

ℓ_j - length of j-th mechanism link

$\vec{\tilde{e}}_j$ - unit vector which coincides with the axis of the j-th joint in j-th reference system

$\vec{\tilde{r}}_{jk}$ - distance vector from k-th joint to center of mass of link in j-th reference system

$\vec{r}_{o1}$ - radius vector of the first link center of mass from the origin of the fixed coordinate system

$\vec{e}_1$ - unit vector which coincides with axis of the first joint in the fixed coordinate system

q^i - generalized coordinate describing angular or linear displacements in the i-th joint

$\dot{q}^i$ - generalized velocity in the i-th joint

i - number of link whose dynamic parameters are changing

$j = 1$

(1)

"Assembling"

$$\vec{r}_{Nj} = \vec{e}_j \times (\vec{r}_{j-1,j} \times \vec{e}_j),\ \vec{\tilde{r}}_{Nj} = \vec{\tilde{e}}_j \times (\vec{\tilde{r}}_{jj} \times \vec{\tilde{e}}_j),\ \vec{a}_j = -\vec{r}_{Nj}/|\vec{r}_{Nj}|,\ \vec{\tilde{a}}_j = \vec{\tilde{r}}_{Nj}/|\vec{\tilde{r}}_{Nj}|$$

$$\vec{b}_j = \vec{e}_j \times \vec{a}_j,\ \vec{\tilde{b}}_j = \vec{\tilde{e}}_j \times \vec{\tilde{a}}_j,\ Q_j^o = [\vec{q}_{j1}^{\,o}\ \vec{q}_{j2}^{\,o}\ \vec{q}_{j3}^{\,o}] = [\vec{a}_j\ \vec{e}_j\ \vec{b}_j][\vec{\tilde{a}}_j\ \vec{\tilde{e}}_j\ \vec{\tilde{b}}_j]^T$$

"Position"

$$\vec{q}_{jk} = [\vec{q}_{jk}^{\,o}\cos q_j + (1-\cos q_j)(\vec{e}_j \vec{q}_{jk}^{\,o})\vec{e}_j + \vec{e}_j \times \vec{q}_{jk}^{\,o}\sin q_j](1-\xi_j) + \vec{q}_{jk}^{\,o}\xi_j, \quad (k=1,2,3)$$

$$Q_j = [\vec{q}_{j1}\ \vec{q}_{j2}\ \vec{q}_{j3}],\ \vec{r}_{jj} = Q_j \vec{\tilde{r}}_{jj} + q_j \vec{e}_j \xi_j,\ \vec{r}_{j,j+1} = Q_j \vec{\tilde{r}}_{j,j+1},\ \vec{e}_{j+1} = Q_j \vec{\tilde{e}}_{j+1}$$

$$\vec{r}_{jk} = \vec{r}_{j-1,k} - \vec{r}_{j-1,j} + \vec{r}_{jj}, \qquad k=1,\dots,j-1$$

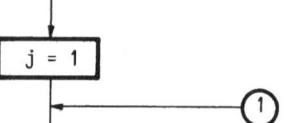

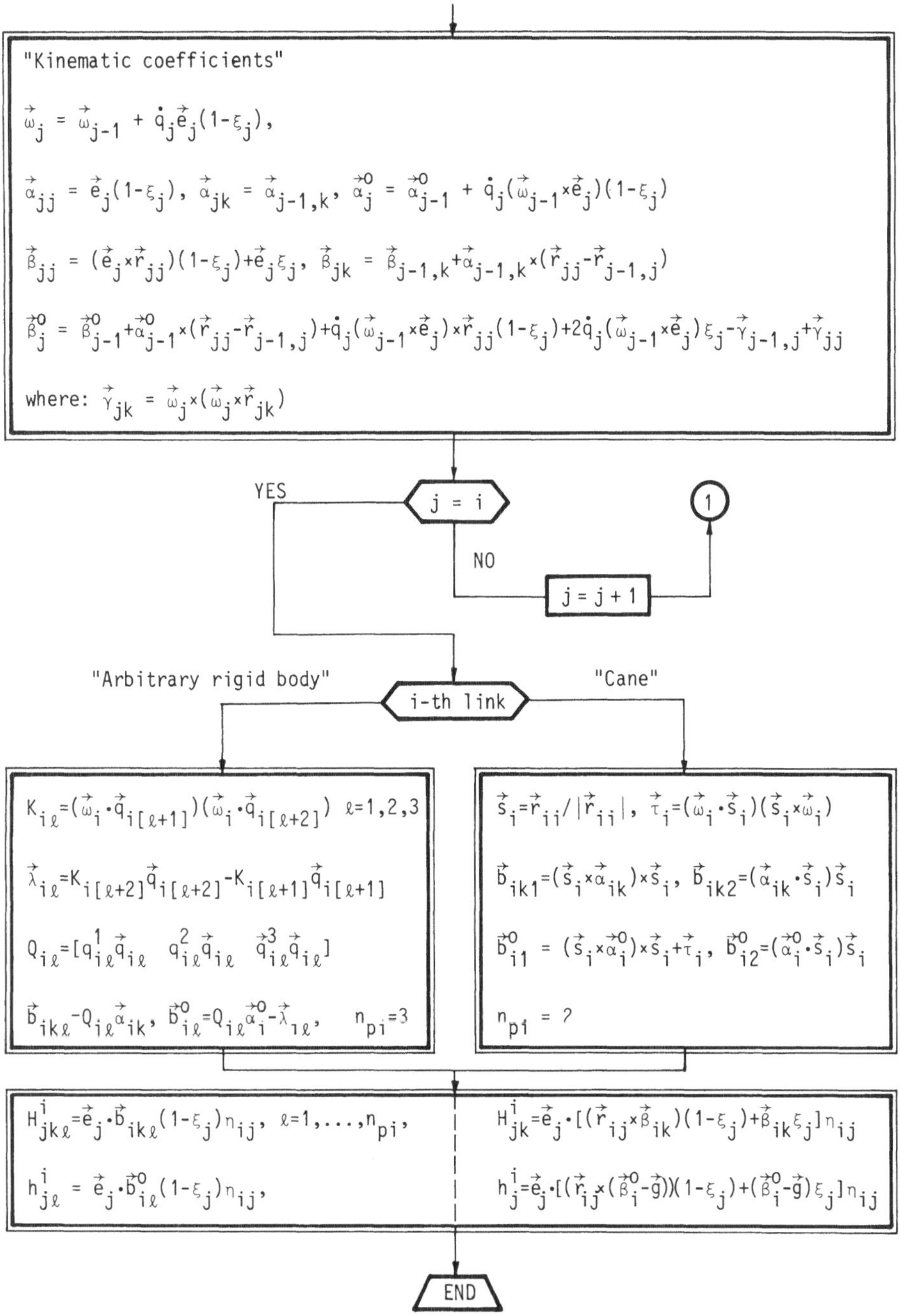

Fig. 3.5. Detailed calculation flow diagram

$$\begin{bmatrix} 0 \\ \text{------} \\ \dfrac{\partial \dot{x}_r}{\partial \theta_{i\ell}} \end{bmatrix} = \begin{bmatrix} A_{q1} & A_{q2} \\ A_{r1} & A_{r2} \end{bmatrix} \begin{bmatrix} 0 \\ \text{------} \\ \dfrac{\partial x_r}{\partial \theta_{i\ell}} \end{bmatrix} + \begin{bmatrix} 0 \\ \text{----} \\ B_r \end{bmatrix} \frac{\partial u}{\partial \theta_{i\ell}} + \begin{bmatrix} F_q \\ \text{----} \\ F_r \end{bmatrix} \frac{\partial P}{\partial \theta_{i\ell}} \tag{3.3.35}$$

Simple transformations give:

$$\frac{\partial u}{\partial \theta_{i\ell}} = \bar{B}_r \bar{A}_{q2} F_q \frac{\partial \dot{P}}{\partial \theta_{i\ell}} - \bar{B}_r (A_{r2} \bar{A}_{q2} F_q + F_r) \frac{\partial P}{\partial \theta_{i\ell}}, \tag{3.3.36}$$

where: $\bar{B}_r = (B_r^T B_r)^{-1} B_r^T$, $\bar{A}_{q2} = (A_{q2}^T A_{q2})^{-1} A_{q2}$.

If expression for $\frac{\partial P}{\partial \theta_{i\ell}}$, (3.3.26) is used, (3.3.36) becomes:

$$\left. \frac{\partial u}{\partial \theta_{i\ell}} \right|_{q=q_o(t)} = \bar{B}_r \bar{A}_{q2} F_q \dot{S}_\ell^i(t, \theta_o) - \bar{B}_r (A_{r2} \bar{A}_{q2} F_q + F_r) S_\ell^i(t, \theta_o), \tag{3.3.37}$$

where $S_\ell^i(t, \theta_o) \equiv S_\ell^i(q^o, \dot{q}^o, \ddot{q}^o, \theta_o)$.

For a special case when $x = x_q$, expression (3.3.37) is greatly simplified:

$$\left. \frac{\partial u}{\partial \theta_{i\ell}} \right|_{q=q^o(t)} = -(B^T B)^{-1} B^T F S_\ell^i(t, \theta_o). \tag{3.3.38}$$

Sensitivity functions of mechanism state variables

In this case it is required to determine $\left. \frac{\partial q}{\partial \theta} \right|_{P=P_o(t)}$. After introducing the states variables $\xi = [\xi_1^T \mid \xi_2^T]^T$, $\xi_1 = q$, $\xi_2 = \dot{q}$, model (3.3.1) can be transformed into canonical form:

$$\dot{\xi} = A_m(\xi, \theta) + B_m(\xi, \theta) P, \quad \xi(t_o) = \xi_o, \tag{3.3.39}$$

where $A_m(\xi, \theta) = [\xi_2^T \mid -(H^{-1} h)^T]^T$, $B_m(\xi, \theta) = [0 \mid (H^{-1})^T]^T$.

Sensitivity function $\frac{\partial \xi}{\partial \theta}$ max be obtained by partial differentiation of the left and right sides of equation (3.3.39) with respect to θ. If $A_M(\xi, \theta) + B_m(\xi, \theta) P$ is denoted by $f(\xi, \theta, P)$, and $\frac{\partial \xi}{\partial \theta}$ by $U(\theta)$, the sensitivity equation becomes [1]:

$$\dot{U}(\theta) = \frac{\partial f}{\partial \xi} U(\theta) + \frac{\partial f}{\partial \theta} \qquad (3.3.40)$$

Calculating matrix $\frac{\partial f}{\partial \xi}$ can be performed by computer linearization of the dynamic model, given in Section 3.2. In order to obtain matrix $\frac{\partial f}{\partial \theta}$ it is required to evaluate $\frac{\partial f}{\partial \theta_\ell}$, $(\ell=1,\ldots,\ell_{max})$ where θ_ℓ is the ℓ-th component of vector of parameters θ.

It is easily shown that:

$$\frac{\partial f}{\partial \theta_\ell} = \begin{bmatrix} 0 \\ \hline -\dfrac{\partial (H^{-1}h)}{\partial \theta_\ell} \end{bmatrix} + \begin{bmatrix} 0 \\ \hline \dfrac{\partial (H^{-1})}{\partial \theta_\ell} \end{bmatrix} P. \qquad (3.3.41)$$

On the other hand from expression (3.3.23) follows that model matrices (3.3.1) can be represented in the form:

$$\begin{aligned} H(\theta_\ell+\Delta\theta_\ell) &= H_o + H_\ell \Delta\theta_\ell , \\ h(\theta_\ell+\Delta\theta_\ell) &= h_o + h_\ell \Delta\theta_\ell , \end{aligned} \qquad (3.3.42)$$

where $H_\ell = \frac{\partial H}{\partial \theta_\ell}$ and $h_\ell = \frac{\partial h}{\partial \theta_\ell}$ are obtained from expression (3.3.20) and (3.3.21). Using expression (3.3.42) one obtains:

$$\frac{\partial H^{-1}}{\partial \theta_\ell} = -H_o^{-1} H_\ell H_o^{-1} \qquad (3.3.43)$$

Substituting (3.3.43) into expression (3.3.41) and taking into account that

$$\frac{\partial (H^{-1}h)}{\partial \theta_\ell} = \frac{\partial H^{-1}}{\partial \theta_\ell} h + H^{-1} \frac{\partial h}{\partial \theta_\ell},$$ it follows:

$$\frac{\partial f}{\partial \theta_\ell} = \begin{bmatrix} 0 \\ \hline H_o^{-1} H_\ell H_o^{-1} h - H^{-1} h_\ell \end{bmatrix} + \begin{bmatrix} 0 \\ \hline -H_o^{-1} H_\ell H_o^{-1} \end{bmatrix} P \qquad (3.3.44)$$

Integration of (3.3.40) gives sensitivity functions $\frac{\partial \xi}{\partial \theta}$. Because ξ is given in vector form $\xi = [q^T \; \dot{q}^T]^T$, it is also possible to obtain sensitivity functions $\frac{\partial q}{\partial \theta}$.

References

[1] Vukobratović M., Kirćanski N., "Computer Assisted Sensitivity Model Generation in Manipulation Robots Dynamics", Mechanism and Machine Theory, No. 1, 1984.

[2] Vukobratović M., D. Stokić, Control of Manipulation Robots: Theory and Application, Springer-Verlag, 1982.

[3] Vukobratović M., Stokić D. and Kirćanski N., Non-Adaptive and Adaptive Control of Manipulation Robots, Springer-Verlag, 1985.

[4] Medvedov V.S., Laskov A.G. and Yuschenko A.S., Control Systems of Manipulation Robots, (in Russian), "Nauka", Moscow, 1978.

[5] Vukobratović M., Kirćanski N., "Computer-Oriented Method for Linearization of Dynamic Models of Active Spatial Mechanisms", Mechanism and Machine Theory, Vol. 17, No. 1, 1982.

[6] Vukobratović M., Stokić D., Applied Control of Manipulation Robots: Analysis, Synthesis and Exercises, Springer-Verlag, 1989.

Appendix 1
Connection Between the Moving and Fixed System

The need for transformation of the values from the internal (moving) into the fixed (absolute) system lies in the requirement to present the kinematics and dynamics of robotic mechanisms in a unique, fixed system of coordinates (Fig. A.1.1).

Therefore let us consider two Cartesian coordinate systems: Oxyz and and O'x'y'z' (Fig. A.1.2). Vector $\vec{r}$ is expressed via its projections onto the axes of both coordinate systems as:

$$\vec{r} = r_x\vec{i} + r_y\vec{j} + r_z\vec{k} \qquad (Oxyz) \qquad (A.1.1)$$

$$\vec{\tilde{r}} = r_{x'}\vec{i}' + r_{y'}\vec{j}' + r_{z'}\vec{k}' \qquad (O'x'y'z') \qquad (A.1.2)$$

Vector $\vec{r}$ is designated by $\vec{\tilde{r}}$, as seen by an observer in the O'x'y'z' coordinate system. The vectors $\vec{i}$, $\vec{j}$, $\vec{k}$; $\vec{i}'$, $\vec{j}'$, $\vec{k}'$ are unit vectors, associated to the corresponding coordinate axes.

Let us express the projections r_x, r_y, r_z via the data of vector $\vec{\tilde{r}}$ in O'x'y'z'. By scalar multiplication of (A.1.1) and (A.1.2) by $\vec{i}$, we obtain:

$$r_x\vec{i}\cdot\vec{i} + r_y\vec{j}\cdot\vec{i} + r_z\vec{k}\cdot\vec{i} = r_{x'}\vec{i}'\cdot\vec{i} + r_{y'}\vec{j}'\cdot\vec{i} + r_{z'}\vec{k}'\cdot\vec{i} \qquad (A.1.3)$$

Using the definition of the dot product of two vectors, and taking into account that $\vec{i}\cdot\vec{i} = 1$, $\vec{j}\cdot\vec{i} = 0$ and $\vec{k}\cdot\vec{i} = 0$, expression (A.1.3) becomes:

$$r_x = r_{x'}\vec{i}'\cdot\vec{i} + r_{y'}\vec{j}'\cdot\vec{i} + r_{z'}\vec{k}'\cdot\vec{i} \qquad (A.1.3a)$$

By analogous multiplication by $\vec{j}$ and $\vec{k}$, we obtain for r_y and r_z:

$$r_y = r_{x'}\vec{i}'\cdot\vec{j} + r_{y'}\vec{j}'\cdot\vec{j} + r_{z'}\vec{k}'\cdot\vec{j} \qquad (A.1.3b)$$

$$r_z = r_{x'}\vec{i}'\cdot\vec{k} + r_{y'}\vec{j}'\cdot\vec{k} + r_{z'}\vec{k}'\cdot\vec{k} \qquad (A.1.3c)$$

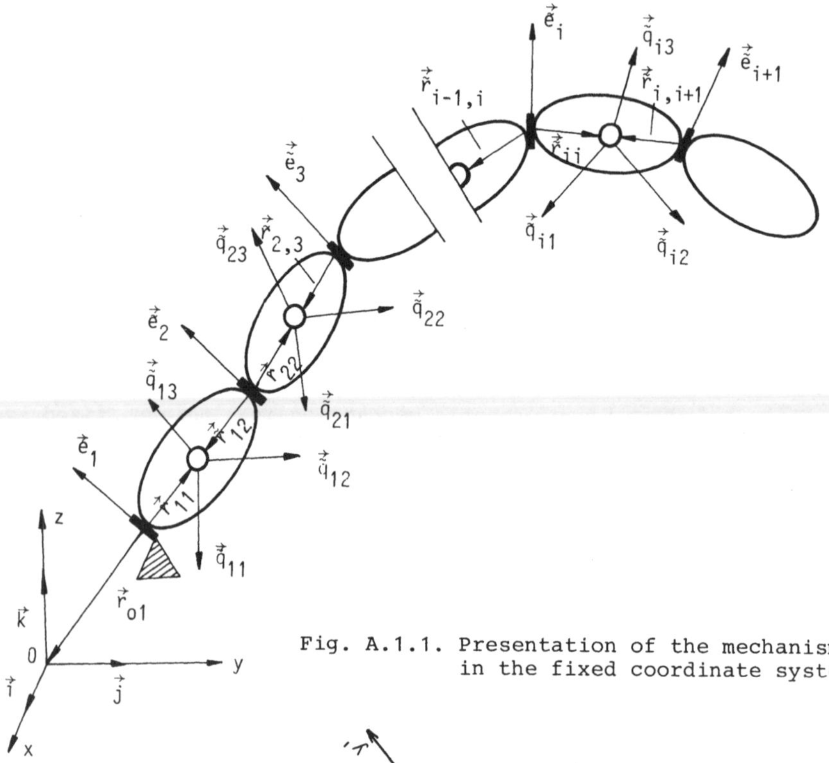

Fig. A.1.1. Presentation of the mechanism in the fixed coordinate system

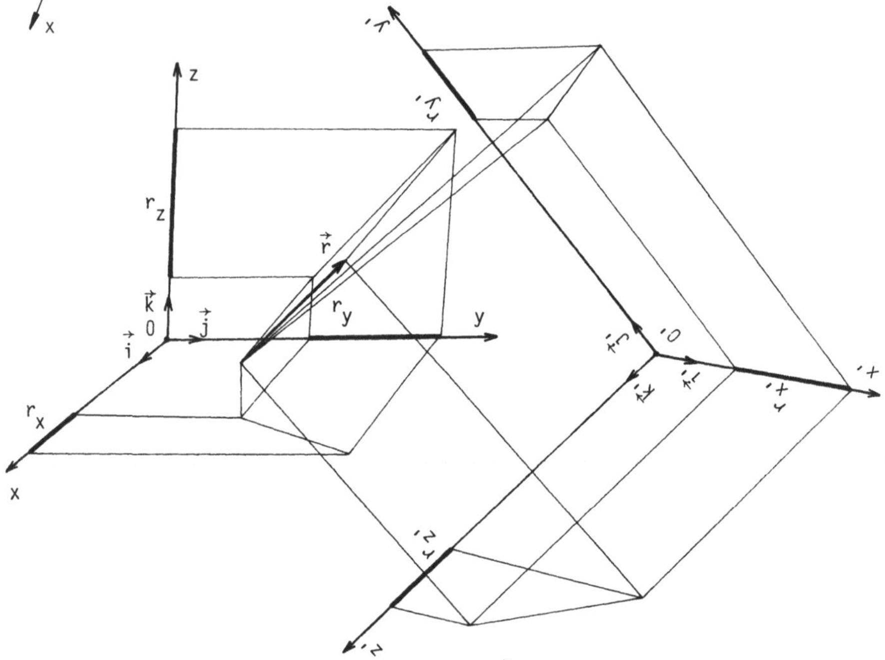

Fig. A.1.2. Presentation of vector $\vec{r}$ in two coordinate systems

Using matrix notation, expressions for r_x, r_y, and r_z can be written in the form:

$$\begin{bmatrix} r_x \\ r_y \\ r_z \end{bmatrix} = \begin{bmatrix} \cos(\vec{i}'\vec{i}) & \cos(\vec{j}'\vec{i}) & \cos(\vec{k}'\vec{i}) \\ \cos(\vec{i}'\vec{j}) & \cos(\vec{j}'\vec{j}) & \cos(\vec{k}'\vec{j}) \\ \cos(\vec{i}'\vec{k}) & \cos(\vec{j}'\vec{k}) & \cos(\vec{k}'\vec{k}) \end{bmatrix} \begin{bmatrix} r_{x'} \\ r_{y'} \\ r_{z'} \end{bmatrix} \qquad \text{(A.1.3d)}$$

or $\vec{r} = Q\vec{\tilde{r}}$,

where Q is the (3×3) cosine transformation matrix, by which the observer from coordinate system Oxyz has to multiply the projections of vector $\vec{\tilde{r}}$ in system O'x'y'z' in order to obtain the values of the vector $\vec{r}$ projections in Oxyz. The stated relations perform connections of the data by which one vector is described in two different coordinate systems.

The columns of matrix Q represent the projections of the unit vectors of the O'x'y'z' system onto the axes of system Oxyz. In order to determine the elements of this matrix of dimensions 3×3, it is evidently necessary to have nine linearly independent equations with respect to q_{ij}; $i,j \in \{1,2,3\}$. We have already seen that knowing the values of one vector's projections in both coordinate systems enables forming three linearly independent equations with respect to q_{ij} (A.1.3d). Consequently, for forming all nine equations with respect to q_{ij} it is necessary to know the values of the projections of three non-colinear, non--zero vectors in both coordinate systems. In that case the system determinant of linear equations w.r. to q_{ij} is different from zero, which enables calculation of the transformation matrix Q (see expr. 2.1 - 2.4).

Let us return to our task of determining the transformation matrices of the robot mechanism. Let us consider Figs. 2.5 and A.1.1. The resting two vectors which are not mutually and also related to the unit vectors of the joint axes $\vec{e}_i$, colinear, are $\vec{a}_i$ and $\vec{b}_i = (\vec{e}_i \times \vec{a}_i)$. From Fig. 2.5 is evident that only for the case $q^i = 0$ the identity $\vec{a}_i \equiv \vec{\tilde{a}}_i$ holds, when it is possible to perform the transformation of the i-th segment coordinate system into the fixed system. Namely, then the vector trihedron $\{\vec{e}_i, \vec{a}_i, \vec{b}_i\}$ and $\{\vec{\tilde{e}}_i, \vec{\tilde{a}}_i, \vec{\tilde{b}}_i\}$ are coinciding in space, so there exist only three non-colinear vectors, the position of which is known in both coordinate systems.

Explanation why the noncolinear unit vectors $\vec{a}$ and $\vec{b}$ were chosen as the other two, lies in the following. Namely, in the general case vectors $\vec{r}_{i-1,i}$ and $\vec{r}_{ii}$ possess no equal modules, so if vector trihedrons of following type were formed:

$$\{\vec{e}_i,\ [-\vec{e}_i\times(\vec{r}_{i-1,i}\times\vec{e}_i)],\ \vec{e}_i\times[-\vec{e}_i\times(\vec{r}_{i-1,i}\times\vec{e}_i)]\},$$

$$\{\tilde{\vec{e}}_i,\ [\tilde{\vec{e}}_i\times(\tilde{\vec{r}}_{ii}\times\tilde{\vec{e}}_i)],\ \tilde{\vec{e}}_i\times[\tilde{\vec{e}}_i\times(\tilde{\vec{r}}_{ii}\times\tilde{\vec{e}}_i)]\},$$

the condition could not be attained that with $q^i = 0$ the corresponding vectors are of same intensity, which is an indispensable prerequisite for calculating the transformation matrix Q_i^o. The stated trihedrons could only coincide in space, which is a necessary, but not sufficient condition for determining the elements of matrix Q_i^o.

The system of equations (2.1) can be represented in expanded form as:

$$
\begin{aligned}
&q_{11}^o\tilde{e}_{i,1} + q_{12}^o\tilde{e}_{i,2} + q_{13}^o\tilde{e}_{i,3} = e_{i,x}\\
&q_{12}^o\tilde{e}_{i,1} + q_{22}^o\tilde{e}_{i,2} + q_{23}^o\tilde{e}_{i,3} = e_{i,y}\\
&q_{31}^o\tilde{e}_{i,1} + q_{32}^o\tilde{e}_{i,2} + q_{33}^o\tilde{e}_{i,3} = e_{i,z}\\
&q_{11}^o\tilde{a}_{i,1} + q_{12}^o\tilde{a}_{i,2} + q_{13}^o\tilde{a}_{i,3} = a_{i,x}\\
&\text{-------------------------------------}\\
&q_{31}^o\tilde{b}_{i,1} + q_{32}^o\tilde{b}_{i,2} + q_{33}\tilde{b}_{i,3} = b_{i,z}
\end{aligned}
\qquad (A.1.4)
$$

where $q_{k,\ell}^o$, $k,\ell\in\{1,2,3\}$ are the elements of matrix Q_i^o; $\tilde{e}_{i,(1,2,3)}$, $\tilde{a}_{i,(1,2,3)}$ and $\tilde{b}_{i,(1,2,3)}$ are the projections of vectors $\tilde{\vec{e}}_i$, $\tilde{\vec{a}}_i$ and $\tilde{\vec{b}}_i$ onto the axes of coordinate system of the i-th segment, $e_{i,(x,y,z)}$, $a_{i,(x,y,z)}$ and $b_{i,(x,y,z)}$ are the projections of vectors $\vec{e}_i$, $\vec{a}_i$, and $\vec{b}_i$ onto the axes of the fixed coordinate system. By grouping the equations with respect to the unknowns $q_{k,1}^o$, $q_{k,2}^o$, $q_{k,3}^o$, $k\in\{1,2,3\}$, we obtain three systems of linear algebraic equations with three unknowns. By applying Kramer's rule we obtain, for instance for q_{11}^o, the following value:

$$q_{11}^{o} = \frac{\begin{vmatrix} e_{i,x} & \tilde{e}_{i,2} & \tilde{e}_{i,3} \\ a_{i,x} & \tilde{a}_{i,2} & \tilde{a}_{i,3} \\ b_{i,x} & \tilde{b}_{i,2} & \tilde{b}_{i,3} \end{vmatrix}}{\Delta}\ ; \qquad \Delta = \begin{vmatrix} \tilde{e}_{i,1} & \tilde{e}_{i,2} & \tilde{e}_{i,3} \\ \tilde{a}_{i,1} & \tilde{a}_{i,2} & \tilde{a}_{i,3} \\ \tilde{b}_{i,1} & \tilde{b}_{i,2} & \tilde{b}_{i,3} \end{vmatrix} .$$

By generalizing the solution of system (A.1.4), the expanded matrix of the equations system is of the form:

$$B = \begin{bmatrix} \tilde{e}_{i,1}\tilde{e}_{i,2}\tilde{e}_{i,3} & 0 & 0 & e_{i,x} \\ 0 & \tilde{e}_{i,1}\tilde{e}_{i,2}\tilde{e}_{i,3} & 0 & e_{i,y} \\ 0 & 0 & \tilde{e}_{i,1}\tilde{e}_{i,2}\tilde{e}_{i,3} & e_{i,z} \\ \tilde{a}_{i,1}\tilde{a}_{i,2}\tilde{a}_{i,3} & 0 & 0 & a_{i,x} \\ 0 & \tilde{a}_{i,1}\tilde{a}_{i,2}\tilde{a}_{i,3} & 0 & a_{i,y} \\ 0 & 0 & \tilde{a}_{i,1}\tilde{a}_{i,2}\tilde{a}_{i,3} & a_{i,z} \\ \tilde{b}_{i,1}\tilde{b}_{i,2}\tilde{b}_{i,3} & 0 & 0 & b_{i,x} \\ 0 & \tilde{b}_{i,1}\tilde{b}_{i,2}\tilde{b}_{i,3} & 0 & b_{i,y} \\ 0 & 0 & \tilde{b}_{i,1}\tilde{b}_{i,2}\tilde{b}_{i,3} & b_{i,z} \end{bmatrix}$$

Let us note that from matrix B the equations of type (A.1.4) can be written directly by selecting the corresponding column.

In order to start the procedure of forming the transformation matrix, it is necessary to know the projections of the fixed vector $\vec{r}_{01}$ in the fixed system (Fig. A.1.1), connecting the coordinate origin to the first mechanism joint. It is also necessary to know the projections of vector $\vec{e}$ in the fixed system and the system, connected to the mass center of the first segment. It should be mentioned that upper index "zero" in the transformation matrix denotes the case of mechanism assembly, when all the relative angular coordinates (q^i) are equal zero (initial configuration of the robot mechanism).

The transformation matrices are calculated in such way, that in each iteration the following segment is added to the robot chain and the corresponding transformation matrix is calculated in recurrent way. Thus Q_i^o is calculated when adding the i-th segment to the chain, whereby Q_{i-1}^o has already been calculated. With Q_{i-1}^o known, $\vec{r}_{i-1,i} = Q_{i-1}^o \tilde{\vec{r}}_{i-1,i}$ and $\vec{e}_i = Q_{i-1}^o \tilde{\vec{e}}_i$ are also known (in general case $\vec{e}_i = Q_{i-1}^o \underset{\sim}{\vec{e}}_i$).

Remark 1: *Non-parallelism of local coordinate systems*

Vectors $\vec{r}_{ii}$ and $\vec{r}_{i,i+1}$ defining the position of joints with respect to the center of mass of the mechanism members are properties of the member itself, so if expressed in the i-th member connected system they are constant, i.e. $\tilde{\vec{r}}_{ii}$ and $\tilde{\vec{r}}_{i,i+1}$ are constant vectors. Furtheron, the rotation axes or translations have a fixed position with respect to the corresponding members, thus the vector $\vec{e}_i$ is constant if expressed in the i-th or (i-1)st member system, i.e. $\tilde{\vec{e}}_i$ and $\underset{\sim}{\vec{e}}_i$ (unit vector of the i-th joint axis in the local coordinate system of the (i-1)st member) are constant vectors. Hence vectors $\tilde{\vec{r}}_{ii}$, $\tilde{\vec{r}}_{i,i+1}$, $\tilde{\vec{e}}_i$, $\underset{\sim}{\vec{e}}_i$, defining the geometry of the members and joints must be given for all the members and mechanism joints.

In that way relation $\vec{e}_{i+1}^{\,o} = Q_i^o \tilde{\vec{e}}_{i+1}$ (2.3) holds only for the case when the local coordinate systems are mutually parallel, so general relation $\vec{e}_{i+1}^{\,o} = Q_i^o \underset{\sim}{\vec{e}}_{i+1}$ is more adequate.

Remark 2: *Colinearity of vectors*

It is necessary to consider the case of vector colinearity (i.e. $\vec{r}_{ii} || \vec{e}_i$ and $\vec{r}_{i-1,i} || \vec{e}_i$). If $\vec{r}_{i-1,i} || \vec{e}_i$, then it is said that member i-1 has "specificity" on upper side and in that case it is necessary to define on that member vector $\vec{r}^{\,A}_{i-1,i}$, perpendicular to $\vec{e}_i$ ($\vec{r}^{\,A}_{i-1,i} \perp \vec{e}_i$) which is used instead of $\vec{r}_{i-1,i}$ when defining the generalized coordinates q^i (Fig. A.1.3).

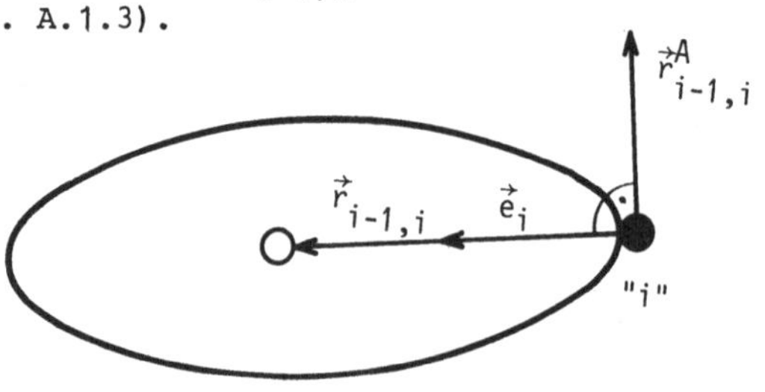

Fig. A.1.3. Specificity of (i-1)st member on upper side

If $\vec{r}_{ii} || \vec{e}_i$, then can be said that the i-th member is "specific" on the lower side and then the unit vector $\vec{r}^A_{ii} \perp \vec{e}_i$ is defined on the member, which is then used instead of $\vec{r}_{ii}$ (Fig. A.1.4).

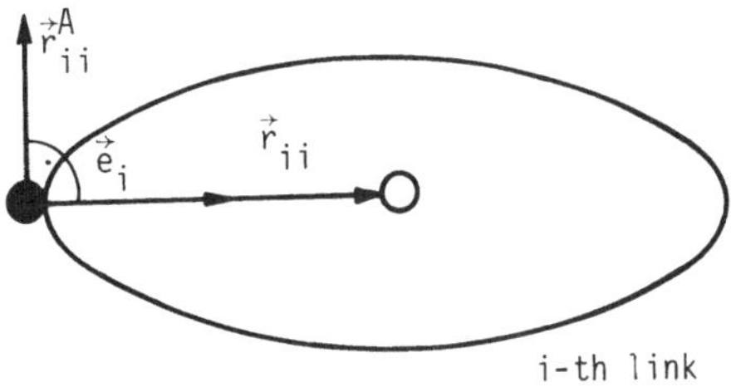

Fig. A.1.4. Specificity on lower side

For the purpose of the efficient application of the proposed algorithm, the problem of specifities can be solved by using the following practical rule: If the specificity of the i-th link occurs on the lower side define the unit vector $\vec{r}^A_{ii}$ in the positive direction of x-axis of the corresponding coordinate system. However, if vectors $\vec{r}_{ii}$ and $\vec{e}_i$ are colinear with x-axis direction, then define $\vec{r}^A_{ii}$ in the positive y-axis direction of the corresponding coordinate system; in the case of the specificity of the (i-1)th link on the upper side, define the unit vector $\vec{r}^A_{i-1,i}$ in the negative x-axis direction or if the vectors $\vec{e}_i$ and $\vec{r}_{i-1,i}$ are colinear with x-axis direction, then define the vector $\vec{r}^A_{i-1,i}$ in the negative y-axis direction. The existence of specificities are solved in the program NOMDYN (Appendix 9) by examining automatically the colinearity and putting corresponding supplementary vectors into the algorithm itself.

[1] Vukobratović M., Potkonjak V., Scientific Fundamentals of Robotic 1: Dynamics of Manipulation Robots, Springer-Verlag, 1982.

[2] Popov E.P., Vereschagin A.F., Zenkevich S.A., Manipulation Robots: Dynamics and Algorithms, (in Russian), "Nauka", Moscow, 1978.

[3] Vukobratović M. (Ed.) Introduction to Robotics, Springer-Verlag, 1988.

Apppendix 2
Manipulator Kinematical Model

Determining the orientation of the manipulator mechanism tip, with respect to a fixed, orthogonal coordinate system has been dealt with in Chapter 2 within the process of computer aided dynamic equation forming, where the derivation of kinematic relations represents a stage in the generation of mechanism dynamics.

In this Appendix, a procedure for describing a simple, open kinematic chain according to the Denavit-Hartenberg approach [1], will be presented.

(a) *Revolute joint* (Fig. A.2.1)

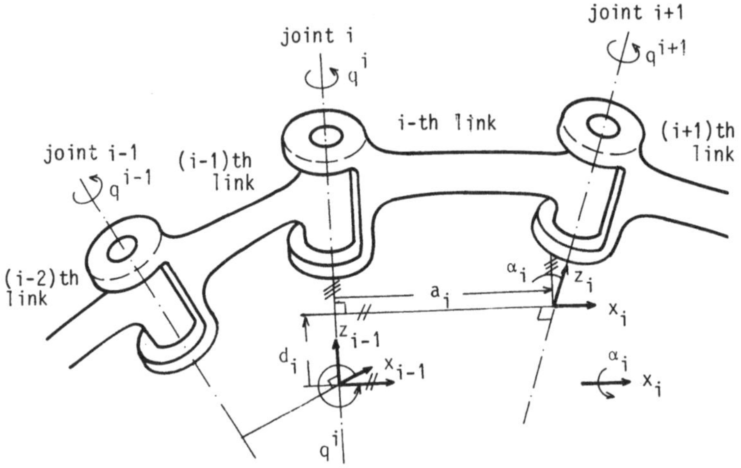

Fig. A.2.1. Denavit-Hartenberg parameters for a revolute joint

The coordinate system $O_i x_i y_i z_i$ for the i-th link in the kinematic chain is adopted such, that axis z_i of the chain coincides with the axis of the (i+1)th joint of the chain, axis x_i is directed along the normal, common to axes of the i-th and (i+1)-th joint, while the coordinate system centre is at the point of intersection of the normal and the axis of (i+1)-th joint. For cases where the neighbouring axes of joints intersect, their point of intersection is taken as the centre. If axes are parallel a centre is chosen such, that the distance along the joint axis to the subsequent common normal, which has been defined, is equal to zero. When the joint axes intersect, the sense of the x axis is taken as that of $\vec{z}_{i-1} \times \vec{z}_i$.

Parameters which characterize the i-th link are:

- a_i - length of the common normal
- α_i - angle formed by axes $\vec{z}_{i-1}$ and $\vec{z}_i$
- d_i - distance between the common normals a_{i-1} and a_i.

The generalized coordinate of the i-th joint is q^i - the angle formed by the common normals a_{i-1} and a_i, measured in a plane normal to the axis of the i-th joint.

(b) *Prismatic joint* (Fig. A.2.2).

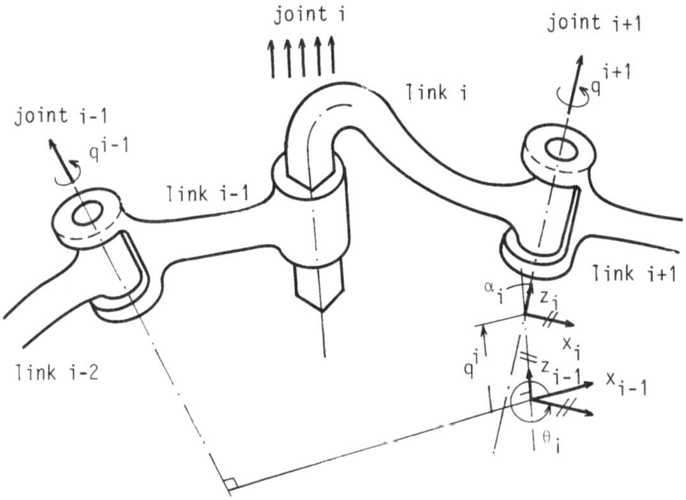

Fig. A.2.2. Denavit-Hartenberg parameters for a prismatic joint

Here, the generalized joint coordinate is the distance d_i. Length a_i is insignificant and is therefore adopted to be zero. The center of the coordinate system for a translational joint is adopted as coinciding with the centre of the subsequent system which has been defined (at joint coordinate $d_i=0$).

Irrespective of whether the kinematic joint is revolute or prismatic, the transformation of the i-th into the (i-1)-th coordinate system can be described by a sequence of successive transformations:

q^i - rotation around $\vec{z}_{i-1}$,

d_i - translation along $\vec{z}_{i-1}$,

a_i translation along the rotated axis $\vec{x}_{i-1}=\vec{x}_i$ and

α_i - rotation around $\vec{x}_i$.

These transformations can be described by the (4×4) homogenous matrix transformations in which the upper (3×3) submatrix describes a rotation of the (i-1)-th with respect to the i-th system and the first three components of the last column represent the translatory displacement of the i-th system centre with respect to the centre of the (i-1)-th system. Thus, the mentioned sequence of rotations and translations gives a transformation matrix of the form:

$$Q^i_{i-1} = \mathrm{Rot}(\vec{z}_{i-1}, q^i) \cdot \mathrm{Trans}(0, 0, d_i) \cdot \mathrm{Trans}(a_i, 0, 0) \cdot \mathrm{Rot}(\vec{x}_i, \alpha_i) \tag{A.2.1}$$

These transformation matrices are:

$$Q^i_{i-1} = \begin{bmatrix} \cos q^i & -\sin q^i & 0 & 0 \\ \sin q^i & \cos q^i & 0 & 0 \\ 0 & 0 & 1 & 0 \\ 0 & 0 & 0 & 1 \end{bmatrix} \begin{bmatrix} 1 & 0 & 0 & a_i \\ 0 & 1 & 0 & 0 \\ 0 & 0 & 1 & d_i \\ 0 & 0 & 0 & 1 \end{bmatrix} \begin{bmatrix} 1 & 0 & 0 & 0 \\ 0 & \cos\alpha_i & -\sin\alpha_i & 0 \\ 0 & \sin\alpha_i & \cos\alpha_i & 0 \\ 0 & 0 & 0 & 1 \end{bmatrix} \tag{A.2.2}$$

Multiplying out gives:

$$Q^i_{i-1} = \begin{bmatrix} \cos q^i & -\sin q^i \cos\alpha_i & \sin q^i \sin\alpha_i & a_i \cos q^i \\ \sin q^i & \cos q^i \cos\alpha_i & -\cos q^i \sin\alpha_i & a_i \sin q^i \\ 0 & \sin\alpha_i & \cos{}_i & d_i \\ 0 & 0 & 0 & 1 \end{bmatrix} \tag{A.2.3}$$

In the case of a prismatic kinematic pair this matrix takes the form:

$$Q^i_{i-1} = \begin{bmatrix} \cos q^i & -\sin q^i \cos q^i & \sin q^i \sin\alpha_i & 0 \\ \sin q^i & \cos q^i \cos\alpha^i & -\cos q^i \sin\alpha_i & 0 \\ 0 & \sin\alpha_i & \cos{}_i & d_i \\ 0 & 0 & 0 & 1 \end{bmatrix} \tag{A.2.4}$$

Transformation between the n-th and the fixed reference system is obtained as:

$$Q_o^n = Q_o^1 \, Q_1^2 \, \cdots \, Q_{n-1}^n \qquad \text{(A.2.5)}$$

The last column of matrix Q_o^n represents manipulator tip position with respect to the fixed coordinate system. The form of these transformation matrices is relatively simple and is applicable for solving in analytic form, both direct and inverse problems of simpler kinematic mechanism structures. Further simplification of matrices is possible in cases where $\alpha_i = \pi/2$ or 0, which is quite frequent with industrial robotic mechanisms.

From the standpoint of kinematic model complexity, Denavit-Hartenberg approach is more suitable than Rodrigue's approach because it provides a simpler transformation matrix. On the other hand, it is clear that the axes of the local link coordinate systems in this approach do not have to coincide with the main axis of the inertia of the link. Furthermore these axes do not represent the central inertial axes (they are not placed in the centre of mass). As a consequence, a more complex dynamic model is obtained because the inertia of the robot link cannot be described completely with only three moments of inertia, so that the complete inertia tensor has to be used.

Amongst the robotic manipulators a frequent case is that at least one link is not completely symmetric, so that the main inertial axes are rotated with respect to the normal, common to the two joints. If it was never-the-less adopted, during the formation of the dynamic model, that one of the main inertial axes coincides with the common normal, an error would occur in the dynamic model which is dependent on mass distribution within the link. These problems are not encountered when using the Rodrigues' theorem though the kinematic model thereby obtained is somewhat more complex. However, in cases where Denavit-Hartenberg coordinate systems also represent the central inertial axes of the link, the use of these coordinates in the dynamic model reduces the number of computational operations required and thus presents a better alternative.

Correlation between Denavit-Hartenberg coordinates and Rodrigue's formula of finite rotations [3]

The two fundamental approaches to the kinematic modelling of active spatial chains are the Denavit-Hartenberg approach and the approach based on Rodrigue's formula of finite rotations, which is developed in the context of the general approach for forming dynamic models of active spatial mechanisms. The basic difference between these two approaches is the result of a different way of adopting the coordinate systems referenced to chain segments (links) and parameters which define the segments (links).

The Denavit-Hartenberg approach has already been presented in this appendix.

In contrast to Denavit-Hartenberg kinematic modelling, the approach using Rodrigue's formula has been developed as a stage in the process of dynamic modelling of active spatial mechanisms. In its basic form (i.e., when performing calculations of transformation matrices and all vectors directly, referencing to the fixed system) this approach is given in Chapter 2, stage 2 (mechanism position) and stage 3 (mechanism kinematics) of forming the dynamic model of open chain manipulation mechanisms.

In order to establish a comparison between the two approaches it is necessary to apply Rodrigue's formula to the case when transformation matrices are calculated beetween two adjoining links.

Transformation matrices Q_i can be represented as successive transformations of the i-th segment coordinate system into (i-1)-th, Q^i_{i-1}, from (i-1)-th into (i-2)-th, Q^{i-1}_{i-2}, e.t.c, from 1st into fixed system Q^1_o, i.e.

$$Q_i \triangleq Q^i_o = Q^1_o Q^2_1 \cdots Q^i_{i-1} = Q^{i-1}_o Q^i_{i-1} \tag{A.2.6}$$

where Q^i_{i-1} is a (3×3) matrix which transforms every vector expressed in $O_i(\vec{q}_{i1}, \vec{q}_{i2}, \vec{q}_{i3})$ coordinate system into the coordinate system the axes of which are parallel to those of $O_{i-1}(\vec{q}_{i-1,1}, \vec{q}_{i-1,2}, \vec{q}_{i-1,3})$ system but the origin of which coincides with O_i (elementary rotation of the coordinate system i in relation to system i-1). Coordinate system referenced to the segments are orthonormal so that:

$$(Q^i_{i-1})^{-1} = (Q^i_{i-1})^T = Q^{i-1}_i \tag{A.2.7}$$

i.e. $Q_i^o = Q_i^{i-1} \cdots Q_1^o$, transformation matrix from the absolute coordinate system into the system of i-th mechanism segment. Matrices $Q_{io}(q^i=0)$ are calculated and memorised, and $Q_{io}^{i-1}(q^i=0)$ are calculated using (A.2.6) and (A.2.7): $Q_{i-1}^i(q^i=0) = (Q_{i-1,o})^{-1}Q_{io}$, $\forall i\in I$. We assume that in the previous step, using (2.4) matrix Q_{i-1}, $i\in I$ has been calculated. Using the memorised initial matrices*) $Q_{i-1,o}$, matrix Q_{io} is calculated after assembling $(q^j \neq 0,\ j=1,\ldots,i-1)$ using (A.2.6) i.e. $Q_{io}(q^j\neq 0,\ j=1,\ldots,i-1,\ q^i=0) = Q_{i-1}(q^j\neq 0,\ j=1,\ldots,i-1)Q_{i-1}^i(q^i=0)$.

A rotation for $q^i\neq 0$ is performed now, using relation (2.4). Thus, instead of doing all calculations in the block 2 (Fig. 3.3) for each joint every time the model is formed, it is sufficient to perform the assembling only once, for joint coordinates $q^i=0$, $\forall i\in I$. Then, when recursive calculations are being performed for $q^i\neq 0$, $i=1,\ldots,n$, using the memorised values of Q_{io}, $Q_{i-1,o}^i$ and matrix Q_{i-1} that has already been calculated for the previous segment (link), calculations of the type (A.2.6) are applied. Thus, the recurrent relations (2.6) can be replaced by analogous relations [4]:

$$Q_i^o\vec{\omega}_i = Q_i^{i-1}[Q_{i-1}^o\vec{\omega}_{i-1} + \vec{\tilde{e}}_i\dot{q}^i\bar{\xi}_i] \tag{A.2.8}$$

where relation $\vec{e}_{i+1} = Q_o^i\vec{\tilde{e}}_{i+1}$ and relations (A.2.6) and (A.2.7) have been used. Additionally:

$$Q_i^o\vec{\varepsilon}_i = Q_i^{i-1}[Q_{i-1}^o\vec{\varepsilon}_{i-1} + (Q_{i-1}^o\vec{\omega}_{i-1})\times(\vec{\tilde{e}}_i\dot{q}^i)\bar{\xi}_i + \vec{\tilde{e}}_i\ddot{q}^i\bar{\xi}_i]$$

$$\begin{aligned} Q_i^o\vec{w}_i = Q_i^{i-1}\{&Q_{i-1}^o\vec{w}_{i-1} + 2(Q_{i-1}^o\vec{\omega}_{i-1})\times\vec{\tilde{e}}_i\dot{q}^i\xi_i - (Q_{i-1}^o\vec{\varepsilon}_{i-1})\times \\ &\times(Q_{i-1}^o\vec{r}_{i-1,i}) - (Q_{i-1}^o\vec{\omega}_{i-1})\times[(Q_{i-1}^o\vec{\omega}_{i-1})\times(Q_{i-1}^o\vec{r}_{i-1,i})] + \ddot{q}^i\vec{\tilde{e}}_i\xi_i\} + \\ &+ (Q_i^o\vec{\varepsilon}_i)\times(Q_i^o\vec{r}_{ii}) + (Q_i^o\vec{\omega}_i)\times[(Q_i^o\vec{\omega}_i)\times(Q_i^o\vec{r}_{ii})] \end{aligned} \tag{A.2.9}$$

Taking into considerations the inverse of relation $\vec{e}_{i+1} = Q_o^i\vec{\tilde{e}}_{i+1}$, $\vec{\tilde{e}}_{i+1} = Q_i^o\vec{e}_{i+1}$ and expressions (A.2.6) and (A.2.7), follows:

$$\vec{\tilde{\varepsilon}}_i = Q_i^o\vec{\varepsilon}_i;\quad \vec{\tilde{\omega}}_i = Q_i^o\vec{\omega}_i;\quad \vec{\tilde{r}}_{ii} = Q_i^o\vec{r}_{ii};\quad \vec{\tilde{r}}_{i,i+1} = Q_i^o\vec{r}_{i,i+1} \tag{A.2.10}$$

---

*) To avoid the possible confusion due to the identical notations of transformation matrices in the mechanism assembling ($q^i=0$) and the matrices, mapping the local system into the absolute system, here, these former matrices were denoted by the subscript o, $Q_{i,o}$, $Q_{i-1,o}$, etc., different from those, presented in (2.2).

for vectors in local coordinate systems relations equivalent to those of (2.6) are obtained [4]:

$$\vec{\tilde{\omega}}_i = Q_i^{i-1}(\vec{\tilde{\omega}}_{i-1} + \vec{\tilde{e}}_i \dot{q}^i \bar{\xi}_i)$$

$$\vec{\tilde{\varepsilon}}_i = Q_i^{i-1}[\vec{\tilde{\varepsilon}}_{i-1} + \vec{\tilde{\omega}}_{i-1} \times \vec{\tilde{e}}_i \dot{q}^i \bar{\xi}_i + \vec{\tilde{e}}_i \ddot{q}^i \bar{\xi}_i]$$

$$\vec{\tilde{w}}_i = Q_i^{i-1}[\vec{\tilde{w}}_{i-1} + 2\vec{\tilde{\omega}}_{i-1} \times \vec{\tilde{e}}_i \dot{q}^i \xi_i - \vec{\tilde{\varepsilon}}_{i-1} \times \vec{\tilde{r}}_{i-1,i} - \vec{\tilde{\omega}}_{i-1} \times (\vec{\tilde{\omega}}_{i-1} \times \vec{\tilde{r}}_{i-1,i}) + \ddot{q}^i \vec{\tilde{e}}_i \xi_i] +$$

$$+ \vec{\tilde{\varepsilon}}_i \times \vec{\tilde{r}}_{ii} + \vec{\tilde{\omega}}_i \times (\vec{\tilde{\omega}}_i \times \vec{\tilde{r}}_{ii}) \qquad \text{(A.2.11)}$$

where, as before, ξ_i is the joint type indicator, and the tilde "~" denotes local coordinates of the system. This enables direct application of relation (2.11) for determining the moments of inertial forces.

In an analogous manner expressions for corresponding coefficients (2.8), i.e. for $\vec{b}_{ij}$, $\vec{b}_i^o$ in (2.17) can be obtained. For example:

$$\vec{\tilde{b}}_{ij} = -J_i \vec{\tilde{\alpha}}_{ij}; \qquad \vec{\tilde{\alpha}}_{ij} = \vec{\tilde{\alpha}}_{i-1,j}$$

$$\vec{\tilde{b}}_i^o = -J_i \vec{\tilde{\theta}}_i; \qquad \vec{\tilde{\theta}}_i = Q_i^{i-1}[\vec{\tilde{\theta}}_{i-1} + \dot{q}^i (\vec{\tilde{\omega}}_{i-1} \times \vec{\tilde{e}}_i) \bar{\xi}_i] \qquad \text{(A.2.12)}$$

$$\vec{\tilde{\lambda}}_i = \vec{\tilde{\omega}}_i \times (J_i \vec{\tilde{\omega}}_i)$$

The effectiveness of recursive relations (A.2.11) and (A.2.12) is evident, and it is also clear that vectors $\vec{\tilde{e}}_i$, $\vec{\tilde{r}}_{i-1,i}$ and $\vec{\tilde{r}}_{ii}$ need not be calculated since they are given as input parameters in defining the structure of the mechanism.

This procedure is analogously applied to remaining phases of the algorithm, until the final expressions for determining the driving forces (moments) in the mechanism joints have been obtained. Relations analogous to (2.30) and (2.31) are derived in a similar way. These relations require the evaluation of vectors,

$$\vec{\tilde{r}}_{ji} = Q_i^{i-1}(\vec{\tilde{r}}_{j,i-1} - \vec{\tilde{r}}_{i-1,i}) + \vec{\tilde{r}}_{ii}, \quad \forall i \in I,\ j=1,\dots,i-1 \qquad \text{(A.2.13)}$$

For forming the transformation matrix Q_{i-1}^i, or Q_i^{i-1} it is possible to directly apply Rodrigues' formula (2.4) to the rotation of the i-th local system in relation to the (i-1)-th local system (Fig. A.2.3):

$$\bar{\vec{q}}_{ij} = \bar{\vec{q}}^o_{ij}\cos q^i + (1-\cos q^i)(\bar{\vec{e}}_i\bar{\vec{q}}^o_{ij})\bar{\vec{e}}_i + \bar{\vec{e}}_i \times \bar{\vec{q}}^o_{ij}\sin q^i$$

$$Q^i_{i-1} = [\bar{\vec{q}}_{i1}\ \bar{\vec{q}}_{i2}\ \bar{\vec{q}}_{i3}] \tag{A.2.14}$$

where $\bar{\vec{q}}^o_{ij}$ are vectors projecting the i-th coordinate system unit axes onto the (i-1)-th system under the condition that the joint coordinate $q^i=0$; $\bar{\vec{e}}_i$ is the unit vector of the i-th joint projected onto the (i-1)-th system. In order to obtain vectors $\bar{\vec{q}}^o_{ij}$, j=1,2,3 directly the projections of unit vectors $\bar{\vec{q}}_{ij}$ under condition $q^i=0$ (i.e., matrix $Q^i_{i-1,o}$), the assembling relations which are valid for zero internal angles (2.1 - 2.3) can be used. According to notation of Fig. A.2.3. these relations become:

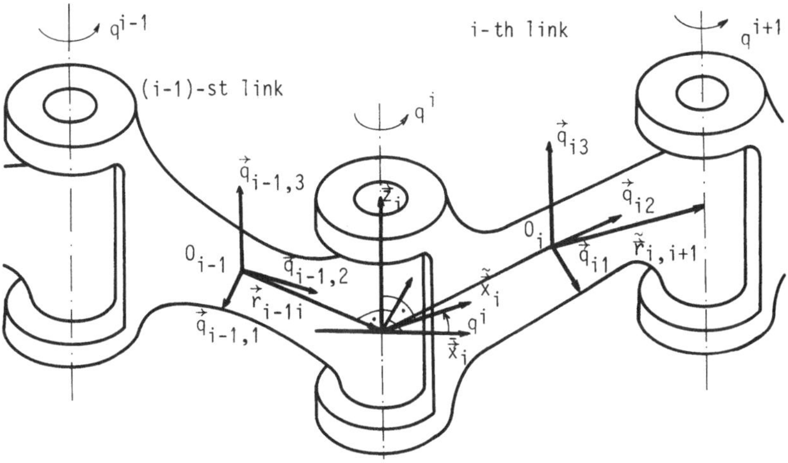

Fig. A.2.3. Vectors which define the kinematic chain where Rodrigues' formula is used

$$\bar{\vec{x}}_i = Q^i_{i-1,o}\tilde{\vec{x}}_i, \quad \bar{\vec{y}}_i = Q^i_{i-1,o}\tilde{\vec{y}}_i, \quad \bar{\vec{z}}_i = Q^i_{i-1}\tilde{\vec{z}}_i \tag{A.2.15}$$

where according to (2.1 - 2.2):

$$\bar{\vec{x}}_i = \bar{\vec{z}}_i \times (\text{ort}\ \bar{\vec{r}}_{i-1,i} \times \bar{\vec{z}}_i), \quad \tilde{\vec{x}}_i = \tilde{\vec{z}}_i \times (\text{ort}\ \tilde{\vec{r}}_{ii} \times \tilde{\vec{z}}_i)$$

$$\bar{\vec{y}}_i = \text{ort}\ \bar{\vec{r}}_{i-1,i} \times \bar{\vec{z}}_i = \bar{\vec{z}}_i \times \bar{\vec{x}}_i, \quad \tilde{\vec{y}}_i = \text{ort}\ \tilde{\vec{r}}_{ii} \times \tilde{\vec{z}}_i = \tilde{\vec{z}}_i \times \tilde{\vec{x}}_i \tag{A.2.16}$$

Relation (2.2) by using the given notation (A.2.15), (A.2.16) becomes:

$$Q^{i}_{i-1,o} = [\bar{\vec{q}}^{o}_{i1}\ \bar{\vec{q}}^{o}_{i2}\ \bar{\vec{q}}^{o}_{i3}] = [\bar{\vec{x}}_i\ \bar{\vec{y}}_i\ \bar{\vec{z}}_i][\tilde{\vec{x}}_i\ \tilde{\vec{y}}_i\ \tilde{\vec{z}}_i]^T \qquad (A.2.17)$$

After multiplying the matrices on the right side of relation (A.2.17) columns $\bar{\vec{q}}^{o}_{ij}$ may be represented in the form

$$\bar{\vec{q}}^{o}_{ij} = \tilde{x}^{(j)}_i \bar{\vec{x}}_i + \tilde{y}^{(j)}_i \bar{\vec{y}}_i + \tilde{z}^{(j)}_i \bar{\vec{z}}_i, \qquad i,j=1,2,3 \qquad (A.2.18)$$

where (j) denotes the j-th projection of the corresponding vector.

In this way the projections of unit vectors $\vec{q}_{ij}$ onto the (i-1)-th coordinate system, under condition $q^i=0$ have been obtained in the form of a linear combination of three orthogonal unit vectors positioned at the centre of i-th joint. The columns of the transformation matrix are obtained by substituting (A.2.18) into (A.2.14). For the case of a rotational joint these columns are given by:

$$\bar{\vec{q}}_{ij} = (\tilde{x}^{(j)}_i \cos q^i - \tilde{y}^{(j)}_i \sin q^i)\bar{\vec{x}}_i + (\tilde{y}^{(j)}_i \cos q^i + \tilde{x}^{(j)}_i \sin q^i)\bar{\vec{y}}_i + \tilde{z}^{(j)}_i \bar{\vec{z}}_i$$

(A.2.19)

Rearranging this expression we obtain:

$$\bar{\vec{q}}_{ij} = \vec{d}^{s}_{ij} \sin q^i + \vec{d}^{c}_{ij} \cos q^i + \vec{d}^{k}_{ij} \qquad (A.2.20)$$

where:

$$\vec{d}^{s}_{ij} = \tilde{x}^{(j)}_i \bar{\vec{y}}_i - \tilde{y}^{(j)}_i \bar{\vec{x}}_i;\quad \vec{d}^{c}_{ij} = \tilde{x}^{(j)}_i \bar{\vec{x}}_i + \tilde{y}^{(j)}_i \bar{\vec{y}}_i;\quad \vec{d}^{k}_{ij} = \tilde{z}^{(j)}_i \bar{\vec{z}}_i. \qquad (A.2.21)$$

In the case of translational (prismatic) joint, the columns of matrix Q^{i}_{i-1} are given by relation (A.2.18).

While forming the dynamic model it is suitable to have the coordinate systems of segments located in their center of mass so that their axis coincide with the central inertial axes of the segment (Fig. A.2.3). The unit axes of the i-th coordinate system in Fig. A.2.3. are denoted by $\vec{q}_{i1}$, $\vec{q}_{i2}$ and $\vec{q}_{i3}$. The vectors describing the i-th mechanism segment in the defining of mechanism configuration are: $\tilde{\vec{r}}_{ii}$ - the distance vector between the centre of mass of i-th segment and the centre of i-th joint, $\tilde{\vec{r}}_{i,i+1}$ - the distance vector between the centre of mass of i-th segment and the centre of (i+1)-th joint, joint axis vectors $\tilde{\vec{z}}_i$ and $\tilde{\vec{z}}_{i+1}$ (all these vectors are described with respect to i-th coordinate system which is denoted by "~"). Characteristic vectors described in relation to (i-1)-th coordinate system are denoted by -. The joint coordinate q^i for a revolute joint is defined as the angle between the projections of vectors $\vec{r}_{i-1,i}$ and $\vec{r}_{ii}$ onto the plane perpendicular to $\vec{z}_i$, while in the case of a prismatic pair, q^i represents the distance between the end point $\vec{r}_{i-1,i}$ and the starting point $\vec{r}_{ii}$.

Transformation matrix Q^i_{i-1}, from i-th to (i-1)-th segment is determined by the application of Rodrigues' formula of finite rotations, and relations (A.2.14-A.2.21). The transformation matrix for zero valued joint coordinates $Q^i_{i-1,o}$ is determined from conditions (A.2.15), i.e. from the condition that at $q^i=0$ unit vectors $\bar{\vec{x}}_i$ and $\vec{x}_i$ which represent vector projections of $\vec{r}_{i-1,i}$ and $\vec{r}_{ii}$ onto the plane perpendicular to z_i joint axis, respectively, coincide, as well as from the condition that $Q^i_{i-1,o}$ mapps the joint axis $\tilde{\vec{z}}_i$ vector from the i-th to the (i-1)-th coordinate system, and the vector $\tilde{\vec{y}}_i$ into vector $\bar{\vec{y}}_i$.

For a general case when $q^i \neq 0$, the columns of transformation matrix $Q^i_{i-1} = [\bar{\vec{q}}_{i1}\ \bar{\vec{q}}_{i2}\ \bar{\vec{q}}_{i3}]$ are obtained in the form (A.2.20) and (A.2.21). Vectors (A.2.21) do not depend on the coordinates q^i, because vectors $\tilde{\vec{x}}_i$, $\tilde{\vec{y}}_i$, $\tilde{\vec{z}}_i$, $\bar{\vec{x}}_i$, $\bar{\vec{y}}_i$, $\bar{\vec{z}}_i$ are constant and defined by the mechanism configuration. In this way the transformation matrix between the i-th and the (i-1)-th coordinate systems is determined, and it is a function of the known vectors and the sine and cosine of the joint angle q^i. For the case of a prismatic joint the columns of matrix Q^i_{i-1} are given by relation (A.2.18). The transformation matrix between the i-th and fixed (reference) coordinate system is obtained using (A.2.6).

Let us now consider a specific case when the vector $\vec{r}_{ii}$ is colinear with the axis $\vec{z}_i$ of the i-th joint. Then, according to (A.2.16), $\vec{x}_i=0$, $\vec{y}_i=0$ so that the transformation matrix (A.2.17) cannot be uniquely determined. In this case, an arbitrary auxiliary vector $\vec{r}^A_{ii}$ which is not colinear with $\vec{r}_{ii}$, is introduced and relations (A.2.16), (A.2.17) applied with the exception that $\vec{r}^A_{ii}$ is used instead of $\vec{r}_{ii}$. The choice of vector $\vec{r}^A_{ii}$ uniquely defines the position of the i-th segment in relation to (i-1)-th at $q^i=0$. The procedure is repeated for the case when the vector $\vec{r}_{i-1,i}$ is colinear with $\vec{z}_i$.

The differences between the mentioned procedures are the result of different ways in which segments are described. While in Denavit-Hartenberg coordinates, the kinematic pair is described by three parameters and a joint coordinate, in the second approach every segment is described by four vectors. Let us consider the necessary constraints on the choice of segment coordinate systems so that the more general description using a set of vectors may be reduced to one using only the three parameters.

In the approach using Rodrigues' formula, the coordinate system is positioned at the centre of mass, with axes colinear with the central inertial axes. However, from the kinematic standpoint, its position and orientation are arbitrary. According to Denavit-Hartenberg notation, the centre is defined by the intersection of the (i+1)-th joint axis and the common normal between i-th and (i+1)-th joint. Its orientation is also fixed by the direction of the (i+1)-th joint axis and the vector of common normal. If in the application of Rodrigues' formula, constraints are introduced, then the coordinate system unit vectors $\vec{q}_{i1}$ and $\vec{q}_{i3}$ coincide with $\vec{x}_i$ and $\vec{z}_i$ from the Denavit-Hartenberg approach and vector $\vec{r}_{ii}$ becomes $\overrightarrow{O_{i-1}O_i}$ (Fig. A.2.4). Vector $\vec{r}_{i,i+1}$ then degenerates into a zero vector. In order to determine the transformation matrix

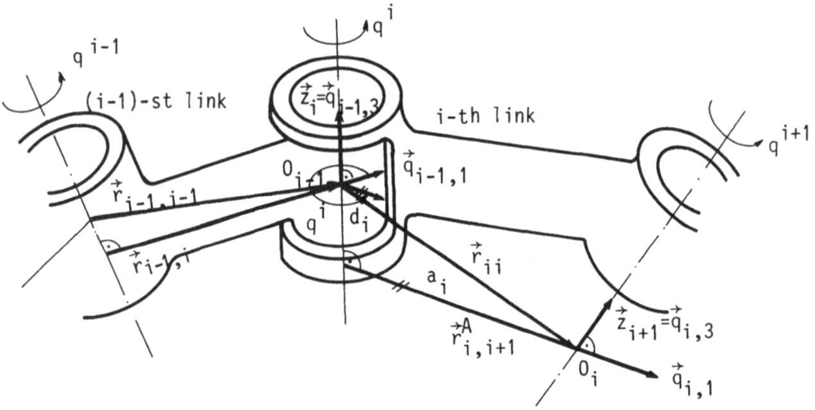

Fig. A.2.4. Segment vectors for the case when the coordinate systems are chosen using Denavit-Hartenberg parameters

$Q^i_{i-1,o}$, according to (A.2.16), (A.2.17), it is necessary to introduce the auxiliary vector $\vec{r}^{A}_{i,i+1}$, in the direction of the common normal of axes $\vec{z}_i$ and $\vec{z}_{i+1}$. Thus, all i-th segment vectors required for the application of Rodrigues' formula have been identified. The transformation matrix Q^i_{i-1}, thus obtained, must be of the same form as that obtained using Denavit-Hartenbergs' approach (A.2.3). In order to demonstrate this we begin from relations (A.2.16), (A.2.20) and (A.2.21). Directly from (A.2.16) is obtained:

$$\vec{x}_i = [1\ 0\ 0]^T, \quad \bar{\vec{x}}_i = [1\ 0\ 0]^T, \quad \bar{\vec{y}}_i = [0\ 1\ 0]^T, \quad \bar{\vec{z}}_i = [0\ 0\ 1]^T \tag{A.2.22}$$

Substituting (A.2.22) into (A.2.21) gives the matrix Q^i_{i-1} of the form:

$$Q^i_{i-1} = \begin{bmatrix} \cos q^i & -\tilde{z}_i^{(3)} \sin q^i & \tilde{z}_i^{(2)} \sin q^i \\ \sin q^i & \tilde{z}_i^{(3)} \cos q^i & -\tilde{z}_i^{(2)} \cos q^i \\ \tilde{z}_i^{(1)} & \tilde{z}_i^{(2)} & \tilde{z}_i^{(3)} \end{bmatrix} \tag{A.2.23}$$

Because the $\vec{q}_{i,1}$ axis is perpendicular to axis $\vec{z}_i$ of the i-th joint we can introduce:

$$\tilde{z}_i^{(1)} = 0, \quad \tilde{z}_i^{(2)} = \sin\alpha_i, \quad \tilde{z}_i^{(3)} = \cos\alpha_i \tag{A.2.24}$$

Substituting (A.2.24) into (A.2.23) gives the transformation matrix in the form:

$$Q^i_{i-1} = \begin{bmatrix} \cos q^i & -\sin q^i \cos\alpha_i & \sin q^i \sin\alpha_i \\ \sin q^i & \cos q^i \cos\alpha_i & -\cos q^i \sin\alpha_i \\ 0 & \sin\alpha_i & \cos\alpha_i \end{bmatrix} \tag{A.2.25}$$

which agrees with the submatrix of matrix (A.2.3).

Position of the i-th segment tip in relation to (i-1)-th coordinate system (distance $O_{i-1}O_i$ in Fig. A.2.4) becomes:

$$\bar{\vec{p}}_i = Q^i_{i-1} \vec{r}_{ii} = Q^i_{i-1} \begin{bmatrix} a_i \\ d_i \sin\alpha_i \\ d_i \cos\alpha_i \end{bmatrix} = \begin{bmatrix} a_i \cos q^i \\ a_i \sin q^i \\ d_i \end{bmatrix} \tag{A.2.26}$$

(where a_i and d_i are parameters defined in the Denavit-Hartenberg approach), and represent the last column in the homogenous transformation matrix (A.2.3).

In this way, by using the formalism of describing the kinematic chain used in the approach by the use of Rodrigues'formula, the transformation matrices and tip position have been obtained from the Denavit-

-Hartenberg approach. In other words, it was shown that this approach represents a special case of Rodrigues' formula application.

From the standpoint of kinematic model complexity it is clear that Denavit-Hartenbergs' approach is more suitable because it produces simpler transformation matrices (A.2.25) in relation to a more general form (A.2.20). On the other hand, it is clear that the axes of local coordinate systems of the segments in Denavit-Hartenbergs' approach do not have to coincide with the main inertial axes of the segment. Furthermore, these axes do not represent the central axes of inertia. The consequence of this is a more complex dynamic model, since the segment inertial characteristics cannot be described by three moments of inertia, but by the inertia tensor containing nine elements.

With industrial robots, it often happens that at least one segment is not completely symmetrical so that the main axes of inertia are rotated in relation to the normal common to the joints.

If, nevertheless, in forming the dynamic model it was adopted that one axis of inertia coincides with the common normal between two segments, a fault would result in dynamic modelling, which would depend on mass distribution within the segment. These problems are not encountered in the approach using Rodrigues' formula, however, in this case the kinematic model is somewhat more complex. However, in the case when the axes of Denavit-Hartenberg coordinate systems also represent the central axes of inertia of the segment, this approach is more suitable, since it requires fewer computational operations.

E X A M P L E S

Kinematic models for two typical manipulation mechanisms will be assembled on the basis of Denavit-Hartenberg coordinates.

Semianthropomorphic mechanism

A kinematic scheme of a semianthropomorphic manipulation robot with coordinates assigned according to Denavit-Hartenberg notation is shown in Fig. A.2.5. which also includes a table of Denavit-Hartenberg parameters for the mechanism in question. In the figure, only the y_6 axis is marked. However, it is clear that the axes are determined such that the coordinate systems $x_i y_i z_i$ are right-hand orientated, $i=1,2,\ldots,6$.

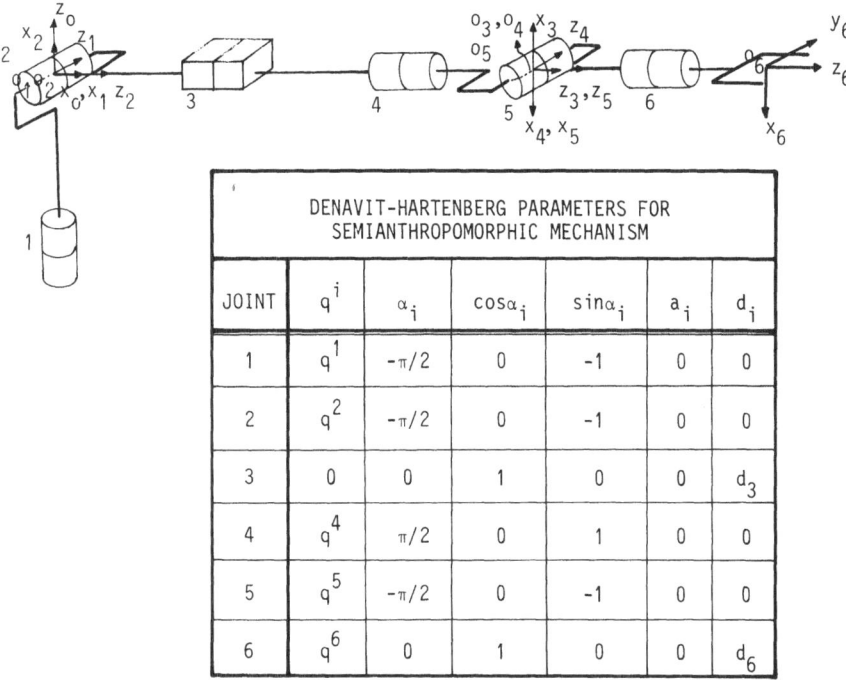

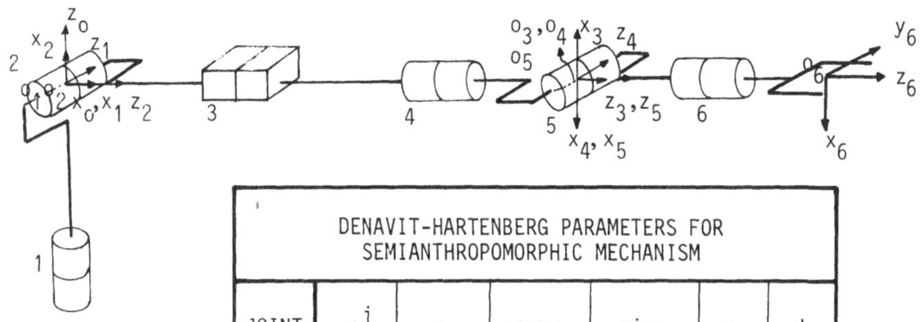

DENAVIT-HARTENBERG PARAMETERS FOR SEMIANTHROPOMORPHIC MECHANISM						
JOINT	q^i	α_i	$\cos\alpha_i$	$\sin\alpha_i$	a_i	d_i
1	q^1	$-\pi/2$	0	-1	0	0
2	q^2	$-\pi/2$	0	-1	0	0
3	0	0	1	0	0	d_3
4	q^4	$\pi/2$	0	1	0	0
5	q^5	$-\pi/2$	0	-1	0	0
6	q^6	0	1	0	0	d_6

Fig. A.2.5. Kinematic scheme of semianthropomorphic manipulator

On the basis of the table shown in Fig. A.2.5. the following transformation matrices are obtained Q^1_{i-1}, $i=1,2,\ldots,6$:

$$Q^1_o = \begin{bmatrix} C1 & 0 & -S1 & 0 \\ S1 & 0 & C1 & 0 \\ 0 & -1 & 0 & 0 \\ 0 & 0 & 0 & 1 \end{bmatrix}, \qquad Q^2_1 = \begin{bmatrix} C2 & 0 & -S2 & 0 \\ S2 & 0 & C2 & 0 \\ 0 & -1 & 0 & 0 \\ 0 & 0 & 0 & 1 \end{bmatrix}$$

$$Q^3_2 = \begin{bmatrix} 1 & 0 & 0 & 0 \\ 0 & 1 & 0 & 0 \\ 0 & 0 & 1 & d_3 \\ 0 & 0 & 0 & 1 \end{bmatrix}, \qquad Q^4_3 = \begin{bmatrix} C4 & 0 & S4 & 0 \\ S4 & 0 & -C4 & 0 \\ 0 & 1 & 0 & 0 \\ 0 & 0 & 0 & 1 \end{bmatrix}$$

$$Q_4^5 = \begin{bmatrix} C5 & 0 & -S5 & 0 \\ S5 & 0 & C5 & 0 \\ 0 & -1 & 0 & 0 \\ 0 & 0 & 0 & 1 \end{bmatrix}, \qquad Q_5^6 = \begin{bmatrix} C6 & -S6 & 0 & 0 \\ S6 & C6 & 0 & 0 \\ 0 & 0 & 1 & d_6 \\ 0 & 0 & 0 & 1 \end{bmatrix}$$

In the above matrices Ci, Si, i=1,2,...,6 denote the sines and cosines of angles q^i, i=1,2,...,6 respectively.

$$Q_o^6 = Q_o^1 \cdot Q_1^2 \cdots Q_5^6$$

$$Q_o^2 = Q_o^1 \cdot Q_1^2 = \begin{bmatrix} C1 & 0 & -S1 & 0 \\ S1 & 0 & C1 & 0 \\ 0 & -1 & 0 & 0 \\ 0 & 0 & 0 & 1 \end{bmatrix} \begin{bmatrix} C2 & 0 & S2 & 0 \\ S2 & 0 & C2 & 0 \\ 0 & -1 & 0 & 0 \\ 0 & 0 & 0 & 1 \end{bmatrix} =$$

$$= \begin{bmatrix} C1 \cdot C2 & S1 & -C1 \cdot S2 & 0 \\ S1 \cdot C2 & -C1 & -S1 \cdot S2 & 0 \\ -S2 & 0 & -C2 & 0 \\ 0 & 0 & 0 & 1 \end{bmatrix}$$

$$Q_o^3 = Q_o^2 \cdot Q_2^3 = \begin{bmatrix} C1 \cdot C2 & S1 & -C1 \cdot S2 & 0 \\ S1 \cdot C2 & -C1 & -S1 \cdot S2 & 0 \\ -S2 & 0 & -C2 & 0 \\ 0 & 0 & 0 & 1 \end{bmatrix} \begin{bmatrix} 1 & 0 & 0 & 0 \\ 0 & 1 & 0 & 0 \\ 0 & 0 & 1 & d_3 \\ 0 & 0 & 0 & 1 \end{bmatrix} =$$

$$= \begin{bmatrix} C1 \cdot C2 & S1 & -C1 \cdot S2 & -C1 \cdot S2 \cdot d_3 \\ S1 \cdot C2 & -C1 & -S1 \cdot S2 & -S1 \cdot S2 \cdot d_3 \\ -S2 & 0 & -C2 & -C2 \cdot d_3 \\ 0 & 0 & 0 & 1 \end{bmatrix}$$

$$Q_o^4 = Q_o^3 \cdot Q_3^4 = \begin{bmatrix} C1\cdot C2 & S1 & -C1\cdot S2 & -C1\cdot S2\cdot d_3 \\ S1\cdot C2 & -C1 & -S1\cdot S2 & -S1\cdot S2\cdot d_3 \\ -S2 & 0 & -C2 & -C2\cdot d_3 \\ 0 & 0 & 0 & 1 \end{bmatrix} \begin{bmatrix} C4 & 0 & S4 & 0 \\ S4 & 0 & -C4 & 0 \\ 0 & 1 & 0 & 0 \\ 0 & 0 & 0 & 1 \end{bmatrix} =$$

$$= \begin{bmatrix} C1\cdot C2\cdot C4+S1\cdot S4 & -C1\cdot S2 & C1\cdot C2\cdot S4-S1\cdot C4 & -C1\cdot S2\cdot d_3 \\ S1\cdot C2\cdot C4-C1\cdot S4 & -S1\cdot S2 & S1\cdot C2\cdot S4+C1\cdot C4 & -S1\cdot S2\cdot d_3 \\ -S2\cdot C4 & -C2 & 0 & -C2\cdot d_3 \\ 0 & 0 & 0 & 1 \end{bmatrix}$$

$$Q_o^5 = Q_o^4 \cdot Q_4^5 = \begin{bmatrix} (C1\cdot C2\cdot C4+S1\cdot S4)\cdot C5-C1\cdot S2\cdot S5 & S1\cdot C4-C1\cdot C2\cdot S4 \\ (S1\cdot C2\cdot C4-C1\cdot S4)\cdot C5-S1\cdot S2\cdot S5 & S1\cdot S2 \\ -S2\cdot C4\cdot C5-C2\cdot C5 & 0 \\ 0 & 0 \end{bmatrix}$$

$$\begin{bmatrix} -(C1\cdot C2\cdot C4+S1\cdot S4)\cdot S5+C1\cdot S2\cdot C5 & -C1\cdot S2\cdot d_3 \\ (C1\cdot S4-S1\cdot C2\cdot C4)\cdot S5+S1\cdot S2\cdot C5 & -S1\cdot S2\cdot d_3 \\ -S2\cdot C4\cdot S5-C2\cdot C5 & -C2\cdot d_3 \\ 0 & 1 \end{bmatrix}$$

$$Q_o^6 = Q_o^5 \cdot Q_5^6 = \begin{bmatrix} a_{11} & a_{12} & a_{13} & a_{14} \\ a_{21} & a_{22} & a_{23} & a_{24} \\ a_{31} & a_{32} & a_{33} & a_{34} \\ a_{41} & a_{42} & a_{43} & a_{44} \end{bmatrix}$$

$$a_{11} = [(C1\cdot C2\cdot C4+S1\cdot S4)\cdot C5-C1\cdot S2\cdot S5]\cdot C6+(S1\cdot C4-C1\cdot C2\cdot S4)\cdot S6$$

$$a_{12} = [C1\cdot S2\cdot S6-(C1\cdot C2\cdot C4+S1\cdot S4)\cdot C5]\cdot S6+(S1\cdot C4-C1\cdot C2\cdot S4)\cdot C6$$

$$a_{13} = C1\cdot S2\cdot C5-(C1\cdot C2\cdot C4+S1\cdot S4)\cdot S5$$

$$a_{14} = [C1\cdot S2\cdot C5-(C1\cdot C2\cdot C4+S1\cdot S4)\cdot S5]\cdot d_6-C1\cdot S2\cdot d_3$$

$$a_{21} = [(S1\cdot C2\cdot C4+C1\cdot S4)\cdot C5-S1\cdot S2\cdot S5)]C6+S1\cdot S2\cdot S6$$

$$a_{22} = [S1\cdot S2\cdot S5-(S1\cdot C2\cdot C4-C1\cdot S4)\cdot C5]S6+S1\cdot S2\cdot C6$$

$$a_{23} = (C1\cdot S4-S1\cdot C2\cdot C4)\cdot S5+S1\cdot S2\cdot C5$$

$$a_{24} = [(C1\cdot S4-S1\cdot C2\ C4)\cdot S5+S1\cdot S2\cdot C5]d_6-S1\cdot S2\cdot d_3$$

$$a_{31} = -(S2\cdot C4\cdot C5+C2\cdot C5)\cdot C6$$

$$a_{32} = (S2\cdot C4\cdot C5+C2\cdot C5)\cdot S6$$

$$a_{33} = S2\cdot C4\cdot S5-C2\cdot C5,$$

$$a_{34} = (S2\cdot C5\cdot S5-C2\cdot C5)d_6-C2\cdot d_3$$

$$a_{41} = 0, \quad a_{42} = 0, \quad a_{43} = 0, \quad a_{44} = 1$$

Arthropoidal mechanism

The kinematic scheme of this manipulation mechanism (the "PUMA" mechanism) is given in Fig. A.2.6. The variable quantities are joint angles q^i, i=1,2,3,...,6 and the table provided gives the values of angles α_i and those values of a_i and d_i which are not zero.

The coordinate systems are chosen according to the aforementioned rules. O_o is positioned at the centre of second joint and z_o is in the direction of the first joint axes. Axis x_o is chosen perpendicular to the axis of the first joint. For reasons of clarity, no axis is drawn in the diagram, though it is clear that in this case y_o coincides with z_1 since $\vec{x}_o\times\vec{y}_o = \vec{z}_o$. O_1 is the intersection of the normal common to the axes of the first and second joint with the axis of the second joint, therefore it coincides with O_o. The remaining coordinate systems are determined analogously. The angles of twist α_i are determined according to the positions of neighbouring coordinate systems: $\alpha_1 = 90^o$, since it is required to perform a rotation for $+90^o$ about the x_o axis so that z_o and z_1 could coincide and $\alpha_2 = 0^o$, since z_1 and z_2 are of a common sense and direction etc. Lengths a_i and d_i are also determined from the relative positions of coordinate systems: $a_1 = 0$, since O_1 and O_2 coincide, while a_2 has the value equal to the distance between O_1 and

O_2. Remaining a_i for this manipulator are 0, as are all d_i except for d_4 and d_6 because origins O_4 and O_6 of joint coordinate systems are translated for distances d_4 and d_6 along z_3 and z_5 axes, respectively.

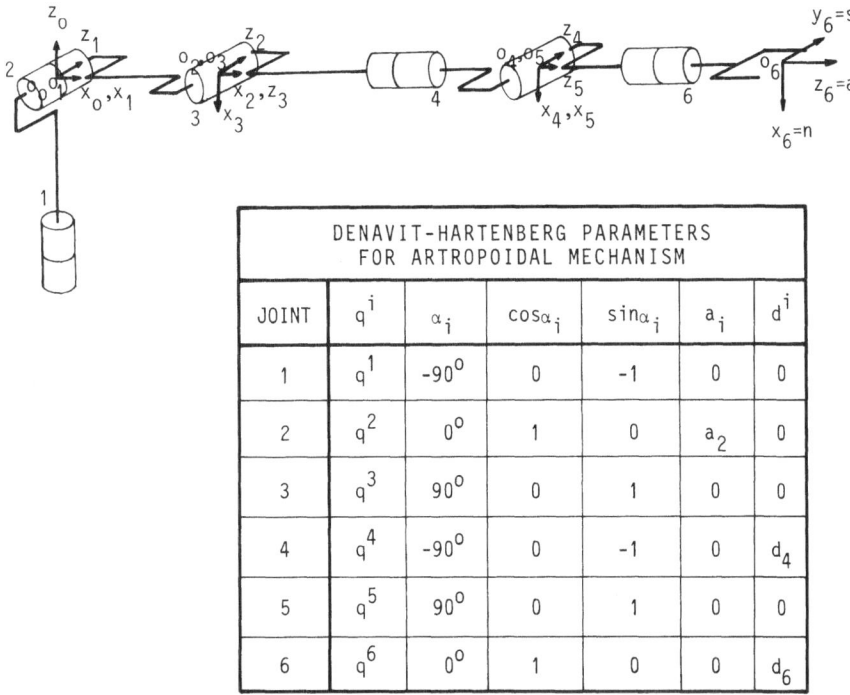

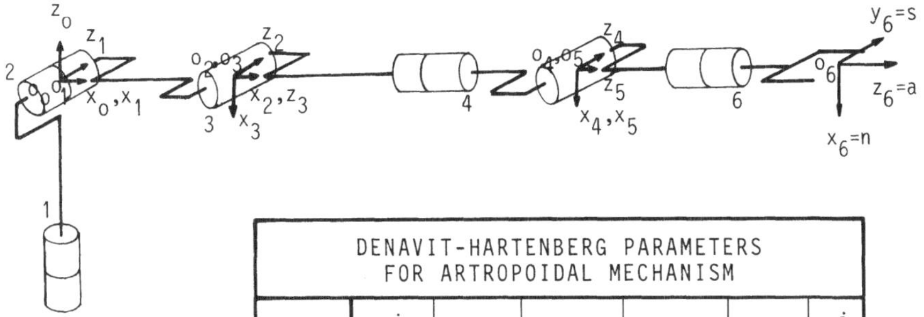

DENAVIT-HARTENBERG PARAMETERS FOR ARTROPOIDAL MECHANISM						
JOINT	q^i	α_i	$\cos\alpha_i$	$\sin\alpha_i$	a_i	d^i
1	q^1	-90^0	0	-1	0	0
2	q^2	0^0	1	0	a_2	0
3	q^3	90^0	0	1	0	0
4	q^4	-90^0	0	-1	0	d_4
5	q^5	90^0	0	1	0	0
6	q^6	0^0	1	0	0	d_6

Fig. A.2.6. Kinematic scheme of an arthropoidal manipulation robot

In the manner described, the tables in Fig. A.2.6. are filled in and according to these the actual transformation matrices are formed.

From the given table it follows that matrices Q^i_{i-1}, i-1,2,...,6 for the considered manipulation robot have the form:

$$Q^1_o = \begin{bmatrix} C1 & 0 & -S1 & 0 \\ S1 & 0 & C1 & 0 \\ 0 & -1 & 0 & 0 \\ 0 & 0 & 0 & 1 \end{bmatrix}, \qquad Q^2_1 = \begin{bmatrix} C2 & -S2 & 0 & a_2C2 \\ S2 & C2 & 0 & a_2S2 \\ 0 & 0 & 1 & 0 \\ 0 & 0 & 0 & 1 \end{bmatrix}$$

$$Q_2^3 = \begin{bmatrix} C3 & 0 & S3 & 0 \\ S3 & 0 & -C3 & 0 \\ 0 & 1 & 0 & 0 \\ 0 & 0 & 0 & 1 \end{bmatrix}, \quad Q_3^4 = \begin{bmatrix} C4 & 0 & -S4 & 0 \\ S4 & 0 & C4 & 0 \\ 0 & -1 & 0 & d_4 \\ 0 & 0 & 0 & 1 \end{bmatrix}$$

$$Q_4^5 = \begin{bmatrix} C5 & 0 & S5 & 0 \\ S5 & 0 & -C5 & 0 \\ 0 & 1 & 0 & 0 \\ 0 & 0 & 0 & 1 \end{bmatrix}, \quad Q_5^6 = \begin{bmatrix} C6 & -S6 & 0 & 0 \\ S6 & C6 & 0 & 0 \\ 0 & 0 & 1 & d_6 \\ 0 & 0 & 0 & 1 \end{bmatrix}$$

Matrix Q_o^6 can be obtained by successive multiplications from the tip to the base and vice versa. Here, we adopt the first way, in which the fourth column of matrix T_o^i represents position of the center of joint i in the external coordinate system. It follows:

$$Q_6^4 = Q_5^4 \cdot Q_6^5 = \begin{bmatrix} C5 \cdot C6 & -C5 \cdot S6 & S5 & S5 \cdot d_6 \\ S5 \cdot C6 & -S5 \cdot S6 & -C5 & -C5 \cdot d_6 \\ S6 & C6 & 0 & 0 \\ 0 & 0 & 0 & 1 \end{bmatrix}$$

$$Q_6^3 = Q_4^3 \cdot Q_6^4 = \begin{bmatrix} C4 \cdot C5 \cdot C6 - S4 \cdot S6 & -C4 \cdot C5 \cdot S6 & C4 \cdot S5 & C4 \cdot S5 \cdot d_6 \\ S4 \cdot C5 \cdot C6 + C4 \cdot S6 & -S4 \cdot C5 \cdot S6 & S4 \cdot S5 & -S4 \cdot C5 \cdot d_6 \\ -S5 \cdot C6 & S5 \cdot S6 & C5 & C5 \cdot d_6 + d_4 \\ 0 & 0 & 0 & 1 \end{bmatrix}$$

$$Q_6^2 = Q_3^2 \cdot Q_6^3 = \left[\begin{matrix} C3 \cdot (C4 \cdot C5 \cdot C6 - S4 \cdot S6) - S3 \cdot S5 \cdot C6 \\ S3 \cdot (C4 \cdot C5 \cdot C6 - S4 \cdot S6) + C3 \cdot S5 \cdot C6 \\ S4 \cdot C5 \cdot C6 + C4 \cdot S6 \\ 0 \end{matrix} \right.$$

$$\begin{array}{cc}
C3\cdot(-C4\cdot C5\cdot S6-S4\cdot C6)+S3\cdot S5\cdot S6 & C3\cdot C4\cdot S5+S3\cdot C5 \\
S3\cdot(-C4\cdot C5\cdot S6-S4\cdot C6)-C3\cdot S5\cdot S6 & S3\cdot C4\cdot S5-C3\cdot C5 \\
-S4\cdot C5\cdot S6+C4\cdot C6 & S4\cdot S5 \\
0 & 0
\end{array}$$

$$\left.\begin{array}{c}
C3\cdot C4\cdot C5\cdot d_6+S3\cdot(C5\cdot d_6+d_4) \\
S3\cdot C4\cdot S5\cdot d_6-C3\cdot(C5\cdot d_6+d_4) \\
-S4\cdot C5\cdot d_6 \\
1
\end{array}\right]$$

$$Q_6^1 = Q_2^1\cdot Q_6^2 = \left[\begin{array}{c}
C23\cdot(C4\cdot C5\cdot C6-S4\cdot S6)-S23\cdot S5\cdot C6 \\
S23\cdot(C4\cdot C5\cdot C6-S4\cdot S6)+C23\cdot S5\cdot C6 \\
S4\cdot C5\cdot C6+C4\cdot S6 \\
0
\end{array}\right.$$

$$\begin{array}{c}
C23\cdot(-C4\cdot C5\cdot S6-S4\cdot C6)+S23\cdot S5\cdot S6 \\
S23\cdot(-C4\cdot C5\cdot S6-S4\cdot C6)-C23\cdot S5\cdot S6 \\
-S4\cdot C5\cdot S6+C4\cdot C6 \\
0
\end{array}$$

$$\begin{array}{c}
C23\cdot C4\cdot S5+S23\cdot C5 \\
S23\cdot C4\cdot S5-C23\cdot C5 \\
S4\cdot S5 \\
0
\end{array}$$

$$\left.\begin{array}{c}
C23\cdot C4\cdot S5\cdot d_6+S23\cdot(C5\cdot d_6+d_4)+a_2\cdot C2 \\
S23\cdot C4\cdot S5\cdot d_6-C23\cdot(C5\cdot d_6+d_4)+a_2\cdot S2 \\
-S4\cdot C5\cdot d_6 \\
1
\end{array}\right]$$

C23 denotes $\cos(q^2+q^3)$ and S23, $\sin(q^2+q^3)$. This simplification is always adopted when the axes of neigbouring joints are parallel.

$$Q_6^o = \begin{bmatrix} a_{11} & a_{12} & a_{13} & a_{14} \\ a_{21} & a_{22} & a_{23} & a_{24} \\ a_{31} & a_{32} & a_{33} & a_{34} \\ a_{41} & a_{42} & a_{43} & a_{44} \end{bmatrix}$$

a_{11} = C1·[C23·(C4·C5·C6−S4·S6)−S23·S5·C6]−S1·(S4·C5·C6+C4·S6)

a_{12} = C1·[C23·(−C4·C5·S6−S4·C6)+S23·(S5·S6)]−S1.

· (−S4·C5·S6+C4·C6)

a_{13} = C1·(C23·C4·S5+S23·C5)−S1·S4·S5

a_{14} = C1·[C23·C4·S5·d_6+S23·(C5·d_6+d_4)+a_2·C2]+S1·S4·C5·d_6

a_{21} = S1·[C43·(C4·C5·C6−S4·S6)−S23·S5·C6]+C1·(S4·C5·C6+C4·S6)

a_{22} = S1·[C23·(−C4·C5·S6−S4·C6)+S23·(S5·S6)]+C1·(−S4·C5·S6+C4·C6)

a_{23} = S1·(C23·C4·S5+S23·C5)+C1·S4·S5

a_{24} = S1·[C23·C4·S5·d_6+S23·(C5·d_6+d_4)+a_2·C2]−C1·S4·C5·d_6

a_{31} = −[S23·(C4·C5·C6−S4·S6)+C23·S5·C6]

a_{32} = −[S23·(−C4·C5·S6−S4·C6)−C23·S5·S6], a_{33}=−(S23·C4·S5−C23·C5)

a_{34} = −[S23·C4·S5·d_6−C23·(C5·d_6+d_4)+a_2·S2]

$a_{41} = 0$, $a_{42} = 0$, $a_{43} = 0$, $a_{44} = 1$

References

[1] Paul R., Robots Manipulators: Mathematics, Programming and Control, The MIT Press, 1981.

[2] Denavit J., Hartenberg R.S., "Kinematic Notation for Lower-Pair Mechanisms Based on Matrices", ASME J. of Applied Mechanics, June 1955, pp. 215-221.

[3] Vukobratović M., Kirćanski M., Scientific Fundamentals of Robotics 3: Kinematics and Trajectory Synthesis of Manipulation Robots, Springer-Verlag, 1985.

[4] Vukobratović M., Cvetković V., "Computer-Oriented Algorithm Modelling of Active Spatial Mechanisms for Application in Robotics", Trans. on Systems, Man and Cybernetics, Vol. SMC-12, No. 6, 1982.

Appendix 3
Determining Velocities and Accelerations

The recurrent kinematic relations, by which the angular and linear velocities and angular and linear accelerations of the mechanism segments are determined applying the basic relations of the rigid body mechanics are derived.

The absolute body velocity equals the sum of its transfer and relative velocity:

$$\vec{\omega} = \vec{\omega}_p + \vec{\omega}_r \tag{A.3.1}$$

For two points A and B of the body follows:

$$\vec{v}_B = \vec{v}_A + \vec{\omega} \times \vec{r}_{AB} \tag{A.3.2}$$

where $\vec{\omega}$ is the angular body velocity, $\vec{r}_{AB}$ is the radius vector from point A to point B.

Let us consider the i-th kinematic pair, consisting of the i-th and (i-1)st mechanism segment (Fig. A.3.1). Motion of the (i-1)st segment will be considered as transfering one, and motion in the i-th joint as relative. Then the following relations hold: for the case of linear (prismatic) kinematic pair:

$$\vec{\omega}_i = \vec{\omega}_{i-1}; \quad \vec{v}_i = \vec{v}_{i-1} - \vec{\omega}_{i-1} \times \vec{r}_{i-1,i} + \dot{q}^i \vec{e}_i \tag{A.3.3}$$

and for the case of revolute kinematic pair:

$$\vec{\omega}_i = \vec{\omega}_{i-1} + \dot{q}^i \vec{e}_i, \quad \vec{v}_i = \vec{v}_{i-1} - \vec{\omega}_{i-1} \times \vec{r}_{i-1,i} + \vec{\omega}_i \times \vec{r}_{ii} \tag{A.3.4}$$

Expressions (A.3.3, A.3.4) can be considered recurrent; if gradually all the mechanism segments are circled, starting from the initial one, the velocities of all segments can be calculated using these expressions.

For the accelerations of points A and B in rigid body free motion, moving with angular velocity $\vec{\omega}$ and angular acceleration $\vec{\varepsilon}$:

$$\vec{\varepsilon} = \vec{\varepsilon}_p+\vec{\varepsilon}_r+\vec{\omega}_p\times\vec{\omega}_r, \qquad \vec{w}_B = \vec{w}_A+\vec{\varepsilon}\times\vec{r}_{AB}+\vec{\omega}\times(\vec{\omega}\times\vec{r}_{AB}) \tag{A.3.5}$$

Considering the coordinate system $Q_{i-1}(q_{i-1,j},\ j=1,2,3)$, connected to the (i-1)st mechanism segment and $Q_i(q_{ij},\ j=1,2,3)$, connected to the i-th mechanism segment, the relation is easily established via relations (A.3.1 - A.3.3), considering motion of each previous (i-1)st segment as transfering and motion of the i-th segment as relative:

$$\begin{aligned} &\vec{v} = \vec{v}_i, \quad \vec{v}_p = \vec{v}_{i-1}, \quad \vec{\omega} = \vec{\omega}_i, \quad \vec{\omega}_p = \vec{\omega}_{i-1} \\ &\vec{\varepsilon} = \vec{\varepsilon}_i, \quad \vec{\varepsilon}_p = \vec{\varepsilon}_{i-1}, \quad \vec{w} = \vec{w}_i, \quad \vec{w}_p = \vec{w}_{i-1} \end{aligned} \tag{A.3.6}$$

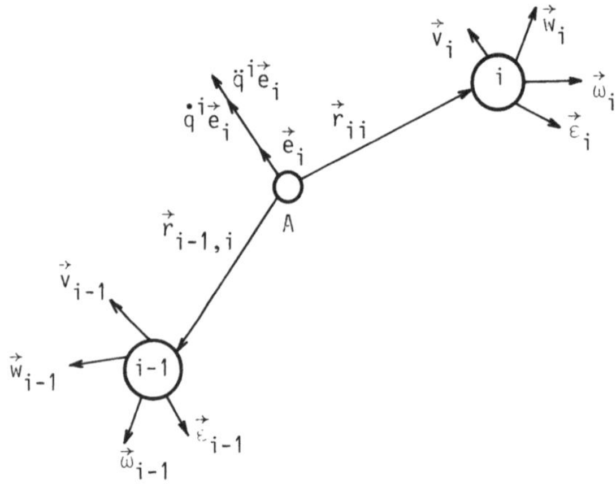

Fig. A.3.1. Velocities and accelerations of kinematic pair

Now the following relations can be written:

- for the prismatic kinematic pair: $\vec{\varepsilon}_i = \vec{\varepsilon}_{i-1}$,
- for the revolute kinematic pair: $\vec{\varepsilon}_i = \vec{\varepsilon}_{i-1}+\ddot{q}^i\vec{e}_i+\dot{q}^i\vec{\omega}_{i-1}\times\vec{e}_i$.

Let us pass to determining the linear accelerations. Denote with "A" the point of the momentary i-th joint center. Acceleration of point of the (i-1)st segment, coinciding with point "A" let us designate with $\vec{w}_{A,i-1}$. Then can be written (Fig. A.3.1):

$$\vec{w}_{A,i-1} = \vec{w}_{i-1} - \vec{\varepsilon}_{i-1} \times \vec{r}_{i-1,i} - \vec{\omega}_{i-1} \times (\vec{\omega}_{i-1} \times \vec{r}_{i-1,i}) \tag{A.3.7}$$

For the i-th segment point, coinciding instantly with point "A" can be written:

- for the case of prismatic kinematic pair: $\vec{w}_{A,i} = w_{A,i-1} + \ddot{q}^i \vec{e}_i + 2\dot{q}^i \vec{\omega}_{i-1} \times \vec{e}_i$,

- for the revolute kinematic pair: $\vec{w}_{A,i} = \vec{w}_{A,i-1}$.

Now can be written $\vec{w}_i = \vec{w}_{A,i} + \vec{\varepsilon}_i \times \vec{r}_{ii} + \vec{\omega}_i \times (\vec{\omega}_i \times \vec{r}_{ii})$.

It should be mentioned that the relative movement q^i in the expressions mentioned above understands both linear and revolute displacements.

References

[1] Whittaker E.T., Analytical Dynamics, London, Cambridge, 1937.

[2] Pars, Analytical Dynamics (in Russian), "Nauka", Moscow, 1971.

Appendix 4
Moment of Momentum Rigid Body with Respect to a Fixed Pole

Let us consider the rigid body in (Fig. A.4.1). Coordinate system $O\xi\eta\zeta$ is firmly connected to the body, and is moving with the same.

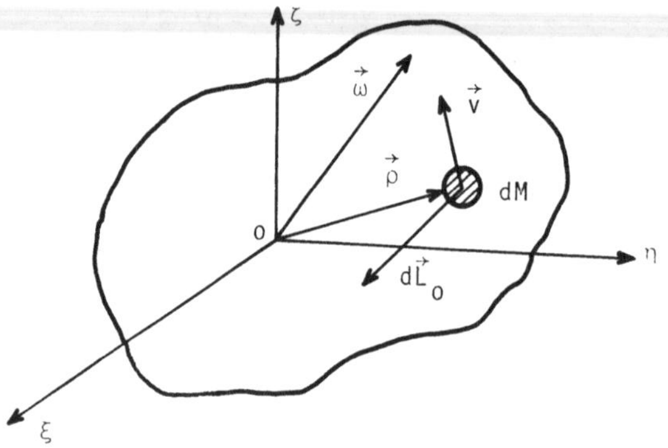

Fig. A.4.1. Moment of momentum of rigid body elementary mass

Moment of momentum of rigid body with respect to fixed pole O is:

$$\vec{L}_O = \int_M \vec{\rho} \times \vec{v} dM = \int_M \vec{\rho} \times (\vec{\omega} \times \vec{\rho}) dM \tag{A.4.1}$$

where vector $\vec{\rho}$ is the position vector of the elementary mass dM with respect to pole O, $\vec{v}$ is the velocity vector of elementary mass, $\vec{\omega}$ is the angular velocity vector of the rigid body. After expanding the double cross product in expression (A.4.1) one obtains:

$$\vec{L}_O = \vec{\omega} \int_M \rho^2 dM - \int_M \vec{\rho} (\vec{\omega} \cdot \vec{\rho}) dM \tag{A.4.2}$$

or,

$$\vec{L}_O = J_O \vec{\omega} - \int_M (\omega_\xi \xi + \omega_\eta \eta + \omega_\zeta \zeta) \vec{\rho} dM \tag{A.4.3}$$

where J_O is the polar moment of inertia for pole O. As the double polar moment of inertia equals the sum of the axial moments of inertia for the coordinate axes, $2J_O = J_\xi + J_\eta + J_\zeta$, the projections of the moment of momentum onto the coordinate axes are:

$$L_\xi = J_\xi \omega_\xi - J_{\xi\eta}\omega_\eta - J_{\xi\zeta}\omega_\zeta$$

$$L_\eta = -J_{\xi\eta}\omega_\xi + J_\eta \omega_\eta - J_{\eta\zeta}\omega_\zeta \qquad (A.4.4)$$

$$L_\zeta = -J_{\xi\zeta}\omega_\xi - J_{\eta\zeta}\omega_\eta + J_\zeta \omega_\zeta$$

If the coordinate axes are the main axes of inertia, the components of the moment of momentum for those axes are:

$$L_1 = J_1\omega_1, \qquad L_2 = J_2\omega_2, \qquad L_3 = J_3\omega_3 \qquad (A.4.5)$$

Euler's dynamic equations

For the rigid body there exists the relation between the rate change of moment of momentum and the moment of external forces:

$$\frac{d\vec{L}_O}{dt} = \vec{M}_O, \qquad (A.4.6)$$

that the time derivative of the moment of momentum for pole O equals the moment of external forces for the same pole as moment point.

If this moment of external forces equals zero, i.e. if external forces are not acting on the body (rotation by inertia), the moment of momentum is constant.

Let us designate by $\vec{L}_O$ the position vector of the terminal point of the moment of momentum (Fig. A.4.2); then the time derivative of this vector represents the absolute velocity of point N.

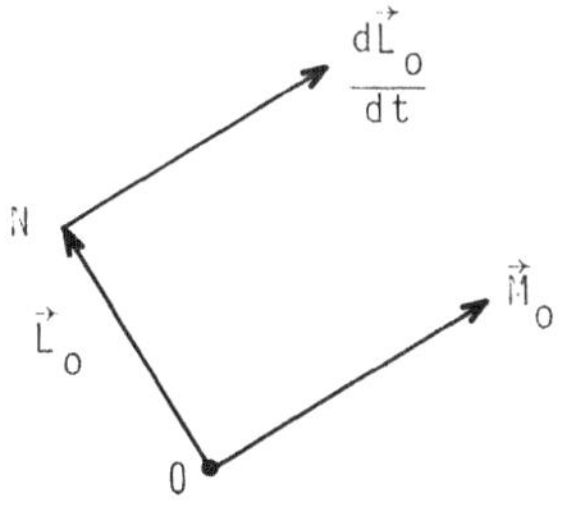

Fig. A.4.2. Moment of momentum rate change

As the absolute velocity consists of the relative and transfer velocity, the rate of change of moment of momentum can be written in the form:

$$\frac{d\vec{L}_o}{dt} = \dot{\vec{L}}_o + \vec{\omega} \times \vec{L}_o = \vec{M}_o \tag{A.4.7}$$

where $\dot{\vec{L}}_o = \dot{L}_\xi \vec{\xi}_o + \dot{L}_\eta \vec{\eta}_o + \dot{L}_\zeta \vec{\zeta}_o$ is the relative time derivative of the position vector $\vec{L}_o$, and $\vec{\xi}_o$, $\vec{\eta}_o$, $\vec{\zeta}_o$ are unit vectors of the ξ, η, ζ axes, respectively.

Equation (A.4.7) represents the Euler's dynamic equation in vector form. To this equation correspond three scalar equations for the coordinate axes of the moving trihedron:

$$\begin{aligned} \dot{L}_\xi + \omega_\eta L_\zeta - \omega_\zeta L_\eta &= M_\xi \\ \dot{L}_\eta + \omega_\zeta L_\xi - \omega_\xi L_\zeta &= M_\eta \\ \dot{L}_\zeta + \omega_\xi L_\eta - \omega_\eta L_\xi &= M_\zeta \end{aligned} \tag{A.4.8}$$

where the moment of momentum components for the coordinate axes are determined by equations (A.4.4). If the coordinate axes are the main axes of inertia, Euler's equations obtain a simpler form:

$$\begin{aligned} J_1\dot{\omega}_1 - (J_2-J_3)\omega_2\omega_3 &= M_1 \\ J_2\dot{\omega}_2 - (J_3-J_1)\omega_3\omega_1 &= M_2 \\ J_3\dot{\omega}_3 - (J_1-J_2)\omega_1\omega_2 &= M_3 \end{aligned} \tag{A.4.9}$$

where J_1, J_2, J_3 are the main moments of inertia, ω_1, ω_2, ω_3 are the projections of the angular velocity vector onto the main axes of inertia and moments M_1, M_2, M_3 are the main moments of all external forces for these axes.

References

[1] Kane T.R. Dynamics, New York: Holt, Reinhart and Winston, 1968.

Appendix 5
Specifities of Lever-Mechanisms Dynamics

A lever, more precisely a cane, is a homogenous cylinder, the diameter of which can be assumed negligible in comparison to its length. Let such cane be characterized by its length 2ℓ, its mass m and moments of inertia, J_N, with respect to an axis, perpendicular to the cane, passing through its center of mass and J_S, with respect to the cane axis. Let us note that, although the cane diameter was neglected, its moment of inertia J_S is considered a finite value, which represents an idealization widely used in mechanics.

Determining the main vector of inertial forces of the cane does not differ from the general case, while for determining the main moment direct use of Euler's equations is not suitable. Hence, different way was proposed by Stepanenko for deriving it (see footnote of Appendix).

The following expression can be written based on (A.4.4) and (A.4.6):

$$\vec{M} = -\frac{d}{dt}(\underline{\underline{J}}\vec{\omega}) = -\underline{\underline{J}}\frac{d\vec{\omega}}{dt} - \frac{d\underline{\underline{J}}}{dt}\vec{\omega} \tag{A.5.1}$$

where $\vec{M}$ and $\vec{\omega}$ are vectors of the moment of inertial forces and angular velocity, respectively, and $\underline{\underline{J}}$ is the body tensor of inertia, having with respect to the fixed coordinate system Oxyz the form:

$$\underline{\underline{J}} = \begin{bmatrix} J_{xx} & -J_{xy} & -J_{xz} \\ -J_{xy} & J_{yy} & -J_{yz} \\ -J_{xz} & -J_{yz} & J_{zz} \end{bmatrix} \tag{A.5.2}$$

The member $\frac{d\underline{\underline{J}}}{dt}\vec{\omega}$ in expression (A.5.1) is unsuitable, while the tensor of inertia derivative has a very complex form in the fixed coordinate system. The usual way of avoiding these difficulties consists in projecting the equations (A.5.1) onto the moving (local) coordinate system and differentiating them. As a result, Euler's dynamic equations (2.11) are obtained.

However, for a cane another way is possible. Second member in expression (A.5.1) can be represented in the form:

$$\frac{d\underline{\underline{J}}}{dt}\vec{\omega} = \underline{\underline{J}}\vec{\tau} \tag{A.5.3}$$

where $\vec{\tau} = (\vec{\omega}\cdot\vec{s})(\vec{s}\times\vec{\omega})$, $\vec{s}$ is the unit vector oriented along the cane axis. Vector $\vec{\tau}$ was nominated as equivalent angular acceleration.

In that case the expression for the moment of inertial forces can be written in the form:

$$\vec{M} = -\underline{\underline{J}}(\vec{\varepsilon}+\vec{\tau}) \tag{A.5.4}$$

where $\vec{\varepsilon}$ is the vector of the body angular acceleration. Furtheron, let vectors $\vec{M}$ and $\vec{\varepsilon}$ be divided into two components each (perpendicular and parallel to the cane axis):

$$\vec{M}_N = (\vec{s}\times\vec{M})\times\vec{s}; \qquad \vec{M}_S = (\vec{M}\cdot\vec{s})\vec{s}$$
$$\vec{\varepsilon}_N = (\vec{s}\times\vec{\varepsilon})\times\vec{s}; \qquad \vec{\varepsilon}_S = (\vec{\varepsilon}\cdot\vec{s})\vec{s} \tag{A.5.5}$$

From the definition of vector $\vec{\tau}$ is concluded, that the same is perpendicular to the cane axis. By the fact that the cane is symmetrical, and that its moment of inertia with respect to some arbitrary axis, perpendicular to s, passing through its center of mass, is J_N, expression (A.5.4) can be written as:

$$\vec{M} = -J_N(\vec{\varepsilon}_N+\vec{\tau}) - J_S\vec{\varepsilon}_S \tag{A.5.6}$$

Equation (A.5.6) determines the moment os inertial forces in the fixed system.

In order to derive the equivalent angular acceleration $\vec{\tau}$, let consider some point of the cane at distance ρ from its center of mass. The acceleration $\vec{w}_i$ of this point is:

$$\vec{w}_i = \rho\vec{\omega}\times(\vec{\omega}\times\vec{s}) \tag{A.5.7}$$

Let decompose $\vec{w}_i$ in two components: perpendicular to the cane - $\vec{w}_{in}$

and parallel to same - $\vec{w}_{is}$ (Fig. A.5.1):

$$\vec{w}_{in} = (\vec{s}\times\vec{w}_i)\times\vec{s}; \qquad \vec{w}_{is} = (\vec{s}\cdot\vec{w}_i)\vec{s} \tag{A.5.8}$$

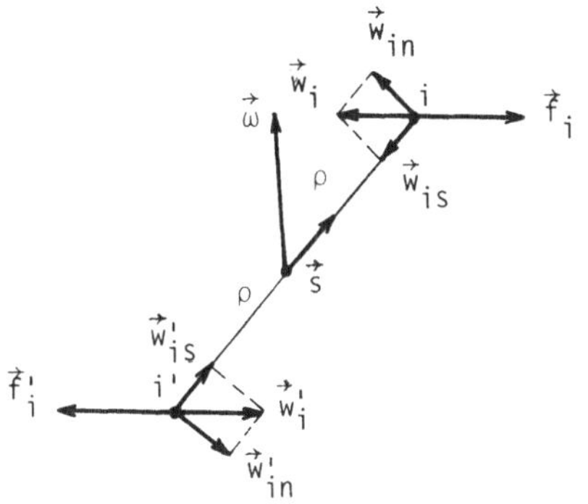

Fig. A.5.1. Cane accelerations

Since the cane is symmetrical, opposite to the center of mass at distance ρ from the same is a point i' the accelerations of which are $\vec{w}'_{in}$ and $\vec{w}'_{is}$:

$$\vec{w}'_{in} = -\vec{w}_{in}; \qquad \vec{w}'_{is} = -\vec{w}_{is} \tag{A.5.9}$$

Elementary inertial forces $\vec{f}_i$ and $\vec{f}'_i$ in points i and i' evidently will be proportional to accelerations. Hereby the components of the inertial forces depending on $\vec{w}_{is}$ and $\vec{w}'_{is}$ will be oriented along the cane axis and its resultant will be zero. Hence, the moment of inertial forces depends on components $\vec{w}_{in}$ and $\vec{w}'_{in}$ only.

Now, such acceleration $\vec{\tau}$ is adopted, that produces in points i and i' accelerations $\vec{w}_{in}$ and $\vec{w}'_{in}$, i.e. $\vec{\tau}$ should satisfy:

$$\rho\vec{\tau}\times\vec{s} = \vec{w}_{in} \tag{A.5.10}$$

For a cane, accelerations of all its points lie in one plane (plane of the vectors $\vec{s}$ and $\vec{\omega}$), so vector $\vec{\tau}$ is perpendicular to this plane. Introduce into (A.5.10) the expressions for $\vec{w}_{in}$ and $\vec{w}_i$, given by (A.5.8) and (A.5.7), respectively. One obtains:

$$\vec{\tau}\times\vec{s} = [\vec{s}\times(\vec{\omega}\times(\vec{\omega}\times\vec{s}))]\times\vec{s}, \tag{A.5.11}$$

wherefrom due to $\vec{\tau}\perp\vec{s}$ follows:

$$\vec{\tau} = \vec{s}\times(\vec{\omega}\times(\vec{\omega}\times\vec{s})). \qquad (A.5.12)$$

Transforming the right side (expanding the double vector product), final form is obtained:

$$\vec{\tau} = (\vec{\omega}\cdot\vec{s})(\vec{s}\times\vec{\omega}) \qquad (A.5.13)$$

The algorithm for calculating the moments of the cane inertial forces follows directly from expressions (A.5.3)-(A.5.6).

Let now the question be considered, in what cases the "cane type" model can be used and what errors are committed at doing it.

Let a homogenous circular cylinder of diameter d and length 2ℓ be considered. Connect to it a coordinate system, base of which is $\vec{q}_1$, $\vec{q}_2$, $\vec{q}_3$, whereby axis q_1 is oriented along the cylinder axis. Let the Euler's equations for the moment of inertial forces be written:

$$M^1 = -J_S\varepsilon^1$$

$$M^2 = -J_N\varepsilon^2 + (J_N-J_S)\omega^1\omega^3 \qquad (A.5.14)$$

$$M^3 = -J_N\varepsilon^3 + (J_S-J_N)\omega^1\omega^2$$

where the upper indexes designate the projection number onto the axes of the moving coordinate system, i.e.:

$$\varepsilon^j = \vec{\varepsilon}\cdot\vec{q}_j; \qquad M^j = \vec{M}\cdot\vec{q}_j; \qquad \omega^j = \vec{\omega}\cdot\vec{q}_j; \qquad j=1,2,3 \qquad (A.5.15)$$

For the sake of simplicity here has been omitted the upper tilde, previously designating the vectors given by projections in the moving coordinate system.

Let it be supposed that the cane has moments of inertia J_N, J_S, mass m and length 2ℓ, like the cylinder. Write for the cane expressions for the moment of inertial forces. In order to compare the result with (A.5.14), write basic equation (A.5.6) by means of the projections onto the moving axes. Then:

$$M^1 = -J_S\varepsilon_S = -J_S\varepsilon^1$$

$$M^2 = -J_N(\vec{\varepsilon}_N \cdot \vec{q}_2) - J_N(\vec{\tau} \cdot \vec{q}_2) \tag{A.5.16}$$

$$M^3 = -J_N(\vec{\varepsilon}_N \cdot \vec{q}_3) - J_N(\vec{\tau} \cdot \vec{q}_3)$$

Using expressions (A.5.13) for $\vec{\tau}$, it can be written:

$$\vec{\tau} \cdot \vec{q}_j = \vec{q}_j \cdot (\vec{\omega} \cdot \vec{q}_1)(\vec{q}_1 \times \vec{\omega}) = \omega^1 \vec{q}_j \cdot (\vec{q}_1 \times \vec{\omega}) = \omega^1 \vec{\omega} \cdot (\vec{q}_j \times \vec{q}_1)$$

Then:

$$\vec{\tau} \cdot \vec{q}_2 = -\omega^1 \vec{\omega} \cdot \vec{q}_3 = -\omega^1 \omega^3$$

$$\vec{\tau} \cdot \vec{q}_3 = \omega^1 \vec{\omega} \cdot \vec{q}_2 = \omega^1 \omega^2$$

Introducing these expressions into (A.5.16) and taking into account that:

$$\vec{\varepsilon}_N \cdot \vec{q}_2 = \varepsilon^2; \qquad \vec{\varepsilon}_N \cdot \vec{q}_3 = \varepsilon^3$$

it is obtained:

$$M^1 = -J_s \varepsilon^1$$

$$M^2 = -J_N \varepsilon^2 + J_N \omega^1 \omega^3 \tag{A.5.18}$$

$$M^3 = -J_N \varepsilon^3 - J_N \omega^1 \omega^2$$

Comparing (A.5.18) with (A.5.14) can be noted, that due to passing to cane, following moment is "lost":

$$\Delta\vec{M} = J_s \omega^1 (\omega^2 \vec{q}_3 - \omega^3 \vec{q}_2) \tag{A.5.19}$$

This moment is perpendicular to the cane, its module being:

$$|\Delta M| = J_s |\omega^1 \vec{\omega}_N| \tag{A.5.20}$$

where $\vec{\omega}_N$ - component of angular velocity, perpendicular to the cane axis:

$$\vec{\omega}_N = (\vec{s} \times \vec{\omega}) \times \vec{s} \tag{A.5.21}$$

From (A.5.20) follows that ΔM is a variable, which equals zero when the vector of angular velocity coincides with the cylinder axis, or when is perpendicular to it; ΔM attains its maximum value when the angle between $\vec{\omega}$ and the cylinder axis is $\pi/4$. It is interesting to note, that ΔM depends on angular velocity only. Let separate from the Euler's equations (A.5.14) the component of the moment, depending on angular velocity and denote it by:

$$\vec{M}_{\omega} = (J_N - J_S)\omega^1(\omega^3\vec{q}_2 - \omega^2\vec{q}_3) \tag{A.5.22}$$

Then the relative error can be determine from relation:

$$\delta = \left|\frac{\Delta M}{M_{\omega}}\right| \tag{A.5.23}$$

Introducing in this ratio the values for ΔM and M_{ω} from (A.5.20) and (A.5.22), and adopting that for the cylinder $J_N = m(\frac{d^2}{16} + \frac{L^2}{12})$, $J_S = \frac{m}{8}d^2$, $L = 2\ell$, the relative error is obtained:

$$\delta = \frac{2}{\frac{4}{3}(\frac{L}{d})^2 - 1} \tag{A.5.24}$$

In Fig. A.5.2 error δ as function of the cylinder length to diameter ratio is given. It follows, that for ratios over 5 the error is smaller than 6%*).

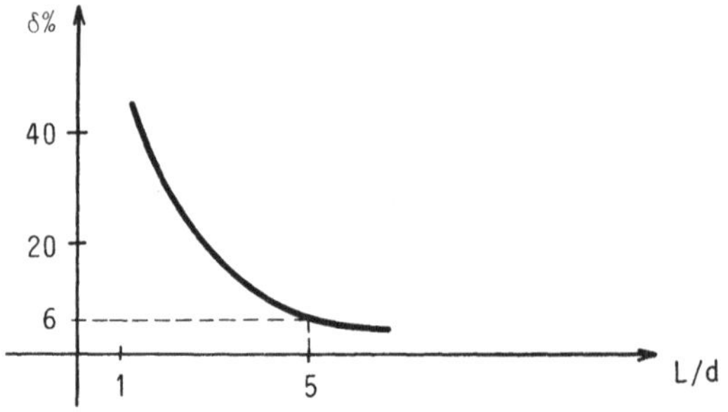

Fig. A.5.2. Error as function of L/D

*) The complete proof in this Appendix is done by Y. Stepanenko in his book Dynamics of Spatial Mechanisms (in Russian), Mathematical Institute, Belgrad, 1974.

Appendix 6
Mathematical Models of Driving Units

For actuators of industrial robots electromechanical, electrohydraulical and electropneumatical servodrives are used.

Electric servomotors used nowadays in practice in most cases are DC permanent-magnet motors and AC motors (induction motors and synchronous motors with permanent-magnet rotor).

Although to-day application of individual types of actuators is changing, it can be concluded that electric servomotors are dominant with robots of up to 100 kg carrying capacity, while electrohydraulical actuators are used above 100 kg. In the future shifting of this border upwards is expected to the benefit of electric drives. Due to evident tendency of practical application, the electrical actuator without reducer, i.e. the so-called Direct-Drive will be presented. Finally, the electropneumatical drive has its chance in industrial robotics, in spite of the fact that for the time being such drive is applied for low-mass manipulation and with short manipulation strokes.

Permanent-magnet DC servomotor

Position controlled servodrives with high dynamic characteristics are today almost exclusively DC motors fed by power converters. The stator field is produced by permanent magnets of rare-earth, so the possibility of field weakening is excluded. The mechanical construction of a DC motor is complicated by the commutator, which sets a limit to 3000 - 4000 min^{-1} on the rotational speed, when powerful units are used. DC motors is sensitive to torque overload at low speed or standstill operations.

There are two types of DC servomotors: the slim drum type motor with radial magnetic field for machine tool applications, and the disk motor with an iron-free armature and axial magnetic field for robotic applications, because of its compact design and short axial length.

Servodrive with permanent-magnet DC servomotor, encloses the DC servomotor and the r.p.m. controller, as well as the servo-controller. On the motor shaft are mounted the tachogenerator, encoder and the electromechanical brake. R.p.m. control is effected by DC voltage control at the output of the controller.

In Fig. A.6.1. the schematic presentation of a permanent-magnet DC servomotor with constant magnetic field is given.

Based on scheme in Fig. A.6.1. the following differential equations can be written:

$$L_R \frac{di_R}{dt} + r_R i_R + C_E N_v \dot{q} = u$$

$$J_M N_v \ddot{q} + \frac{F_v \dot{q}}{N_m} + \frac{M^*}{N_m} = C_M i_R$$

(A.6.1)

where: i_R - rotor control current [A]

u - rotor control voltage [V]

r_R - rotor resistance [Ohm]

L_R - rotor winding inductance [H]

q - output angle of reducer output shaft [rad]

$\dot{q}$ - output angular velocity of output motor shaft [rad/s]

$\ddot{q}$ - output angular acceleration of output motor shaft [rad/s^2]

C_M - torque constant [Nm/A]

C_E - electromotor force constant [V/rad/s]

F_v - viscous damping constant [Nm/rad/s]

J_M - rotor moment of inertia [kgm^2]

J_R - rotor moment of inertia reduced to output shaft
$J_R = J_M N_v N_m$ [kgm^2]

N_v - speed reduction ratio

N_m - torque multiplier ratio

M^* - load torque [Nm].

By rearranging equation system (A.6.1) the linear third order mathematical model of a DC motor can be obtained in the form of state space presentation:

$$\dot{x}^i = A^i x^i + b^i N(u^i) + f^i M_i^* \tag{A.6.2}$$

where: x^i - state vector, $x^i = [q^i, \dot{q}^i, i_R^i]^T$,

u^i - scalar input of control signal

$N(u^i)$ - nonlinearity of amplitude saturation type (motor voltage)

$$N(u^i) = \begin{cases} u_m^i & \text{for} \quad u^i \geq u_m^i \\ u^i, & \text{for} \quad -u_m^i < u^i < u_m^i \\ -u_m^i, & \text{for} \quad u^i \leq -u_m^i \end{cases} \tag{A.6.3}$$

M_i^* - driving torque (load) acting on i-th actuator

A^i - subsystem matrix of dimension (3×3)

b^i - input distribution vector of dimension (3×1)

f^i - load distribution vector of dimension (3×1).

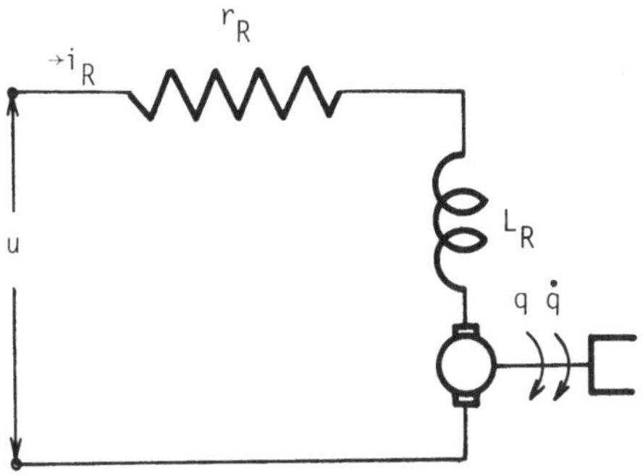

Fig. A.6.1. Simplified scheme of DC motor

Matrix A^i and the vectors f^i and b^i can be written in following form:

$$A^i = \begin{bmatrix} 0 & 1 & 0 \\ 0 & -F_v/J_R & \dfrac{C_M N_m}{J_R} \\ 0 & -\dfrac{C_E N_v}{L_R} & -\dfrac{r_R}{L_R} \end{bmatrix}, \quad f^i = \begin{bmatrix} 0 \\ -1/J_R \\ 0 \end{bmatrix}, \quad b^i = \begin{bmatrix} 0 \\ 0 \\ 1/L_R \end{bmatrix} \tag{A.6.4}$$

In the case when viscous friction can be neglected, element a_{22} of matrix A^i vanishes.

The mathematical model of the DC motor, which is in the general case of third order can be reduced to second order, under the condition that rotor inductance can be neglected. Then matrix A^i and vectors f^i and b^i can be given as:

$$A^i = \begin{bmatrix} 0 & 1 \\ 0 & -\dfrac{C_E C_M N_m N_V}{J_R r_R} \end{bmatrix}, \quad f^i = \begin{bmatrix} 0 \\ -1/J_R \end{bmatrix}, \quad b^i = \begin{bmatrix} 0 \\ \dfrac{C_M N_m}{J_R r_R} \end{bmatrix} \qquad (A.6.5)$$

Since in the considered mathematical model the angular motor speed is used, the electromotor force constant C_E is of the same nature and value as the torque constant C_M.

AC servomotor

Under AC servomotors the induction (asynchronous) motors and synchronous motors (SM) with permanent-magnet rotor will be understood. High performance AC servomotors can be made with pulse-width modulated transistor (thyristor) inverters and microcomputer reprogrammable control. Both types of AC servomotors have the possibility of field weakening and four quadrant operation. Also, there are two types of AC servomotors by construction: slim drum motors and disk motors, for same applications as in the case of DC servomotors.

Induction motors (IM). Induction motors with squirrel cage rotor are very interesting like servodrives, even in regard to SM, because of their rugged mechanical construction and low cost. The main problem in applying induction motors consists in the balance of the flux level to the nominal value, down to zero frequency and up to base speed, and then constant power beyond it. In the constant power region, the flux is programmed to be inversely proportional to the motor speed. A servodrive with an induction motor encloses the motor with tachogenerator, encoder and brake and the r.p.m. regulator.

To realize the mathematical model of IM, we would suppose the following:

1. Stator phase windings are identical and mutually displaced per perihel of machine for 120° electrical,

2. Magnetomotive force (MMF) of windings, and flux too, are sinusoidally placed per perihel of air-gap,

3. Air-gap is uniform, more exactly, the stator and the rotor are of cylindrical form,

4. Rotor cage is displaced in that manner, so we can understand the rotor MMF is sinusoidally placed per perihel, with the same pole number as the stator MMF,

5. Winding resistances of stator and rotor, and all dissipated inductances are constant,

6. The phenomenon of eddy currents and hysteresis can be ignored,

7. All electromagnet energy is based on the magnet circuit of the machine (all parasite capacitances are ignored),

8. Magnet characteristic is linear, the machine is not saturated,

9. Resistance changes versus temperature and skin effect, and changes of dissipated inductances versus saturation and position of rotor, as well as higher spatial harmonics are ignored.

According to the way of flux maintenance, two control methods can be distinguished [5, 12]:

(a) Method of "vector - oriented field"

(b) "Volt - Hertz" method.

(a) Basic system of the equations of the mathematical model of the induction motor (IM) is given in matrix form in the component "d-q" coordinate system, synchronized with the rotating magnetic field:

$$\begin{bmatrix} u_{ds} \\ u_{qs} \\ 0 \\ 0 \end{bmatrix} = \begin{bmatrix} r_s+sL_s & -\omega_s L_s & sL_m & -\omega_s L_m \\ \omega_s L_s & r_s+sL_s & \omega_s L_m & sL_m \\ sL_m & -\omega_{s\ell} L_M & r_R+sL_R & -\omega_{s\ell} L_R \\ \omega_{s\ell} L_m & sL_m & \omega_{s\ell} L_R & r_R+sL_R \end{bmatrix} \begin{bmatrix} i_{ds} \\ i_{qs} \\ i_{dr} \\ i_{qr} \end{bmatrix} \quad \text{(A.6.6)}$$

The algebraic equation of the electric torque is:

$$M_e = L_m(i_{qs}\, i_{dr} - i_{ds}\, i_{qr}) \quad \text{(A.6.7)}$$

as well as the differential equation of dynamic equilibrium is:

$$J_M N_v \ddot{q} + \frac{F_v \dot{q}}{N_m} + \frac{M^*}{N_m} = n\pi M_e \quad (A.6.8)$$

u_{ds}, u_{qs} - stator voltages in the d-q component coordinate system [V]

i_{ds}, i_{qs} - stator currents in the d-q component coordinate system [A]

i_{dr}, i_{qr} - rotor currents in the d-q component coordinate system [A]

J_M - rotor moment of inertia [kgm^2]

J_R - rotor moment of inertia reduced to output shaft [kgm^2]

L_s - stator windings inductance [H]

L_R - rotor windings inductance [H]

L_M - magnetizing inductance [H]

$u^i = [u_{ds}, u_{qs}]^T$ - control vector

n - number of phases

π - number of pole pairs

M_e - motor electric torque [Nm]

M^* - mechanical load torque [Nm]

s - Laplace operator

r_s - resistance of stator windings [Ω]

r_R - resistance of rotor windings [Ω]

ω_s - angular speed of stator rotational magnetic field [rad/s]

$\omega_r = \dot{q}^m$ - rotor angular velocity [rad/s]

$\omega_{s\ell} = \omega_s - \omega_r$ - slippage angular velocity [rad/s]

$q^m = \theta_r$ - rotor position with respect to d-axis [rad]

θ_s - stator position with respect to d-axis [rad]

σ - dissipation factor

ψ_r - rotor flux [Vs]

All rotor values were reduced to the stator.

By suitable transformation the mathematical model of the induction motor (IM) can be presented as a fifth-order nonlinear model in the x - state space, with element $x_i \in X$

$$x = [x_1\ x_2\ x_3\ x_4\ x_5]^T \quad (A.6.9)$$

where: $x_1 = q^m$ - position of the IM shaft

$x_2 = \dot{q}^m$ - IM angular velocity, $x_3 = M_e$ - electrical torque

$x_4 = \psi_r = (\psi_r^T \psi_r)^{1/2}$ - module of rotor flux vector

$x_5 = \dot{\psi}_r$,

in the following form [5, 12]:

$$\dot{x}^i = A^i(x^i) + b^i(x^i)N(u^i) + f^i M_i^* \tag{A.6.10}$$

i.e.

$$s\begin{bmatrix} x_1 \\ x_2 \\ x_3 \\ x_4 \\ x_5 \end{bmatrix} = \begin{bmatrix} 0 & x_2 & 0 & 0 & 0 \\ 0 & -\frac{F_v}{J_R}x_2 & \frac{n\pi N_m}{J_R}x_3 & 0 & 0 \\ 0 & -(\frac{x_4^2}{\sigma L_R} + \frac{x_4 x_5}{r_R})N_v x_2 & -(\frac{r_s}{\sigma L_s} + \frac{r_R}{\sigma L_R})x_3 & 0 & 0 \\ 0 & 0 & 0 & 0 & x_5 \\ 0 & r_R \frac{x_2 x_3}{x_4} N_v & r_R^2 \frac{x_3^2}{x_4^3} & -\frac{r_s r_R}{\sigma L_R L_s} x_4 & -(\frac{r_s}{\sigma L_s} + \frac{r_R}{\sigma L_R})x_5 \end{bmatrix}$$

$$+ \begin{bmatrix} 0 & 0 \\ 0 & 0 \\ 0 & \frac{L_M}{\sigma L_s L_R} x_4 \\ 0 & 0 \\ \frac{r_R L_M}{\sigma L_s L_R} & 0 \end{bmatrix} \begin{bmatrix} u_{ds} \\ u_{qs} \end{bmatrix} + \begin{bmatrix} 0 \\ -1/J_R \\ 0 \\ 0 \\ 0 \end{bmatrix} M^* \tag{A.6.11}$$

where: $A^i(x^i)$ - subsystem matrix of dimension (5×5)

$b^i(x^i)$ - input distribution matrix of dimension (5×2)

f^i - load distribution vector of dimension (5×1)

$N(u^i)$ - nonlinearity of amplitude saturation type.

The IM is controlled in a coordinate system, which is synchronized with the rotational magnetic field and is shown schematically in Fig. A.6.2.

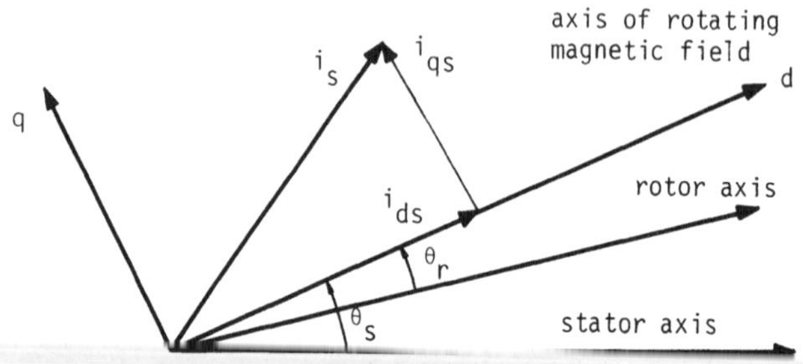

Fig. A.6.2. Schematic presentation of rotating coordinate system of IM

With IM the rotor flux is a time-variant value and it is necessary to maintain it on its nominal value, i.e. in some narrow region about the nominal flux $\psi_r \in [\psi_{rN} - \frac{\Delta\psi}{2}, \psi_{rN} + \frac{\Delta\psi}{2}]$.

The fifth-order nonlinear mathematical model of IM, according to (A.6.11) can be represented as a linear time-varying model of third order and introduced into the complete dynamic manipulator model, along with added calculation from the two resting equations of the time-varying rotor flux.

Such mathematical model, regarded from the motor side is still a fifth-order nonlinear model and it will be nominated as a fifth-order model or a model "three plus two", in the following matrix form:

$$\dot{x}^i = A^i(t)x^i + b^i(t)N(u^i) + f^i M_i^* \tag{A.6.12}$$

where:

$$A^i(t) = \begin{bmatrix} 0 & 1 & 0 \\ 0 & -F_V/J_R & n\pi N_m/J_R \\ 0 & -N_V\left(\frac{\psi_r^2}{\sigma L_R} + \frac{\psi_r \dot{\psi}_r}{r_R}\right) & -\left(\frac{R_s}{\sigma L_s} + \frac{r_R}{\sigma L_R}\right) \end{bmatrix},$$

$$b^i(t) = \begin{bmatrix} 0 \\ 0 \\ \frac{L_M}{\sigma L_s L_R}\psi_r \end{bmatrix}, \qquad f^i = \begin{bmatrix} 0 \\ -1/J_R \\ 0 \end{bmatrix} \tag{A.6.13}$$

The time-varying flux is calculated from the two resting model equations (A.6.11).

If the rotor flux variations are neglected and regarded as always equal to the nominal flux, the IM mathematical model can be represented as a third-order linear model in following form:

$$\dot{x}^i = A^i x^i + b^i N(u^i) + f^i M_i^* \tag{A.6.14}$$

where:

$$A^i = \begin{bmatrix} 0 & 1 & 0 \\ 0 & -F_V/J_R & n\pi N_m/J_R \\ 0 & -N_V \frac{\psi_{rN}^2}{\sigma L_R} & -\left(\frac{r_s}{\sigma L_s} + \frac{r_R}{\sigma L_R}\right) \end{bmatrix},$$

$$b^i = \begin{bmatrix} 0 \\ 0 \\ \frac{L_M}{\sigma L_s L_R}\psi_{rN} \end{bmatrix}, \qquad f^i = \begin{bmatrix} 0 \\ -1/J_R \\ 0 \end{bmatrix} \tag{A.6.15}$$

(b) The "Volt - Hertz" - method basically maintains the flux on the nominal level by keeping the ratio U_s/ω_s constant, so the mathematical model reduces to a second-order linear model in the state space in following form:

$$\dot{x}^i = A^i x^i + b^i N(F(U_s)) + f^i M_i^* \tag{A.6.16}$$

$$A^i = \begin{bmatrix} 0 & 1 \\ 0 & -F_V/J_R \end{bmatrix}, \quad b^i = \begin{bmatrix} 0 \\ n\pi N_m/J_R \end{bmatrix}, \quad f^i = \begin{bmatrix} 0 \\ -1/J_R \end{bmatrix} \tag{A.6.17}$$

U_s - motor phase voltage [V]

$F(U_s)=M_e$ - motor electric torque [Nm]

$$U_s = \frac{r_s}{L_s}\psi_{sN}\sqrt{\frac{[(\omega_s-\dot{q}^m)\frac{L_R}{r_R}+\omega_s\frac{L_s}{r_s}]^2+[1-(\omega_s-\dot{q}^m)\omega_s\sigma\frac{L_R}{r_R}\frac{L_s}{r_s}]^2}{1+[(\omega_s-\dot{q}^m)\sigma\frac{L_R}{r_R}]^2}} \tag{A.6.18}$$

$$\omega_s = \dot{q}^m + \frac{r_R}{\psi_r^2}\frac{M_e}{n\pi} \tag{A.6.19}$$

$$F(U_s)=M_e = \frac{U_s^2}{r_s}\frac{L_s}{r_s}\frac{(1-\sigma)(\omega_s-\dot{q}^m)\frac{L_R}{r_R}}{[(\omega_s-\dot{q}^m)\frac{L_R}{r_R}+\omega_s\frac{L_s}{r_s}]^2+[1-(\omega_s-\dot{q}^m)\omega_s\sigma\frac{L_R}{r_R}\frac{L_s}{r_s}]^2} \tag{A.6.20}$$

Expressions (A.6.18 - A.6.20) are recurrent for calculating the equivalent control $F(U_s)$.

Synchronous motors with permanent magnet rotor (SM)

SM fed from inverters are becoming increasingly interesting in a wide range of speed control applications, particularly following the recent introduction of commercial neodymium-ironboron (Nd-Fe-B) magnet material with a relative permeability very near unity. The magnet appears to the stator magnetomotive force as an equivalent air gap. Maximum torque per ampere of stator current is achieved when the stator field is displaced by $\delta = 90^o$ angulary from the magnet field of the rotor.

Cross section of four-pole SM is given in Fig. A.6.3b, and sinusoidal dependance of torque on the load angle δ is shown in Fig. A.6.3c.

Servodrives with synchronous motors with a permanent magnet rotor consists of SM with tachogenerator, encoder and brake and an r.p.m. controller, as well as a servocontroller. In the forming of SM mathematical models, two methods exist, too:

a) Method of "vector-oriented field", and
b) "Volt - Hertz method".

a) Method of "vector-oriented field". Initial mathematical model equations of SM, with the same suppositions like in the case of IM, is given in matrix form in the "d-q" coordinate system, which is synchronized with the rotor coordinate system:

$$\begin{bmatrix} u_{ds} \\ u_{qs} \end{bmatrix} = \begin{bmatrix} r_s+sL_s & -\omega_s L_s & 0 \\ \omega_s L_s & r_s+sL_s & \omega_s L_m \end{bmatrix} \begin{bmatrix} i_{ds} \\ i_{qs} \end{bmatrix} \tag{A.6.21}$$

The algebraic equation of the electric torque for SM is

$$M_e = \psi_F i_{sq} \tag{A.6.22}$$

where ψ_F is the flux of permanent magnet rotor.

The notations used are the same as in the case of IM. The equation (A.6.8), is valid also for SM. Then, in the case of SM, the angular velocity of the rotational magnetic field and that of the rotor are identical, $\omega_s = \omega_r$.

By rearranging the previous equations (A.6.8, A.6.21, A.6.22) analogously, as done in the case of IM, the mathematical model of SM is obtained in the state space X, with elements $x^i \epsilon X$:

$$x = [x_1 \; x_2 \; x_3 \; x_4]^T \tag{A.6.23}$$

where: $x_1 = q^m$ - SM shaft position; $x_2 = \dot{q}^m$ - SM angular velocity; $x_3 = M_e$ - SM electric torque; $x_4 = \delta$ - load angle (spatial angle between the stator and magnet fields)

in the following form:

$$\dot{x}^i = A^i(x^i) + b^i(x^i)N(u^i) + f^i M_i^* \tag{A.6.24}$$

i.e.:

$$s \begin{bmatrix} x_1 \\ x_2 \\ x_3 \\ x_4 \end{bmatrix} = \begin{bmatrix} 0 & x_2 & 0 & 0 \\ 0 & -\frac{F_V}{J_R} x_2 & \frac{\mu\pi N_m}{J_R} x_3 & 0 \\ 0 & -N_V \frac{\psi_F^2}{L_s} x_2 & -(\frac{r_s}{L_s} + N_V x_2 \operatorname{ctg} x_4) x_3 & 0 \\ 0 & -N_V(1 + \frac{\psi_F^2}{L_s} \frac{1}{x_3} \frac{\sin 2x_4}{2}) & 0 & 0 \end{bmatrix} +$$

$$+ \begin{bmatrix} 0 & 0 \\ 0 & 0 \\ 0 & \frac{\psi_F}{L_s} \\ -\frac{\psi_F}{L_s} \frac{\sin^2 x_4}{x_3} & \frac{\psi_F}{L_s} \frac{\sin 2x_4}{2x_3} \end{bmatrix} \begin{bmatrix} u_{ds} \\ u_{qs} \end{bmatrix} + \begin{bmatrix} 0 \\ -1/J_R \\ 0 \\ 0 \end{bmatrix} M^* \tag{A.6.25}$$

The SM mathematical model is of fourth order and nonlinear. The SM is being controlled in a coordinate system, synchronized with the rotor coordinate system, presented schematically in Fig. A.6.3a.

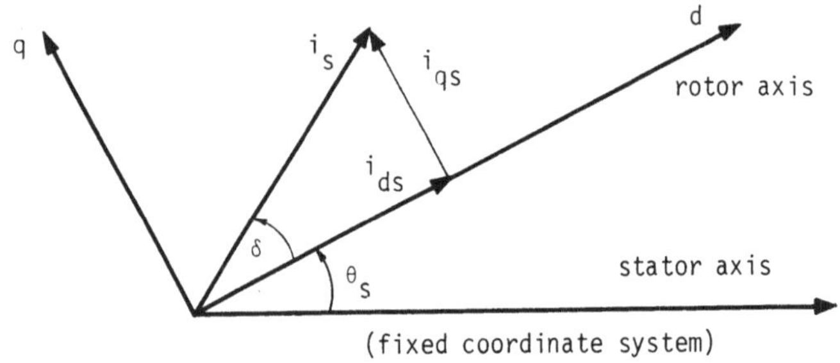

Fig. A.6.3a. Schematic presentation of the rotating coordinate systems of SM

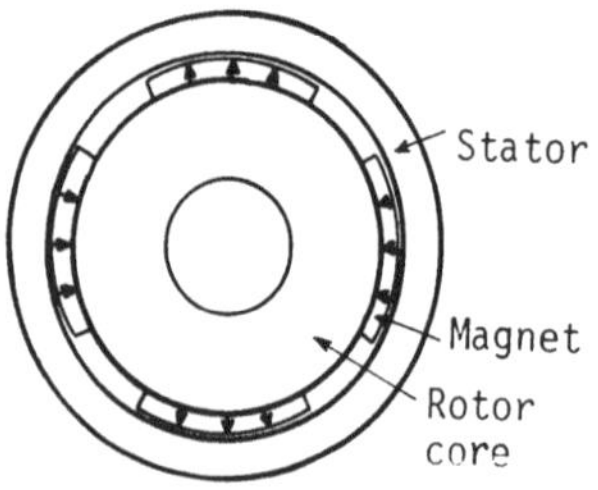

Fig. A.6.3b. Cross section of four-pole SM

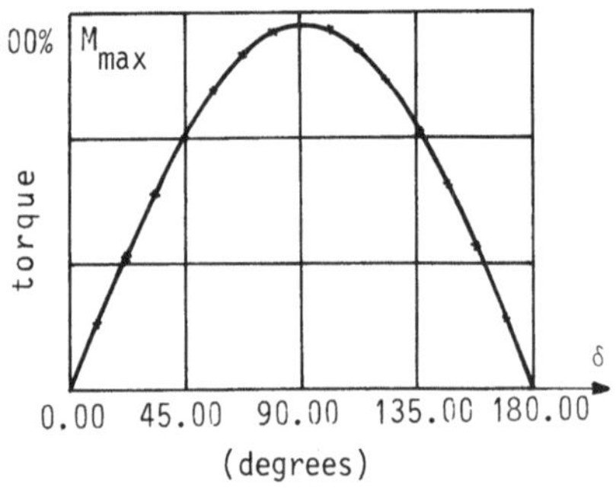

Fig. A.6 .3c. Shape of the maximal torque versus load angle δ

With SM, the variable parameter is the load angle (δ) and it is necessary to keep it constant at the value of $\delta = \pi/2$, because in that case the developed electric moment is maximal, i.e. in some region around $\pi/2$, $\delta \in [\pi/2 - \frac{\Delta\delta}{2}, \pi/2 + \frac{\Delta\delta}{2}]$. For maintaining this function same methods as for IM are used. The SM nonlinear mathematical model of fourth order via (A.6.25) can be represented as a linear time-varying third-order model and this model is used for forming the complete manipulator dynamic system model, along with added calculation of the load angle δ from the resting equation. This mathematical model, regarded from the motor side, is still a fourth-order nonlinear model, or a model "three plus one" and is given in following matrix form:

$$\dot{x}^i = A^i(t)x^i + b^i N(u^i) + f^i M_i^* \tag{A.6.26}$$

where:

$$A^i(t) = \begin{bmatrix} 0 & 1 & 0 \\ 0 & -F_v/J_R & n\pi N_m/J_R \\ 0 & -N_v \frac{\psi_F^2}{L_s} & -(\frac{r_s}{L_s} + \omega_s ctg\delta) \end{bmatrix}$$

$$b^i = \begin{bmatrix} 0 \\ 0 \\ \frac{\psi_F}{L_s} \end{bmatrix}, \qquad f^i = \begin{bmatrix} 0 \\ -1/J_R \\ 0 \end{bmatrix} \tag{A.6.27}$$

and the time-varying load angle δ is calculated from the fourth equation.

If we neglect the load angle variations and consider $\delta = \pi/2$, the mathematical model of SM can be represented as a linear time invariant third-order model in the form:

$$\dot{x}^i = A^i x^i + b^i N(u^i) + f^i M_i^* \tag{A.6.28}$$

where

$$A^i = \begin{bmatrix} 0 & 1 & 0 \\ 0 & -F_v/J_R & n\pi N_m/J_R \\ 0 & -N_v \frac{\psi_F^2}{L_s} & -r_s/L_s \end{bmatrix}, \quad b^i = \begin{bmatrix} 0 \\ 0 \\ \frac{\psi_F}{L_s} \end{bmatrix}, \quad f^i = \begin{bmatrix} 0 \\ -1/J_R \\ 0 \end{bmatrix} \tag{A.6.29}$$

b) <u>The "Volt - Hertz" method</u>. Using this method for the control of SM, the mathematical model reduces to a second-order linear model in the state space, given in following form:

$$\dot{x}^i = A^i x^i + b^i N(F(U_s)) + f^i M_i^* \tag{A.6.30}$$

where:

$$A^i = \begin{bmatrix} 0 & 1 \\ 0 & -F_v/J_R \end{bmatrix}, \quad b^i = \begin{bmatrix} 0 \\ N_m/J_R \end{bmatrix}, \quad f^i = \begin{bmatrix} 0 \\ -1/J_R \end{bmatrix} \tag{A.6.31}$$

U_s - effective value of the phase voltage of stator,

$F(U_s)=M_e$ - motor electric torque [Nm]:

$$U_s = \frac{M_e}{C_M} \cdot \sqrt{(\omega_s L_s)^2 + (C_e \omega_s + r_s \frac{M_e}{C_M})^2} \tag{A.6.32}$$

$$\omega_s = \dot{q}^m = N_V \cdot \dot{q} \tag{A.6.33}$$

$$M_e = C_M \cdot I_s \tag{A.6.34}$$

C_M - torque constant [Nm/A]

C_e - electromotor force constant [V/rad/s].

Relations (A.6.32 - A.6.34) represent recurrent relations for the calculation of the control $F(U_s)$.

Direct-drive motors

Actuators, being used for direct-drive of robots, can be classified in two basic groups:

1. *Electric servomotors with continual output.*

2. *Electric servomotors with discrete output.*

In the first group belong servomotors, on the input of which a continual command signal is applied and a continual output (e.g. angle) is obtained, while in the second group belong those on the input of which a continual signal is applied and at their output some discrete value is obtained (e.g. the rotor rotates by some angle and maintains that value during the duration of the command signal). For the second group of motors the term *step motors* is used.

Due to the fact that the direct-drive poses high demands, notably concerning torque, it is necessary to carry out some modifications of the conventional DC or AC motors. To-day, direct-drives (although still in initial applications) are mainly based on DC motors, so they will be treated in this book.

Let us limit to two subgroups only:

(I) Modified Torque DC Servomotors

(II) Brushless DC Servomotors

Into the second servomotor group of the step motors can be included:

(a) Permanent-magnet step motors

(b) Variable reluctance step motors

(c) Hybrid motors

(d) Step motors with rotating magnetic field.

I *Modified DC Torque Servomotors*

Basic difference between the standard and the modified DC servomotors, used in direct drive is in the fact, that with modified DC servomotors the output torque is maximized, while with the standard ones, the power.

Output torque is given by expression [7]:

$$P^m = C_A \cdot AD_r = C_A \ell_r D_r^2 \qquad (A.6.35)$$

where C_A is a constant, defined by the winding magnetic field, ℓ_r is the rotor length, D_r is the motor diameter and A is the rotor cross section.

Equation of voltage equilibrium of the modified DC servomotor is the same as with the standard one:

$$u = r_R i_R + e + L_R \dot{i}_R \qquad (A.6.36)$$

where u is the rotor voltage, r_R is the resistance of the rotor circuit, i_R is the rotor current, e is the back EMF, L_R is the inductivity of rotor. Back EMF is given by:

$$e = C_M \dot{q} \qquad (A.6.37)$$

where C_M is the torque constant, or back EMF constant, when IS of units are used ($C_M = C_E$) and $\dot{q}$ is the angular velocity.

If analysis of the energetic balance is performed and the input electric power ei_R is equalized with the output mechanical power Q^m, one obtains

$$Q^m = ei_R \qquad (A.6.38)$$

or

$$ei_R = M_e\dot{q} \tag{A.6.39}$$

By substituting equation (A.6.37) into (A.6.39) it is obtained

$$M_e = C_M i_R \tag{A.6.40}$$

C_M, the torque constant is only constant for some ideal torque motor, while, in practical conditions it varies with the rotor position.

In the case when the rotor rotates with constant angular velocity,

$$\dot{q} = \text{const} \tag{A.6.41}$$

$$M_e = \frac{C_M}{r_R} u - \frac{C_M^2}{r_R} \dot{q} \tag{A.6.42}$$

When the rotor rotates with constant angular velocity $\dot{q}$ = const, on the output shaft the mechanical power Q^m is obtained

$$Q^m = M_e\dot{q} = (\frac{C_M}{r_R} u - \frac{C_M^2}{r_R} \dot{q})\dot{q} \tag{A.6.43}$$

The term of the square motor constant resulting from the back EMF, has the role of a damping factor (see the curve in Fig. A.6.4).

By varying the torque with respect to angular velocity, the standard and most characteristic diagram is given by the linear function in Fig. A.6.5.

Using equations (A.6.36) and (A.6.40) dependence of motor torque as function of the rotor voltage u is obtained and if the Laplace transformation is performed of the expression for torque, it is obtained:

$$M_e(s) = \frac{\frac{C_M}{r_R} U(s) - \frac{C_M^2}{r_R} \Omega(s)}{1 + T_e s} \tag{A.6.44}$$

where, $\Omega(s)$ is Laplace transform of angular velocity and T_e is the electrical time constant, given by:

$$T_e = \frac{L_R}{r_R} \tag{A.6.45}$$

The electrical time constant T_e is small for standard DC motors, but for this version of modified motor, it should be introduced. In order to generate a high torque at the motor output shaft, the direct drive motor must possess a large diameter and windings producing large inductivity. Variation of the electrical time constant is of great influence on system stability (augmenting T_e produces a stabilizing effect), so that it should be notably taken care of.

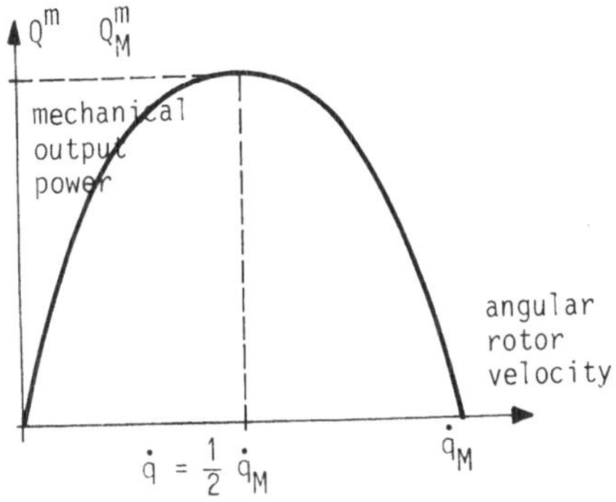

Fig. A.6.4. Mechanical power w.r. to angular velocity

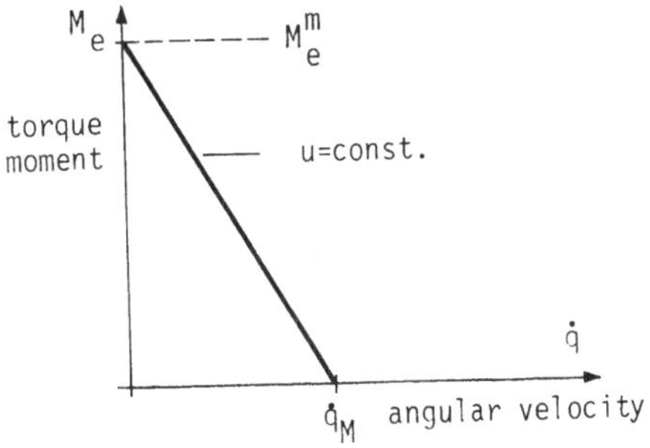

Fig. A.6.5. Torque w.r. to angular velocity

Basic deficience of these motors are large current values, transferred to the rotor via brushes, producing powerful sparking, which is harmful.

Mathematical model of the torque DC servomotor

The mathematical model of the torque DC servomotor can also be represented sufficiently correctly by the same linear differential equations as in the case of conventional DC motors [7]:

$$
\begin{aligned}
&L_R \dot{i}_R + r_R i_R + C_E \dot{q} = u \\
&J_R \ddot{q} + F_V \dot{q} + M^* = C_M i_R
\end{aligned}
\qquad (A.6.46)
$$

where J_R is the rotor moment of inertia, M^* is the torque of external motor load, while other designations were given in previous text.

Here can be pointed out, too, that due to the fact that direct-drive motor are in question, the viscous damping constant F_V can be neglected, because the viscous friction of the rotor itself is significantly smaller than the viscous friction of the motor-reducer unit.

If the actuator model is now given in the state space as:

$$\dot{x} = Ax + bN(u) + fM^{*} \tag{A.6.47}$$

where x is the state vector, A is the system matrix, b is the input distribution vector, f is the load distribution vector, N(u) is the voltage saturation, M^{*} is the external load torque.

If equations (A.6.46) are solved with respect to $\dot{i}_R$ and $\ddot{q}$:

$$\dot{i}_R = -\frac{C_E}{L_R}\dot{q} - \frac{r_R}{L_R} i_R + \frac{1}{L_R} u \tag{A.6.48}$$

$$\ddot{q} = \frac{C_M}{J_R} i_R - \frac{1}{J_R} M^{*} \tag{A.6.49}$$

and the state vector $x = x(q, \dot{q}, i_R)^T$ is adopted, expression for A, b and f are obtained:

$$A = \begin{bmatrix} 0 & 1 & 0 \\ 0 & 0 & \frac{C_M}{J_R} \\ 0 & -\frac{C_E}{L_R} & -\frac{r_R}{L_R} \end{bmatrix}, \qquad b = \begin{bmatrix} 0 \\ 0 \\ \frac{1}{L_R} \end{bmatrix}, \qquad f = \begin{bmatrix} 0 \\ -\frac{1}{J_R} \\ 0 \end{bmatrix} \tag{A.6.50}$$

If the rotor inductivity is neglected ($L_r \approx 0$), the third-order actuator model can be reduced to a second-order model. In equation (A.6.46) the first member becomes zero so one obtains:

$$i_R = -\frac{C_E}{r_R}\dot{q} + \frac{1}{r_R} u \tag{A.6.51}$$

When expression (A.6.51) is substituted in (A.6.49) it becomes:

$$\ddot{q} = -\frac{C_M C_E}{J_R r_R}\dot{q} + \frac{C_M}{J_R r_R} u - \frac{1}{J_R} M^{*} \tag{A.6.52}$$

Now A, b and f are:

$$A = \begin{bmatrix} 0 & 1 \\ 0 & -\frac{C_M C_E}{J_R r_R} \end{bmatrix}, \qquad b = \begin{bmatrix} 0 \\ \frac{C_M}{J_R r_R} \end{bmatrix}, \qquad f = \begin{bmatrix} 0 \\ -\frac{1}{J_R} \end{bmatrix} \qquad \text{(A.6.53)}$$

Brushless DC servomotors

To the good properties of the DC servomotors belong: expressed linearity of the torque characteristic, great starting torque, possibility of simple speed control, while the AC servomotors possess simplicity of design, long service life, absence of sparking (with brushless motors).

In order to profit of the advantages of one and the other servomotors, the brushless DC servomotor has been adopted. First solution which imposes itself is to substitute mechanical commutation with the DC servomotors, by electronic commutation and thus avoid wearing up, friction and sparking at the brushes. However, such direct substitution is expensive due to the great number of switching elements to be installed, so a more economical solution was searched for. It was found partially in the design with permanent magnets in the rotor and windings on the stator. The windings of the stator, with the aim to reduce the number of switching elements, were divided in two, maximally four phases and the magnetic field was realized so that on a soft iron rotor permanent magnet poles were mounted. These poles are mostly produced from ferrites and rare earth elements, samarium and other. Use of rare earth materials produces higher torque, better torque to inertia ratio and a size reduction by 20-30% of the whole motor, compared to the ferrite solution. Torque to inertia ratio for ferrite magnets is about 3000 and for rare earth magnets attains even 13000. Speed attained with rare earth magnets in about 46% greater as compared with ferrite ones. Temperature stability is also better with rare earth magnets, because their temperature coefficient is smaller than with ferrites almost by order of magnitude.

These advantages of rare earth compared with ferrites reflects in price, too, so they are used only in cases which are economically justified. Compared with conventional AC and DC servomotors, the ratio of torque to inertia is in the case of ferrite brushless motors about three times greater and in the case of rare earth motors this ratio is up to twelve times greater.

Let the work of a three-phased motor be considered, i.e. a motor in which the stator windings are divided in three parts, conditionally named phases. When electric current flows through phase 1, the stator winding will produce a magnetic field, which will attract the permanent magnet on the rotor, so rotation will start.

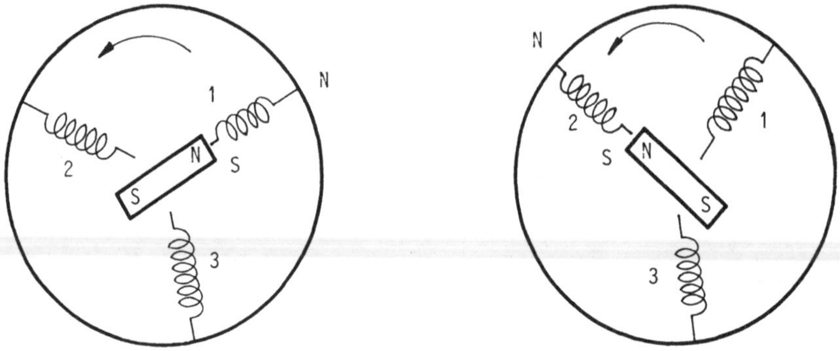

Fig. A.6.6. Three-phased DC brushless motor scheme

If in the appropriate moment the supply is switched from phase 1 to phase 2, the magnetic field will rotate in the positive sence by 120° and thus the rotor will continue to rotate. If this switching is continued in same sequence, in the motor is generated a rotating magnetic field, which atract the rotor and rotation continues. Reversing the switching sequence, i.e. if instead of sequence 1-2-3-1 phases are switched in sequence 1-3-2-1, the electromagnetic field and the rotor will rotate in the opposite sense. For synchronizing the supply switching from one phase to another the corresponding position sensor is used. Usually, it is a Hall's sensor placed in the motor housing, coupled with a small permanent magnet being on the rotor shaft. Each time one of the magnet poles passes past the Hall's sensor, a signal is generated which triggers the commutator logic unit to perform the switching. Exact position of the sensor on the housing is important, because this influences the commutation correctness. In that way, the synchronously switched phase current coupled with the permanent magnet field produces a torque, proportional to the current and the magnetic flux intensity. The rotor is accelerated up to the point in which the applied voltage is reduced by the value of the produced back EMF, which limits the current to the value needed to carry the load. Hence, the higher the applied voltage, the higher will be the speed at some determined external load, i.e. the higher the load, higher current values will be needed.

In Fig. A.6.7. variation of torque depending on the phase and angular position was given for a three-phase DC motor.

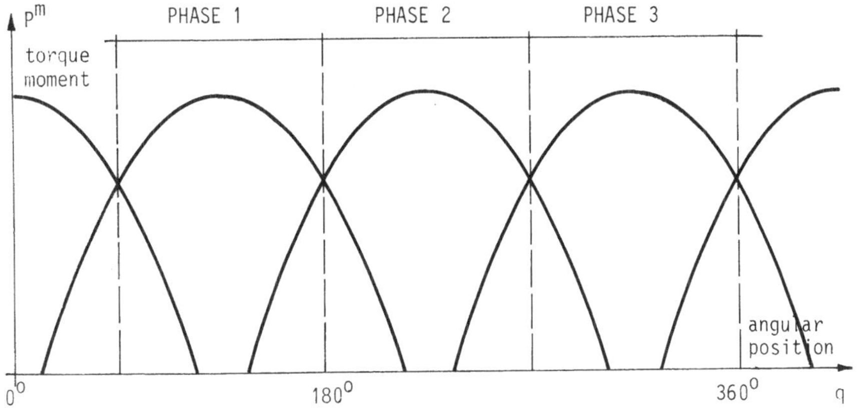

Fig. A.6.7. Variation of torque depending on phase and angular position

Here should be pointed out, that the brushless DC motors are nothing else but synchronous motors with permanent magnets in the rotor (SM), in a special (more expensive) version, due to the Hall's sensors integrated into the motor. From other side the brushless DC motors have a simplified control structure, very similar to that of stepping motors.

The described operation of the brushless motors is simple and easily understood and was taken here for describing the working principle of these motors. However, in the majority of practical realizations, two phases are energized simultaneously, because in that way greater torque is produced. The windings of the stator are divided in three phases and star-connected, two of each are energized in same time. Synchronization of the overlap is achieved by means of a small position magnet on the rotor and three Hall's sensors distributed on the stator at 60° between them (Fig. A.6.8). Excitated by the position magnet, the sensors produce signals: logical zeroes and unities, which are conducted to the inputs of the logical unit with binary decoder and a series of "or"-elements. Therefrom they are conducted to bipolar transistors or FET - transistors, switching the phases to the DC supply. When the binary decoder is actuated by the input signal, it transmits depending on the signal from the Hall's sensors, one address, for instance Nr. 6 (Fig. A.6.8). This number passes through the corresponding "or"-elements and switches on units 1 and 6, thus, conduction the current from the DC supply

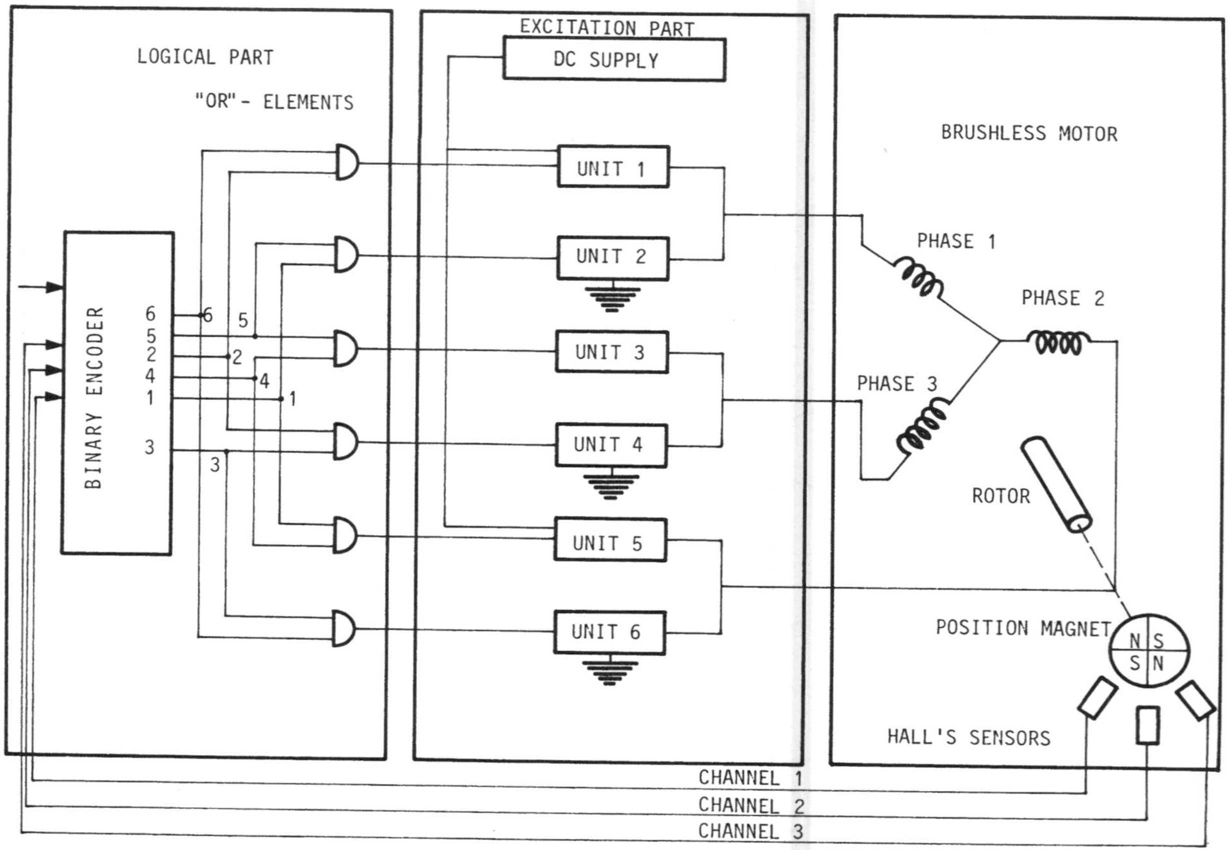

Fig. A.6.8. Electronic commutator of brushless DC motor with two-phase excitation

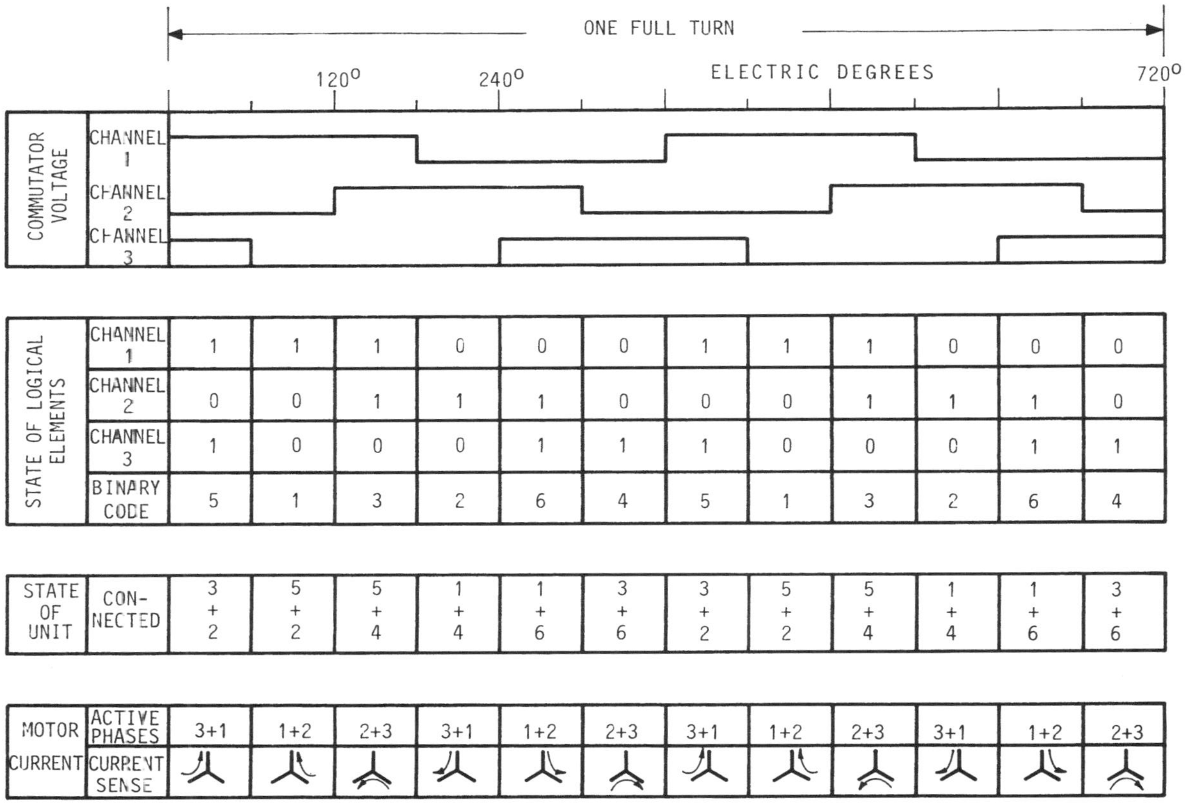

Fig. A.6.9. Illustration of logical elements states and current and voltage values in the motor and commutator after Fig. A.6.8.

through unit 1, phases 1 and 2, then through unit 6 to ground. When the rotor rotates by 60°, signal from the Hall's sensors changes the signal at the input of the binary decoder, so that it transmits the address 4. This number switches units 3 and 6 together and the current flows through unit 3, phase 3, phase 2 and unit 6 to ground (Fig. A.6.8).

By further rotation of the rotor for 60° the third Hall's sensor is activated, connecting the corresponding phases according to the table in Fig. A.6.9. and thus rotation continues until the activating signal from the input of the binary decoder is annulled (Fig. A.6.8).

Mathematical model of brushless DC servomotors

The mathematical model of brushless DC motors is similar to the one of conventional DC motors and the linearity of the speed-torque characteristic is also very similar. Let start from the two basic differential equations, describing the voltage and dynamic equilibrium, respectively:

$$L_s \dot{i}_s + r_s i_s + C_E \dot{q} = u \qquad (A.6.54)$$

$$J_R \ddot{q} + F_V \dot{q} + M^*(q, \dot{q}) = M_e(q, i_s) \qquad (A.6.55)$$

where: L_s is the inductivity of stator winding [H]

r_s is the resistance of stator winding [Ohm]

$\dot{q}$ is the rotor angular velocity [rad/s]

C_E is the motor voltage constant [V/rad/s]

J_R is the motor moment of inertia [kgm^2]

F_V is viscous damping constant [Nm/rad/s]

$M_e(q, i_s)$ is the torque developed by motor [Nm]

$M^*(q, \dot{q})$ is the external load torque [Nm]

i_s is the stator winding current [A]

u is the stator winding voltage [V]

It is notably necessary to pay attention to the member M_e, the developed motor torque, which depends on the stator winding current and rotor position, and can be expressed as:

$$M_e(q, i_s) = C_M \cdot i_s(t) \tag{A.6.56}$$

Since a three phased motor is considered, in each winding flows the individual current:

$$i_j = i_s \cdot \sin q_j, \qquad j=1,2,3 \tag{A.6.57}$$

or:

$$\begin{aligned} i_1 &= i_s \cdot \sin q_1, \\ i_2 &= i_s \cdot \sin q_2, \\ i_3 &= i_s \cdot \sin q_3 \end{aligned} \tag{A.6.58}$$

If (A.6.58) is substituted in (A.6.56), one obtains:

$$M_e(q, i_s) = C_M[i_1 \sin q_1 + i_2 \sin q_2 + i_3 \sin q_3] \tag{A.6.59}$$

or

$$M_e(q, i_s) = C_M \sum_{j=1}^{3} i_j \sin q_j \tag{A.6.60}$$

and finally:

$$M_e(q, i_s) = C_M i_s \sum_{j=1}^{3} \sin^2 q_j \tag{A.6.61}$$

Let now write the model in the state space:

$$\begin{aligned} \dot{q} &= \dot{q} \\ \ddot{q} &= -\frac{F_v}{J_R}\dot{q} + \frac{C_M i_s}{J_R} \sum_{j=1}^{3} \sin^2 q_j - \frac{1}{J_R} M^* \\ \dot{i}_s &= -\frac{C_E}{L_s}\dot{q} - \frac{r_s}{L_s} i_s + \frac{1}{L_s} u \end{aligned} \tag{A.6.62}$$

This system of equations, as a multivariable nonlinear system with variable coefficients, can be represented in the state space as:

$$\dot{x} = A(t)x + bN(u) + fM^* \tag{A.6.63}$$

If the position is expressed as time function:

$$q = q(t) \tag{A.6.64}$$

it is obtained:

$$A(q) = \begin{bmatrix} 0 & 1 & 0 \\ 0 & -\frac{F_V}{J_R} & \frac{C_M}{J_R}\sum_{j=1}^{3}\sin^2 q_j \\ 0 & -\frac{C_E}{L_s} & -\frac{r_s}{L_s} \end{bmatrix}, \quad b = \begin{bmatrix} 0 \\ 0 \\ \frac{1}{L_s} \end{bmatrix}, \quad f = \begin{bmatrix} 0 \\ -\frac{1}{J_R} \\ 0 \end{bmatrix} \tag{A.6.65}$$

It can be noted, that the mathematical model of brushless DC motors from the preceding expression (A.6.65) is almost identical to that of SM given by expression (A.6.29), in which the constants are written in somewhat different form:

$$C_E = \psi_F, \text{ electromotor force constant} \tag{A.6.66}$$

$$C_M = n\pi\psi_F, \text{ torque constant} \tag{A.6.67}$$

The mathematical model (A.6.65) takes care about the step-by-step control structure, so it is given positionally dependent, introducing the time dependence in the DC brushless motor mathematical model and makes the complete model of the manipulation robot more complex. A SM controlled in that way renders the following results: more expensive motor, simpler control structure of the motor itself and a more complex robot control as a whole. In this case the viscous damping constant F_V can be neglected for the same reason as in the preceding example of the DC torque servomotor.

In this case, too, when the stator inductivity is neglected ($L_s \approx 0$), the third-order model is reduced to a second-order model. Namely, in equation (A.6.54) the member $L_s \dot{i}_s$ becomes zero, and one obtains:

$$i_s = -\frac{C_E}{r_s}\dot{q} + \frac{1}{r_s}u \tag{A.6.68}$$

When expression (A.6.68) is substituted in (A.6.62) we get:

$$\ddot{q} = -\frac{C_M}{J_R r_s} C_E \sum_{j=1}^{3}\sin^2 q_j \cdot \dot{q} + \frac{C_M}{J_R r_s}\sum_{j=1}^{3}\sin^2 q_j \cdot u - \frac{1}{J_R} M^* \tag{A.6.69}$$

Finally, the third-order model is reduced to a second-order one, where A(q), b(q) and f are:

$$A(q) = \begin{bmatrix} 0 & 1 \\ 0 & -\dfrac{C_M C_E}{J_R r_s} \sum_{j=1}^{3} \sin^2 q_j \end{bmatrix}, \quad b(q) = \begin{bmatrix} 0 \\ \dfrac{C_M}{J_R r_s} \sum_{j=1}^{3} \sin^2 q_j \end{bmatrix}, \quad f = \begin{bmatrix} 0 \\ -\dfrac{1}{J_R} \end{bmatrix} \qquad (A.6.70)$$

Electrohydraulic actuators

Electrohydraulic actuators are currently the sole powering units in industrial robotics for manipulation of working objects with a mass over 100 kg. It is also important to stress the possibility of direct generation of translational and rotational motion by hydraulic components without the use of gears. High specific power (pressures over 200 bar) enables the generation of significant forces at relatively small volumes. Such driving systems exhibit short response times, high rigidity and simple connection with electronic components. The electrohydraulic actuator*) consists of a four-way servovalve and a hydraulic cylinder or a hydraulic vane motor. Control signal from the controller after the inevitable amplification in the electrical amplifier serve as inputs to the servovalve, where a low power electrical signal is converted into an hydraulic signal of high power level which controls the hydraulic actuators. Hydraulic actuators power particular degrees of freedom and can in principle be linked to the segments of the mechanism via some power transfer unit although they are in most practical cases directly coupled to mechanical d.o.f. of the robotic mechanism. The servovalve is the most important and the most precise component of an electrohydraulic system. Different types of servovalves exist, though we shall confine ourselves to that most often applied in robotics: two-stage proportional servovalve. The schematic presentation of a two-stage proportional electrohydraulic servovalve is given in Fig. A.6.10. This type of servovalve consists of an electric torque motor and two stages of hydraulic amplification:

First stage - hydraulic preamplifier consists of the flapper and the nozzle, which follows the flapper motion at all times.

Second stage - hydraulic four-way spool valve.

In forming the mathematical model we shall use the common assumptions from literature [1, 2, 4].

*) Most of the presented material on electrohydraulic servovalves relies on H. Merrit [1] - fundamental textbook in this domain.

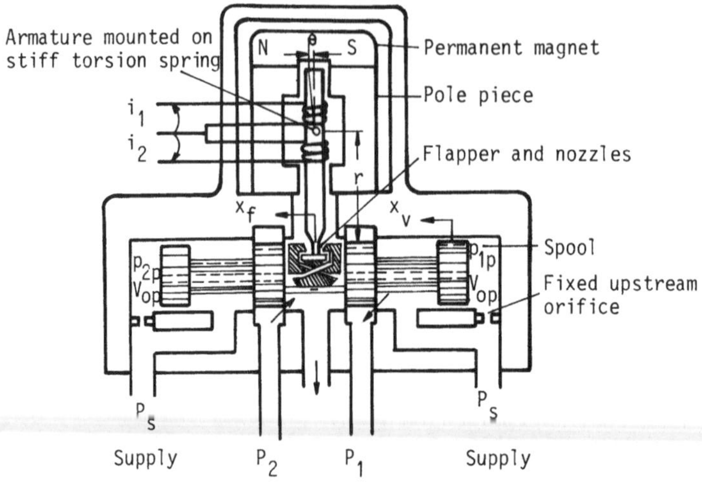

Fig. A.6.10. Schematic of a two-stage electrohydraulic servovalve

Permanent magnet torque motor consists of the permanent magnet and the armature mounted onto the shaft with a torsion spring. The armature carries two sets of windings.

Balancing the forces acting on the armature, we obtain a differential equation,

$$K_t \Delta i = J_a \ddot{\theta} + B_a \dot{\theta} + (K_a - K_m)\theta + T_L \qquad \text{(A.6.71)}$$

where K_t - motor torque constant [Nm/A], $\Delta i = i_1 - i_2$ - difference in currents flowing through windings [A], J_a - armature and load inertia, B_a - armature viscous friction coefficient [Nms], K_m - magnetic spring constant of motor [Nm/rad], θ - angular displacement of armature [rad], K_a - mechanical torsion spring constant of armature pivot [Nm/rad], T_L - load torque. In further text Δi will be denoted by i.

Under the assumption that J_a is negligible in relation to the inertia of other system elements (assumption 1), and that B_a is also negligible (assumption 2), (A.6.71) becomes:

$$K_t \cdot i = (K_a - K_m)\theta + T_L \qquad \text{(A.6.72)}$$

A three-land-four-way spool valve is shown in Fig. A.6.11.

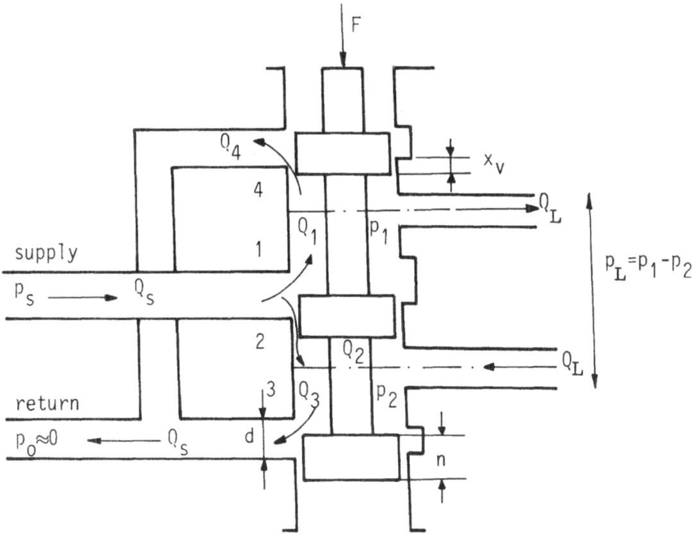

Fig. A.6.11. A three-land-four-way spool valve

We will make the following assumptions:

- four-way valve is zero lapped (land width identical to port width) i.e. it has a critical center, (assumption 3).
- Valve displacement x_v never reaches its limits, (assumption 4).
- Valve chamber is fixed (assumption 5).
- Flow change due to fluid compressibility is negligible, since we are interested in static characteristics, (assumption 6).

On the basis of continuity equation it follows,

$$Q_L = Q_1 - Q_4, \qquad Q_L = Q_3 - Q_2 \tag{A.6.73}$$

and the load pressure is defined by

$$p_L = p_1 - p_2 \tag{A.6.74}$$

The flows through particular valving orifices are given by,

$$Q_1 = C_d A_1 \sqrt{\frac{2}{\rho}(p_s - p_1)}, \qquad Q_2 = C_d A_2 \sqrt{\frac{2}{\rho}(p_s - p_2)} \tag{A.6.75}$$

$$Q_3 = C_d A_3 \sqrt{\frac{2}{\rho} p_2}, \qquad Q_4 = C_d A_4 \sqrt{\frac{2}{\rho} p_1} \qquad (A.6.76)$$

Orifice areas are function of spool valve displacement,

$$A_1 = A_1(x_v), \quad A_2 = (-x_v), \quad A_3 = A_3(x_v), \quad A_4 = A_4(-x_v) \qquad (A.6.77)$$

where Q_L - load pressure flow $[m^3/s]$, p_L - load pressure $[N/m^2]$, C_d - dimensionless damping coefficient, ρ - hydraulic flow density $[kg/m^3]$, p_s - supply pressure, x_v - spool-valve piston displacement [m].

Equations (A.6.73)-(A.6.77) provide the basis for looking at the flow as a nonlinear function of valve position and flow pressure:

$$Q_L = Q_L(x_v, p_L) \qquad (A.6.78)$$

We make the following assumptions:

- valving orifices are matched and symmetrical and all four orificies are equal at the piston rod neutral position, (assumption 7),

- orifice areas are linear functions of displacement x_v (assumption 8).

On the basis of these assumptions only one orifice area (A) has to be defined. The single most important parameter for defining an orifice area is the area gradient (w) (the rate of change of orifice area with stroke). Under assumption 7, and a further assumption that return pressure is negligible, it follows:

$$Q_1 = Q_3, \qquad Q_2 = Q_4 \qquad (A.6.79)$$

Substituting (A.6.75), (A.6.76) into (A.6.79) and under assumption 7, it follows

$$p_s = p_1 + p_2 \qquad (A.6.80)$$

which using (A.6.74) gives

$$p_1 = \frac{p_s + p_L}{2}, \qquad p_2 = \frac{p_s - p_L}{2} \qquad (A.6.81)$$

Under the assumption that the fluid is incompressible (assumption 9) using assumptions 7, 8, equations (A.6.73), (A.6.75), (A.6.81) and that Q_2 and $Q_4(x_v>0)$, or Q_1 and $Q_3(x_v<0)$ are equal to zero, the load pressure flow for critical center spool valves are given by expression*)

$$Q_L = C_d|A|\frac{x_v}{|x_v|}\sqrt{\frac{1}{\rho}(p_s - \frac{x_v}{|x_v|}\,p_L)} \qquad (A.6.82)$$

or for valves with rectangular orifice

$$Q_L = C_d w x_v\sqrt{\frac{1}{\rho}(p_s - \frac{x_v}{|x_v|}\,p_L)} \qquad (A.6.83)$$

Expression (A.6.82) represents a nonlinear static characteristic (Fig. A.6.12) of the servovalve.

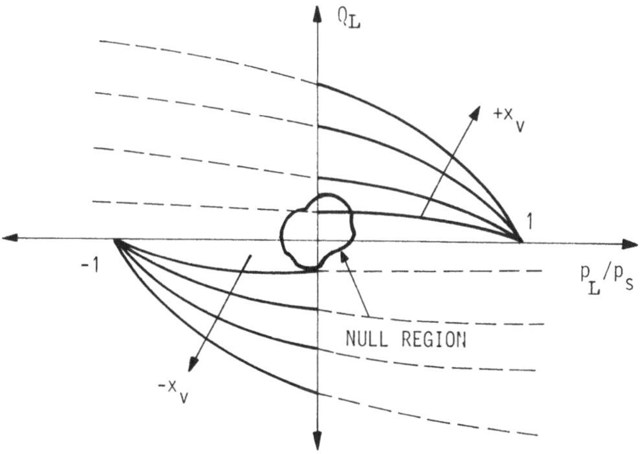

Fig. A.6.12. Steady-state characteristic of the servo-valve

The static characteristic (A.6.83) exists in quadrants II and IV only in transient conditions such as a fast change in x_v, fluid or load inertias can not result in the instantaneous change of direction of fluid flow. With the aim of obtaining a linear system, the nonlinear steady-state characteristic (A.6.82) will be linearized by expansion into a Taylor's series about a particular operating point $Q_L = Q_{L1}$, i.e.,

$$Q_L - Q_{L1} = \Delta Q_L = \left.\frac{\partial Q_L}{\partial x_v}\right|_1 \Delta x_v + \left.\frac{\partial Q_L}{\partial p_L}\right|_1 \Delta p_L \qquad (A.6.84)$$

or

---

*) Expression $x_v/|x_v|$ denotes function sign of this value.

$$\Delta Q_L = K_q \Delta x_v - K_c \Delta p_L \tag{A.6.85}$$

where K_q and K_c - flow gain and flow-pressure coefficient respectively.

Mathematical model of servovalve controlled hydraulic cylinder

Consider the servovalve controlled hydraulic cylinder (Fig. A.6.13).

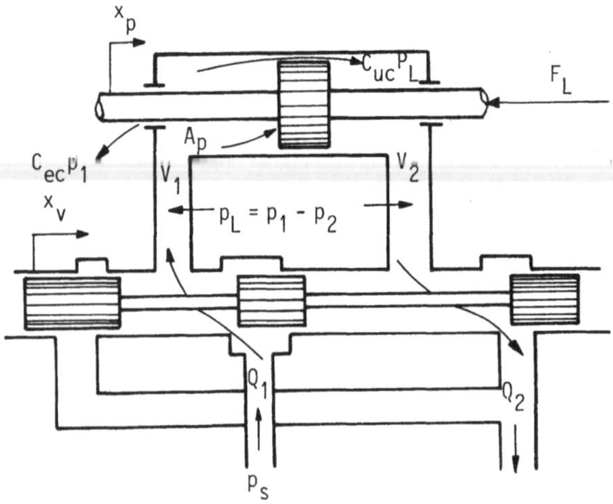

Fig. A.6.13. Servovalve controlled hydraulic cylinder

Because the servovalve orifices can be assumed matched and symmetrical, coefficients K_q and K_c for the branch valving towards the cylinder and the return branch are the same. Under the assumption of constant supply pressure, linearized flow equations are:

$$\begin{aligned} Q_1 &= K_q x_v - 2K_c p_1 \\ Q_2 &= K_q x_v + 2K_c p_2 \end{aligned} \tag{A.6.86}$$

Defining the load pressure flow as

$$Q_L = \frac{Q_1 + Q_2}{2} \tag{A.6.87}$$

and using (A.6.74) we obtain:

$$Q_L = K_q x_v - K_c p_L \tag{A.6.88}$$

Introducing assumptions that:

- Internal or cross-part leakage of piston is proportional to the load pressure (assumption 10), $Q_{uc} = C_{uc}p_L$, C_{uc} - internal leakage coefficient of piston $[m^5/sN]$, Q_{us} - flow of internal leakage $[m^3/s]$.

- External leakage in every chamber of the hydraulic cylinder is proportional to the pressure in that chamber (assumption 11), $Q_{ec1} = C_{ec}p_1$, $Q_{ec2} = C_{ec}p_2$, Q_{ec1}, Q_{ec2} - flow of external leakage in corresponding chambers, C_{ec} - external leakage coefficient.

Applying the continuity equation to each of the piston chambers yields

$$Q_1 - C_{uc}(p_1-p_2) - C_{ec}p_1 = \frac{dV_1}{dt} + \frac{V_1}{\beta}\frac{dp_1}{dt} \tag{A.6.89}$$

$$C_{uc}(p_1-p_2) - C_{ec}p_2 - Q_2 = \frac{dV_2}{dt} + \frac{V_2}{\beta}\frac{dp_2}{dt} \tag{A.6.90}$$

where V_1 and V_2 are the volumes of respective chambers, β - fluid compressibility coefficient $[N/m^2]$. The volumes of piston chambers are

$$V_1 = V_{o1} + A_p x_p, \qquad V_2 = V_{o2} - A_p x_p \tag{A.6.91}$$

where A_p - piston area, x_p - piston displacement V_{o1}, V_{o2} - initial volumes of respective cylinder chambers.

We make the following assumption:

- The piston is positioned such that the initial volumes in each chamber are equal (assumption 12) i.e.,

$$V_{o1} = V_{o2} = V_o \tag{A.6.92}$$

On the basis of (A.6.91), (A.6.92) the total volume of fluid under pressure in both chambers is

$$V_u = V_1+V_2 = V_{o1}+V_{o2} = 2V_o \tag{A.6.93}$$

On the basis of (A.6.74), (A.6.87), (A.6.89 - A.6.93) and the assumption that $A_p x_p << V_o$, the flow of load pressure can be expressed as,

$$Q_L = A_p \dot{x}_p + C_{tc} p_L + \frac{V_u}{4\beta} \dot{p}_L \tag{A.6.94}$$

where $C_{tc} = C_{uc} + C_{ec}/2$ - coefficient of total piston leakage $[m^5/sN]$.

By balancing the forces, the force acting on the piston can be determined, i.e.,

$$F_p = A_p p_L = m_p \ddot{x}_p + B_c \dot{x}_p + kx_p + F_L \tag{A.6.95}$$

where: F_p - force generated by piston [N], m_p - total mass of piston and load referred to the piston, B_c - viscous friction coefficient [Ns/m], k - load spring gradient, F_L - external load force on piston [N].

Mathematical model of torque motors is analogous to the model on the basis of equations (A.6.94), (A.6.95) i.e.,

$$Q_L = D_m \dot{\theta}_m + C_{tc} p_L + \frac{V_u}{4\beta} \dot{p}_L, \quad P^m = D_m p_L = I_m \ddot{\theta}_m + B_m \dot{\theta}_m + M^* \tag{A.6.96}$$

where: D_m - motor volume displacement $[m^3]$, θ_m - angular displacement at motor shaft [rad], P^m - torque developed by motor [Nm], I_m - moment of inertia of rotating parts $[kgm^2]$, B_m - viscous friction coefficient [Nms], M^* - external load moment [Nm].

On the basis of equations (A.6.88), (A.6.94), (A.6.95) along with the assumption that spring load is negligible (k=0), where $(K_c+C_{tc})B_c/A_p^2<<1$ with a further mention that load forces F_L are often neglected in the process of system construction, the transfer function will be:

$$W(s) = \frac{x_p(s)}{x_v(s)} = \frac{K_q/A_p}{s\left(\frac{s^2}{\omega_h^2} + \frac{2\delta_h}{\omega_h} s + 1\right)} \tag{A.6.97}$$

where: $\omega_h = \sqrt{\frac{4\beta A_p^2}{V_u m_p}}$ - hydraulic natural frequency,

$$\delta_h = \frac{K_c+C_{tc}}{A_p}\sqrt{\frac{\beta m_p}{V_u}} + \frac{B_c}{4A_p}\sqrt{\frac{V_u}{\beta m_p}}$$ - damping ratio.

Behaviour of the first stage, the preamplifier, can be described similarly i.e., as a relation between flapper displacement x_f and piston displacement x_v, with x_v having a feedback loop

$$\frac{x_v(s)}{x_f(s) - x_v(s)} = \frac{K_{qp}/\omega_f A_v}{(\frac{s}{\omega_f} + 1)(\frac{s^2}{\omega_{hp}^2} + \frac{2\delta_{hp}}{\omega_{hp}} s + 1)} \tag{A.6.98}$$

where $\omega_{hp} = \sqrt{\frac{2\beta A_v^2}{V_{op} M_v}}$ - hydraulic natural frequency of the first stage,

$\delta_{hp} = \frac{\omega_{hp} K_{cp} M_v}{2A_v^2}$ - nondimensional damping of the first stage,

$\omega_f = \frac{0.43 w p_s K_{cp}}{2A_v^2}$ - break frequency resulting from action of flow force on the piston,

K_{qp}, K_{cp} - coefficients of the linearized flow equations for the first stage,

A_v - piston area,

M_v - piston mass,

V_{op} - volume of parts behind piston ends,

P_s - supply pressure, p_L - load pressure.

Under the assumption that the break frequency is low, behaviour of the first stage in "s" domain (A.6.98) can be described by the 3-rd order differential equation

$$\frac{K_{qp}}{A_v} x_f = \frac{1}{\omega_{hp}^2} \dddot{x}_v + \frac{2\delta_{hp}}{\omega_{hp}} \ddot{x}_v + \dot{x}_v + \frac{K_{qp}}{A_v} x_v \tag{A.6.99}$$

On the basis of forces balance which act on the piston one obtains

$$p_{Lp} A_v = M_v \ddot{x}_v + 0.43 w (p_s - p_L) x_v \tag{A.6.100}$$

where $p_{Lp} = p_{1p} - p_{2p}$ - load pressure at first stage (pilot valve load pressure) and $0.43\, w(p_s - p_L)$ - gradient of flow spring load.

Linearizing equation (A.6.100) around $p_{Lo} = 0$, we obtain:

$$A_v \Delta p_{Lp} = M_v \Delta \ddot{x}_v + 0.43 w p_s \Delta x_v - 0.43 w x_{vo} \Delta p_L \tag{A.6.101}$$

where x_{vo} - initial piston displacement.

The external load moment acting on the armature, is given by expression

$$T_L = rA_N p_{Lp} + r(8\pi C_{df}^2 p_s x_{fo}) x_v \qquad (A.6.102)$$

where r - length of the rotating part of the armature, A_N- nozzle area, C_{df} - flow coefficient, x_{fo} - initial flapper displacement.

Unifying equations (A.6.71), (A.6.99 - A.6.102), and using relation $x_f = r\theta$ (Fig. A.6.10) along with assumption that (A.6.99) reduces to 2-nd order equation due to $\frac{1}{\omega_{hp}^2}$ multiplying $\dddot{x}_v$ being small, one obtains a differential equation describing the behaviour of the electric motor with the first stage hydraulic amplification

$$K_t \cdot i = K_{p1}\ddot{x}_v + K_{p2}\dot{x}_v + K_{p3}x_v - K_{p4}p_L \qquad (A.6.103)$$

where: $K_{p1} = \frac{(K_a - K_m)A_v}{rK_{qp}\omega_{hp}^2} + \frac{rA_N M_v}{A_v}$,

$$K_{p2} = \frac{2\delta_{hp}A_v(K_a - K_m)}{r\omega_{hp}K_{qp}}$$

$$K_{p3} = \frac{0.43 rA_N p_s w}{A_v} + \frac{(K_a - K_m)(A_v + K_{qp})}{rK_{qp}} + 8r\pi C_{df}^2 p_s x_{fo}$$

$$K_{p4} = \frac{0.43 rA_N w x_{vo}}{A_v}$$

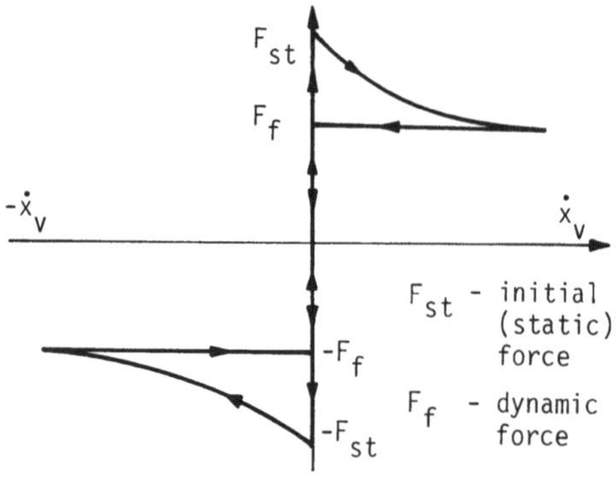

Fig. A.6.14. Dry friction force

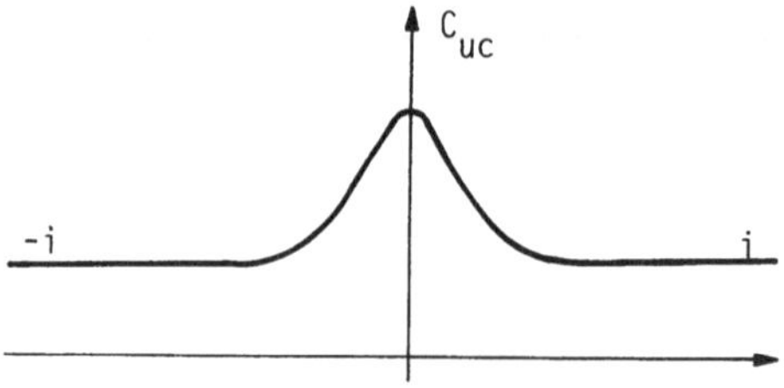

Fig. A.6.15. Internal leakage

Under the assumption that the pressure feedback loop can be neglected because displacement $x_f \approx 0$, which implies a relatively small load pressure change p_{Lp}, it follows

$$K_t \cdot i = K_{p1}\ddot{x}_v + K_{p2}\dot{x}_v + K_{p3}x_v \qquad (A.6.104)$$

Equations (A.6.83), (A.6.94), (A.6.95) (or A.6.96) and equation (A.6.104) represent the basic equations of the electrohydraulic actuator model. Electrohydraulic actuators can, beside the basic nonlinearity (A.6.83) also possess other nonlinearities which can be introduced into the mathematical model in the aim of obtaining its realistic description.

Dry friction in hydraulic actuators produces a friction force which is independent of rate of change of displacement $\dot{x}_v$, but alternates in sense (Fig. A.6.14).

Internal leakage varies with input current and reaches a maximum at its null value (Fig. A.6.15).

Linearity denotes the deviation degree from straight line when all the remaining variables are kept constant (A.6.72). *Symmetry* denotes the equality degree between the flow gains of both polarity. *Hysteresis* (Fig. A.6.16) is an important nonlinearity of the system.

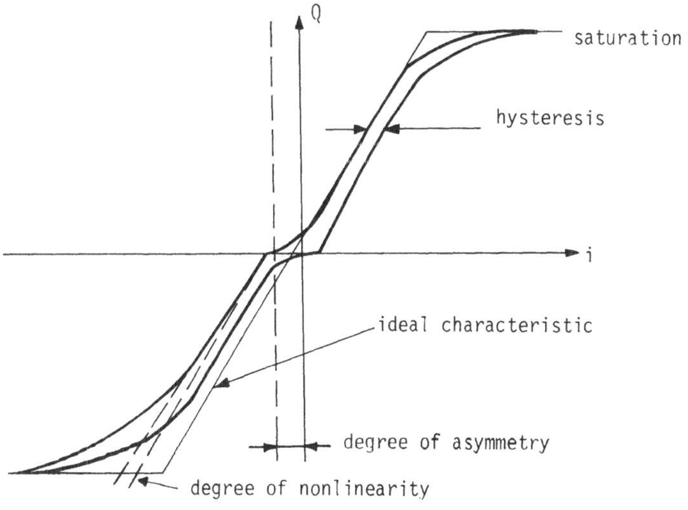

Fig. A.6.16. Servovalve nonlinearities

Various mathematical models of actuators

In a general case the mathematical model of electrohydraulic actuators (A.6.84, A.6.94, A.6.95, A.6.104) is a nonlinear system [4, 11]:

$$S^i: \quad \dot{x}^i = g^i(x^i, \theta^i, u^i, M_i^*), \qquad y^i = G^i x^i \tag{A.6.105}$$

where x^i - subsystem state vector*), θ^i - i-th system parameters vector, u^i - subsystem input, M_i^* - driving moment or force in i-th joint, y^i - output of subsystem, G^i - output subsystem matrix. Values of u^i are amplitude constrained by saturation type nonlinearities:

$$N(u^i) = \begin{cases} -u_m^i & u^i \leq -u_m^i \\ u^i & -u_m^i \leq u_i \leq u_m^i \\ u_m^i & u^i \geq u_m^i \end{cases}$$

Adopting a subsystem state vector of the form $x^i = [x_v, \dot{x}_v, x_p, \dot{x}_p, p_L]^T$ the mathematical model of (A.6.105) can be written in matrix form:

$$S^i: \quad \dot{x}^i = g^i(x^i, \theta^i) + b^i(\theta)N(u^i) + f^i(\theta)M_i^* \tag{A.6.106}$$

where

$$g^i(x^i, \theta^i) = \begin{bmatrix} \dot{x}_v \\ -(K_{p3}/K_{p1})x_v - (K_{p2}/K_{p1})\dot{x}_v \\ \dot{x}_p \\ \dfrac{A_p}{m_p} p_L - \dfrac{B_c}{m_p}\dot{x}_p \\ \dfrac{4\beta C_d w x_v}{V_u}\sqrt{\dfrac{1}{\rho}\left(p_s - \dfrac{x_v}{|x_v|}p_L\right)} - \dfrac{4\beta A_p \dot{x}_p}{V_u} - \dfrac{4\beta C_{tc} p_L}{V_u} \end{bmatrix}$$

---

*) Subsystem implies one degree of freedom of the robot.

$$b^i(\theta^i) = \begin{bmatrix} 0 \\ K_t/K_{p1} \\ 0 \\ 0 \\ 0 \end{bmatrix}, \quad f^i(\theta^i) = \begin{bmatrix} 0 \\ 0 \\ 0 \\ -1/m_p \\ 0 \end{bmatrix}, \quad M_i^* = F_L \qquad (A.6.107)$$

Mathematical model defined by (A.6.106, A.6.107) represents a fifth order nonlinear model in state space. With the aim of obtaining a linearised model a standard procedure is to linearize the servovalve static characteristic (A.6.83) around the nominal operating point. On the basis of (A.6.88), coefficients K_{q1} - flow gain $[m^3/smA]$ and K_{c1} - flow/pressure coefficient $[m^5/sN]$ are calculated as partial derivatives of static characteristic about the nominal operating point

$$K_{q1} = \left.\frac{\partial Q_L}{\partial x_v}\right|^o = C_d w \sqrt{\frac{1}{\rho}\left(p_s - \frac{x_v}{|x_v|} p_L\right)} \qquad (A.6.108)$$

$$K_{c1} = \left.\frac{\partial Q_L}{\partial p_L}\right|^o = \frac{C_d w |x_v| \sqrt{\frac{1}{\rho}\left(p_s - \frac{x_v}{|x_v|} p_L\right)}}{2\left(p_s - \frac{x_v}{|x_v|} p_L\right)} \qquad (A.6.109)$$

On the basis of (A.6.107) a linear stationary mathematical model of actuators in state space (linear system with respect to state) results:

$$S^i: \quad \dot{x}^i = A^i(\theta^i)x^i + b^i(\theta^i)N(u^i) + f^i(\theta^i)M_i^* \qquad (A.6.110)$$

where

$$A^i = \begin{bmatrix} 0 & 1 & 0 & 0 & 0 \\ -K_1 & -K_2 & 0 & 0 & 0 \\ 0 & 0 & 0 & 1 & 0 \\ 0 & 0 & 0 & -B_c/m_p & A_p/m_p \\ \frac{4\beta K_{q1}}{V_u} & 0 & 0 & -\frac{4\beta A_p}{V_u} & -\frac{4\beta(C_{tc}+K_{c1})}{V_u} \end{bmatrix} \qquad (A.6.111)$$

where $K_1 = K_{p3}/K_{p1}$, $K_2 = K_{p2}/K_{p1}$.

Matrices b^i and f^i and the subsystem state vector are the same as in the previous case. Such a mathematical model represents a linear, time invariant model because the parameters vector θ^i does not depend on time. However, it is of extreme importance to examine the influence of the change in coefficients K_{q1} and K_{c1} at various operating points since such models may become inadequate in representing system behaviour when wide variations in the state vector occur (eq. change in working pressure) which is often the case with robotic mechanisms. Also worth mentioning is that with robotic mechanisms, hydraulic pistons having a connecting rod on one side only are used, which results in a variable volume along the motion. Therefore, taking that $V_1=V_{o1}+A_1x_p$, $V_2=V_{o2}-A_2x_p$ and substituting these in (A.6.89, A.6.90), on the basis of (A.6.81) and (A.6.87) the following model equations are obtained*):

$$\begin{aligned} K_t \cdot i &= K_{p1}\ddot{x}_v + K_{p2}\dot{x}_v + K_{p3}x_v, \\ Q_L &= A_{K1}\dot{x}_p + C_{tc}p_L + \frac{V_u}{4\beta}\dot{p}_L \\ {}^{*)}Q_L &= K_{q1}x_v - K_{c1}p_L \\ F_p &= A_{K1}p_L = m_p\ddot{x}_p + B_c\dot{x}_p + F_L - A_{K2}p_s \end{aligned} \tag{A.6.112}$$

where coefficients K_{q1} and K_{c1} are always calculated along the movement in accordance with (A.6.108, A.6.109) and the variable volume according to

$$V_u = V_{up} + 2A_{K2}x_p \tag{A.6.113}$$

where $A_{K1} = (A_1+A_2)/2$, $A_{K2} = (A_1-A_2)/2$, A_1 - piston area on the side without the piston-rod, A_2 - piston area on piston-rod side, V_{up} - initial volume of hydraulic cylinder.

On the basis of equations denoted by (A.6.112) a linear time varying mathematical model in state space can be formed [4, 11]:

$$S^i\colon \dot{x}^i = A^i(x^{io}(t),\ \theta^i)x^i+b^i(\theta^i)N(u^i)+f^i(\theta^i)M_i^*+t^i(\theta) \tag{A.6.114}$$

where:

---

*) It is assumed that the performed approximation has no significant affect on the calculation of load flow.

$$A^i(x^{io}(t),\theta^i) = \begin{bmatrix} 0 & 0 & 0 & 0 & 0 \\ -K_1 & -K_2 & 0 & 0 & 0 \\ 0 & 0 & 0 & 1 & 0 \\ 0 & 0 & 0 & -B_c/m_p & A_{K1}/m_p \\ \dfrac{4\beta K_{q1}(x^{io}(t))}{V_u(x^{io}(t))} & 0 & 0 & \dfrac{4\beta A_{K1}}{V_u(x^{io}(t))} & \dfrac{4\beta(K_{c1}(x^{io}(t))+C_{tc})}{V_u(x^{io}(t))} \end{bmatrix}$$

$$t^i(\theta^i) = \begin{bmatrix} 0 \\ 0 \\ 0 \\ \dfrac{A_{K2}P_s}{m_p} \\ 0 \end{bmatrix}, \quad K_1 = K_{p3}/K_{p1}, \quad K_2 = K_{p2}/K_{p1} \tag{A.6.115}$$

Matrices $b^i(\theta^i)$ and $f^i(\theta^i)$ remain unchanged according to (A.6.107).

It is clear that the A^i subsystem matrix contains elements which vary along trajectory $x^{io}(t)$ due to time variant coefficients K_{q1}, K_{c1}, V_u. Column matrix $t^i(\theta^i)$ represents a constant perturbation arising from nonsymmetry of the piston surface.

The order of the mathematical model in the modelling of electrohydraulic actuators is of great importance since it directly affects the complexity of microcomputer implementations of control laws from the memory space standpoint and especially as regards the execution times. It has been shown that for most purposes in industrial robotics a third order mathematical model is a sufficiently true description of its behaviour.

Under the assumption that valve piston mass is negligible, frequency ω_{hp} is high, which justifies transforming equation (A.6.104) into the form,

$$K_t \cdot i = K_{p3} x_v \tag{A.6.116}$$

On the basis of this equation, which represents the behaviour of proportional servovalves, we obtain a third order model. On the basis of (A.6.116) static characteristic (A.6.83) can be represented as:

$$Q_L = C_d w \frac{K_t}{K_{p3}} i \sqrt{\frac{1}{\rho}(p_s - \frac{i}{|i|} p_L)}, \quad \text{i.e.}$$

$$Q_L = K_H i \sqrt{p_s - \frac{i}{|i|} p_L} \tag{A.6.117}$$

where $K_H = C_d w \frac{K_t}{K_{p3}} \sqrt{\frac{1}{\rho}}$ - servovalve constant.

Adopting the state variables in the form $x^i = [x_p, \dot{x}_p, p_L]^T$ and taking into account corrections for variations in volume of electrohydraulic cylinder, the nonlinear mathematical model will be of the form [4, 11]:

$$S^i: \quad \dot{x}^i = A^i(x^{io}, \theta^i)x^i + d^i(x^{io}, x^i, \theta^i)N(u^i) + f^i(\theta^i)M_i^* + t^i(\theta^i) \tag{A.6.118}$$

$$\text{where: } A^i(x^{io}, \theta^i) = \begin{bmatrix} 0 & 1 & 0 \\ 0 & -B_c/m_p & A_{K1}/m_p \\ 0 & -\dfrac{4\beta A_{K1}}{V_u(x^{io})} & -\dfrac{4\beta C_{tc}}{V_u(x^{io})} \end{bmatrix} \tag{A.6.119}$$

$$d^i(x^{io}, x^i, \theta^i) = \begin{bmatrix} 0 \\ 0 \\ \dfrac{4\beta H_K \sqrt{p_s - \frac{i}{|i|} p_L}}{V_u(x^{io})} \end{bmatrix}, \quad f^i(\theta^i) = \begin{bmatrix} 0 \\ -1/m_p \\ 0 \end{bmatrix},$$

$$t^i(\theta^i) = \begin{bmatrix} 0 \\ \dfrac{A_{K2} p_s}{m_p} \\ 0 \end{bmatrix} \tag{A.6.120}$$

Matrices of linear time-invariant mathematical model (A.6.110) for the third order case are

$$A^i(\theta^i) = \begin{bmatrix} 0 & 1 & 0 \\ 0 & -B_c/m_p & A_p/m_p \\ 0 & -\dfrac{4\beta A_p}{V_u} & -\dfrac{4\beta(K_c+C_{tc})}{V_u} \end{bmatrix}, \qquad b^i(\theta^i) = \begin{bmatrix} 0 \\ 0 \\ \dfrac{4\beta K_q}{V_u} \end{bmatrix},$$

$$f^i(\theta^i) = \begin{bmatrix} 0 \\ -1/m_p \\ 0 \end{bmatrix} \qquad \text{(A.6.121)}$$

where coefficients K_q i K_c are calculated as

$$K_q = \frac{\partial Q_L}{\partial i}\bigg|^{o} = K_H\sqrt{p_s - \frac{i}{|i|}\, p_L},$$

$$K_c = \frac{\partial Q_L}{\partial p_L}\bigg|^{o} = \frac{K_H i\sqrt{p_s - \frac{i}{|i|}\, p_L}}{2(p_s - \frac{i}{|i|}\, p_L)} \qquad \text{(A.6.122)}$$

For the case of linear non-stationary mathematical model, the model in state space has the form

$$S^i:\ \dot{x}^i = A^i(x^{io},\ \theta^i)x^i + b^i(x^{io},\ \theta^i)N(u^i) +$$

$$+ f^i(\theta^i)M_i^* + f^i(\theta^i) \qquad \text{(A.6.123)}$$

where: $$A^i(x^{io},\ \theta^i) = \begin{bmatrix} 0 & 1 & 0 \\ 0 & -\dfrac{B_c}{m_p} & \dfrac{A_{K1}}{m_p} \\ 0 & -\dfrac{4\beta A_{k1}}{V_u(x^{io})} & -\dfrac{4\beta(K_c(x^{io})+C_{tc})}{V_u(x^{io})} \end{bmatrix}$$

$$b^i(x^{io}, \theta^i) = \begin{bmatrix} 0 \\ 0 \\ \dfrac{4\beta K_q(x^{io})}{V_u(x^{io})} \end{bmatrix}$$

and coefficients K_q and K_c are calculated according to (A.6.122) along the total movement, while V_u is calculated according to (A.6.113). Matrices $f^i(\theta^i)$ and $t^i(\theta^i)$ are given by expression (A.6.120).

If the effect of fluid compressibility can be neglected, the mathematical model can be reduced to second order whose state vector $x^i = [x_p, \dot{x}_p]^T$. In this case the nonlinear mathematical model acquires the the form

$$S^i: \quad \dot{x}^i = g^i(x^i, u^i, \theta^i, M_i^*)$$

where

$$g^i = (x^i, u^i, \theta^i, M_i^*) = \begin{bmatrix} \dot{x}_p \\ g_2^i(x^i, u^i, \theta^i, P_i) \end{bmatrix} \qquad (A.6.124)$$

where g_2^i represents a complex function which is obtained by solving equations w.r.t $\ddot{x}_p$, on the basis of (A.6.83), II and IV equations (A.6.112) and (A.6.117)

$$K_H i \sqrt{p_s - \frac{i}{|i|}\frac{m_p}{A_{K1}}\ddot{x}_p - \frac{i}{|i|}\frac{B_c}{A_{K1}}\dot{x}_p - \frac{i}{|i|}\frac{F_L}{A_{K1}} + \frac{i}{|i|}\frac{A_{K2}p_s}{A_{K1}}} =$$

$$= A_{K1}\dot{x}_p + C_{tc}\left(\frac{m_p}{A_{k1}}\ddot{x}_p + \frac{B_c}{A_{K1}}\dot{x}_p + \frac{F_L}{A_{K1}} - \frac{A_{K2}p_s}{A_{K1}}\right) \qquad (A.6.125)$$

In this case, the matrices of the linear, time invariant model (A.6.110) become:

$$A^i(\theta^i) = \begin{bmatrix} 0 & 1 \\ 0 & -\dfrac{1}{m_p}\left(B_c + \dfrac{A_p^2}{(K_c + C_{tc})}\right) \end{bmatrix}, \qquad b^i(\theta^i) = \begin{bmatrix} 0 \\ \dfrac{A_p K_q}{m_p(C_{tc} + K_c)} \end{bmatrix},$$

$$f^i(\theta^i) = \begin{bmatrix} 0 \\ -\frac{1}{m_p} \end{bmatrix}$$

where coefficients K_q and K_c are calculated according to (A.6.122).

Linear, time varying second order model has the form according to (A.6.123) while the model matrices are

$$A^i(x^{io}, \theta^i) = \begin{bmatrix} 0 & 1 \\ 0 & -\frac{1}{m_p}\left(B_c - \frac{A_{K1}^2}{K_c(x^{io})+C_{tc}}\right) \end{bmatrix}$$

$$b^i(x^{io}, \theta^i) = \begin{bmatrix} 0 \\ \frac{A_{K1}K_q(x^{io})}{m_p(C_{tc}+K_c(x^{io}))} \end{bmatrix}, \qquad f^i(\theta^i) = \begin{bmatrix} 0 \\ -\frac{1}{m_p} \end{bmatrix}$$

$$t^i(\theta^i) = \begin{bmatrix} 0 \\ \frac{A_{K2}P_s}{m_p} \end{bmatrix}$$

If the manipulator d.o.f. are rotational and linear actuators are used, the connection between state coordinates of the actuators model x^i and the state coordinates of the model representing the mechanical part of the system, $(q^i, \dot{q}^i)$ is not linear and is most often expressed using the cosine formula. In a similar manner the relation between load M_i^* acting on actuator S^i and the driving torque or force P_i can be established. In a general case it can be written

$$P_i = f(M_i^*, q^i) \tag{A.6.126}$$

where f - some nonlinear function M_i^* and q^i. As already mentioned when a rotational mechanism joint is driven by linear actuators, the relation between the load on actuator M_i^* and the torque developed in the joint P_i can be described by a cosine formula or some other relation which is dependent on joint position. For example, if a rotational joint is driven by a D.C. electric motor or vane hydraulic actuator, relation (A.6.126) may be trivial: $P_i^M = M_i^*$ as has been stressed in the preceding text. Relation (A.6.126) can be simply written as

$$P_i = z^i(q^i)M_i^* \qquad (A.6.127)$$

where z^i - nonlinear function q^i.

Thus, for the case of manipulation robot GORO-101 (Fig. A.6.17) these relations have the following form.

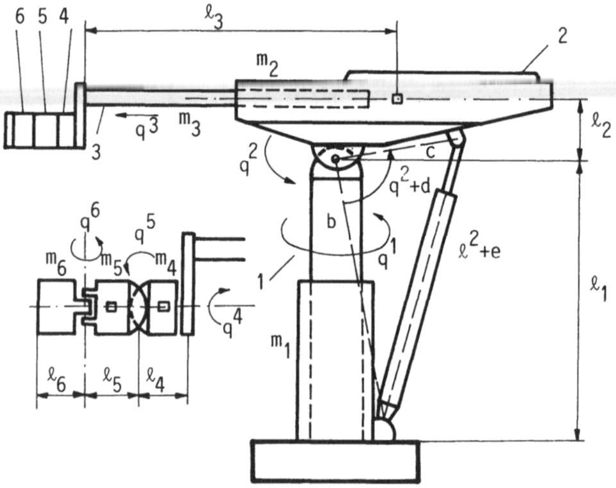

Fig. A.6.17. Spray-painting robot GORO-101

Connections between manipulator angles q^i and piston positions ℓ^i are linear, $q^i = c_i\ell^i$ for all d.o.f. except for i=2

$$\ell^1 = aq^1, \quad \ell^2 = (b^2+c^2-2bc\cdot\cos(q^2+d))^{1/2}-e \qquad (A.6.128)$$

where a, b, c, d and e - geometric parameters

Because the loading of linear actuators is realised through forces, and the generalised forces P_i in rotational joints are moments, the load distribution vectors for the first two d.o.f. are

$$f^1 = [0,\ 1/(m_p^1 a),\ 0]^T, \quad f^2 = [0,\ 1/(m_p^2 c\ \cos\ q^2),\ 0]^T \qquad (A.6.129)$$

The remaining four d.o.f. are powered by actuators having the same motion type as the corresponding d.o.f.

Using relation (2.36) or (A.6.128) and (A.6.129) expression (2.37) for generalised forces of the mechanical part of manipulation robot system can be easily derived.

Electropneumatic actuators

An electropneumatic actuator consists of an electropneumatic servovalve and a pneumatic servomotor (Fig. A.6.18).

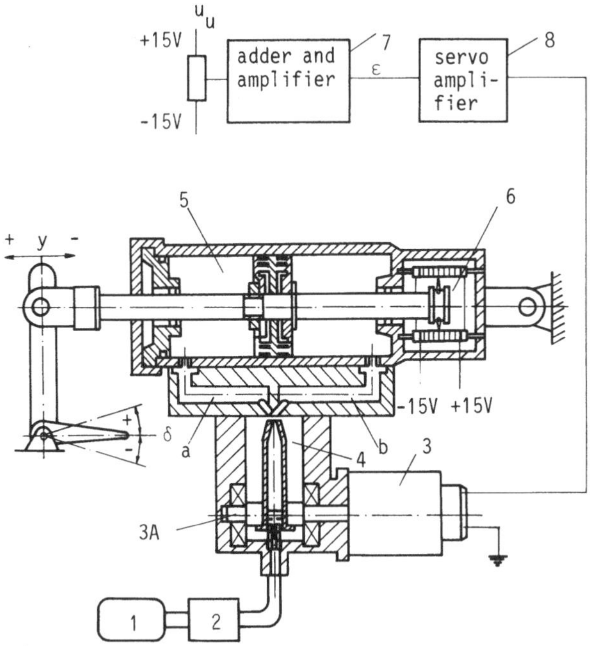

Fig. A.6.18. Electropneumatic actuator

In Fig. A.6.18. an independent source of energy (gas under pressure) with a valving and pressure reduction group (1 and 2) is given. The electromechanical transformer (3) transforms the electrical signal from the servoamplifier into a corresponding turning angle of the shaft and nozzle. (3A) Mechanical - pneumatic transformer (4) provides for different pressures in chambers a and b i.e., flow proportional to the angle through which the nozzle is turned. Electro-mechanical transformer (3) and mechanical-pneumatic transformer (4) together form the servovalve. Pneumatic servomotor (5) realises the change of position of the loaded con-rod (x or δ) and the displacement on the sliding potentiometer of the feedback loop (6). Electronic adder and amplifier (7) use the voltage across the controlled potentiometer (u_u) and across the feedback potentiometer (u_{ps}) to form the error signal ε. The error signal is amplified with respect to the power in the servoamplifier (8) and sent to the electro-mechanical transformer (3). With the aim of improving dynamic characteristics and positional accuracy, it is possible to introduce additional feedback loops and compensators.

The basic mathematical model of the electro-pneumatic actuator is presented by the following system of equations:

- current through electro-mechanical transformer:

$$L_n \frac{di}{dt} + R_n i = u_p \tag{A.6.130}$$

where: L_n - mean value inductance of the electro-pneumatic servovalve (EPSR), i - current through EPSR, R_n - mean value resistance of EPSR, u_p - output voltage of the servovalve,

- turning, active torque on the nozzle:

$$M_i = K_{m,i} i \tag{A.6.131}$$

where: M_i - electromagnetic moment acting on the nozzle, $K_{m,i}$ - coefficient of moment amplification with respect to current,

- turning angle of the nozzle:

$$I_m \frac{d^2\alpha}{dt^2} + f_m \frac{d\alpha}{dt} + c_m \alpha + c_{tm} \operatorname{sign}\alpha = (M_i - M_{\Delta Q} - M_{\Delta p}) \tag{A.6.132}$$

where: I_m - moment of inertia of moving parts (EPSR), α - nozzle turning angle, f_m - viscous friction coefficient (EPSR), c_m - positional stiffness of EPSR, C_{tm} - dry friction force of EPSR, $M_{\Delta Q}$ - nozzle torque due to different flows, $M_{\Delta p}$ - nozzle torque due to different pressures

$$\alpha_{min} < \alpha < \alpha_{max}, \quad M_{\Delta Q} = K_{M,\Delta Q} \Delta Q, \quad M_{\Delta p} = K_{M,\Delta p} \Delta p$$

- flow through mechanical-pneumatic transformer:

$$\Delta Q = K_{\Delta Q,\alpha} \alpha - K_{\Delta Q,\Delta p} \Delta p = f(\alpha, \Delta p) \tag{A.6.133}$$

where $K_{\Delta Q,\alpha}$ - flow gain coefficient with respect to the turning angle,

$K_{\Delta Q,\Delta p}$ - flow gain coefficient with respect to pressure, Δp - pressure difference between the EPSM chambers, $\Delta p = p_2 - p_1$, $K_{M,\Delta p}$, $K_{M,\Delta Q}$ - corresponding gain coefficients,

- flow through pneumatic servomotor:

$$\frac{d\Delta p}{dt} = \frac{Kp_o}{V_o} \{\Delta Q - 2A \frac{dy}{dt}\} \tag{A.6.134}$$

where: K - polythropic exponent, $p_o = p_n/2$ - mean chamber pressure, V_o - volume of one half of EPSM, p_n - input pressure to electropneumatic servosystem (EPSS), A - piston area,

- unloaded servomotor:

$$m \frac{d^2y}{dt^2} + f \frac{dy}{dt} + cy + F_{tr} \text{signy} = A\Delta p \tag{A.6.135}$$

where: m - mass of moving parts of EPSM, y - output (con-rod position), c - positional stiffness of EPSM, F_{tr} - dry friction force of EPSM, f - viscous friction coefficient of EPSM, $A\Delta p = P$ - driving force of unloaded actuator.

Mathematical model of electropneumatic actuator in state space

From the given dynamic equations of the electropneumatic actuator it may be concluded that a six th-order model is in question. In order to represent the model in state space, the following state quantities are adopted, x_i, i = 1,2,...,6:

$x_1 = y$ - linear displacement of con-rod

$x_2 = \dot{y}$ - con-rod motion velocity

$x_3 = \Delta p$ - pressure difference between the chambers

$x_4 = \alpha$ - nozzle turning angle

$x_5 = \dot{\alpha}$ - angular velocity of nozzle turning

$x_6 = i$ - current through EPSR

The model in state space becomes:

$$\dot{x}_1 = x_2$$

$$\dot{x}_2 = -\frac{c}{m}x_1 - \frac{F_{tr}}{m}\,\text{sign}\,x_1 - \frac{f}{m}x_2 + \frac{1}{m}p$$

$$\dot{x}_3 = -\frac{2AKp_o}{V_o}x_2 - \frac{Kp_oK_{\Delta Q,\Delta p}}{V_o}x_3 + \frac{Kp_oK_{\Delta Q,\alpha}}{V_o}x_4$$

$$\dot{x}_4 = x_5$$

$$\dot{x}_5 = -\frac{K_{M,\Delta p}-K_{M,\Delta Q}K_{\Delta Q,\Delta p}}{I_m}x_3 - \frac{c_m+K_{M,\Delta Q}K_{\Delta Q,\alpha}}{I_m}x_4 - \frac{c_{tm}}{I_m}\,\text{sign}\,x_4 - \frac{f_m}{I_m}x_5 + \frac{K_{m,i}}{I_m}x_6$$

$$\dot{x}_6 = -\frac{R_n}{L_n}x_6 + \frac{1}{L_n}u_p$$

Mathematical model of electropneumatic servovalve in matrix form

Equations (A.6.136) can be written in matrix form as

$$\dot{x} = Ax + \tilde{A}(x) + bN(u_p) + fM^* \tag{A.6.137}$$

where: $x = [x_1\ x_2\ x_3\ x_4\ x_5\ x_6]^T$ - state vector,

$$A = \begin{bmatrix} 0 & 1 & 0 & 0 & 0 & 0 \\ -\frac{c}{m} & -\frac{f}{m} & 0 & 0 & 0 & 0 \\ 0 & -\frac{2AKp_o}{V_o} & -\frac{Kp_oK_{\Delta Q,\Delta p}}{V_o} & \frac{Kp_oK_{\Delta Q,\alpha}}{V_o} & 0 & 0 \\ 0 & 0 & 0 & 0 & 1 & 0 \\ 0 & 0 & -\frac{K_{M,\Delta p}-K_{M,\Delta Q}K_{\Delta Q,\Delta p}}{I_m} & -\frac{c_m+K_{M,\Delta Q}K_{\Delta Q,\alpha}}{I_m} & -\frac{f_m}{I_m} & \frac{K_{mi}}{I_m} \\ 0 & 0 & 0 & 0 & 0 & -\frac{R_n}{L_n} \end{bmatrix}$$

$$\tilde{A}(x) = [0\quad -\frac{F_{tr}}{m}\,\text{sign}x_1\quad 0\quad 0\quad -\frac{c_{tm}}{I_m}\,\text{sign}x_4\quad 0]^T$$

$$b = [0\ \ 0\ \ 0\ \ 0\ \ 0\ \ 1/L_n]^T, \qquad f = [0\ \ \frac{1}{m}\ \ 0 \cdots 0]^T$$

$N(u_p)$ - nonlinearity of the amplitude saturation type.

Simplified mathematical model

Instead of the differential equations describing the nozzle turning angle, we shall adopt a linear dependency between the nozzle turning angle and the current through the transformer,

$$\alpha = K_{\alpha,i} i \tag{A.6.139}$$

In this manner the model is reduced to fourth order. The equations become:

- current through electromechanical transformer

$$L_n \frac{di}{dt} + R_n i = u_p \tag{A.6.140}$$

where: L_n - mean inductance, R_n - mean resistance of EPSR,

- turning angle of nozzle

$$\alpha = K_{\alpha,i} i \tag{A.6.141}$$

where: $K_{\alpha,i}$ - coefficient of proportionality,

- flow through mechanical-pneumatic transformer

$$\Delta Q = f(\alpha, \Delta p) = K_{\Delta Q,\alpha}\alpha + K_{\Delta Q,\Delta p}\Delta p \tag{A.6.142}$$

- flow through pneumatic servomotor

$$\Delta Q = \frac{M p_n \zeta A_K}{R T_n} \frac{dy}{dt} + \frac{M V_o}{K R T_n} \frac{d\Delta p}{dt} \tag{A.6.143}$$

where M - molecular mass of gas, p_n - supply pressure, ζ - pressure loss coefficient, R - universal gas constant, T_n - supply temperature, A_K - active piston area, K - polytropic exponent, V_o - total volume of EPSM (piston in mid-way position),

- unloaded servomotor

$$m\frac{d^2y}{dt^2} + f\frac{dy}{dt} + cy + F_{tr}\text{sign}y = A_K\Delta p \qquad (A.6.144)$$

If we want to obtain at the output the angular displacement, substituting $y = r_\delta\delta$, (A.6.144) becomes:

$$mr_\delta\frac{d^2\delta}{dt^2} + fr_\delta\frac{d\delta}{dt} + cr_\delta\delta + F_{tr}\text{sign}\delta = A_K\Delta p \qquad (A.6.145)$$

$$J_K\frac{d^2\delta}{dt^2} + c_{VT}\frac{d\delta}{dt} + c_{SM}\delta + c_t\text{sign}\delta = P^m \qquad (A.6.146)$$

$$c_t = F_{tr}r_\delta, \quad c_{SM} = cr_\delta^2, \quad c_{VT} = fr_\delta^2, \quad J_K = mr_\delta^2, \quad P^m = A_K\Delta pr_\delta$$

In the case that the actuator is subjected to external load, the above expression becomes:

$$J_K\frac{d^2\delta}{dt^2} + c_{VT}\frac{d\delta}{dt} + c_{SM}\delta + c_t\text{sign}\delta = P^m - M^* \qquad (A.6.147)$$

where: M^* - external load, r_δ - radius of the working force, J_K - moment of inertia of moving parts taken about axis of rotation, c_{VT} - coefficient of viscous friction, c_{SM} - hinge moment coefficient, c_t - dry friction moment, P^m - actuator driving torque.

Mathematical model in state space

For the adopted state quantities

$x_1 = \delta$ - piston-rod position

$x_2 = \dot{\delta}$ - piston-rod speed

$x_3 = \Delta p$ - difference in chamber pressures of EPSM

$x_4 = i$ - current through EPSR

model in state space can be presented in the form:

$$\dot{x}_1 = x_2$$

$$\dot{x}_2 = -\frac{c_{SM}}{J_K}x_1 - \frac{c_t}{J_K}\text{sign}x_1 - \frac{c_{VT}}{J_K}x_2 + \frac{A_K r_\delta}{J_K}x_3$$

$$\dot{x}_3 = -\frac{Kp_n\zeta A_K r_\delta}{V_o} x_2 + \frac{KRT_n K_{\Delta Q,\Delta p}}{MV_o} x_3 + \frac{KRT_n K_{\Delta Q,\alpha} \cdot K_{\alpha,i}}{MV_o} x_4 \quad \text{(A.6.148)}$$

$$\dot{x}_4 = -\frac{R_n}{L_n} x_4 + \frac{1}{L_n} u_p$$

Mathematical model in matrix form

(a) For the case when the driving torque is joined to matrix A along the state vector: $P^m = A_K r_\delta x_3$

$$\dot{x} = Ax + \tilde{A}(x) + bN(u_p)$$

$$x = [x_1\ x_2\ x_3\ x_4]^T \quad \text{(A.6.149)}$$

$$A = \begin{bmatrix} 0 & 1 & 0 & 0 \\ -\frac{c_{SM}}{J_K} & -\frac{c_{VT}}{J_K} & \frac{A_K r_\delta}{J_K} & 0 \\ 0 & -\frac{Kp_n\zeta A_K r_\delta}{V_o} & \frac{KRT_n K_{\Delta Q,\Delta p}}{MV_o} & \frac{KRT_n K_{\Delta Q,\alpha} K_{\alpha,i}}{MV_o} \\ 0 & 0 & 0 & -\frac{R_n}{L_n} \end{bmatrix}$$

$$\tilde{A}(x) = [0\ -\frac{c_t}{J_K}\ \mathrm{sign}x_1\ 0\ 0]^T, \quad b = [0\ 0\ 0\ 1/L_n] \quad \text{(A.6.150)}$$

(b) For the case when driving torque is separated:

$$\dot{x} = Ax + \tilde{A}(x) + bN(u_p) + fP^m$$

$$\tilde{A}(x) = [0\ -\frac{c_t}{J_K}\ \mathrm{sign}x_1\ 0\ 0]^T \quad \text{(A.6.151)}$$

$$b = [0\ 0\ 0\ \frac{1}{L_n}]^T, \quad f = [0\ \frac{1}{J_K}\ 0\ 0]^T$$

$$A = \begin{bmatrix} 0 & 1 & 0 & 0 \\ -\frac{c_{SM}}{J_K} & -\frac{c_{VT}}{J_K} & 0 & 0 \\ 0 & -\frac{Kp_n \zeta A_K r_\delta}{V_o} & \frac{K \cdot K_{\Delta Q,\Delta p} RT_n}{MV_o} & \frac{K \cdot K_{\Delta Q\alpha} K_{\alpha,i} RT_n}{MV_o} \\ 0 & 0 & 0 & -\frac{R_n}{L_n} \end{bmatrix} \tag{A.6.152}$$

Third order mathematical model

Model of electro-pneumatic actuators is further simplified for cases when coil inductance of the electromechanical transformer can be neglected. In that case, the equations used for modelling the electro-pneumatic actuator reduces to:

- current through electromechanical transformer:

$$R_n \cdot i = u_p \tag{A.6.153}$$

- nozzle turning angle:

$$\alpha = K_{\alpha,i} \cdot i \tag{A.6.154}$$

- flow through mechanical-pneumatic transformer:

$$\Delta Q = f(\alpha, \Delta p) = K_{\Delta Q,\alpha} \cdot \alpha + K_{\Delta Q,\Delta p} \cdot \Delta p \tag{A.6.155}$$

- flow through pneumatic servomotor:

$$Q = \frac{Mp_n \zeta A_K}{RT_n} \cdot \frac{dy}{dt} + \frac{MV_o}{KRT_n} \frac{d\Delta p}{dt} \tag{A.6.156}$$

- unloaded servomotor:

$$m \frac{d^2 y}{dt^2} + f \frac{dy}{dt} + cy + F_{tr} \operatorname{sign} y = A_K \Delta p = F_P \tag{A.6.157}$$

$$y = r_\delta \cdot \delta, \qquad \dot{y} = r_\delta \cdot \dot{\delta}, \qquad \ddot{y} = r_\delta \cdot \ddot{\delta}$$

$$\underbrace{mr_\delta^2}_{J_K}\ddot{\delta} + \underbrace{fr_\delta^2}_{c_{VT}}\dot{\delta} + \underbrace{cr_\delta^2}_{c_{SM}}\delta + \underbrace{F_{tr}r\delta}_{c_t} \,\mathrm{sign}\delta = \underbrace{F_P r_\delta}_{P^m} \tag{A.6.158}$$

Mathematical model in state space

Let us adopt the state values:

$x_1 = \delta$ - piston rod position

$x_2 = \dot{\delta}$ - piston rod speed

$x_3 = \Delta p$ - pressure difference between chambers.

Model in state space has the following form:

$$\dot{x}_1 = x_2$$

$$\dot{x}_2 = -\frac{c_{SM}}{J_K}x_1 - \frac{c_{VT}}{J_K}x_2 - \frac{c_t}{J_K}\,\mathrm{sign}x_1 + \frac{1}{J_K}P^m \tag{A.6.159}$$

$$\dot{x}_3 = -\frac{Kp_n\zeta A_K r_\delta}{V_o}x_2 + \frac{KRT_n K_{\Delta Q,\Delta p}}{MV_o}x_3 + \frac{K_{\Delta Q,\alpha}K_{\alpha,i}KRT_n}{R_n MV_o}u_p$$

Mathematical model in matrix form

$$\dot{x} = Ax + \tilde{A}(x) + bN(u_p) + fP^m \tag{A.6.160}$$

where:

$$A = \begin{bmatrix} 0 & 0 & 0 \\ -\frac{c_{SM}}{J_K} & -\frac{c_{VT}}{J_K} & 0 \\ 0 & -\frac{Kp_n\zeta A_K r_\delta}{V_o} & \frac{KRT_n K_{\Delta q,\Delta p}}{MV_o} \end{bmatrix}$$

$$\tilde{A}(x) = [0 \quad -\frac{c_t}{J_K}\,\mathrm{sign}x_1 \quad 0]^T \tag{A.6.161}$$

$$b = [0 \quad 0 \quad \frac{K_{\Delta Q,\alpha}K_{\alpha,i}KRT_n}{R_n MV_o}]^T, \qquad f = [0 \quad \frac{1}{J_K} \quad 0]^T$$

If dry friction is neglected and state variables chosen as y, $\dot{y}$, Δp, equations of the model are:

$$\dot{x} = Ax + bN(u_p) + fF_p \tag{A.6.162}$$

$$\dot{x}_1 = x_2$$

$$\dot{x}_2 = -\frac{c}{m} x_1 - \frac{f}{m} x_2 + \frac{1}{m} p^m \tag{A.6.163}$$

$$\dot{x}_3 = -\frac{Kp_n \zeta A_K}{V_o} x_2 + \frac{K \cdot K_{\Delta Q,\Delta p} R \cdot T_n}{MV_o} x_3 + \frac{K \cdot K_{\Delta Q,\alpha} K_{\alpha,i} RT_n}{MV_o R_n} u_p$$

where:

$$m = \frac{J_K}{r_\delta^2}, \quad f = \frac{c_{VT}}{r_\delta^2}, \quad c = \frac{c_{SM}}{r_\delta^2} \tag{A.6.164}$$

Symbols used has been explained in preceding text.

On the basis of (A.6.162, A.6.163) matrices of the third order model are:

$$A = \begin{bmatrix} 0 & 1 & 0 \\ -\frac{c}{m} & -\frac{f}{m} & 0 \\ 0 & -\frac{Kp_n \zeta A_K}{V_o} & \frac{K \cdot K_{\Delta Q,\Delta p} R \cdot T_n}{MV_o} \end{bmatrix}$$

$$b = \left[0 \quad 0 \quad \frac{K \cdot K_{\Delta Q,\alpha} K_{\alpha,i} R \cdot T_n}{MV_o R_n}\right]^T \tag{A.6.165}$$

$$f = \left[0 \quad -\frac{1}{m} \quad 0\right]^T$$

References

[1] Merrit E.H., Hydraulic Control Systems, John Wiley & Sons, 1967.

[2] Vukobratović M., Borovac B., Stokić D., "Influence Analysis of the Actuator Model Complexity on Manipulator Control Synthesis", Mechanism and Machine Theory, Vol. 18, No 2, 1983.

[3] Vukobratović M., Stokić D., Kirćanski N., Scientific Fundamentals of Robotics 5: Non-Adaptive and Adaptive Control of Manipulation Robots, Springer-Verlag, 1985.

[4] Vukobratović M., Katić D., Potkonjak V., "Computer-Assisted Choice of Electrohydraulic Servosystems for Manipulation Robots Using Complete Mathematical Models", Mechanism and Machine Theory, Vol. 22, No 5, 1987.

[5] Lessmeier R., Schumacher W., Leonard W., "Microprocessor-Controlled AC - Servo Drives with Synchronous or Induction Motors: Which is Preferable?" IEEE Trans. on Industry Applications, Vol. IA-22, No 5, 1985.

[6] Koyama M., Yano M., Kamiyama I., Yano S., "Microprocessor-Based Vector Control System for Induction Motor Drives with Rotor Time Constant Identification Function" IEEE Trans. on Industry Applications, Vol. IA-22, No 3, 1986.

[7] Asada H., Youcef T., Direct-Drive Robots, MIT, London, 1987.

[8] Asada H., Kanade T., "Design of Direct-Drive Mechanical Arms", ASME Journal of Vibration, Acoustics, Stress and Reliability in Design, Vol. 105, No 3, 1983.

[9] Cheng R.M.H., Fahim A.E.F., "Characteristics of a Novel Pneumatic Stepping Motor", Fluidics Quartely, Vol. 10, No 3, 1978.

[10] Lakota N.A., Fundamentals of Design of Tracking Systems, (in Russian), "Mashinostroenie", Moscow, 1980.

[11] Katić D., Vukobratović M., "The Influence of Actuator Model Complexity on Control Synthesis for High Performance Robot Trajectory Tracking", Proceedings of the IFAC/IFIP/IMACS International Symposium Theory of Robots, Decembar 1986, Vienna, Austria, pp. 217-222.

[12] Akamatsu N., Ikeda K., Tomei H., Yano S., "High Performance IM Drive by Coordinate Control Using a Controlled Current Inverter", IEEE Trans. on Industry Application, Vol. IA-18, No. 4, 1982.

[13] Šabanović A., Izosimov D., "Application of Sliding Modes in Induction Motor Control", IEEE Trans. on Industry Applications, Vol. IA-17, No. 1, 1981.

Appendix 7
Automatic Forming of Dynamic Models

Example 1: "Cylindrical" Mechanism (Basic Configuration)

ϕ - rotation of column around vertical axis

z - vertical translation (heave)

x - horizontal translation (reach)

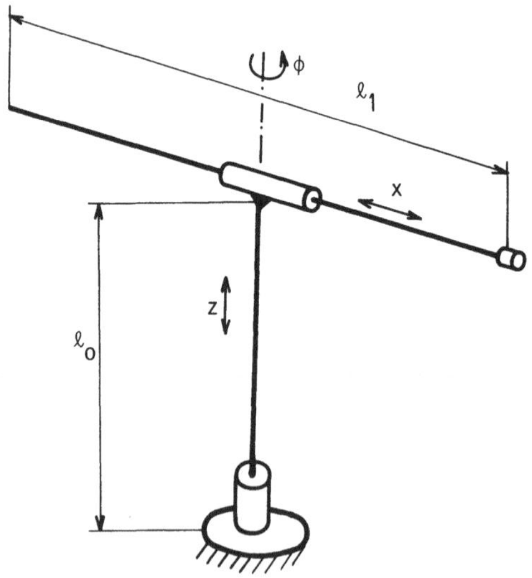

Fig. A.7.1. Scheme of "cilindrical" mechanism (basic configuration)

1. Introduction

The mechanism presented in Fig. A.7.1. can be presented schematically for some adopted initial position in the absolute coordinate system, as illustrated in Fig. A.7.2.

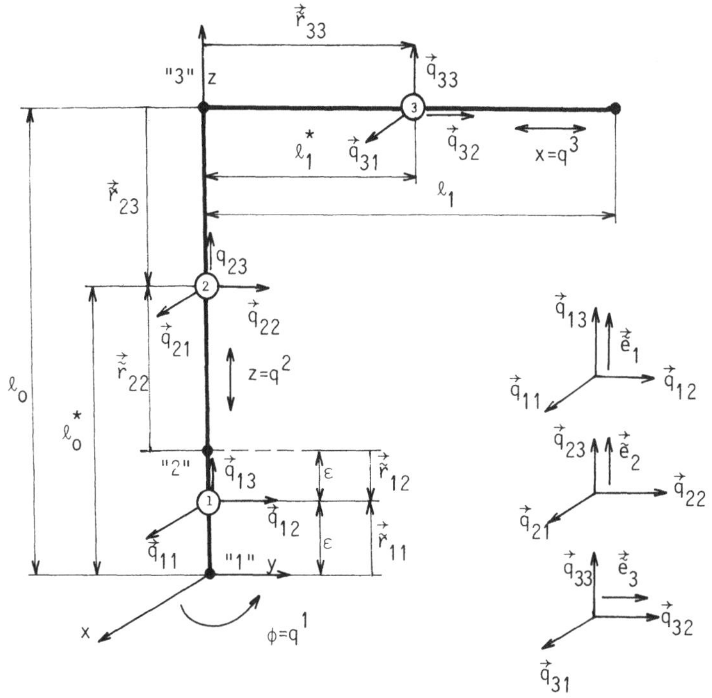

Fig. A.7.2. Initial position of mechanism

Mechanism in Fig. A.7.1. has three degrees of freedom:

- I - rotation of column around vertical axis,

- II - vertical translation (heave) z,

- III - horizontal translation (reach) x.

The cylindrical joint of IV-th class connecting the mechanism to the basis is divided in two V-th class joints:

- revolute joint "1",

- prismatic joint "2",

between which the zero-length segment "1" was introduced, having a zero-mass too (first segment joined to basis), i.e.

$$m_1 = 0$$

A right-oriented coordinate system was adopted.

For initial mechanism position the relative coordinates are zero, i.e.

$$\phi^o = 0$$

$$z^o = 0$$

$$x^o = 0$$

2. Segment "1"

Joint is revolute, $\xi_1 = 0$.

Segment is considered as a body (J_{11}, J_{12}, J_{13}). The segment is "specific" on lower side because $\tilde{\vec{e}}_1 || \tilde{\vec{r}}_{11}$:

$$\vec{e}_1 = (0,\ 0,\ 1)^T$$

$$\vec{r}_{o1} = (-1,\ 0,\ 0)$$

$$\tilde{\vec{e}}_1 = (0,\ 0,\ 1)^T \qquad \tilde{\vec{e}}_1 || \tilde{\vec{r}}_{11}$$

$$\tilde{\vec{r}}_{11} = (0,\ 0,\ \varepsilon)^T$$

$$\tilde{\vec{r}}_{12} = (0,\ 0,\ -\varepsilon)^T \qquad \varepsilon \approx 0$$

$$\tilde{\vec{r}}^A_{11} = (\varepsilon,\ 0,\ 0)^T$$

2.1. *Assembly*

$$\vec{r}_{N1} = \vec{e}_1 \times (\vec{r}^{\,A}_{o1} \times \vec{e}_1) = (0,\ 0,\ 1)^T \times [(-1,\ 0,\ 0)^T \times (0,\ 0,\ 1)^T] = (-1,\ 0,\ 0)^T$$

$$\overset{\rightrightarrows}{r}_{N1} = \overset{\rightrightarrows}{e}_1 \times (\overset{\rightrightarrows}{r}^{A}_{11} \times \overset{\rightrightarrows}{e}_1) = (0,\ 0,\ 1)^T \times [(\varepsilon,\ 0,\ 0)^T \times (0,\ 0,\ 1)^T] = (\varepsilon,\ 0,\ 0)^T$$

$$\vec{a}_1 = \frac{-\vec{r}_{N1}}{|\vec{r}_{N1}|} = (1,\ 0,\ 0)^T, \qquad \overset{\rightrightarrows}{a}_1 = \frac{+\overset{\rightrightarrows}{r}_{N1}}{|\overset{\rightrightarrows}{r}_{N1}|} = (1,\ 0,\ 0)^T$$

$$\vec{e}_1 = (0,\ 0,\ 1)^T, \qquad \overset{\rightrightarrows}{e}_1 = (0,\ 0,\ 1)^T$$

$$\vec{b}_1 = \vec{e}_1 \times \vec{a}_1 = (0,\ 1,\ 0)^T, \qquad \overset{\rightrightarrows}{b}_1 = \overset{\rightrightarrows}{e}_1 \times \overset{\rightrightarrows}{a}_1 = (0,\ 1,\ 0)^T$$

Transformation matrix before rotation

$$Q_1^o = [\vec{e}_1\ \vec{a}_1\ \vec{b}_1][\overset{\rightrightarrows}{e}_1\ \overset{\rightrightarrows}{a}_1\ \overset{\rightrightarrows}{b}_1]^T = \begin{bmatrix} 0 & 1 & 0 \\ 0 & 0 & 1 \\ 1 & 0 & 0 \end{bmatrix} \begin{bmatrix} 0 & 0 & 1 \\ 1 & 0 & 0 \\ 0 & 1 & 0 \end{bmatrix}$$

$$Q_1^o = I = \begin{bmatrix} 1 & 0 & 0 \\ 0 & 1 & 0 \\ 0 & 0 & 1 \end{bmatrix} = [\vec{q}^{\,o}_{11}\ \vec{q}^{\,o}_{12}\ \vec{q}^{\,o}_{13}]$$

2.2. *Position after rotation (changes of generalized coordinate ϕ)*

Rodrigues formula

$$\vec{q}_{1j} = [\vec{q}^{\,o}_{1j}\cos\phi + (1-\cos\phi)(\vec{e}_1 \cdot \vec{q}^{\,o}_{1j})\vec{e}_1 + {}$$

$$+ \vec{e}_1 \times \vec{q}^{\,o}_{1j}\sin\phi](1-\xi_1) + \vec{q}^{\,o}_{1j}\xi_1, \qquad (j=1,2,3)$$

$$\vec{q}_{11} = (\cos\phi,\ \sin\phi,\ 0)^T$$

$$\vec{q}_{12} = (-\sin\phi,\ \cos\phi,\ 0)^T$$

$$\vec{q}_{13} = (0,\ 0,\ 1)^T$$

Transformation matrix after rotation

$$Q_1 = [\vec{q}_{11}\ \vec{q}_{12}\ \vec{q}_{13}] = \begin{bmatrix} \cos\phi & -\sin\phi & 0 \\ \sin\phi & \cos\phi & 0 \\ 0 & 0 & 1 \end{bmatrix}$$

$$\vec{r}_{11} = Q_1\vec{\tilde{r}}_{11} = (0,\ 0,\ \varepsilon) = (0,\ 0,\ 0)^T$$

$$, \quad \varepsilon \approx 0$$

$$\vec{r}_{12} = Q_1\vec{\tilde{r}}_{12} = (0,\ 0,\ -\varepsilon) = (0,\ 0,\ 0)^T$$

$$\vec{\tilde{e}}_2 = (0,\ 0,\ 1)^T, \qquad \vec{e}_2 = Q_1\cdot\vec{\tilde{e}}_2 = (0,\ 0,\ 1)^T$$

2.3. *Angular and linear velocity*

$$\vec{\omega}_1 = \vec{\omega}_o + \dot{\phi}\vec{e}_1(1-\xi_1) = (0,\ 0,\ \dot{\phi})^T$$

$$\vec{v}_1 = \vec{v}_o - \vec{\omega}_o \times \vec{r}_{o1} + \vec{\omega}_1 \times \vec{r}_{11} + \dot{\phi}\vec{e}_1\xi_1 = (0,\ 0,\ 0)^T$$

$$\vec{\omega}_o = (0,\ 0,\ 0)^T$$

2.4. *Angular and linear acceleration**)

Vector coefficients in the expression for the angular acceleration are:

$$\vec{\alpha}_{11} = \vec{e}_1(1-\xi_1) = (0,\ 0,\ 1)^T$$

$$\vec{\theta}_o = \vec{\alpha}_1^o = \vec{\alpha}_o^o + \dot{\phi}(\vec{\omega}_o \times \vec{e}_1)(1-\xi_1) = (0,\ 0,\ 0)^T,\ \vec{\varepsilon}_o = (0,\ 0,\ 0)^T$$

Angular acceleration

$$\vec{\varepsilon}_1 = \vec{\alpha}_{11}\ddot{\phi} + \vec{\alpha}_1^o = (0,\ 0,\ \ddot{\phi})^T$$

Vector coefficients of linear acceleration

$$\vec{\beta}_{11} = (\vec{e}_1 \times \vec{r}_{11})(1-\xi_1) + \vec{e}_1\xi_1 = (0,\ 0,\ 0)^T$$

$$\vec{\eta}_o = \vec{\beta}_1^o = \vec{\beta}_o^o + \vec{\alpha}_o^o \times (\vec{r}_{11} - \vec{r}_{o1}) + \dot{\phi}(\vec{\omega}_o \times \vec{e}_1) \times \vec{r}_{11}(1-\xi_1) +$$

$$+ 2\dot{\phi}(\vec{\omega}_o \times \vec{e}_1)\xi_1 - \vec{\omega}_o \times (\vec{\omega}_o \times \vec{r}_{o1}) + \vec{\omega}_1 \times (\vec{\omega}_1 \times \vec{r}_{11}) = (0,\ 0,\ 0)^T,\ \vec{w}_o = (0,\ 0,\ 0)^T$$

*) In the examples the vector coefficients $\vec{\theta}_i$ and $\vec{\eta}_i$ (see 2.7 and 2.8) for angular and linear accelerations, respectively, were denoted by $\vec{\alpha}_i^o$ and $\vec{\beta}_i^o$, in flow-chart (block 3, page 169) and NOMDYN (Appendix 9).

Linear acceleration

$$\vec{w}_1 = \vec{\beta}_{11}\ddot{\phi}+\vec{\beta}_1^o = (0,\ 0,\ 0)^T$$

2.5. *Inertial force*

$$\vec{a}_{11} = -m_1\vec{\beta}_{11} = (0,\ 0,\ 0)^T$$

$$\vec{a}_1^o = -m_1\vec{\beta}_1^o = (0,\ 0,\ 0)^T$$

$$\vec{F}_1 = -m_1\vec{w}_1 = \vec{a}_{11}\ddot{\phi} + \vec{a}_1^o = (0,\ 0,\ 0)^T$$

2.6. *Moment of inertial forces*

Vector $\vec{\lambda}_i$ in the expression for moment of inertial forces

$$\vec{\lambda}_1 = Q_1 \begin{bmatrix} (J_{12}-J_{13})(\vec{\omega}_1\cdot\vec{q}_{12})(\vec{\omega}_1\cdot\vec{q}_{13}) \\ (J_{13}-J_{11})(\vec{\omega}_1\cdot\vec{q}_{13})(\vec{\omega}_1\cdot\vec{q}_{11}) \\ (J_{11}-J_{12})(\vec{\omega}_1\cdot\vec{q}_{11})(\vec{\omega}_1\cdot\vec{q}_{12}) \end{bmatrix} = \begin{bmatrix} 0 \\ 0 \\ 0 \end{bmatrix}$$

$$T_1^{jk} = \sum_{\ell=1}^{3} q_{1\ell}^k\ q_{1\ell}^j\ J_{1\ell}$$

$$T_1 = \left[T_1^{jk}\right]_{3,3} = \begin{bmatrix} \cos^2\phi J_{11}+\sin^2\phi J_{12} & \sin\phi\cos\phi(J_{11}-J_{12}) & 0 \\ \sin\phi\cos\phi(J_{11}-J_{12}) & \sin^2\phi J_{11}+\cos^2\phi J_{12} & 0 \\ 0 & 0 & J_{13} \end{bmatrix}$$

$$b_1^o = -T_1\vec{\alpha}_1^o + \vec{\lambda}_1 = (0,\ 0,\ 0)^T$$

$$b_{11} = -T_1\vec{\alpha}_{11} = (0,\ 0,\ -J_{13})^T$$

Moment of inertial forces

$$\vec{M}_1 = \vec{b}_{11}\ddot{\phi} + \vec{b}_1^o = (0,\ 0,\ -J_{13}\ddot{\phi})^T$$

3. Segment "2"

Joint is prismatic $\xi_2 = 1$.

Consider the segment as a cane (J_S, J_N). Segments 1 and 2 are "specific" from the upper and lower side, respectively because $\vec{\tilde{e}}_2 || \vec{\tilde{r}}_{22}$ and $\vec{e}_2 || \vec{r}_{12}$

$$\vec{e}_2 = (0,\ 0,\ 1)^T, \qquad \vec{\tilde{e}}_2 = (0,\ 0,\ 1)^T$$

$$\vec{r}_{12} = (0,\ 0,\ -\varepsilon)^T \qquad \vec{\tilde{r}}_{22} = (0,\ 0,\ \ell_o^*)^T$$

$$\vec{r}^{\,A}_{12} = (-\varepsilon\sin\phi,\ \varepsilon\cos\phi,\ 0)^T \qquad \vec{\tilde{r}}_{23} = (0,\ 0,\ -(\ell_o - \ell_o^*))^T$$

$$\vec{\tilde{r}}^{\,A}_{22} = (0,\ -\ell_o^*,\ 0)^T$$

$$\vec{\tilde{e}}_3 = (0,\ 1,\ 0)^T$$

3.1. *Assembly*

$$\vec{r}_{N2} = \vec{e}_2 \times (\vec{r}^{\,A}_{12} \times \vec{e}_2) = (0,\ 0,\ 1) \times ((-\varepsilon\sin\phi,\ +\varepsilon\cos\phi,\ 0) \times (0,\ 0,\ 1)) =$$

$$= (-\varepsilon\sin\phi,\ +\varepsilon\cos\phi,\ 0)^T$$

$$\vec{\tilde{r}}_{N2} = \vec{\tilde{e}}_2 \times (\vec{\tilde{r}}^{\,A}_{22} \times \vec{\tilde{e}}_2) = (0,\ 0,\ 1) \times ((0,\ -\ell_o^*,\ 0) \times (0,\ 0,\ 1)) =$$

$$= (0,\ -\ell_o^*,\ 0)^T$$

$$\vec{a}_2 = -\frac{\vec{r}_{N2}}{|\vec{r}_{N2}|} = (\sin\phi,\ -\cos\phi,\ 0)^T, \qquad \vec{\tilde{a}}_2 = \frac{\vec{\tilde{r}}_{N2}}{|\vec{\tilde{r}}_{N2}|} = (0,\ -1,\ 0)^T$$

$$\vec{e}_2 = (0,\ 0,\ 1)^T, \qquad \vec{\tilde{e}}_2 = (0,\ 0,\ 1)^T$$

$$\vec{b}_2 = \vec{e}_2 \times \vec{a}_2 = (\cos\phi,\ \sin\phi,\ 0)^T \qquad \vec{\tilde{b}}_2 = \vec{\tilde{e}}_2 \times \vec{\tilde{a}}_2 = (1,\ 0,\ 0)^T$$

Transformation matrix before changing the generalized coordinate z

$$Q_2^o = [\vec{e}_2\ \vec{a}_2\ \vec{b}_2][\vec{\tilde{e}}_2\ \vec{\tilde{a}}_2\ \vec{\tilde{b}}_2]^T = \begin{bmatrix} 0 & \sin\phi & \cos\phi \\ 0 & -\cos\phi & \sin\phi \\ 1 & 0 & 0 \end{bmatrix} \begin{bmatrix} 0 & 0 & 1 \\ 0 & -1 & 0 \\ 1 & 0 & 0 \end{bmatrix} =$$

$$= \begin{bmatrix} \cos\phi & -\sin\phi & 0 \\ \sin\phi & \cos\phi & 0 \\ 0 & 0 & 1 \end{bmatrix} = Q_1 = [\vec{q}^{\,o}_{21}\ \vec{q}^{\,o}_{22}\ \vec{q}^{\,o}_{23}]$$

3.2. *Position after changing the generalized coordinate z*

Rodrigues formula

$$\vec{q}_{2j} = \vec{q}^{\,o}_{2j}, \qquad j=1,2,3$$

$$\vec{q}_{21} = \vec{q}^{\,o}_{21} = [\cos\phi,\ \sin\phi,\ 0]^T$$

$$\vec{q}_{22} = \vec{q}^{\,o}_{22} = [-\sin\phi,\ \cos\phi,\ 0]^T$$

$$\vec{q}_{23} = \vec{q}^{\,o}_{23} = [0,\ 0,\ 1]^T$$

Transformation matrix after changing the generalized coordinate z

$$Q_2 = [\vec{q}_{21}\ \vec{q}_{22}\ \vec{q}_{23}] = \begin{bmatrix} \cos\phi & -\sin\phi & 0 \\ \sin\phi & \cos\phi & 0 \\ 0 & 0 & 1 \end{bmatrix}$$

$$\vec{r}_{22} = Q_2\tilde{\vec{r}}_{22} + z\vec{e}_2 = (0,\ 0,\ \ell^*_o+z)^T$$

$$\vec{r}_{23} = Q_2\tilde{\vec{r}}_{23} = (0,\ 0,\ -(\ell_o-\ell^*_o))^T$$

$$\vec{r}_{21} = \vec{r}_{11} - \vec{r}_{12} + \vec{r}_{22} = (0,\ 0,\ \ell^*_o + z)$$

$$\vec{e}_3 = Q_2\tilde{\vec{e}}_3 = (-\sin\phi,\ \cos\phi,\ 0)^T$$

3.3. *Angular and linear velocity*

$$\vec{\omega}_2 = \vec{\omega}_1 + \dot{z}e_2(1-\xi_2) = \vec{\omega}_1 = (0,\ 0,\ \dot{\phi})^T$$

$$\vec{v}_2 = \vec{v}_1 - \vec{\omega}_1\times\vec{r}_{12} + \vec{\omega}_2\times\vec{r}_{22} + \dot{z}\vec{e}_2\xi_2 = (0,\ 0,\ \dot{z})^T$$

3.4. *Angular and linear acceleration*

Vector coefficients of angular acceleration

$$\vec{\alpha}_{22} = \vec{e}_2(1-\xi_2) = (0,\ 0,\ 0)^T$$

$$\vec{\alpha}_{21} = \vec{\alpha}_{11} = (0,\ 0,\ 1)^T$$

$$\vec{\alpha}_2^o = \vec{\alpha}_1^o + \dot{z}(\vec{\omega}_1 \times \vec{e}_2)(1-\xi_2) = (0,\ 0,\ 0)^T$$

Angular acceleration

$$\vec{\varepsilon}_2 = \vec{\alpha}_{21}\ddot{\phi} + \vec{\alpha}_{22}\ddot{z} + \vec{\alpha}_2^o = (0,\ 0,\ \ddot{\phi})^T$$

Vector coefficients of linear acceleration

$$\vec{\beta}_{22} = (\vec{e}_2 \times \vec{r}_{22})(1-\xi_2) + \vec{e}_2\xi_2 = (0,\ 0,\ 1)^T$$

$$\vec{\beta}_{21} = \vec{\beta}_{11} + \vec{\alpha}_{11} \times (\vec{r}_{22} - \vec{r}_{12}) = (0,\ 0,\ 0)^T$$

$$\vec{\beta}_2^o = \vec{\beta}_1^o + \vec{\alpha}_1^o \times (\vec{r}_{22} - \vec{r}_{12}) + \dot{z}(\vec{\omega}_1 \times \vec{e}_2) \times \vec{r}_{22}(1-\xi_2) +$$

$$+ 2\dot{z}(\vec{\omega}_1 \times \vec{e}_2)\xi_2 - \vec{\omega}_1 \times (\vec{\omega}_1 \times \vec{r}_{12}) + \vec{\omega}_2 \times (\vec{\omega}_2 \times \vec{r}_{22}) = (0,\ 0,\ 0)^T$$

Linear acceleration

$$\vec{w}_2 = \vec{\beta}_{21}\ddot{\phi} + \vec{\beta}_{22}\ddot{z} + \vec{\beta}_2^o = (0,\ 0,\ \ddot{z})^T$$

3.5. *Inertial force*

$$\vec{a}_{21} = -m_2\vec{\beta}_{21} = (0,\ 0,\ 0)^T$$

$$\vec{a}_{22} = -m_2\vec{\beta}_{22} = (0,\ 0,\ -m_2)^T$$

$$\vec{a}_2^o = -m_2\vec{\beta}_2^o = (0,\ 0,\ 0)^T$$

$$\vec{F}_2 = -m_2\vec{w}_2 = \vec{a}_{21}\ddot{\phi} + \vec{a}_{22}\ddot{z} + \vec{a}_2^o = (0,\ 0,\ -m_2\ddot{z})^T$$

3.6. *Moment of inertial force*

Unit vector along cane's axis

$$\vec{s}_2 = \frac{\vec{r}_{22}}{|\vec{r}_{22}|} = (0,\ 0,\ 1)^T$$

Vector coefficients

$$\vec{b}_{21} = -J_{N2}(\vec{s}_2 \times \vec{\alpha}_{21}) \times \vec{s}_2 - J_{s2}(\vec{\alpha}_{21} \cdot \vec{s}_2)\vec{s}_2 = (0,\ 0,\ -J_{s2})^T$$

$$\vec{b}_{22} = -J_{N2}(\vec{s}_2 \times \vec{\alpha}_{22}) \times \vec{s}_2 - J_{s2}(\vec{\alpha}_{22} \cdot \vec{s}_2)\vec{s}_2 = (0,\ 0,\ 0)^T$$

$$\vec{\tau}_2 = (\vec{\omega}_2 \cdot \vec{s}_2)(\vec{s}_2 \times \vec{\omega}_2) = (0,\ 0,\ 0)^T$$

$$\vec{b}_2^{\,o} = -J_{N2}[(\vec{s}_2 \times \vec{\alpha}_2^{\,o}) \times \vec{s}_2 + \vec{\tau}_2] - J_{s2}(\vec{\alpha}_2^{\,o} \cdot \vec{s}_2)\vec{s}_2 = (0,\ 0,\ 0)^T$$

Moment of inertial force

$$\vec{M}_2 = \vec{b}_{21}\ddot{\phi} + \vec{b}_{22}\ddot{z} + \vec{b}_2^{\,o} = (0,\ 0,\ -J_{s2}\ddot{\phi})^T$$

4. Segment "3"

Joint is prismatic $\xi_3 = 1$.

Consider the segment as a cane (J_s, J_N). The segment is "specific" from lower side because $\vec{e}_3 || \tilde{\vec{r}}_{33}$

$$\vec{e}_3 = (-\sin\phi,\ \cos\phi,\ 0)^T \qquad \tilde{\vec{e}}_3 = (0,\ 1,\ 0)^T$$

$$\vec{r}_{23} = (0,\ 0,\ -(\ell_o - \ell_o^*))^T \qquad \tilde{\vec{r}}_{33} = (0,\ \ell_1^*,\ 0)^T$$

$$\tilde{\vec{r}}_{33}^{\,A} = (0,\ 0,\ \ell_1^*)^T$$

4.1. *Assembly*

$$\vec{r}_{N3} = \vec{e}_3 \times (\vec{r}_{23} \times \vec{e}_3) = (-\sin\phi,\ \cos\phi,\ 0) \times ((0,\ 0,\ -(\ell_1 - \ell_1^*)) \times (-\sin\phi,\ \cos\phi,\ 0)) =$$

$$= (0,\ 0,\ -(\ell_1 - \ell_1^*))^T$$

$$\tilde{\vec{r}}_{N3} = \tilde{\vec{e}}_3 \times (\tilde{\vec{r}}_{33}^{\,A} \times \tilde{\vec{e}}_3) = (0,\ 1,\ 0) \times ((0,\ 0,\ \ell_1^*) \times (0,\ 1,\ 0)) = (0,\ 0,\ \ell_1^*)^T$$

$$\vec{a}_3 = \frac{-\vec{r}_{N3}}{|\vec{r}_{N3}|} = (0,\ 0,\ 1)^T \qquad \vec{\tilde{a}}_3 = \frac{\vec{\tilde{r}}_{N3}}{|\vec{\tilde{r}}_{N3}|} = (0,\ 0,\ 1)^T$$

$$\vec{e}_3 = (-\sin\phi,\ \cos\phi,\ 0)^T \qquad \vec{\tilde{e}}_3 = (0,\ 1,\ 0)^T$$

$$\vec{b}_3 = \vec{e}_3 \times \vec{a}_3 = (\cos\phi,\ \sin\phi,\ 0)^T \qquad \vec{\tilde{b}}_3 = \vec{\tilde{e}}_3 \times \vec{\tilde{a}}_3 = (1,\ 0,\ 0)^T$$

Transformation matrix before changing the generalized coordinate x

$$Q_3^o = [\vec{e}_3\ \vec{a}_3\ \vec{b}_3][\vec{\tilde{e}}_3\ \vec{\tilde{a}}_3\ \vec{\tilde{b}}_3]^T = \begin{bmatrix} -\sin\phi & 0 & \cos\phi \\ \cos\phi & 0 & \sin\phi \\ 0 & 1 & 0 \end{bmatrix} \begin{bmatrix} 0 & 1 & 0 \\ 0 & 0 & 1 \\ 1 & 0 & 0 \end{bmatrix} =$$

$$= \begin{bmatrix} \cos\phi & -\sin\phi & 0 \\ \sin\phi & \cos\phi & 0 \\ 0 & 0 & 1 \end{bmatrix} = Q_2 = [\vec{q}^{\,o}_{31}\ \vec{q}^{\,o}_{32}\ \vec{q}^{\,o}_{33}]$$

4.2. *Position after changing the generalized coordinate x*

Rodrigues formula

$$\vec{q}_{3j} = \vec{q}^{\,o}_{3j}, \qquad j=1,2,3$$

$$\vec{q}_{31} = (\cos\phi,\ \sin\phi,\ 0)^T$$

$$\vec{q}_{32} = (-\sin\phi,\ \cos\phi,\ 0)^T$$

$$\vec{q}_{33} = (0,\ 0,\ 1)^T$$

Transformation matrix after changing the generalized coordinate x

$$Q_3 = [\vec{q}_{31}\ \vec{q}_{32}\ \vec{q}_{33}] = \begin{bmatrix} \cos\phi & -\sin\phi & 0 \\ \sin\phi & \cos\phi & 0 \\ 0 & 0 & 1 \end{bmatrix}$$

$$\vec{r}_{33} = Q_3\vec{\tilde{r}}_{33} + x\vec{e}_3 = (-(\ell_1^*+x)\sin\phi,\ (\ell_1^*+x)\cos\phi,\ 0)^T$$

$$\vec{r}_{31} = \vec{r}_{21} - \vec{r}_{23} + \vec{r}_{33} = (-(\ell_1^*+x)\sin\phi,\ (\ell_1^*+x)\cos\phi,\ \ell_o+z)^T$$

$$\vec{r}_{32} = \vec{r}_{22} - \vec{r}_{23} + \vec{r}_{33} = (-(\ell_1^*+x)\sin\phi,\ (\ell_1^*+x)\cos\phi,\ \ell_o+z)^T$$

4.3. *Angular and linear velocity*

$$\vec{\omega}_3 = \vec{\omega}_2 + \dot{x}\vec{e}_3(1-\xi_3) = (0,\ 0,\ \dot{\phi})^T$$

$$\vec{v}_3 = \vec{v}_2 - \vec{\omega}_2 \times \vec{r}_{23} + \vec{\omega}_3 \times \vec{r}_{33} + \dot{x}\vec{e}_3\xi_3 =$$

$$= (-\dot{\phi}(\ell_1^* + x)\cos\phi - \dot{x}\sin\phi,\ -\dot{\phi}(\ell_1^* + x)\sin\phi + \dot{x}\cos\phi,\ \dot{z})^T$$

4.4. *Angular and linear acceleration*

Vector coefficient of angular acceleration

$$\vec{\alpha}_{31} = \vec{\alpha}_{21} = (0,\ 0,\ 1)^T$$

$$\vec{\alpha}_{32} = \vec{\alpha}_{22} = (0,\ 0,\ 0)^T$$

$$\vec{\alpha}_{33} = \vec{e}_3(1-\xi_3) = (0,\ 0,\ 0)^T$$

$$\vec{\alpha}_3^o = \vec{\alpha}_2^o + \dot{x}(\vec{\omega}_2 \times \vec{e}_3)(1-\xi_3) = (0,\ 0,\ 0)^T$$

Angular acceleration

$$\vec{\varepsilon}_3 = \vec{\alpha}_{31}\ddot{\phi} + \vec{\alpha}_{32}\ddot{z} + \vec{\alpha}_{33}\ddot{x} + \vec{\alpha}_3^o = (0,\ 0,\ \ddot{\phi})^T$$

Vector coefficients of linear acceleration

$$\vec{\beta}_{31} = \vec{\beta}_{21} + \vec{\alpha}_{21} \times (\vec{r}_{33} - \vec{r}_{23}) = (-(\ell_1^* + x)\cos\phi,\ -(\ell_1^* + x)\sin\phi,\ 0)^T$$

$$\vec{\beta}_{32} = \vec{\beta}_{22} + \vec{\alpha}_{22} \times (\vec{r}_{33} - \vec{r}_{23}) = (0,\ 0,\ 1)^T$$

$$\vec{\beta}_{33} = (\vec{e}_3 \times \vec{r}_{33})(1-\xi_3) + \vec{e}_3\xi_3 = (-\sin\phi,\ \cos\phi,\ 0)^T$$

$$\vec{\beta}_3^o = \vec{\beta}_2^o + \vec{\alpha}_2^o \times (\vec{r}_{33} - \vec{r}_{23}) + \dot{x}(\vec{\omega}_2 \times \vec{e}_3) \times \vec{r}_{33}(1-\xi_3) +$$

$$+ 2\dot{x}(\vec{\omega}_2 \times \vec{e}_3)\xi_3 - \vec{\omega}_2 \times (\vec{\omega}_2 \times \vec{r}_{23}) + \vec{\omega}_3 \times (\vec{\omega}_3 \times \vec{r}_{33}) =$$

$$= (\dot{\phi}^2(\ell_1^* + x)\sin\phi - 2\dot{x}\dot{\phi}\cos\phi,\ -\dot{\phi}^2(\ell_1^* + x)\cos\phi - 2\dot{x}\dot{\phi}\sin\phi,\ 0)^T$$

Linear acceleration

$$\vec{w}_3 = \vec{\beta}_{31}\ddot{\phi} + \vec{\beta}_{32}\ddot{z} + \vec{\beta}_{33}\ddot{x} + \vec{\beta}_3^{o} =$$

$$= \begin{bmatrix} -\ddot{\phi}(\ell_1^*+x)\cos\phi - \ddot{x}\sin\phi + \dot{\phi}^2(\ell_1^*+x)\sin\phi - 2\dot{x}\dot{\phi}\cos\phi \\ -\ddot{\phi}(\ell_1^*+x)\sin\phi - \ddot{x}\cos\phi - \dot{\phi}^2(\ell_1^*+x)\cos\phi - 2\dot{x}\dot{\phi}\sin\phi \\ \ddot{z} \end{bmatrix}$$

4.5. *Inertial force*

Vector coefficients

$$\vec{a}_{31} = -m_3\vec{\beta}_{31} = [m_3(\ell_1^*+x)\cos\phi,\ m_3(\ell_1^*+x)\sin\phi,\ 0]^T$$

$$\vec{a}_{32} = -m_3\vec{\beta}_{32} = (0,\ 0,\ -m_3)^T$$

$$\vec{a}_{33} = -m_3\vec{\beta}_{33} = (m_3\sin\phi,\ -m_3\cos\phi,\ 0)^T$$

$$\vec{a}_3^{o} = -m_3\vec{\beta}_3^{o} = \begin{bmatrix} -m_3\dot{\phi}^2(\ell_1^*+x)\sin\phi + 2m_3\dot{x}\dot{\phi}\cos\phi \\ m_3\dot{\phi}^2(\ell_1^*+x)\cos\phi + 2m_3\dot{x}\dot{\phi}\sin\phi \\ 0 \end{bmatrix}$$

Inertial force

$$\vec{F}_3 = -m_3\vec{w}_3 = \vec{a}_{31}\ddot{\phi} + \vec{a}_{32}\ddot{z} + \vec{a}_{33}\ddot{x} + \vec{a}_3^{o} =$$

$$= \begin{bmatrix} m_3\ddot{\phi}(\ell_1^*+x)\cos\phi + m_3\ddot{x}\sin\phi - m_3\dot{\phi}^2(\ell_1^*+x)\sin\phi + 2m_3\dot{x}\dot{\phi}\cos\phi \\ m_3\ddot{\phi}(\ell_1^*+x)\sin\phi - m_3\ddot{x}\cos\phi + m_3\dot{\phi}^2(\ell_1^*+x)\cos\phi + 2m_3\dot{x}\dot{\phi}\sin\phi \\ -m_3\ddot{z} \end{bmatrix}$$

4.6. *Moment of inertial force*

Unit vector along cane's axis

$$\vec{s}_3 = \frac{\vec{r}_{33}}{|\vec{r}_{33}|} = (-\sin\phi,\ \cos\phi,\ 0)^T$$

Vector coefficients

$$\vec{b}_{31} = -J_{N3}(\vec{s}_3\times\vec{\alpha}_{31})\times\vec{s}_3 - J_{s3}(\vec{\alpha}_{31}\cdot\vec{s}_3)\vec{s}_3 = (0,\ 0,\ -J_{N3})^T$$

$$\vec{b}_{32} = -J_{N3}(\vec{s}_3\times\vec{\alpha}_{32})\times\vec{s}_3 - J_{s3}(\vec{\alpha}_{32}\cdot\vec{s}_3)\vec{s}_3 = (0,\ 0,\ 0)^T$$

$$\vec{b}_{33} = -J_{N3}(\vec{s}_3\times\vec{\alpha}_{33})\times\vec{s}_3 - J_{s3}(\vec{\alpha}_{33}\cdot\vec{s}_3)\vec{s}_3 = (0,\ 0,\ 0)^T$$

$$\vec{\tau}_3 = (\vec{\omega}_3\cdot\vec{s}_3)(\vec{s}_3\times\vec{\omega}_3) = (0,\ 0,\ 0)^T$$

$$\vec{b}_3^{o} = -J_{N3}[(\vec{s}_3\times\vec{\alpha}_3^{o})\times\vec{s}_3+\vec{\tau}_3] - J_{s3}(\vec{\alpha}_3^{o}\cdot\vec{s}_3)\vec{s}_3 = (0,\ 0,\ 0)^T$$

Moment of inertial force

$$\vec{M}_3 = \vec{b}_{31}\ddot{\phi} + \vec{b}_{32}\ddot{z} + \vec{b}_{33}\ddot{x} + \vec{b}_3^{o} = (0,\ 0,\ -J_{N3}\ \ddot{\phi})^T$$

5. Forming of differential equations

Model

$$H(q)\ddot{q} + h(q,\ \dot{q}) = P$$

5.1. *Elements of H-matrix*

$$\xi_1 = 0 \quad H_{ik} = -\vec{e}_i \sum_{j=\max(i,k)}^{n} (\vec{b}_{jk}+\vec{r}_{ji}\times\vec{a}_{jk}) \qquad i=1,\ k=1,2,3$$

$$H_{11} = -\vec{e}_1\cdot\sum_{j=1}^{3}(\vec{b}_{j1}+\vec{r}_{j1}\times\vec{a}_{j1}) = -\vec{e}_1\cdot(\vec{b}_{11}+\vec{r}_{11}\times\vec{a}_{11}+\vec{b}_{21}+\vec{r}_{21}\times\vec{a}_{21}+$$
$$+\ \vec{b}_{31}+\vec{r}_{31}\times\vec{a}_{31}) = m_3(\ell_1^*+x)^2+J_{13}+J_{s2}+J_{N3}$$

$$H_{12} = -\vec{e}_1\cdot\sum_{j=2}^{3}(\vec{b}_{j2}+\vec{r}_{j1}\times\vec{a}_{j2}) = -\vec{e}_1\cdot(\vec{b}_{22}+\vec{r}_{21}\times\vec{a}_{22}+\vec{b}_{32}+\vec{r}_{31}\times\vec{a}_{32})=0$$

$$H_{13} = -\vec{e}_1\cdot(\vec{b}_{33}+\vec{r}_{31}\times\vec{a}_{33}) = 0$$

$$\xi_2 = 1 \quad H_{ik} = -\vec{e}_i \sum_{j=\max(i,k)}^{n} \vec{a}_{jk} \qquad i=2,\ k=1,2,3$$

$$H_{21} = -\vec{e}_2 \cdot \sum_{j=2}^{3} \vec{a}_{j1} = -\vec{e}_2 \cdot (\vec{a}_{21}+\vec{a}_{31}) = 0$$

$$H_{22} = -\vec{e}_2 \cdot \sum_{j=2}^{3} \vec{a}_{j2} = -\vec{e}_2 \cdot (\vec{a}_{22}+\vec{a}_{32}) = m_2 + m_3$$

$$H_{23} = -\vec{e}_2 \cdot \vec{a}_{33} = 0$$

$$\xi_3 = 1 \quad H_{ik} = -\vec{e}_i \cdot \sum_{j=\max(i,k)}^{n} \vec{a}_{jk} \qquad i=3,\ k=1,2,3$$

$$H_{31} = -\vec{e}_3 \cdot \sum_{j=3}^{3} \vec{a}_{j1} = -\vec{e}_3 \cdot \vec{a}_{31} = 0$$

$$H_{32} = -\vec{e}_3 \cdot \sum_{j=3}^{3} \vec{a}_{j2} = -\vec{e}_3 \cdot \vec{a}_{32} = 0$$

$$H_{33} = -\vec{e}_3 \cdot \vec{a}_{33} = m_3$$

5.2. *Elements of h-matrix*

$$G_1 = (0,\ 0,\ -m_1 g)^T, \quad G_2 = (0,\ 0,\ -m_2 g)^T, \quad G_3 = (0,\ 0,\ -m_3 g)^T$$

$$\xi_1 = 0 \quad h^i = -\vec{e}_i \cdot \sum_{j=i}^{n} (\vec{r}_{ji} \times (\vec{a}_j^{\,o}+\vec{G}_j) + \vec{b}_j^{\,o}) \qquad i=1$$

$$h_1 = -\vec{e}_1 \cdot \sum_{j=1}^{3} [\vec{r}_{j1} \times (\vec{a}_j^{\,o}+\vec{G}_j) + \vec{b}_j^{\,o}] = -\vec{e}_1 \cdot [\vec{r}_{11} \times (\vec{a}_1^{\,o}+\vec{G}_1) + \vec{b}_1^{\,o} + \vec{r}_{21} \times (\vec{a}_2^{\,o}+\vec{G}_2) +$$

$$+ \vec{b}_2^{\,o} + \vec{r}_{31} \times (\vec{a}_3^{\,o}+\vec{G}_3) + \vec{b}_3^{\,o}] = 2m_3 \dot{x}\dot{\phi}(\ell_3^* + x)$$

$$\xi_2 = 1 \quad h^i = -\vec{e}_i \cdot \sum_{j=i}^{n} (\vec{a}_j^{\,o}+\vec{G}_j) \qquad i=2$$

$$h_2 = -\vec{e}_2 \cdot \sum_{j=2}^{3} (\vec{a}_j^{\,o}+\vec{G}_j) = -\vec{e}_2 \cdot (\vec{a}_2^{\,o}+\vec{G}_2+\vec{a}_3^{\,o}+\vec{G}_3) = (m_2+m_3)g$$

$$\xi_3 = 1 \quad h^i = -\vec{e}_i \cdot \sum_{j=i}^{n} (\vec{a}_j^{\,o}+\vec{G}_j) \qquad i=3$$

$$h_3 = -\vec{e}_3 \cdot (\vec{a}_3^{\,o}+\vec{G}_3) = -m_3 \dot{\phi}^2 (\ell_1^* + x)$$

$$P = H(q)\ddot{q} + h(q,\ \dot{q})$$

$$\begin{bmatrix} P_1^M \\ P_2^F \\ P_3^F \end{bmatrix} = \begin{bmatrix} m_3(\ell_1^*+x)^2+J_{s2}+J_{N3} & 0 & 0 \\ 0 & m_2+m_3 & 0 \\ 0 & 0 & m_3 \end{bmatrix} \begin{bmatrix} \ddot{\phi} \\ \ddot{z} \\ \ddot{x} \end{bmatrix} + \begin{bmatrix} 2m_3\dot{x}\dot{\phi}(\ell_1^*+x) \\ (m_2+m_3)g \\ -m_3\dot{\phi}^2(\ell_1^*+x) \end{bmatrix}$$

6. Differential equations of motion

$$P_1^M = [m_3(\ell_1^*+x)^2+J_{13}+J_{s2}+J_{N3}]\ddot{\phi}+2m_3\dot{x}\dot{\phi}(\ell_1^*+x)$$

$$P_2^F = (m_2+m_3)\ddot{z} + (m_2+m_3)g$$

$$P_3^F = m_3\ddot{x} - m_3\dot{\phi}^2(\ell_1^*+x)$$

Example 2: "Stanford" Manipulator (Basic Configuration)

ϕ - rotation of column around vertical axis

ψ - rotation around horizontal axis

x - horizontal translation

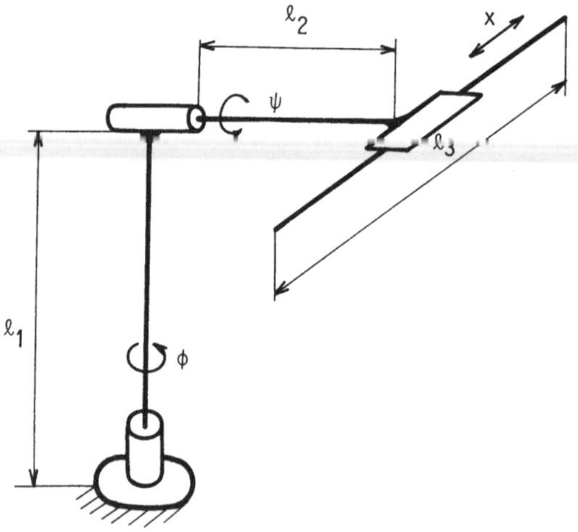

Fig. A.7.3. Mechanical scheme of "Stanford" manipulator (basic configuration)

1. Introduction

The mechanism presented in Fig. A.7.3. can be schematically presented for some adopted initial position, in the absolute coordinate system as given in Fig. A.7.4.

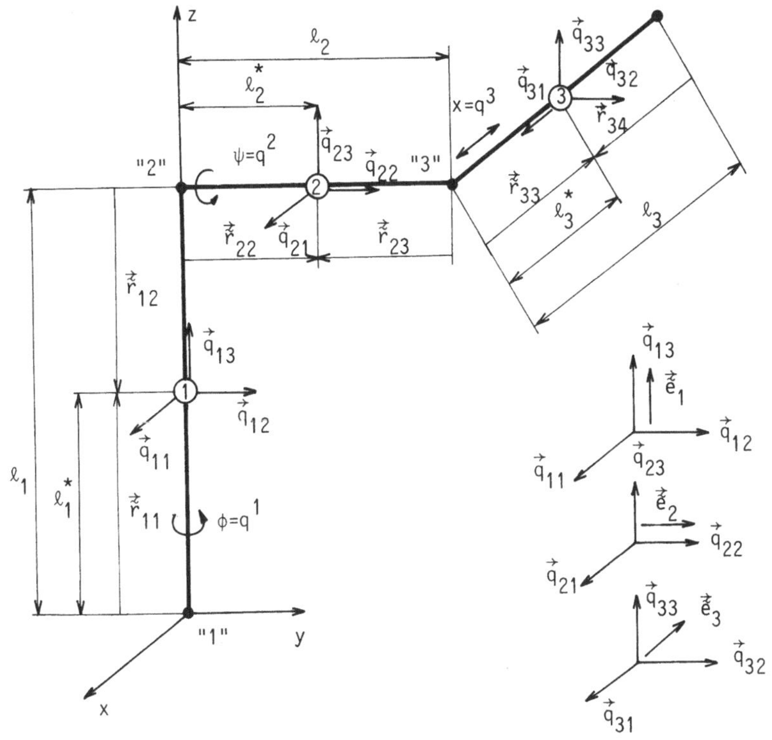

Fig. A.7.4. Initial position of mechanism

The mechanism has three degrees of freedom:

- rotation of column around vertical axis: ϕ
- rotation around horizontal axis: ψ
- horizontal translation: x.

For initial mechanism position the relative coordinates are zero, i.e.

$$\phi^{o} = 0, \quad \psi^{o} = 0, \quad x^{o} = 0$$

2. Segment "1"

Joint is revolute, $\xi_1 = 0$.

Consider the segment as a body (J_{11}, J_{12}, J_{13}). The segment is "specific" from the lower side because $\vec{\bar{r}}_{11} || \vec{\bar{e}}_1$.

$$\vec{e}_1 = (0,\ 0,\ 1)^T \qquad \vec{\bar{e}}_1 = (0,\ 0,\ 1)^T$$

$$\vec{r}^{\,A}_{o1} = (-1,\ 0,\ 0)^T \qquad \vec{\bar{r}}_{11} = (0,\ 0,\ \ell_1^*)^T$$

$$\vec{\bar{r}}^{\,A}_{11} = (\ell_1^*,\ 0,\ 0)^T$$

2.1. *Assembly*

$$\vec{r}_{N1} = \vec{e}_1 \times (\vec{r}^{\,A}_{o1} \times \vec{e}_1) = (0,\ 0,\ 1)^T \times ((-1,\ 0,\ 0)^T \times (0,\ 0,\ 1)^T) = (-1,\ 0,\ 0)^T$$

$$\vec{\bar{r}}_{N1} = \vec{\bar{e}}_1 \times (\vec{\bar{r}}^{\,A}_{11} \times \vec{\bar{e}}_1) = (0,\ 0,\ 1)^T \times ((\ell_1^*,\ 0,\ 0)^T \times (0,\ 0,\ 1)^T) = (\ell_1^*,\ 0,\ 0)^T$$

$$\vec{a}_1 = -\frac{\vec{r}_{N1}}{|\vec{r}_{N1}|} = (1,\ 0,\ 0)^T \qquad \vec{\bar{a}}_1 = \frac{\vec{\bar{r}}_{N1}}{|\vec{\bar{r}}_{N1}|} = (1,\ 0,\ 0)^T$$

$$\vec{e}_1 = (0,\ 0,\ 1)^T \qquad \vec{\bar{e}}_1 = (0,\ 0,\ 1)^T$$

$$\vec{b}_1 = \vec{e}_1 \times \vec{a}_1 = (0,\ 1,\ 0)^T \qquad \vec{\bar{b}}_1 = \vec{\bar{e}}_1 \times \vec{\bar{a}}_1 = (0,\ 1,\ 0)^T$$

Transformation matrix before rotation

$$Q_1^o = [\vec{e}_1\ \vec{a}_1\ \vec{b}_1][\vec{\bar{e}}_1\ \vec{\bar{a}}_1\ \vec{\bar{b}}_1]^T = \begin{bmatrix} 0 & 1 & 0 \\ 0 & 0 & 1 \\ 1 & 0 & 0 \end{bmatrix} \begin{bmatrix} 0 & 0 & 1 \\ 1 & 0 & 0 \\ 0 & 1 & 0 \end{bmatrix} =$$

$$= \begin{bmatrix} 1 & 0 & 0 \\ 0 & 1 & 0 \\ 0 & 0 & 1 \end{bmatrix} = [\vec{q}^{\,o}_{11}\ \vec{q}^{\,o}_{12}\ \vec{q}^{\,o}_{13}]$$

2.2. *Position after rotation (changes of generalized coordinate ϕ)*

$$\vec{q}_{1j} = \vec{q}^{\,o}_{1j}\cos\phi + (1-\cos\phi)(\vec{e}_1 \cdot \vec{q}_{1j}) \cdot \vec{e}_1 + \vec{e}_1 \times \vec{q}_{1j}\sin\phi, \qquad j=1,2,3$$

$$\vec{q}_{11} = \vec{q}^{\,o}_{11}\cos\phi + (1-\cos\phi)(\vec{e}_1 \cdot \vec{q}^{\,o}_{11}) \cdot \vec{e}_1 + \vec{e}_1 \times \vec{q}^{\,o}_{11}\sin\phi =$$

$$= \vec{q}^{\,o}_{11}\cos\phi+\vec{e}_1\times\vec{q}^{\,o}_{11}\sin\phi=\cos\phi\begin{bmatrix}1\\0\\0\end{bmatrix}+\sin\phi\begin{bmatrix}0\\0\\1\end{bmatrix}\times\begin{bmatrix}1\\0\\0\end{bmatrix}=\begin{bmatrix}\cos\phi\\\sin\phi\\0\end{bmatrix}$$

$$\vec{q}_{12} = \vec{q}^{\,o}_{12}\cos\phi+(1-\cos\phi)(\vec{e}_1\cdot\vec{q}^{\,o}_{12})\cdot\vec{e}_1+\vec{e}_1\times\vec{q}^{\,o}_{12}\sin\phi =$$

$$= \vec{q}^{\,o}_{12}\cos\phi+\vec{e}_1\times\vec{q}^{\,o}_{12}\sin\phi=\cos\phi\begin{bmatrix}0\\1\\0\end{bmatrix}+\sin\phi\begin{bmatrix}0\\0\\1\end{bmatrix}\times\begin{bmatrix}0\\1\\0\end{bmatrix}=\begin{bmatrix}-\sin\phi\\\cos\phi\\0\end{bmatrix}$$

$$\vec{q}_{13} = \vec{q}^{\,o}_{13}\cos\ +(1-\cos\phi)(\vec{e}_1\cdot\vec{q}^{\,o}_{13})\cdot\vec{e}_1+\vec{e}_1\times\vec{q}^{\,o}_{13}\sin\phi =$$

$$= \vec{q}^{\,o}_{13}\cos\phi+\vec{e}_1(1-\cos\phi) = [0,0,1]^T$$

Transformation matrix after rotation

$$Q_1 = [\vec{q}_{11}\ \vec{q}_{12}\ \vec{q}_{13}] = \begin{bmatrix}\cos\phi & -\sin\phi & 0\\ \sin\phi & \cos\phi & 0\\ 0 & 0 & 1\end{bmatrix}$$

$$\vec{r}_{11} = Q_1\vec{\bar{r}}_{11} = Q_1[0,0,\ell_1^*]^T = [0,0,\ell_1^*]^T$$

$$\vec{r}_{12} = Q_1\vec{\bar{r}}_{12} = Q_1[0,0,-(\ell_1-\ell_1^*)]^T = [0,0,-(\ell_1-\ell_1^*)]^T$$

$$\vec{e}_2 = Q_1\vec{\bar{e}}_2 = Q_1[0,1,0]^T = [-\sin\phi,\cos\phi,0]^T$$

2.3. *Angular and linear velocity*

$$\vec{\omega}_1 = \vec{\omega}_o + \dot{\phi}\vec{e}_1(1-\xi_1); \qquad \vec{\omega}_o = [0,\ 0,\ 0]^T$$

$$\vec{\omega}_1 = \dot{\phi}\vec{e}_1 = [0,0,\dot{\phi}]^T$$

$$\vec{v}_1 = \vec{v}_o - \vec{\omega}_o\times\vec{r}_{o1} + \vec{\omega}_1\times\vec{r}_{11} + \dot{\phi}\vec{e}_1\xi_1$$

$$\vec{v}_1 = \vec{\omega}_1\times\vec{r}_{11} = [0,0,\dot{\phi}]^T\times[0,0,\ell_1^*]^T = [0,0,0]^T$$

2.4. *Angular and linear acceleration*

$$\vec{\alpha}_{11} = \vec{e}_1(1-\xi_1)$$

$$\vec{\alpha}_{11} = \vec{e}_1 = [0, 0, 1]^T$$

$$\vec{\alpha}_1^o = \vec{\alpha}_o^o + \dot{\phi}(\vec{\omega}_o \times \vec{e}_i)(1-\xi_1)$$

$$\vec{\omega}_o = [0, 0, 0]^T$$

$$\vec{\alpha}_1^o = [0, 0, 0]^T$$

$$\vec{\beta}_{11} = (\vec{e}_1 \times \vec{r}_{11})(1-\xi_1) + \vec{e}_1 \xi_1$$

$$\vec{\beta}_{11} = \vec{e}_1 \times \vec{r}_{11} = [0, 0, 1]^T \times [0, 0, \ell_1^*]^T = [0, 0, 0]^T$$

$$\vec{\gamma}_{o1} = \vec{\omega}_o \times (\vec{\omega}_o \times \vec{r}_{o1}) = [0, 0, 0]^T$$

$$\vec{\gamma}_{11} = \vec{\omega}_1 \times (\vec{\omega}_1 \times \vec{r}_{11}) = [0, 0, \dot{\phi}] \times \{[0, 0, \dot{\phi}] \times [0, 0, \ell_1^*]\} = [0, 0, 0]^T$$

$$\vec{\beta}_1^o = \vec{\beta}_o^o + \vec{\alpha}_o^o \times (\vec{r}_{11} - \vec{r}_{o1}) + \dot{\phi}(\vec{\omega}_o \times \vec{e}_1) \times \vec{r}_{11} + 2\dot{\phi}(\vec{\omega}_o \times \vec{e}_1) - (\vec{\gamma}_{o1}) + \vec{\gamma}_{11}$$

$$\vec{\beta}_1^o = \vec{\gamma}_{11} = [0, 0, 0]^T$$

Angular acceleration

$$\vec{\varepsilon}_1 = \vec{\alpha}_{11}\ddot{\phi} + \vec{\alpha}_{1o} = [0, 0, \ddot{\phi}]^T$$

Linear acceleration

$$\vec{w}_1 = \vec{\beta}_{11}\ddot{\phi} + \vec{\beta}_1^o = [0, 0, 0]^T$$

2.5. *Inertial force*

$$\vec{a}_{11} = -m_1 \vec{\beta}_{11} = [0, 0, 0]^T$$

$$\vec{a}_1^o = -m_1 \vec{\beta}_1^o = [0, 0, 0]^T$$

$$\vec{F}_1 = \vec{a}_{11}\ddot{\phi} + \vec{a}_1^o = [0, 0, 0]^T$$

2.6. *Moment of inertial force*

$$T_1^{jk} = \sum_{\ell=1}^{3} q_{1\ell}^{j}\, q_{1\ell}^{k}\, J_{1\ell}$$

$$T_1^{11} = \sum_{\ell=1}^{3} q_{1\ell}^{1}\, q_{1\ell}^{1}\, J_{1\ell} = \cos^2\phi J_{11} + \sin^2\phi J_{12}$$

$$T_1^{12} = \sum_{\ell=1}^{3} q_{1\ell}^{1}\, q_{1\ell}^{2}\, J_{1\ell} = \sin\phi\cos\phi(J_{11}-J_{12})$$

$$T_1^{13} = \sum_{\ell=1}^{3} q_{1\ell}^{1}\, q_{1\ell}^{3}\, J_{1\ell} = 0$$

$$T_1^{21} = T_1^{12} = \sin\phi\cos\phi(J_{11}-J_{12})$$

$$T_1^{22} = \sum_{\ell=1}^{3} q_{1\ell}^{2}\, q_{1\ell}^{2}\, J_{1\ell} = \sin^2\phi J_{11} + \cos^2\phi J_{12}$$

$$T_1^{23} = \sum_{\ell=1}^{3} q_{1\ell}^{2}\, q_{1\ell}^{3}\, J_{1\ell} = 0$$

$$T_1^{31} = T_1^{13} = 0$$

$$T_1^{32} = T_1^{23} = 0$$

$$T_1^{33} = \sum_{\ell=1}^{3} q_{1\ell}^{3}\, q_{1\ell}^{3}\, J_{1\ell} = J_{13}$$

$$T_1 = \begin{bmatrix} \cos^2\phi J_{11}+\sin^2\phi J_{12} & \sin\phi\cos\phi(J_{11}-J_{12}) & 0 \\ \sin\phi\cos\phi(J_{11}-J_{12}) & \sin^2\phi J_{11}+\cos^2\phi J_{12} & 0 \\ 0 & 0 & J_{13} \end{bmatrix}$$

$$\vec{\omega}_1\vec{q}_{11} = [0, 0, \dot{\phi}][\cos\phi, \sin\phi, 0]^T = 0$$

$$\vec{\omega}_1\vec{q}_{12} = [0, 0, \dot{\phi}][-\sin\phi, \cos\phi, 0]^T = 0$$

$$\vec{\omega}_1\vec{q}_{13} = [0, 0, \dot{\phi}][0, 0, 1]^T = \dot{\phi}$$

$$\vec{\lambda}_1^j = q_{11}^j(J_{12}-J_{13})(\vec{\omega}_1\vec{q}_{12})(\vec{\omega}_1\vec{q}_{13})+q_{12}^j(J_{13}-J_{11})(\vec{\omega}_1\vec{q}_{13})(\vec{\omega}_1\vec{q}_{11}) +$$

$$+ q_{13}^j(J_{11}-J_{12})(\vec{\omega}_1\vec{q}_{11})(\vec{\omega}_1\vec{q}_{12}) = 0, \qquad j=1,2,3;$$

$$\vec{\lambda}_1 = [0\ 0\ 0]^T$$

$$\vec{b}_{11} = -T_1\vec{\alpha}_{11} = -T_1[0, 0, 1]^T = -[T_{13}, T_{23}, T_{33}]^T = [0, 0, -J_{13}]^T$$

$$\vec{b}_1^o = -T_1\vec{\alpha}_1^o + \vec{\lambda}_1 = [0, 0, 0]^T$$

$$\vec{M}_1 = \vec{b}_{11}\ddot{\phi} + \vec{b}_1^o = [0, 0, -J_{13}\ddot{\phi}]^T$$

3. Segment "2"

Joint is revolute, $\xi_2 = 0$

Consider the segment as a body (J_{21}, J_{22}, J_{23}). The segment is "specific" from the lower side, $\vec{\tilde{r}}_{22} || \vec{\tilde{e}}_2$.

$$\vec{e}_2 = (-\sin\phi, \cos\phi, 0)^T \qquad \vec{\tilde{e}}_2 = (0, 1, 0)^T$$

$$\vec{r}_{12} = (0, 0, -(\ell_1-\ell_1^*))^T \qquad \vec{\tilde{r}}_{22} = (0, \ell_2^*, 0)^T$$

$$\vec{\tilde{r}}_{22}^A = (0, 0, \ell_2^*)^T$$

3.1. *Assembly*

$$\vec{r}_{N2} = \vec{e}_2\times(\vec{r}_{12}\times\vec{e}_2) = [-\sin\phi, \cos\phi, 0]^T\times\{[0, 0, -(\ell_1-\ell_1^*)]^T$$

$$\times[-\sin\phi, \cos\phi, 0]^T\} = [-\sin\phi, \cos\phi, 0]^T$$

$$\times[(\ell_1-\ell_1^*)\cos\phi, (\ell_1-\ell_1^*)\sin\phi, 0]^T = [0, 0, -(\ell_1-\ell_1^*)]^T$$

$$\vec{a}_2 = -\vec{r}_{N2}/|\vec{r}_{N2}| = [0, 0, 1]^T$$

$$\vec{\tilde{r}}_{N2} = \vec{\tilde{e}}_2\times(\vec{\tilde{r}}_{22}^A\times\vec{\tilde{e}}_2) = [0, 1, 0]^T\times\{[0, 0, \ell_2^*]^T\times[0\ 1\ 0]^T\} = [0, 0, \ell_2^*]^T$$

$$\vec{\tilde{a}}_2 = +\vec{\tilde{r}}_{N2}/|\vec{\tilde{r}}_{N2}| = [0, 0, 1]^T$$

$$\vec{e}_2 = [-\sin\phi, \cos\phi, 0]^T \qquad \vec{\tilde{e}}_2 = [0, 1, 0]^T$$

$$\vec{b}_2 = \vec{e}_2\times\vec{a}_2 = [-\sin\phi, \cos\phi, 0]^T\times[0, 0, 1]^T = [\cos\phi, \sin\phi, 0]^T$$

$$\vec{\tilde{b}}_2 = \vec{\tilde{e}}_2\times\vec{\tilde{a}}_2 = [0, 1, 0]^T\times[0, 0, 1]^T = [1, 0, 0]^T$$

Transformation matrix before rotation

$$Q_2^o = [\vec{q}_{21}^{\,o}\ \vec{q}_{22}^{\,o}\ \vec{q}_{23}^{\,o}] = [\vec{e}_2\ \vec{a}_2\ \vec{b}_2][\vec{\bar{e}}_2\ \vec{\bar{a}}_2\ \vec{\bar{b}}_2]^T =$$

$$= \begin{bmatrix} -\sin\phi & 0 & \cos\phi \\ \cos\phi & 0 & \sin\phi \\ 0 & 1 & 0 \end{bmatrix} \times \begin{bmatrix} 0 & 1 & 0 \\ 0 & 0 & 1 \\ 1 & 0 & 0 \end{bmatrix} = \begin{bmatrix} \cos\phi & -\sin\phi & 0 \\ \sin\phi & \cos\phi & 0 \\ 0 & 0 & 1 \end{bmatrix}$$

3.2. *Position after rotation (changes of generalized coordinate)*

$$\vec{q}_{2j} = \vec{q}_{2j}^{\,o}\cos\psi + (1-\cos\psi)(\vec{e}_2\cdot\vec{q}_{2j}^{\,o})\vec{e}_2 + \vec{e}_2\times\vec{q}_{2j}^{\,o}\sin\psi, \qquad j=1,2,3$$

$$\vec{q}_{21} = \vec{q}_{21}^{\,o}\cos\psi + (1-\cos\psi)(\vec{e}_2\cdot\vec{q}_{21}^{\,o})\vec{e}_2 + \vec{e}_2\times\vec{q}_{21}^{\,o}\sin\psi = \vec{q}_{21}^{\,o}\cos\psi +$$

$$+ \vec{e}_2\times\vec{q}_{21}^{\,o}\sin\psi = \cos\psi\begin{bmatrix} \cos\phi \\ \sin\phi \\ 0 \end{bmatrix} + \sin\psi\begin{bmatrix} -\sin\phi \\ \cos\phi \\ 0 \end{bmatrix} \times \begin{bmatrix} \cos\phi \\ \sin\phi \\ 0 \end{bmatrix} = \begin{bmatrix} \cos\phi\cos\psi \\ \sin\phi\cos\psi \\ -\sin\psi \end{bmatrix}$$

$$\vec{q}_{22} = \vec{q}_{22}^{\,o}\cos\psi + (1-\cos\psi)(\vec{e}_2\cdot\vec{q}_{22}^{\,o})\vec{e}_2 + \vec{e}_2\times\vec{q}_{22}^{\,o}\sin\psi = \vec{q}_{22}^{\,o} = \begin{bmatrix} -\sin\phi \\ \cos\phi \\ 0 \end{bmatrix}$$

$$\vec{q}_{23} = \vec{q}_{23}^{\,o}\cos\psi + (1-\cos\psi)(\vec{e}_2\cdot\vec{q}_{23}^{\,o})\vec{e}_2 + \vec{e}_2\times\vec{q}_{23}^{\,o}\sin\psi = \vec{q}_{23}^{\,o}\cos\psi + \vec{e}_2\times\vec{q}_{23}^{\,o}\sin\psi =$$

$$= \cos\psi\begin{bmatrix} 0 \\ 0 \\ 1 \end{bmatrix} + \sin\psi\begin{bmatrix} -\sin\phi \\ \cos\phi \\ 0 \end{bmatrix} \times \begin{bmatrix} 0 \\ 0 \\ 1 \end{bmatrix} = \begin{bmatrix} \cos\phi\sin\psi \\ \sin\phi\sin\psi \\ \cos\psi \end{bmatrix}$$

Transformation matrix after rotation

$$Q_2 = [\vec{q}_{21}\ \vec{q}_{22}\ \vec{q}_{23}] = \begin{bmatrix} \cos\phi\cos\psi & -\sin\phi & \cos\phi\sin\psi \\ \sin\phi\cos\psi & \cos\phi & \sin\phi\sin\psi \\ -\sin\psi & 0 & \cos\psi \end{bmatrix}$$

$$\vec{r}_{22} = Q_2\vec{\bar{r}}_{22} = Q_2[0,\ \ell_2^*,\ 0]^T = [-\ell_2^*\sin\phi,\ \ell_2^*\cos\phi,\ 0]^T$$

$$\vec{r}_{23} = Q_2\vec{\bar{r}}_{23} = Q_2[0,\ -(\ell_2-\ell_2^*),\ 0]^T = [(\ell_2-\ell_2^*)\sin\phi,\ -(\ell_2-\ell_2^*)\cos\phi,\ 0]^T$$

$$\vec{e}_3 = Q_2\vec{\bar{e}}_3 = Q_2[-1,\ 0,\ 0]^T = [-\cos\phi\cos\psi,\ -\sin\phi\cos\psi,\ \sin\psi]^T$$

$$\vec{r}_{21} = \vec{r}_{11}-\vec{r}_{12}+\vec{r}_{22}=[0,\ 0,\ \ell_1^*]^T-[0,\ 0,\ -(\ell_1-\ell_1^*)]^T+$$

$$+\ [-\ell_2^*\sin\phi,\ \ell_2^*\cos\phi,\ 0]^T = [-\ell_2^*\sin\phi,\ \ell_2^*\cos\phi,\ \ell_1]^T$$

3.3. *Angular and linear velocity*

$$\vec{\omega}_2 = \vec{\omega}_1+\dot{\psi}\vec{e}_2(1-\xi_2)=[0,\ 0,\ \dot{\phi}]^T+\dot{\psi}[-\sin\phi,\ \cos\phi,\ 0]^T=[-\dot{\psi}\sin\phi,\ \dot{\psi}\cos\phi,\dot{\phi}]^T$$

$$\vec{v}_2 = \vec{v}_1-\vec{\omega}_1\times\vec{r}_{12}+\vec{\omega}_2\times\vec{r}_{22}+\dot{\psi}\vec{e}_2\xi_2=-[0\ 0\ \dot{\phi}]^T\times[0,\ 0,\ -(\ell_1-\ell_1^*)]^T +$$

$$+[-\dot{\psi}\sin\phi,\ \dot{\psi}\cos\phi,\ \dot{\phi}]^T\times[-\ell_2^*\sin\phi,\ \ell_2^*\cos\phi,\ 0]^T =$$

$$=[-\dot{\phi}\ell_2^*\cos\phi,\ -\dot{\phi}\ell_2^*\sin\phi,\ 0]^T$$

3.4. *Angular and linear acceleration*

$$\vec{\alpha}_{21} = \vec{\alpha}_{11} = [0,\ 0,\ 1]^T$$

$$\vec{\alpha}_{22} = \vec{e}_2(1-\xi_2) = [-\sin\phi,\ \cos\phi,\ 0]^T$$

$$\vec{\alpha}_2^o = \vec{\alpha}_1^o+\dot{\psi}\vec{\omega}_1\times\vec{e}_2(1-\xi_2)=\dot{\psi}[0,\ 0,\ \dot{\phi}]^T\times[-\sin\phi,\ \cos\phi,\ 0]^T =$$

$$= [-\dot{\phi}\dot{\psi}\cos\phi,\ -\dot{\phi}\dot{\psi}\sin\phi,\ 0]^T$$

$$\vec{\beta}_{21} = \vec{\beta}_{11}+\vec{\alpha}_{11}\times(\vec{r}_{22}-\vec{r}_{12})=[0,\ 0,\ 1]^T\times\{[-\ell_2^*\sin\phi,\ \ell_2^*\cos\phi,\ 0]^T -$$

$$-[0,\ 0,\ -(\ell_1-\ell_1^*)]^T\}=[0,\ 0,\ 1]^T\times[-\ell_2^*\sin\phi,\ \ell_2^*\cos\phi,\ \ell_1-\ell_1^*]^T =$$

$$=[-\ell_2^*\cos\phi,\ -\ell_2^*\sin\phi,\ 0]^T$$

$$\vec{\beta}_{22} = (\vec{e}_2\times\vec{r}_{22})(1-\xi_2)+\vec{e}_2\vec{\xi}_2=[-\sin\phi,\ \cos\phi,\ 0]^T\ \times$$

$$\times\ [-\ell_2^*\sin\phi,\ \ell_2^*\cos\phi,\ 0]^T = [0,\ 0,\ 0]^T$$

$$\vec{\gamma}_{12} = \vec{\omega}_1\times(\vec{\omega}_1\times\vec{r}_{12})=[0,\ 0,\ \dot{\phi}]^T\times\{[0,\ 0,\ \dot{\phi}]^T\times[0,\ 0,\ -(\ell_1-\ell_1^*)]^T\} =$$

$$= [0,\ 0,\ 0]^T$$

$$\vec{\gamma}_{22} = \vec{\omega}_2\times(\vec{\omega}_2\times\vec{r}_{22})=[-\dot{\psi}\sin\phi,\ \dot{\psi}\cos\phi,\ \dot{\phi}]^T\times\{[-\dot{\psi}\sin\phi,\ \dot{\psi}\cos\phi,\ \dot{\phi}]^T\ \times$$

$$\times\ [-\ell_2^*\sin\phi,\ \ell_2^*\cos\phi,\ 0]^T\} = [-\dot\psi\sin\phi,\ \dot\psi\cos\phi,\ \dot\phi]^T \times$$

$$\times\ [-\dot\phi\ell_2^*\cos\phi,\ -\dot\phi\ell_2^*\sin\phi,\ 0] = [\dot\phi^2\ell_2^*\sin\phi,\ -\dot\phi^2\ell_2^*\cos\phi,\ \dot\phi\dot\psi\ell_2^*]^T$$

$$\vec\beta_2^o = \vec\beta_1^o + \vec\alpha_1^o \times (\vec r_{22} - \vec r_{12}) + \dot\psi(\vec\omega_1 \times \vec e_2) \times \vec r_{22} - \vec\gamma_{12} + \vec\gamma_{22} + 2\dot\psi(\vec\omega_1 \times \vec e_2)\xi_2 =$$

$$= \dot\psi\{[0,\ 0,\ \dot\phi]^T \times [-\sin\phi,\ \cos\phi,\ 0]^T\} \times [-\ell_2^*\sin\phi,\ \ell_2^*\cos\phi,\ 0]^T +$$

$$+ [\dot\phi^2\ell_2^*\sin\phi,\ -\dot\phi^2\ell_2^*\cos\phi,\ \dot\phi\dot\psi\ell_2^*]^T =$$

$$= [-\dot\phi\dot\psi\cos\phi,\ -\dot\phi\dot\psi\sin\phi,\ 0]^T \times [-\ell_2^*\sin\phi,\ \ell_2^*\cos\phi,\ 0]^T +$$

$$+ [\dot\phi^2\ell_2^*\sin\phi,\ -\dot\phi^2\ell_2^*\cos\phi,\ \dot\phi\dot\psi\ell_2^*]^T =$$

$$= [\dot\phi^2\ell_2^*\sin\phi,\ -\dot\phi^2\ell_2^*\cos\phi,\ 0]^T$$

Angular acceleration

$$\vec\varepsilon_2 = \vec\alpha_{21}\ddot\phi + \vec\alpha_{22}\ddot\psi + \vec\alpha_2^o = \ddot\phi[0,\ 0,\ 1]^T + \ddot\psi[-\sin\phi,\ \cos\phi,\ 0]^T +$$

$$+ [-\dot\phi\dot\psi\cos\phi,\ -\dot\phi\dot\psi\sin\phi,\ 0]^T = [-\ddot\psi\sin\phi - \dot\phi\dot\psi\cos\phi,\ \ddot\psi\cos\phi - \dot\phi\dot\psi\sin\phi,\ \ddot\phi]^T$$

Linear acceleration

$$\vec w_2 = \vec\beta_{21}\ddot\phi + \vec\beta_{22}\ddot\psi + \vec\beta_2^o = \ddot\phi[-\ell_2^*\cos\phi,\ -\ell_2^*\sin\phi,\ 0]^T +$$

$$+ [\dot\phi^2\ell_2^*\sin\phi,\ -\dot\phi^2\ell_2^*\cos\phi,\ 0]^T = [-\ddot\phi\ell_2^*\cos\phi + \dot\phi^2\ell_2^*\sin\phi,$$

$$-\ddot\phi\ell_2^*\sin\phi - \dot\phi^2\ell_2^*\cos\phi,\ 0]^T$$

3.5. *Inertial force*

$$\vec a_{21} = -m_2\vec\beta_{21} = [m_2\ell_2^*\cos\phi,\ m_2\ell_2^*\sin\phi,\ 0]^T$$

$$\vec a_{22} = -m_2\vec\beta_{22} = [0,\ 0,\ 0]^T$$

$$\vec a_2^o = -m_2\vec\beta_2^o = [-\dot\phi^2 m_2\ell_2^*\sin\phi,\ \dot\phi^2 m_2\ell_2^*\cos\phi,\ 0]^T$$

$$\vec F_2 = \vec a_{21}\ddot\phi + \vec a_{22}\ddot\psi + \vec a_2^o = [m_2\ell_2^*(\ddot\phi\cos\phi - \dot\phi^2\sin\phi),\ m_2\ell_2^*(\ddot\phi\sin\phi + \dot\phi^2\cos\phi),\ 0]^T$$

3.6. *Moment of inertial force*

$$T_2^{jk} = \sum_{\ell=1}^{3} q_{2\ell}^{j}\, q_{2\ell}^{k}\, J_{2\ell}$$

$$T_2^{11} = \sum_{\ell=1}^{3} q_{2\ell}^{1}\, q_{2\ell}^{1}\, J_{2\ell} = \cos^2\phi(J_{21}\cos^2\psi+J_{23}\sin^2\psi)+J_{22}\sin^2\phi$$

$$T_2^{12} = \sum_{\ell=1}^{3} q_{2\ell}^{1}\, q_{2\ell}^{2}\, J_{2\ell} = \sin\phi\cos\phi(J_{21}\cos^2\psi-J_{22}+J_{23}\sin^2\psi)$$

$$T_2^{13} = \sum_{\ell=1}^{3} q_{2\ell}^{1}\, q_{2\ell}^{3}\, J_{2\ell} = \cos\phi\sin\psi\cos\psi(-J_{21}+J_{23})$$

$$T_2^{21} = T_2^{12} = \sin\phi\cos\phi(J_{21}\cos^2\psi-J_{22}+J_{23}\sin^2\psi)$$

$$T_2^{22} = \sum_{\ell=1}^{3} q_{2\ell}^{2}\, q_{2\ell}^{2}\, J_{2\ell} = \sin^2\phi(J_{21}\cos^2\psi+J_{23}\sin^2\psi)+J_{22}\cos^2\phi$$

$$T_2^{23} = \sum_{\ell=1}^{3} q_{2\ell}^{2}\, q_{2\ell}^{3}\, J_{2\ell} = \sin\phi\sin\psi\cos\psi(-J_{21}+J_{23})$$

$$T_2^{31} = T_2^{13} = \cos\phi\sin\psi\cos\psi(-J_{21}+J_{23})$$

$$T_2^{32} = T_2^{23} = \sin\phi\sin\psi\cos\psi(-J_{21}+J_{23})$$

$$T_2^{33} = \sum_{\ell=1}^{3} q_{2\ell}^{3}\, q_{2\ell}^{3}\, J_{2\ell} = J_{21}\sin^2\psi + J_{23}\cos^2\psi$$

$$\vec{\omega}_2\vec{q}_{21} = [-\dot{\psi}\sin\phi,\ \dot{\psi}\cos\phi,\ \dot{\phi}][\cos\phi\cos\psi,\ \sin\phi\cos\psi,\ -\sin\psi]^T=-\dot{\phi}\sin\psi$$

$$\vec{\omega}_2\vec{q}_{22} = [-\dot{\psi}\sin\phi,\ \dot{\psi}\cos\phi,\ \dot{\phi}][-\sin\phi,\ \cos\phi,\ 0]^T = \dot{\psi}$$

$$\vec{\omega}_2\vec{q}_{23} = [-\dot{\psi}\sin\phi,\ \dot{\psi}\cos\phi,\ \dot{\phi}][\cos\phi\sin\psi,\ \sin\phi\sin\psi,\ \cos\psi] = \dot{\phi}\cos\psi$$

$$\vec{\lambda}_2 = Q_2 \begin{bmatrix} (J_{22}-J_{23})(\vec{\omega}_2\cdot\vec{q}_{22})(\vec{\omega}_2\cdot\vec{q}_{23}) \\ (J_{23}-J_{21})(\vec{\omega}_2\cdot\vec{q}_{23})(\vec{\omega}_2\cdot\vec{q}_{21}) \\ (J_{21}-J_{22})(\vec{\omega}_2\cdot\vec{q}_{21})(\vec{\omega}_2\cdot\vec{q}_{22}) \end{bmatrix} =$$

$$= \begin{bmatrix} \cos\phi\cos\psi & -\sin\phi & \cos\phi\sin\psi \\ \sin\phi\cos\psi & \cos\phi & \sin\phi\sin\psi \\ -\sin\psi & 0 & \cos\psi \end{bmatrix} \begin{bmatrix} \dot{\phi}\dot{\psi}(J_{22}-J_{23})\cos\psi \\ -\dot{\phi}^2(J_{23}-J_{21})\sin\psi\cos\psi \\ -\dot{\phi}\dot{\psi}(J_{21}-J_{22})\sin\psi \end{bmatrix} =$$

$$= \begin{bmatrix} \dot{\phi}^2(J_{23}-J_{21})\sin\phi\sin\psi\cos\psi+\dot{\phi}\dot{\psi}\cos\phi(-J_{21}\sin^2\psi+J_{22}-J_{23}\cos^2\psi) \\ -\dot{\phi}^2(J_{23}-J_{21})\cos\phi\sin\psi\cos\psi+\dot{\phi}\dot{\psi}\sin\phi(-J_{21}\sin^2\psi+J_{22}-J_{23}\cos^2\psi) \\ \dot{\phi}\dot{\psi}(J_{23}-J_{21})\sin\psi\cos\psi) \end{bmatrix}$$

$$\vec{b}_{21} = -T_2\vec{\alpha}_{21} = -T_2[0,\ 0,\ 1]^T = [-T_2^{13},\ -T_2^{23},\ -T_2^{33}]^T =$$

$$= \begin{bmatrix} -(J_{23}-J_{21})\cos\phi\sin\psi\cos\psi \\ -(J_{23}-J_{21})\sin\phi\sin\psi\cos\psi \\ -J_{21}\sin^2\psi-J_{23}\cos^2\psi \end{bmatrix}$$

$$\vec{b}_{22} = -T_2\vec{\alpha}_{22} = -T_2[-\sin\phi,\ \cos\phi,\ 0]^T = \begin{bmatrix} T_2^{11}\sin\phi-T_2^{12}\cos\phi \\ T_2^{21}\sin\phi-T_2^{22}\cos\phi \\ T_2^{31}\sin\phi-T_2^{32}\cos\phi \end{bmatrix} =$$

$$= -\begin{bmatrix} -\sin\phi\cos^2\phi(J_{21}\cos^2\psi+J_{23}\sin^2\psi)- \\ \quad -J_{22}\sin^3\phi+\sin\phi\cos^2\phi(J_{21}\cos^2\psi-J_{22}+J_{23}\sin^2\psi) \\ -\sin^2\phi\cos\phi(J_{21}\cos^2\psi-J_{22}+J_{23}\sin^2\psi)\ + \\ \quad +\sin^2\phi\cos\phi(J_{21}\cos^2\psi+J_{23}\sin^2\psi)+J_{22}\cos^3\phi \\ -\sin\phi\cos\phi\sin\psi\cos\psi(J_{23}-J_{21})\ + \\ \quad +\sin\phi\cos\phi\sin\psi\cos\psi(J_{23}-J_{21}) \end{bmatrix} =$$

$$= [J_{22}\sin\phi,\ -J_{22}\cos\phi,\ 0]^T$$

$$\vec{b}_2^o = -T_2\vec{\alpha}_2^o + \vec{\lambda}_2 = -T_2[-\dot{\phi}\dot{\psi}\cos\phi,\ -\dot{\phi}\dot{\psi}\sin\phi,\ 0]^T + \vec{\lambda}_2 =$$

$$= \begin{bmatrix} T_2^{11}\dot{\phi}\dot{\psi}\cos\phi+T_2^{12}\dot{\phi}\dot{\psi}\sin\phi \\ T_2^{21}\dot{\phi}\dot{\psi}\cos\phi+T_2^{22}\dot{\phi}\dot{\psi}\sin\phi \\ T_2^{31}\dot{\phi}\dot{\psi}\cos\phi+T_2^{32}\dot{\phi}\dot{\psi}\sin\phi \end{bmatrix} + \vec{\lambda}_2$$

$$b_2^{o(1)} = \dot{\phi}\dot{\psi}\cos^3\phi(J_{21}\cos^2\psi+J_{23}\sin^2\psi)+\dot{\phi}\dot{\psi}J_{22}\sin^2\phi\cos\phi+$$

$$+ \dot{\phi}\dot{\psi}\sin^2\phi\cos\phi(J_{21}\cos^2\psi-J_{22}+J_{23}\sin^2\psi)+$$

$$+ \dot{\phi}^2(J_{23}-J_{21})\sin\phi\sin\psi\cos\psi+\dot{\phi}\dot{\psi}\cos\phi(-J_{21}\sin^2\psi+J_{22}-J_{23}\cos^2\psi)=$$

$$= \dot{\phi}^2(J_{23}-J_{21})\sin\phi\sin\psi\cos\psi+\dot{\phi}\dot{\psi}\cos\phi[J_{22}+(J_{23}-J_{21})(\sin^2\psi-\cos^2\psi)]$$

$$b_2^{o(2)} = \dot{\phi}\dot{\psi}\sin\phi\cos^2\phi(J_{21}\cos^2\psi-J_{22}+J_{23}\sin^2\psi)+\dot{\phi}\dot{\psi}\sin^3\phi(J_{21}\cos^2\psi+$$

$$+ J_{23}\sin^2\psi)+J_{22}\sin\phi\cos^2\phi-\dot{\phi}^2(J_{23}-J_{21})\cos\phi\sin\psi\cos\psi+$$

$$+ \dot{\phi}\dot{\psi}\sin\phi(-J_{21}\sin^2\psi+J_{22}-J_{23}\cos^2\psi)=-\dot{\phi}^2(J_{23}-$$

$$- J_{21})\cos\phi\sin\psi\cos\psi+\dot{\phi}\dot{\psi}\sin\phi[J_{22}+(J_{23}-J_{21})(\sin^2\psi-\cos^2\psi)]$$

$$b_2^{o(3)} = \dot{\phi}\dot{\psi}\cos^2\phi\sin\psi\cos\psi(J_{23}-J_{21})+\dot{\phi}\dot{\psi}\sin^2\phi\sin\psi\cos\psi(J_{23}-J_{21})+$$

$$+ \dot{\phi}\dot{\psi}(J_{23}-J_{21})\sin\psi\cos\psi=2\dot{\phi}\dot{\psi}(J_{23}-J_{21})\sin\psi\cos\psi$$

$$\vec{b}_2^o = \begin{bmatrix} b_2^{o(1)} \\ b_2^{o(2)} \\ b_2^{o(3)} \end{bmatrix} =$$

$$= \begin{bmatrix} \dot{\phi}^2(J_{23}-J_{21})\sin\phi\sin\psi\cos\psi+\dot{\phi}\dot{\psi}\cos\phi[J_{22}+(J_{23}-J_{21})(\sin^2\psi-\cos^2\psi)] \\ -\dot{\phi}^2(J_{23}-J_{21})\cos\phi\sin\psi\cos\psi+\dot{\phi}\dot{\psi}\sin\phi[J_{22}+(J_{23}-J_{21})(\sin^2\psi-\cos^2\psi)] \\ 2\dot{\phi}\dot{\psi}(J_{23}-J_{21})\sin\psi\cos\psi \end{bmatrix}$$

Moment of inertial force

$$M_2 = \vec{b}_{21}\ddot{\phi} + \vec{b}_{22}\ddot{\psi} + \vec{b}_2^{o} = [M_2^1\ M_2^2\ M_2^3]^T =$$

$$= \begin{bmatrix} -\ddot{\phi}(J_{23}-J_{21})\cos\phi\sin\psi\cos\psi+\ddot{\psi}J_{22}\sin\phi+\dot{\phi}^2(J_{23}-J_{21})\sin\phi\sin\psi\cos\psi+ \\ +\dot{\phi}\dot{\psi}\cos\phi[J_{22}+(J_{23}-J_{21})(\sin^2\psi-\cos^2\psi)] \\ -\ddot{\phi}(J_{23}-J_{21})\sin\phi\sin\psi\cos\psi-\ddot{\psi}J_{22}\cos\phi-\dot{\phi}^2(J_{23}-J_{21})\cos\phi\sin\psi\cos\psi+ \\ +\dot{\phi}\dot{\psi}\sin\phi[J_{22}+(J_{23}-J_{21})(\sin^2\psi-\cos^2\psi)] \\ -\ddot{\phi}(J_{21}\sin^2\psi+J_{23}\cos^2\psi)+2\dot{\phi}\dot{\psi}(J_{23}-J_{21})\sin\psi\cos\psi \end{bmatrix}$$

4. Segment "3"

The joint is prismatic $\xi_3=1$.

Consider the segment as a body (J_{31}, J_{32}, J_{33}). The segment is "specific" from the lower side, $\tilde{\vec{r}}_{33} || \tilde{\vec{e}}_3$

$$\vec{e}_3 = [-\cos\phi\cos\psi,\ -\sin\phi\cos\psi,\ \sin\psi]^T$$

$$\vec{r}_{23} = (-(\ell_2-\ell_2^*)\cos\phi\sin\psi,\ -(\ell_2-\ell_2^*)\sin\phi\sin\psi,\ -(\ell_2-\ell_2^*)\cos\psi)^T$$

$$\tilde{\vec{e}}_3 = (-1,\ 0,\ 0)^T \qquad \tilde{\vec{r}}_{33} = (-\ell_3^*,\ 0,\ 0)^T$$

$$\tilde{\vec{r}}_{33}^{A} = (0,\ \ell_3^*,\ 0)^T$$

4.1. *Assembly*

$$\vec{r}_{N3} = \vec{e}_3\times(\vec{r}_{23}\times\vec{e}_3) = \vec{e}_3\times\{[(\ell_2-\ell_2^*)\sin\phi,\ -(\ell_2-\ell_2^*)\cos\phi,\ 0]^T \times$$

$$\times [-\cos\phi\cos\psi,\ -\sin\phi\cos\psi,\ \sin\psi]^T\} = [-\cos\phi\cos\psi,\ -\sin\phi\cos\psi,\ \sin\psi]^T \times$$

$$\times [-(\ell_2-\ell_2^*)\cos\phi\sin\psi,\ -(\ell_2-\ell_2^*)\sin\phi\sin\psi,\ -(\ell_2-\ell_2^*)\cos\psi] =$$

$$= [(\ell_2-\ell_2^*)\sin\phi,\ -(\ell_2-\ell_2^*)\cos\phi,\ 0]^T$$

$$\vec{a}_3 = -\vec{r}_{N3}/|\vec{r}_{N3}| = [-\sin\phi,\ \cos\phi,\ 0]^T$$

$$\vec{\tilde{r}}_{N3} = \vec{\tilde{e}}_3 \times (\vec{\tilde{r}}^A_{33} \times \vec{\tilde{e}}_3) = [-1,\ 0,\ 0]^T \times \{[0,\ \ell_3^*,\ 0] \times [-1,\ 0,\ 0]^T\} =$$

$$= [0,\ \ell_3^*,\ 0]^T$$

$$\vec{\tilde{a}}_3 = \frac{\vec{\tilde{r}}_{N3}}{|\vec{\tilde{r}}_{N3}|} = [0,\ 1,\ 0]^T$$

$$\vec{\tilde{e}}_3 = [-1,\ 0,\ 0]^T \qquad \vec{e}_3 = [-\cos\phi\cos\psi,\ -\sin\phi\cos\psi,\ \sin\psi]^T$$

$$\vec{b}_3 = \vec{e}_3 \times \vec{a}_3 = [-\cos\phi\cos\psi,\ -\sin\phi\cos\psi,\ \sin\psi]^T \times [-\sin\phi,\ \cos\phi,\ 0]^T =$$

$$= [-\cos\phi\sin\psi,\ -\sin\phi\sin\psi,\ -\cos\psi]$$

$$\vec{\tilde{b}}_3 = \vec{\tilde{e}}_3 \times \vec{\tilde{a}}_3 = [-1,\ 0,\ 0]^T \times [0,\ 1,\ 0]^T = [0,\ 0,\ -1]^T$$

Transformation matrix before translation x

$$Q_3^o = [\vec{q}^{\,o}_{31}\ \vec{q}^{\,o}_{32}\ \vec{q}^{\,o}_{33}] = [\vec{e}_3\ \vec{a}_3\ \vec{b}_3][\vec{\tilde{e}}_3\ \vec{\tilde{a}}_3\ \vec{\tilde{b}}_3]^T =$$

$$= \begin{bmatrix} -\cos\phi\cos\psi & -\sin\phi & -\cos\phi\sin\psi \\ -\sin\phi\cos\psi & \cos\phi & -\sin\phi\sin\psi \\ \sin\psi & 0 & -\cos\psi \end{bmatrix} \begin{bmatrix} -1 & 0 & 0 \\ 0 & 1 & 0 \\ 0 & 0 & -1 \end{bmatrix} =$$

$$= \begin{bmatrix} \cos\phi\cos\psi & -\sin\phi & \cos\phi\sin\psi \\ \sin\phi\cos\psi & \cos\phi & \sin\phi\sin\psi \\ -\sin\psi & 0 & \cos\psi \end{bmatrix}$$

4.2. *Position after translation (changes of relative coordinate x)*

$$\vec{q}_{3j} = \vec{q}^{\,o}_{3j} \qquad j=1,2,3$$

Transformation matrix after translation

$$Q_3 = [\vec{q}_{31}\ \vec{q}_{32}\ \vec{q}_{33}] = Q_3^o = \begin{bmatrix} \cos\phi\cos\psi & -\sin\phi & \cos\phi\sin\psi \\ \sin\phi\cos\psi & \cos\phi & \sin\phi\sin\psi \\ -\sin\psi & 0 & \cos\psi \end{bmatrix}$$

$$\vec{r}_{33} = Q_3\vec{\tilde{r}}_{33} + x\vec{e}_3 = Q_3[-\ell_3^*,\ 0,\ 0]^T + x[-\cos\phi\cos\psi,\ -\sin\phi\cos\psi,\ \sin\psi]^T =$$

$$= [-(x+\ell_3^*)\cos\phi\cos\psi,\ -(x+\ell_3^*)\sin\phi\cos\psi,\ (x+\ell_3^*)\sin\psi]^T$$

$$\vec{r}_{34} = Q_3\bar{\vec{r}}_{34} = Q_3[(\ell_3-\ell_3^*),\ 0,\ 0]^T =$$

$$= [(\ell_3-\ell_3^*)\cos\phi\cos\psi,\ (\ell_3-\ell_3^*)\sin\phi\cos\psi,\ -(\ell_3-\ell_3^*)\sin\psi]^T$$

$$\vec{r}_{31} = \vec{r}_{21}-\vec{r}_{23}+\vec{r}_{33} = [-\ell_2^*\sin\phi,\ \ell_2^*\cos\phi,\ \ell_1]^T-[(\ell_2-\ell_2^*)\sin\phi,$$

$$-(\ell_2-\ell_2^*)\cos\phi,\ 0]^T+[-(x+\ell_3^*)\cos\phi\cos\psi,\ -(x+\ell_3^*)\sin\phi\cos\psi,$$

$$(x+\ell_3^*)\sin\psi]^T = [-\ell_2\sin\phi-(x+\ell_3^*)\cos\phi\cos\psi,$$

$$\ell_2\cos\phi-(x+\ell_3^*)\sin\phi\cos\psi,\ \ell_1+(x+\ell_3^*)\sin\psi]^T$$

$$\vec{r}_{32} = \vec{r}_{22}-\vec{r}_{23}+\vec{r}_{33} = [-\ell_2^*\sin\phi,\ \ell_2^*\cos\phi,\ 0]^T -$$

$$-[(\ell_2-\ell_2^*)\sin\phi,\ -(\ell_2-\ell_2^*)\cos\phi,\ 0]^T +$$

$$+[-(x+\ell_3^*)\cos\phi\cos\psi,\ -(x+\ell_3^*)\sin\phi\cos\psi,\ (x+\ell_3^*)\sin\psi]^T =$$

$$= [-\ell_2\sin\phi-(x+\ell_3^*)\cos\phi\cos\psi,\ \ell_2\cos\phi-(x+\ell_3^*)\sin\phi\cos\psi,\ (x+\ell_3^*)\sin\psi]^T$$

4.3. *Angular and linear velocity*

$$\vec{\omega}_3 = \vec{\omega}_2+\dot{x}\vec{e}_3(1-\xi_3) = [-\dot{\psi}\sin\phi,\ \dot{\psi}\cos\phi,\ \dot{\phi}]^T$$

$$\vec{v}_3 = \vec{v}_2-\vec{\omega}_2\times\vec{r}_{23}+\vec{\omega}_3\times\vec{r}_{33}+\dot{x}\vec{e}_3 = [-\dot{\phi}\ell_2^*\cos\phi,\ -\dot{\phi}\ell_2^*\sin\phi,\ 0]^T -$$

$$-[-\dot{\psi}\sin\phi,\ \dot{\psi}\cos\phi,\ \dot{\phi}]^T\times[(\ell_2-\ell_2^*)\sin\phi,\ -(\ell_2-\ell_2^*)\cos\phi,\ 0]^T +$$

$$+[-\dot{\psi}\sin\phi,\ \dot{\psi}\cos\phi,\ \dot{\phi}]^T\times[-(x+\ell_3^*)\cos\phi\cos\psi,\ -(x+\ell_3^*)\sin\phi\cos\psi,$$

$$(x+\ell_3^*)\sin\psi]^T+\dot{x}[-\cos\phi\cos\psi,\ -\sin\phi\cos\psi,\ \sin\psi]^T =$$

$$= \begin{bmatrix} -\dot{\phi}\ell_2^*\cos\phi-\dot{\phi}(\ell_2-\ell_2^*)\cos\phi+\dot{\psi}(x+\ell_3^*)\cos\phi\sin\psi+\dot{\phi}(x+\ell_3^*)\sin\phi\cos\psi-\dot{x}\cos\phi\cos\psi \\ -\dot{\phi}\ell_2^*\sin\phi-\dot{\phi}(\ell_2-\ell_2^*)\sin\phi-\dot{\phi}(x+\ell_3^*)\cos\phi\cos\psi+\dot{\psi}(x+\ell_3^*)\sin\phi\sin\psi-\dot{x}\sin\phi\cos\psi \\ \dot{\psi}(x+\ell_3^*)\sin^2\phi\cos\psi+\dot{\psi}(x+\ell_3^*)\cos^2\phi\cos\psi+\dot{x}\sin\psi \end{bmatrix} =$$

$$= \begin{bmatrix} -\dot{\phi}[\ell_2\cos\phi-(x+\overset{*}{\ell}_3)\sin\phi\cos\psi]+\dot{\psi}(x+\overset{*}{\ell}_3)\cos\phi\sin\psi-\dot{x}\cos\phi\cos\psi \\ -\dot{\phi}[\ell_2\sin\phi+(x+\overset{*}{\ell}_3)\cos\phi\cos\psi]+\dot{\psi}(x+\overset{*}{\ell}_3)\sin\phi\sin\psi-\dot{x}\sin\phi\cos\psi \\ \dot{\psi}(x+\overset{*}{\ell}_3)\cos\psi+\dot{x}\sin\psi \end{bmatrix}$$

4.4. *Angular and linear acceleration*

$$\vec{\alpha}_{31} = \vec{\alpha}_{21} = [0,\ 0,\ 1]^T$$

$$\vec{\alpha}_{32} = \vec{\alpha}_{22} = [-\sin\phi,\ \cos\phi,\ 0]^T$$

$$\vec{\alpha}_{33} = \vec{e}_3(1-\xi_3) = [0,\ 0,\ 0]^T$$

$$\vec{\alpha}^o_3 = \vec{\alpha}^o_2 + \dot{x}(\vec{\omega}_2\times\vec{e}_3)(1-\xi_3) = [-\dot{\phi}\dot{\psi}\cos\phi,\ -\dot{\phi}\dot{\psi}\sin\phi,\ 0]^T$$

$$\vec{\beta}_{31} = \vec{\beta}_{21}+\vec{\alpha}_{21}\times(\vec{r}_{33}-\vec{r}_{23}) = [-\overset{*}{\ell}_2\cos\phi,\ -\overset{*}{\ell}_2\sin\phi,\ 0]^T+[0,\ 0,\ 1]^T \times$$

$$\times\ \{[-(x+\overset{*}{\ell}_3)\cos\phi\cos\psi,\ -(x+\overset{*}{\ell}_3)\sin\phi\cos\psi,\ (x+\overset{*}{\ell}_3)\sin\psi]^T -$$

$$-\ [(\ell_2-\overset{*}{\ell}_2)\sin\phi,\ -(\ell_2-\overset{*}{\ell}_2)\cos\phi,\ 0]^T\} =$$

$$= -\overset{*}{\ell}_2\begin{bmatrix}\cos\phi\\ \sin\phi\\ 0\end{bmatrix} - (x+\overset{*}{\ell}_3)\begin{bmatrix}-\sin\phi\cos\psi\\ \cos\phi\cos\psi\\ 0\end{bmatrix} - (\ell_2-\overset{*}{\ell}_2)\begin{bmatrix}\cos\phi\\ \sin\phi\\ 0\end{bmatrix} =$$

$$= \begin{bmatrix}-\ell_2\cos\phi+(x+\overset{*}{\ell}_3)\sin\phi\cos\psi\\ -\ell_2\sin\phi-(x+\overset{*}{\ell}_3)\cos\phi\cos\psi\\ 0\end{bmatrix}$$

$$\vec{\beta}_{32} = \vec{\beta}_{22} + \vec{\alpha}_{22}\times(\vec{r}_{33}-\vec{r}_{23}) = [-\sin\phi,\ \cos\phi,\ 0]^T \times$$

$$\times\ \{[-(x+\overset{*}{\ell}_3)\cos\phi\cos\psi,\ -(x+\overset{*}{\ell}_3)\sin\phi\cos\psi,\ (x+\overset{*}{\ell}_3)\sin\psi]^T -$$

$$-\ [(\ell_2-\overset{*}{\ell}_2)\sin\phi,\ -(\ell_2-\overset{*}{\ell}_2)\cos\phi,\ 0]^T\} =$$

$$= [(x+\overset{*}{\ell}_3)\cos\phi\sin\psi,\ (x+\overset{*}{\ell}_3)\sin\phi\sin\psi,\ (x+\overset{*}{\ell}_3)\cos\psi]^T$$

$$\vec{\beta}_{33} = \vec{e}_3\xi_3 + (\vec{e}_3\times\vec{r}_{33})(1-\xi_3) = [-\cos\phi\cos\psi,\ -\sin\phi\cos\psi,\ \sin\psi]^T$$

$$\vec{\gamma}_{23} = \vec{\omega}_2 \times (\vec{\omega}_2 \times \vec{r}_{23}) = \vec{\omega}_2 \times \{[-\dot{\psi}\sin\phi,\ \dot{\psi}\cos\phi,\ \dot{\phi}]^T \times$$

$$\times\ [(\ell_2-\overset{*}{\ell}_2)\sin\phi,\ -(\ell_2-\overset{*}{\ell}_2)\cos\phi,\ 0]^T\} =$$

$$= [-\dot{\psi}\sin\phi,\ \dot{\psi}\cos\phi,\ \dot{\phi}]^T \times [\dot{\phi}(\ell_2-\overset{*}{\ell}_2)\cos\phi,\ \dot{\phi}(\ell_2-\overset{*}{\ell}_2)\sin\phi,\ 0]^T =$$

$$= [-\dot{\phi}^2(\ell_2-\overset{*}{\ell}_2)\sin\phi,\ \dot{\phi}^2(\ell_2-\overset{*}{\ell}_2)\cos\phi,\ -\dot{\phi}\dot{\psi}(\ell_2-\overset{*}{\ell}_2)]^T$$

$$\vec{\gamma}_{33} = \vec{\omega}_3 \times (\vec{\omega}_3 \times \vec{r}_{33}) = \vec{\omega}_3 \times \{[-\dot{\psi}\sin\phi,\ \dot{\psi}\cos\phi,\ \dot{\phi}]^T \times$$

$$\times\ [-(x+\overset{*}{\ell}_3)\cos\phi\cos\psi,\ -(x+\overset{*}{\ell}_3)\sin\phi\cos\psi,\ (x+\overset{*}{\ell}_3)\sin\psi]^T\} =$$

$$= [-\dot{\psi}\sin\phi,\ \dot{\psi}\cos\phi,\ \dot{\phi}]^T \times [\dot{\psi}(x+\overset{*}{\ell}_3)\cos\phi\sin\psi\ +$$

$$+\ \dot{\phi}(x+\overset{*}{\ell}_3)\sin\phi\cos\psi,\ -\dot{\phi}(x+\overset{*}{\ell}_3)\cos\phi\cos\psi+$$

$$+\ \dot{\psi}(x+\overset{*}{\ell}_3)\sin\phi\sin\psi,\ \dot{\psi}(x+\overset{*}{\ell}_3)\cos\psi]^T =$$

$$= \begin{bmatrix} (\dot{\phi}^2+\dot{\psi}^2)(x+\overset{*}{\ell}_3)\cos\phi\cos\psi-\dot{\phi}\dot{\psi}(x+\overset{*}{\ell}_3)\sin\phi\sin\psi \\ (\dot{\phi}^2+\dot{\psi}^2)(x+\overset{*}{\ell}_3)\sin\phi\cos\psi+\dot{\phi}\dot{\psi}(x+\overset{*}{\ell}_3)\cos\phi\sin\psi \\ -\dot{\psi}^2(x+\overset{*}{\ell}_3)\sin\psi \end{bmatrix}$$

$$\vec{\beta}_3^o = \vec{\beta}_2^o+\vec{\alpha}_2^o \times (\vec{r}_{33}-\vec{r}_{23})+2\dot{x}\vec{\omega}_2 \times \vec{e}_3-\vec{\gamma}_{23}+\vec{\gamma}_{33}+\ddot{x}(\vec{\omega}_2 \times \vec{e}_3) \times \vec{r}_{33}(1-\xi_3) =$$

$$= \vec{\beta}_2^o+[-\dot{\phi}\dot{\psi}\cos\phi,\ -\dot{\phi}\dot{\psi}\sin\phi,\ 0]^T \times$$

$$\times\ \{[-(x+\overset{*}{\ell}_3)\cos\phi\cos\psi,\ -(x+\overset{*}{\ell}_3)\sin\phi\cos\psi,\ (x+\overset{*}{\ell}_3)\sin\psi]^T\ -$$

$$-\ [(\ell_2-\overset{*}{\ell}_2)\sin\phi,\ -(\ell_2-\overset{*}{\ell}_2)\cos\phi,\ 0]^T\}+2\dot{x}[-\dot{\psi}\sin\psi,\ \dot{\psi}\cos\phi,\ \dot{\phi}]^T \times$$

$$\times\ [-\cos\phi\cos\psi,\ -\sin\phi\cos\psi,\ \sin\psi]^T-\vec{\gamma}_{23}+\vec{\gamma}_{33}=[\beta_3^{o(1)}\ \ \beta_3^{o(2)}\ \ \beta_3^{o(3)}]^T$$

$$\beta_3^{o(1)} = \dot{\phi}^2\overset{*}{\ell}_2\sin\phi-\dot{\phi}\dot{\psi}(x+\overset{*}{\ell}_3)\sin\phi\sin\psi+2\dot{\psi}\dot{x}\cos\phi\sin\psi+2\dot{\phi}\dot{x}\sin\phi\cos\psi+$$

$$+\ \dot{\phi}^2(\ell_2-\overset{*}{\ell}_2)\sin\phi+(\dot{\phi}^2+\dot{\psi}^2)(x+\overset{*}{\ell}_3)\cos\phi\cos\psi-\dot{\phi}\dot{\psi}(x+\overset{*}{\ell}_3)\sin\phi\sin\psi=$$

$$= \dot{\phi}^2[\ell_2\sin\phi+(x+\overset{*}{\ell}_3)\cos\phi\cos\psi]+\dot{\psi}^2(x+\overset{*}{\ell}_3)\cos\phi\cos\psi-$$

$$-\ 2\dot{\phi}\dot{\psi}(x+\overset{*}{\ell}_3)\sin\phi\sin\psi+2\dot{\phi}\dot{x}\sin\phi\cos\psi+2\dot{\psi}\dot{x}\cos\phi\sin\psi$$

$$\beta_3^{o(2)} = -\dot{\phi}^2\ell_2^*\cos\phi+\dot{\phi}\dot{\psi}(x+\ell_3^*)\cos\phi\sin\psi-2\dot{\phi}\dot{x}\cos\phi\cos\psi+2\dot{\psi}\dot{x}\sin\phi\sin\psi-$$

$$-\dot{\phi}^2(\ell_2-\ell_2^*)\cos\phi+(\dot{\phi}^2+\dot{\psi}^2)(x+\ell_3^*)\sin\phi\cos\psi+\dot{\phi}\dot{\psi}(x+\ell_3^*)\cos\phi\sin\psi =$$

$$= -\dot{\phi}^2[\ell_2\cos\phi-(x+\ell_3^*)\sin\phi\cos\psi]+\dot{\psi}^2(x+\ell_3^*)\sin\phi\cos\psi +$$

$$+ 2\dot{\phi}\dot{\psi}(x+\ell_3^*)\cos\phi\sin\psi-2\dot{\phi}\dot{x}\cos\phi\cos\psi+2\dot{\psi}\dot{x}\sin\phi\sin\psi$$

$$\beta_3^{o(3)} = -\dot{\phi}\dot{\psi}(\ell_2-\ell_2^*)+2\dot{\psi}\dot{x}\cos\psi+\dot{\phi}\dot{\psi}(\ell_2-\ell_2^*)-\dot{\psi}^2(x+\ell_3^*)\sin\psi =$$

$$= -\dot{\psi}^2(x+\ell_3^*)\sin\psi+2\dot{\psi}\dot{x}\cos\psi$$

Angular acceleration

$$\vec{\varepsilon}_3 = \vec{\alpha}_{31}\ddot{\phi}+\vec{\alpha}_{32}\ddot{\psi}+\vec{\alpha}_{33}\ddot{x}+\vec{\alpha}_3^o = \begin{bmatrix} -\ddot{\psi}\sin\phi-\dot{\phi}\dot{\psi}\cos\phi \\ \ddot{\psi}\cos\phi-\dot{\phi}\dot{\psi}\sin\phi \\ \ddot{\phi} \end{bmatrix}$$

Linear acceleration

$$\vec{w}_3 = \vec{\beta}_{31}\ddot{\phi} + \vec{\beta}_{32}\ddot{\psi} + \vec{\beta}_{33}\ddot{x} + \vec{\beta}_3^o =$$

$$= \begin{bmatrix} -\ddot{\phi}[\ell_2\cos\phi-(x+\ell_3^*)\sin\phi\cos\psi]+\ddot{\psi}(x+\ell_3^*)\cos\phi\sin\psi-\ddot{x}\cos\phi\cos\psi \\ -\ddot{\phi}[\ell_2\sin\phi+(x+\ell_3^*)\cos\phi\cos\psi]+\ddot{\psi}(x+\ell_3^*)\sin\phi\sin\psi-\ddot{x}\sin\phi\cos\psi \\ \ddot{\psi}(x+\ell_3^*)\cos\psi+\ddot{x}\sin\psi \end{bmatrix} +$$

$$+ \begin{bmatrix} \dot{\phi}^2[\ell_2\sin\phi+(x+\ell_3^*)\cos\phi\cos\psi]+\dot{\psi}^2(x+\ell_3^*)\cos\phi\cos\psi \\ -\dot{\phi}^2[\ell_2\cos\phi-(x+\ell_3^*)\sin\phi\cos\psi]+\dot{\psi}^2(x+\ell_3^*)\sin\phi\cos\psi \\ -\dot{\psi}^2(x+\ell_3^*)\sin\psi \end{bmatrix} +$$

$$+ \begin{bmatrix} -2\dot{\phi}\dot{\psi}(x+\ell_3^*)\sin\phi\sin\psi+2\dot{\phi}\dot{x}\sin\phi\cos\psi+2\dot{\psi}\dot{x}\cos\phi\sin\psi \\ 2\dot{\phi}\dot{\psi}(x+\ell_3^*)\cos\phi\sin\psi-2\dot{\phi}\dot{x}\cos\phi\cos\psi+2\dot{\psi}\dot{x}\sin\phi\sin\psi \\ 2\dot{\psi}\dot{x}\cos\psi \end{bmatrix}$$

4.5. *Inertial force*

$$\vec{a}_{31} = -m_3\vec{\beta}_{31} = \begin{bmatrix} m_3\ell_2\cos\phi - m_3(x+\ell_3^*)\sin\phi\cos\psi \\ m_3\ell_2\sin\phi + m_3(x+\ell_3^*)\cos\phi\cos\psi \\ 0 \end{bmatrix}$$

$$\vec{a}_{32} = -m_3\vec{\beta}_{32} = \begin{bmatrix} -m_3(x+\ell_3^*)\cos\phi\sin\psi \\ -m_3(x+\ell_3^*)\sin\phi\sin\psi \\ -m_3(x+\ell_3^*)\cos\psi \end{bmatrix}$$

$$\vec{a}_{33} = -m_3\vec{\beta}_{33} = \begin{bmatrix} m_3\cos\phi\cos\psi \\ m_3\sin\phi\cos\psi \\ -m_3\sin\psi \end{bmatrix}$$

$$\vec{a}_3^{\,o} = -m_3\vec{\beta}_3^{\,o} = \begin{bmatrix} -\dot\phi^2 m_3[\ell_2\sin\phi + (x+\ell_3^*)\cos\phi\cos\psi] - \dot\psi^2 m_3(x+\ell_3^*)\cos\phi\cos\psi + \\ +2\dot\phi\dot\psi m_3(x+\ell_3^*)\sin\phi\sin\psi - 2\dot\phi\dot x m_3\sin\phi\cos\psi - 2\dot\psi\dot x m_3\cos\phi\sin\psi \\ \dot\phi^2 m_3[\ell_2\cos\phi - (x+\ell_3^*)\sin\phi\cos\psi] - \dot\psi^2 m_3(x+\ell_3^*)\sin\phi\cos\psi - \\ -2\dot\phi\dot\psi m_3(x+\ell_3^*)\cos\phi\sin\psi + 2\dot\phi\dot x m_3\cos\phi\cos\psi - 2\dot\psi\dot x m_3\sin\phi\sin\psi \\ \dot\psi^2 m_3(x+\ell_3^*)\sin\psi - 2\dot\psi\dot x m_3\cos\psi \end{bmatrix}$$

$$\vec{F}_3 = \vec{a}_{31}\ddot\phi + \vec{a}_{32}\ddot\psi + \vec{a}_{33}\ddot x + \vec{a}_3^{\,o} =$$

$$= \begin{bmatrix} \ddot\phi m_3[\ell_2\cos\phi - (x+\ell_3^*)\sin\phi\cos\psi] - \ddot\psi m_3(x+\ell_3^*)\cos\phi\sin\psi + \ddot x m_3\cos\phi\cos\psi \\ \ddot\phi m_3[\ell_2\sin\phi + (x+\ell_3^*)\cos\phi\cos\psi] - \ddot\psi m_3(x+\ell_3^*)\sin\phi\sin\psi + \ddot x m_3\sin\phi\cos\psi \\ -\ddot\psi m_3(x+\ell_3^*)\cos\psi - \ddot x m_3\sin\psi \end{bmatrix} +$$

$$+ \begin{bmatrix} -\dot\phi^2 m_3[\ell_2\sin\phi - (x+\ell_3^*)\cos\phi\cos\psi] - \dot\psi^2 m_3(x+\ell_3^*)\cos\phi\cos\psi \\ \dot\phi^2 m_3[\ell_2\cos\phi - (x+\ell_3^*)\sin\phi\cos\psi] - \dot\psi^2 m_3(x+\ell_3^*)\sin\phi\cos\psi \\ \dot\psi^2 m_3(x+\ell_3^*)\sin\psi \end{bmatrix} +$$

$$= \begin{bmatrix} 2\dot{\phi}\dot{\psi}m_3(x+\ell_3^*)\sin\phi\sin\psi-2\dot{\phi}\dot{x}m_3\sin\phi\cos\psi-2\dot{\psi}\dot{x}m_3\cos\phi\sin\psi \\ -2\dot{\phi}\dot{\psi}m_3(x+\ell_3^*)\cos\phi\sin\psi+2\dot{\phi}\dot{x}m_3\cos\phi\cos\psi-2\dot{\psi}\dot{x}m_3\sin\phi\sin\psi \\ -2\dot{\psi}\dot{x}m_3\cos\psi \end{bmatrix}$$

4.6. *Moment of inertial force*

$$T_3^{jk} = \sum_{\ell=1}^{3} q_{3\ell}^{j}\, q_{3\ell}^{k}\, J_{3\ell}$$

$$T_3^{11} = \sum_{\ell=1}^{3} q_{3\ell}^{1}\, q_{3\ell}^{1}\, J_{3\ell} = \cos^2\phi(J_{31}\cos^2\psi+J_{33}\sin^2\psi)+J_{32}\sin^2\phi$$

$$T_3^{12} = \sum_{\ell=1}^{3} q_{3\ell}^{1}\, q_{3\ell}^{2}\, J_{3\ell} = \sin\phi\cos\psi(J_{31}\cos^2\psi-J_{32}+J_{33}\sin^2\psi)$$

$$T_3^{13} = \sum_{\ell=1}^{3} q_{3\ell}^{1}\, q_{3\ell}^{3}\, J_{3\ell} = \cos\phi\sin\psi\cos\psi(J_{33}-J_{31})$$

$$T_3^{21} = T_3^{12} = \sin\phi\cos\phi(J_{31}\cos^2\psi-J_{32}+J_{33}\sin^2\psi)$$

$$T_3^{22} = \sum_{\ell=1}^{3} q_{3\ell}^{2}\, q_{3\ell}^{2}\, J_{3\ell} = \sin^2\phi(J_{31}\cos^2\psi+J_{33}\sin^2\psi)+J_{32}\cos^2\phi$$

$$T_3^{23} = \sum_{\ell=1}^{3} q_{3\ell}^{2}\, q_{3\ell}^{3}\, J_{3\ell} = \sin\phi\sin\psi\cos\psi(J_{33}-J_{31})$$

$$T_3^{31} = T_3^{13} = \cos\phi\sin\psi\cos\psi(J_{33}-J_{31})$$

$$T_3^{32} = T_3^{23} = \sin\phi\sin\psi\cos\psi(J_{33}-J_{31})$$

$$T_3^{33} = \sum_{\ell=1}^{3} q_{3\ell}^{3}\, q_{3\ell}^{3}\, J_{3\ell} = J_{31}\sin^2\psi+J_{33}\cos^2\psi$$

$$\vec{\omega}_3\vec{q}_{31} = [-\dot{\psi}\sin\phi,\ \dot{\psi}\cos\phi,\ \dot{\phi}][\cos\phi\cos\psi,\ \sin\phi\cos\psi,\ -\sin\psi]^T=-\dot{\phi}\sin\psi$$

$$\vec{\omega}_3\vec{q}_{32} = [-\dot{\psi}\sin\phi,\ \dot{\psi}\cos\phi,\ \dot{\phi}][-\sin\phi,\ \cos\phi,\ 0]^T = \dot{\psi}$$

$$\vec{\omega}_3\vec{q}_{33} = [-\dot{\psi}\sin\phi,\ \dot{\psi}\cos\phi,\ \dot{\phi}][\cos\phi\sin\psi,\ \sin\phi\sin\psi,\ \cos\psi]^T=\dot{\phi}\cos\psi$$

$$\vec{\lambda}_3 = Q_3 \begin{bmatrix} (J_{32}-J_{33})(\vec{\omega}_3\vec{q}_{32})(\vec{\omega}_3\vec{q}_{33}) \\ (J_{33}-J_{31})(\vec{\omega}_3\vec{q}_{33})(\vec{\omega}_3\vec{q}_{31}) \\ (J_{31}-J_{32})(\vec{\omega}_3\vec{q}_{31})(\vec{\omega}_3\vec{q}_{32}) \end{bmatrix} =$$

$$= \begin{bmatrix} \cos\phi\cos\psi & -\sin\phi & \cos\phi\sin\psi \\ \sin\phi\cos\psi & \cos\phi & \sin\phi\sin\psi \\ -\sin\psi & 0 & \cos\psi \end{bmatrix} \begin{bmatrix} \dot{\phi}\dot{\psi}(J_{32}-J_{33})\cos\psi \\ -\dot{\phi}^2(J_{33}-J_{31})\sin\psi\cos\psi \\ -\dot{\phi}\dot{\psi}(J_{31}-J_{32})\sin\psi \end{bmatrix} =$$

$$= \begin{bmatrix} \dot{\phi}^2(J_{33}-J_{31})\sin\phi\sin\psi\cos\psi+\dot{\phi}\dot{\psi}(-J_{31}\sin^2\psi+J_{32}-J_{33}\cos^2\psi)\cos\psi \\ -\dot{\phi}^2(J_{33}-J_{31})\cos\phi\sin\psi\cos\psi+\dot{\phi}\dot{\psi}(-J_{31}\sin^2\psi+J_{32}-J_{33}\cos^2\psi)\sin\phi \\ \dot{\phi}\dot{\psi}(J_{33}-J_{31})\sin\psi\cos\psi \end{bmatrix}$$

$$\vec{b}_{31} = -T_3\vec{\alpha}_{31} = -T_3[0,\ 0,\ 1]^T = [-T_3^{13},\ -T_3^{23},\ -T_3^{33}]^T =$$

$$= \begin{bmatrix} -(J_{33}-J_{31})\cos\phi\sin\psi\cos\psi \\ -(J_{33}-J_{31})\sin\phi\sin\psi\cos\psi \\ -J_{31}\sin^2\psi-J_{33}\cos^2\psi \end{bmatrix}$$

$$\vec{b}_{32} = -T_3\vec{\alpha}_{32} = -T_2[-\sin\phi,\ \cos\phi,\ 0]^T = \begin{bmatrix} T_3^{11}\sin\phi-T_3^{12}\cos\phi \\ T_3^{21}\sin\phi-T_3^{22}\cos\phi \\ T_3^{31}\sin\phi-T_3^{32}\cos\phi \end{bmatrix} =$$

$$= \begin{bmatrix} -\sin\phi\cos^2\phi(J_{31}\cos^2\psi+J_{33}\sin^2\psi)-J_{32}\sin^3\phi\ + \\ +\sin\phi\cos^2\phi(J_{31}\cos^2\psi-J_{32}+J_{33}\sin^2\psi) \\ -\sin^2\phi\cos\phi(J_{31}\cos^2\psi-J_{32}+J_{33}\sin^2\psi)\ + \\ +\sin^2\phi\cos\phi(J_{31}\cos^2\psi+J_{33}\sin^2\psi)+J_{32}\cos^3\phi \\ -\sin\phi\cos\phi\sin\psi\cos\psi(J_{33}-J_{31})+\sin\phi\cos\phi\sin\psi\cos\psi(J_{33}-J_{31}) \end{bmatrix} =$$

$$= [J_{32}\sin\phi,\ -J_{32}\cos\phi,\ 0]^T$$

$$\vec{b}_{33} = -T_3\vec{\alpha}_{33} = -T_3[0,\ 0,\ 0]^T = [0,\ 0,\ 0]^T$$

$$\vec{b}_3^o = -T_3\vec{\alpha}_3^o+\vec{\lambda}_3 = -T_3[-\dot{\phi}\dot{\psi}\cos\phi,\ -\dot{\phi}\dot{\psi}\sin\phi,\ 0]^T + \vec{\lambda}_3 =$$

$$= \begin{bmatrix} T_3^{11}\dot{\phi}\dot{\psi}\cos\phi+T_3^{12}\dot{\phi}\dot{\psi}\sin\phi \\ T_3^{21}\dot{\phi}\dot{\psi}\cos\phi+T_3^{22}\dot{\phi}\dot{\psi}\sin\phi \\ T_3^{31}\dot{\phi}\dot{\psi}\cos\phi+T_3^{32}\dot{\phi}\dot{\psi}\sin\phi \end{bmatrix} + \vec{\lambda}_3$$

$$= \begin{bmatrix} \dot{\phi}^2(J_{33}-J_{31})\sin\phi\sin\psi\cos\psi+\dot{\phi}\dot{\psi}\cos\phi[J_{32}+ \\ \qquad +(J_{33}-J_{31})(\sin^2\psi-\cos^2\psi)] \\ -\dot{\phi}^2(J_{33}-J_{31})\cos\phi\sin\psi\cos\psi+\dot{\phi}\dot{\psi}\sin\phi[J_{32}+ \\ \qquad +(J_{33}-J_{31})(\sin^2\psi-\cos^2\psi)] \\ 2\dot{\phi}\dot{\psi}(J_{33}-J_{31})\sin\psi\cos\psi \end{bmatrix}$$

Moment of inertial force

$$\vec{M}_3 = \vec{b}_{31}\ddot{\phi}+\vec{b}_{32}\ddot{\psi}+\vec{b}_{33}\ddot{x}+\vec{b}_3^o = [M_3^1\ M_3^2\ M_3^3]^T =$$

$$= \begin{bmatrix} -\ddot{\phi}(J_{33}-J_{31})\cos\phi\sin\psi\cos\psi+\ddot{\psi}J_{32}\sin\phi+\dot{\phi}^2(J_{33}-J_{31})\sin\phi\sin\psi\cos\psi+ \\ +\dot{\phi}\dot{\psi}[J_{32}+(J_{33}-J_{31})(\sin^2\psi-\cos^2\psi)]\cos\phi \\ -\ddot{\phi}(J_{33}-J_{31})\sin\phi\sin\psi\cos\psi-\ddot{\psi}J_{32}\cos\phi-\dot{\phi}^2(J_{33}-J_{31})\cos\phi\sin\psi\cos\psi+ \\ +\dot{\phi}\dot{\psi}[J_{32}+(J_{33}-J_{31})(\sin^2\psi-\cos^2\psi)]\sin\phi \\ -\ddot{\phi}(J_{31}\sin^2\psi+J_{33}\cos^2\psi)+2\dot{\phi}\dot{\psi}(J_{33}-J_{31})\sin\psi\cos\psi \end{bmatrix}$$

5. Forming of differential equations

5.1. *Elements of H-matrix*

$$\xi_1 = 0 \quad H_{ik} = -\vec{e}_i\cdot\sum_{j=\max(i,k)}^{n}(\vec{b}_{jk}+\vec{r}_{ji}\times\vec{a}_{jk}) \qquad i=1, \quad k=1,2,3$$

$$H_{111}=-\vec{e}_1\cdot(\vec{b}_{11}+\vec{r}_{11}\times\vec{a}_{11})=-\vec{e}_1\cdot\vec{b}_{11}=-[0,\ 0,\ 1]^T[0,\ 0,\ -J_{13}]^T=J_{13}$$

$$H_{112} = -\vec{e}_1\cdot(\vec{b}_{21}+\vec{r}_{21}\times\vec{a}_{21}) = -\vec{e}_1\cdot\{\vec{b}_{21} + \begin{bmatrix} -\ell_2^* \sin\phi \\ \ell_2^* \cos\phi \\ \ell_1 \end{bmatrix} \times \begin{bmatrix} m_2\ell_2^*\cos\phi \\ m_2\ell_2^*\sin\phi \\ 0 \end{bmatrix}\} =$$

$$= -[0,\ 0,\ 1]\{\begin{bmatrix} -(J_{23}-J_{21})\cos\phi\sin\psi\cos\psi \\ -(J_{23}-J_{21})\sin\phi\sin\psi\cos\psi \\ -J_{21}\sin^2\psi-J_{23}\cos^2\psi \end{bmatrix} + \begin{bmatrix} -m_2\ell_1\ell_2^*\sin\phi \\ m_2\ell_1\ell_2^*\cos\phi \\ -m_2\ell_2^{*2} \end{bmatrix} =$$

$$= m_2\ell_2^{*2} + J_{21}\sin^2\psi + J_{23}\cos^2\psi$$

$$H_{113} = -\vec{e}_1\cdot(\vec{b}_{31}+\vec{r}_{31}\times\vec{a}_{31}) = -\vec{e}_1\cdot\{\vec{b}_{31} + \begin{bmatrix} -\ell_2\sin\phi-(x+\ell_3^*)\cos\phi\cos\psi \\ \ell_2\cos\phi-(x+\ell_3^*)\sin\phi\cos\psi \\ \ell_1+(x+\ell_3^*)\sin\psi \end{bmatrix} \times$$

$$\times \begin{bmatrix} m_3[\ell_2\cos\phi-(x+\ell_3^*)\sin\phi\cos\psi] \\ m_3[\ell_2\sin\phi+(x+\ell_3^*)\cos\phi\cos\psi] \\ 0 \end{bmatrix} =$$

$$= -[0,\ 0,\ 1] \begin{bmatrix} -(J_{33}-J_{31})\cos\phi\sin\psi\cos\psi \\ -(J_{33}-J_{31})\sin\phi\sin\psi\cos\psi \\ -J_{31}\sin^2\psi-J_{33}\cos^2\psi \end{bmatrix} +$$

$$+ \begin{bmatrix} -m_3[\ell_1+(x+\ell_3^*)\sin\psi][\ell_2\sin\phi+(x+\ell_3^*)\cos\phi\cos\psi] \\ m_3[\ell_1+(x+\ell_3^*)\sin\psi][\ell_2\cos\phi-(x+\ell_3^*)\sin\phi\cos\psi] \\ -m_3[\ell_2^2+(x+\ell_3^*)^2\cos^2\psi] \end{bmatrix} \} =$$

$$= m_3[\ell_2^2+(x+\ell_3^*)^2\cos^2\psi]+J_{31}\sin^2\psi+J_{33}\cos^2\psi$$

$$H_{122}=-\vec{e}_1\cdot(\vec{b}_{22}+\vec{r}_{21}\times\vec{a}_{22})=-\vec{e}_1\cdot\vec{b}_{22}=-[0,\ 0,\ 1][J_{22}\sin\phi,\ -J_{22}\cos\phi,\ 0]^T=0$$

$$H_{123} = -\vec{e}_1\cdot(\vec{b}_{32}+\vec{r}_{31}\times\vec{a}_{32}) = -\vec{e}_1\cdot\{\vec{b}_{32} + \begin{bmatrix} -\ell_2\sin\phi-(x+\ell_3^*)\cos\phi\cos\psi \\ \ell_2\cos\phi-(x+\ell_3^*)\sin\phi\cos\psi \\ \ell_1+(x+\ell_3^*)\sin\psi \end{bmatrix} \times$$

$$\times \begin{bmatrix} -m_3(x+\ell_3^*)\cos\phi\sin\psi \\ -m_3(x+\ell_3^*)\sin\phi\sin\psi \\ m_3(x+\ell_3^*)\cos\psi \end{bmatrix} = -[0,\ 0,\ 1]^T \begin{bmatrix} J_{32}\sin\phi \\ -J_{32}\cos\phi \\ 0 \end{bmatrix} +$$

$$+\begin{bmatrix} m_3(x+\ell_3^*)[\ell_1\sin\phi\sin\psi-\ell_2\cos\phi\cos\psi+(x+\ell_3^*)\sin\phi] \\ -m_3(x+\ell_3^*)[\ell_1\cos\phi\sin\psi+\ell_2\sin\phi\cos\psi+(x+\ell_3^*)\cos\phi] \\ m_3\ell_2(x+\ell_3^*)\sin\psi \end{bmatrix}\} =$$

$$= -m_3\ell_2(x+\ell_3^*)\sin\psi$$

$$H_{133} = -\vec{e}_1\cdot(\vec{b}_{33}+\vec{r}_{31}\times\vec{a}_{33}) = -\vec{e}_1\cdot\{ \begin{bmatrix} -\ell_2\sin\phi-(x+\ell_3^*)\cos\phi\cos\psi \\ \ell_2\cos\phi-(x+\ell_3^*)\sin\phi\cos\psi \\ \ell_1+(x+\ell_3^*)\sin\psi \end{bmatrix} \times$$

$$\times \begin{bmatrix} m_3\cos\phi\cos\psi \\ m_3\sin\phi\cos\psi \\ -m_3\sin\psi \end{bmatrix}\} = -[0,\ 0,\ 1] \begin{bmatrix} -m_3(\ell_1\sin\phi\cos\psi+\ell_2\cos\phi\sin\psi) \\ m_3(\ell_1\cos\phi\cos\psi-\ell_2\sin\phi\sin\psi) \\ -m_3\ell_2\cos\psi \end{bmatrix} =$$

$$= m_3\ell_2\cos\psi$$

$$\xi_2 = 0 \quad H^{ik} = -\vec{e}_i\cdot \sum_{j=\max(i,k)}^{n} (\vec{b}_{jk}+\vec{r}_{ji}\times\vec{a}_{jk}) \quad i=2,\quad k=1,2,3$$

$$H_{212} = -\vec{e}_2\cdot(\vec{b}_{21}+\vec{r}_{22}\times\vec{a}_{21}) = -\vec{e}_2\cdot\{\vec{b}_{21} + \begin{bmatrix} -\ell_2^*\sin\phi \\ \ell_2^*\cos\phi \\ 0 \end{bmatrix} \times \begin{bmatrix} m_2\ell_2^*\cos\phi \\ m_2\ell_2^*\sin\phi \\ 0 \end{bmatrix}\} =$$

$$= -[-\sin\phi,\ \cos\phi,\ 0]\{\begin{bmatrix} -(J_{23}-J_{21})\cos\phi\sin\psi\cos\psi \\ -(J_{23}-J_{21})\sin\phi\sin\psi\cos\psi \\ -J_{21}\sin^2\psi-J_{23}\cos^2\psi \end{bmatrix} +$$

$$+ \begin{bmatrix} 0 \\ 0 \\ -m_2\ell_2^{*2} \end{bmatrix}\} = 0$$

$$H_{213} = -\vec{e}_2\cdot(\vec{b}_{31}+\vec{r}_{32}\times\vec{a}_{31}) = -\vec{e}_2\cdot\{\vec{b}_{31} + \begin{bmatrix} -\ell_2\sin\phi-(x+\ell_3^*)\cos\phi\cos\psi \\ \ell_2\cos\phi-(x+\ell_3^*)\sin\phi\cos\psi \\ (x+\ell_3^*)\sin\psi \end{bmatrix} \times$$

$$\times \begin{bmatrix} m_3[\ell_2\cos\phi-(x+\ell_3^*)\sin\phi\cos\psi] \\ m_3[\ell_2\sin\phi+(x+\ell_3^*)\cos\phi\cos\psi] \\ 0 \end{bmatrix}\} =$$

$$= -[-\sin\phi,\ \cos\phi,\ 0]\ \{\begin{bmatrix} -(J_{33}-J_{31})\cos\phi\sin\psi\cos\psi \\ -(J_{33}-J_{31})\sin\phi\sin\psi\cos\psi \\ -J_{31}\sin^2\psi-J_{33}\cos^2\psi \end{bmatrix} +$$

$$+ \begin{bmatrix} -m_3(x+\ell_3^*)[\ell_2\sin\phi+(x+\ell_3^*)\cos\phi\cos\psi]\sin\psi \\ m_3(x+\ell_3^*)[\ell_2\cos\phi-(x+\ell_3^*)\sin\phi\cos\psi]\sin\psi \\ -m_3(x+\ell_3^*)^2\cos^2\psi \end{bmatrix}\} = -m_3\ell_2(x+\ell_3^*)\sin\psi$$

$$H_{222} = -\vec{e}_2\cdot(\vec{b}_{22}+\vec{r}_{22}\times\vec{a}_{22}) = -\vec{e}_2\cdot\vec{b}_{22} =$$

$$= -[-\sin\phi,\ \cos\phi,\ 0][J_{22}\sin\phi,\ -J_{22}\cos\phi,\ 0]^T = J_{22}$$

$$H_{223} = -\vec{e}_2\cdot(\vec{b}_{32}+\vec{r}_{32}\times\vec{a}_{32}) = -\vec{e}_2\cdot\{\vec{b}_{32} + \begin{bmatrix} -\ell_2\sin\phi-(x+\ell_3^*)\cos\phi\cos\psi \\ \ell_2\cos\phi-(x+\ell_3^*)\sin\phi\cos\psi \\ (x+\ell_3^*)\sin\psi \end{bmatrix} \times$$

$$\times \begin{bmatrix} -m_3(x+\ell_3^*)\cos\phi\sin\psi \\ -m_3(x+\ell_3^*)\sin\phi\sin\psi \\ -m_3(x+\ell_3^*)\cos\psi \end{bmatrix} \} = -[-\sin\phi,\ \cos\psi,\ 0]\{ \begin{bmatrix} J_{32}\sin\phi \\ -J_{32}\cos\phi \\ 0 \end{bmatrix} +$$

$$+ \begin{bmatrix} m_3(x+\ell_3^*)[-\ell_2\cos\phi\cos\psi+(x+\ell_3^*)\sin\phi] \\ -m_3(x+\ell_3^*)[\ell_2\sin\phi\cos\psi+(x+\ell_3^*)\cos\phi] \\ m_3\ell_2(x+\ell_3^*)\sin\psi \end{bmatrix} \} = m_3(x+\ell_3^*)^2+J_{32}$$

$$H_{233} = -\vec{e}_2\cdot(\vec{b}_{33}+\vec{r}_{32}\times\vec{a}_{33}) = -\vec{e}_2\cdot\{ \begin{bmatrix} -\ell_2\sin\phi-(x+\ell_3^*)\cos\phi\cos\psi \\ \ell_2\cos\phi-(x+\ell_3^*)\sin\phi\cos\psi \\ (x+\ell_3^*)\sin\psi \end{bmatrix} \times$$

$$\times \begin{bmatrix} m_3\cos\phi\cos\psi \\ m_3\sin\phi\cos\psi \\ -m_3\sin\psi \end{bmatrix} \} = -[-\sin\phi,\ \cos\phi,\ 0] \begin{bmatrix} -m_3\ell_2\cos\phi\sin\psi \\ -m_3\ell_2\sin\phi\sin\psi \\ -m_3\ell_2\cos\psi \end{bmatrix} = 0$$

$$\xi_3 = 1 \quad H^{ik} = -\vec{e}_i\cdot \sum_{j=\max(i,k)}^{n} \vec{a}_{jk} \qquad i=3, \quad k=1,2,3$$

$$H_{313} = -\vec{e}_3\cdot\vec{a}_{31} = -[-\cos\phi\cos\psi,\ -\sin\phi\cos\psi,\ \sin\psi]\cdot$$

$$\cdot \begin{bmatrix} m_3[\ell_2\cos\phi-(x+\ell_3^*)\sin\phi\cos\psi] \\ m_3[\ell_2\sin\phi+(x+\ell_3^*)\cos\phi\cos\psi] \\ 0 \end{bmatrix} = m_3\ell_2\cos\psi$$

$$H_{323} = -\vec{e}_3\cdot\vec{a}_{32} = -[-\cos\phi\cos\psi,\ -\sin\phi\cos\psi,\ \sin\psi]\cdot$$

$$\cdot \begin{bmatrix} -m_3(x+\ell_3^*)\cos\phi\sin\psi \\ -m_3(x+\ell_3^*)\sin\phi\sin\psi \\ -m_3(x+\ell_3^*)\cos\psi \end{bmatrix} = 0$$

$$H_{333} = -\vec{e}_3 \cdot \vec{a}_{33} = -[-\cos\phi\cos\psi, -\sin\phi\cos\psi, \ \sin\psi] \begin{bmatrix} m_3\cos\phi\cos\psi \\ m_3\sin\phi\cos\psi \\ -m_3\sin\psi \end{bmatrix} = m_3$$

$$H_{11} = H_{111}+H_{112}+H_{113} = m_2\ell_2^{*2}+m_3[\ell_2^2+(x+\ell_3^*)^2\cos^2\psi] +$$

$$+ J_{13}+(J_{21}+J_{31})\sin^2\psi+(J_{23}+J_{33})\cos^2\psi$$

$$H_{12} = H_{122}+H_{123} = -m_3\ell_2(x+\ell_3^*)\sin\psi$$

$$H_{13} = H_{133} = m_3\ell_2\cos\psi$$

$$H_{21} = H_{212} + H_{213} = -m_3\ell_2(x+\ell_3^*)\sin\psi$$

$$H_{22} = H_{222}+H_{223} = m_3(x+\ell_3^*)^2+J_{22}+J_{32}$$

$$H_{23} = H_{233} = 0$$

$$H_{31} = H_{313} = m_2\ell_2\cos\psi$$

$$H_{32} = H_{323} = 0$$

$$H_{33} = H_{333} = m_3$$

H - matrix

$$H = \begin{bmatrix} \begin{matrix} m_2\ell_2^{*2}+m_3[\ell_2^2+(x+\ell_3^*)^2\cos^2\psi]+J_{13}+ \\ +(J_{21}+J_{31})\sin^2\psi+(J_{23}+J_{33})\cos^2\psi \end{matrix} & -m_3\ell_2(x+\ell_3^*)\sin\psi & m_3\ell_2\cos\psi \\ -m_3\ell_2(x+\ell_3^*)\sin\psi & m_3(x+\ell_3^*)^2+J_{22}+J_{32} & 0 \\ m_3\ell_2\cos\psi & 0 & m_3 \end{bmatrix}$$

6. Elements of h-matrix

$$G_1 = (0, \ 0, \ -m_1 g)^T$$

$$G_2 = (0, \ 0, \ -m_2 g)^T$$

$$G_3 = (0,\ 0,\ -m_3 g)^T$$

$$\xi_1 = 0 \qquad h^i = -\vec{e}_i \cdot \sum_{j=i}^{n} (\vec{r}_{ji} \times (\vec{a}_j^o + \vec{G}_j) + \vec{b}_j^o) \qquad i=1$$

$$h_{11} = -\vec{e}_1 \cdot [\vec{r}_{11} \times (\vec{a}_1^o + \vec{G}_1) + \vec{b}_1^o] = -\vec{e}_1 \cdot (\vec{r}_{11} \times \vec{G}_1) =$$

$$= -\vec{e}_1 \cdot \{[0,\ 0,\ \ell_1^*]^T \times [0,\ 0,\ -m_1 g]^T\} = 0$$

$$h_{12} = -\vec{e}_1 \cdot [\vec{r}_{21} \times (\vec{a}_2^o + \vec{G}_2) + \vec{b}_2^o] = -\vec{e}_1 \cdot \{ \begin{bmatrix} -\ell_2^* \sin\phi \\ \ell_2^* \cos\phi \\ \ell_1 \end{bmatrix} \times (\begin{bmatrix} -\dot{\phi}^2 m_2 \ell_2^* \sin\phi \\ \dot{\phi}^2 m_2 \ell_2^* \cos\phi \\ 0 \end{bmatrix} +$$

$$+ \begin{bmatrix} 0 \\ 0 \\ -m_2 g \end{bmatrix}) + \vec{b}_2^o \} = -[0,\ 0,\ 1]\ \{ \begin{bmatrix} -\dot{\phi}^2 m_2 \ell_1 \ell_2^* \cos\phi - m_2 \ell_2^* g \cos\phi \\ -\dot{\phi}^2 m_2 \ell_1 \ell_2^* \sin\phi - m_2 \ell_2^* g \sin\phi \\ 0 \end{bmatrix} +$$

$$+ \begin{bmatrix} \dot{\phi}^2 (J_{23}-J_{21}) \sin\phi \sin\psi \cos\psi + \dot{\phi}\dot{\psi} \cos\phi [J_{22} + (J_{23}-J_{21})(\sin^2\psi - \cos^2\psi)] \\ -\dot{\phi}^2 (J_{23}-J_{21}) \cos\phi \sin\psi \cos\psi + \dot{\phi}\dot{\psi} \sin\phi [J_{22} + (J_{23}-J_{21})(\sin^2\psi - \cos^2\psi)] \\ 2\dot{\phi}\dot{\psi} (J_{23}-J_{21}) \sin\psi \cos\psi \end{bmatrix} \} =$$

$$= -2\dot{\phi}\dot{\psi} (J_{23}-J_{21}) \sin\psi \cos\psi$$

$$h_{13} = -\vec{e}_1 \cdot [\vec{r}_{31} \times (\vec{a}_3^o + \vec{G}_3) + \vec{b}_3^o] = -\vec{e}_1 \cdot \{ \begin{bmatrix} -\ell_2 \sin\phi - (x+\ell_3^*) \cos\phi \cos\psi \\ \ell_2 \cos\phi - (x+\ell_3^*) \sin\phi \cos\psi \\ \ell_1 + (x+\ell_3^*) \sin\psi \end{bmatrix} \times$$

$$\times (\begin{bmatrix} -\dot{\phi}^2 m_3 [\ell_2 \sin\phi + (x+\ell_3^*) \cos\phi \cos\psi] - \dot{\psi}^2 m_3 (x+\ell_3^*) \cos\phi \cos\psi + \\ +2\dot{\phi}\dot{\psi} m_3 (x+\ell_3^*) \sin\phi \sin\psi - 2\dot{\phi}\dot{x} m_3 \sin\phi \cos\psi - 2\dot{\psi}\dot{x} m_3 \cos\phi \sin\psi \\ \dot{\phi}^2 m_3 [\ell_2 \cos\phi - (x+\ell_3^*) \sin\phi \cos\psi] - \dot{\psi}^2 m_3 (x+\ell_3^*) \sin\phi \cos\psi - \\ -2\dot{\phi}\dot{\psi} m_3 (x+\ell_3^*) \cos\phi \sin\psi + 2\dot{\phi}\dot{x} m_3 \cos\phi \cos\psi - 2\dot{\psi}\dot{x} m_3 \sin\phi \sin\psi \\ \dot{\psi}^2 m_3 (x+\ell_3^*) \sin\psi - 2\dot{\psi}\dot{x} m_3 \cos\psi \end{bmatrix} +$$

$$+\begin{bmatrix} 0 \\ 0 \\ -m_3 g \end{bmatrix}) + \vec{b}_3^o\} =$$

$$= -[0,\ 0,\ 1] \begin{bmatrix} -\dot{\phi}^2 m_3[\ell_2\cos\phi-(x+\ell_3^*)\sin\phi\cos\psi][\ell_1+(x+\ell_3^*)\sin\psi]+ \\ +\dot{\psi}^2 m_3(x+\ell_3^*)(\ell_1\sin\phi\cos\psi+\ell_2\cos\phi\sin\psi)+2\dot{\phi}\dot{\psi}m_3(x+\ell_3^*)[\ell_1+ \\ +(x+\ell_3^*)\sin\psi]\cos\phi\sin\psi-2\dot{\phi}\dot{x}m_3[\ell_1+(x+\ell_3^*)\sin\psi]\cos\phi\cos\psi+ \\ +2\dot{\psi}\dot{x}m_3[\ell_1\sin\phi\sin\psi-\ell_2\cos\phi\cos\psi+(x+\ell_3^*)\sin\phi]-m_3g[\ell_2\cos\phi- \\ -(x+\ell_3^*)\sin\psi\cos\psi]+\dot{\phi}^2(J_{33}-J_{31})\sin\phi\sin\psi\cos\psi+\dot{\phi}\dot{\psi}\cos\phi[J_{32}+ \\ +(J_{33}-J_{31})(\sin^2\psi-\cos^2\psi)] \\ \\ -\dot{\phi}^2 m_3[\ell_2\sin\phi+(x+\ell_3^*)\cos\phi\cos\psi][\ell_1+(x+\ell_3^*)\sin\psi]- \\ -\dot{\psi}^2 m_3(x+\ell_3^*)(\ell_1\cos\phi\cos\psi-\ell_2\sin\phi\sin\psi)+2\dot{\phi}\dot{\psi}m_3(x+\ell_3^*)[\ell_1+ \\ +(x+\ell_3^*)\sin\psi]\sin\phi\sin\psi-2\dot{\phi}\dot{x}m_3[\ell_1+(x+\ell_3^*)\sin\psi]\sin\phi\cos\psi- \\ -2\dot{\psi}\dot{x}m_3[\ell_1\cos\phi\sin\psi+\ell_2\sin\phi\cos\psi+(x+\ell_3^*)\cos\phi]-m_3g[\ell_2\sin\phi+ \\ +(x+\ell_3^*)\cos\phi\cos\psi]-\dot{\phi}^2(J_{33}-J_{31})\cos\phi\sin\psi\cos\psi+\dot{\phi}\dot{\psi}\sin\phi[J_{32}+ \\ +(J_{33}-J_{31})(\sin^2\psi-\cos^2\psi)] \\ \\ \dot{\psi}^2 m_3\ell_2(x+\ell_3^*)\cos\psi+2\dot{\phi}\dot{\psi}m_3(x+\ell_3^*)^2\sin\psi\cos\psi- \\ -2\dot{\phi}\dot{x}m_3(x+\ell_3^*)\cos^2\psi+2\dot{\psi}\dot{x}m_3\ell_2\sin\psi+2\dot{\phi}\dot{\psi}(J_{33}-J_{31})\sin\psi\cos\psi \end{bmatrix} =$$

$$= -\dot{\psi}^2 m_3\ell_2(x+\ell_3^*)\cos\psi-2\dot{\phi}\dot{\psi}m_3(x+\ell_3^*)^2\sin\psi\cos\psi+2\dot{\phi}\dot{x}m_3(x+\ell_3^*)\cos^2\psi-2\dot{\psi}\dot{x}m_3\ell_2\sin\psi-$$
$$-2\dot{\phi}\dot{\psi}(J_{33}-J_{31})\sin\psi\cos\psi$$

$$\xi_2 = 0 \qquad h^i = -\vec{e}_i \sum_{j=i}^{n} (\vec{r}_{ji} \times (\vec{a}_j^o+\vec{G}_j)+\vec{b}_j^o), \qquad i=2$$

$$h_{22} = -\vec{e}_2\cdot[\vec{r}_{22}\times(\vec{a}_2^{\,o}+\vec{G}_2)+\vec{b}_2^{\,o}] = -\vec{e}_2\cdot\left\{ \begin{bmatrix}-\ell_2^*\sin\phi\\ \ell_2^*\cos\phi\\ 0\end{bmatrix} \times \left(\begin{bmatrix}-\dot\phi^2 m_2\ell_2^*\sin\phi\\ \dot\phi^2 m_2\ell_2^*\cos\phi\\ 0\end{bmatrix} + \right.\right.$$

$$\left.\left. + \begin{bmatrix}0\\ 0\\ -m_2 g\end{bmatrix}\right) + b_2^o\right\} = -[-\sin\phi,\ \cos\phi,\ 0]\left\{ \begin{bmatrix}-m_2 g\ell_2^*\cos\phi\\ -m_2 g\ell_2^*\sin\phi\\ 0\end{bmatrix} + \right.$$

$$\left. + \begin{bmatrix}\dot\phi^2(J_{23}-J_{21})\sin\phi\sin\psi\cos\psi+\dot\phi\dot\psi\cos\phi[J_{22}+(J_{23}-J_{21})(\sin^2\psi-\cos^2\psi)]\\ -\dot\phi^2(J_{23}-J_{21})\cos\phi\sin\psi\cos\psi+\dot\phi\dot\psi\sin\phi[J_{22}+(J_{23}-J_{21})(\sin^2\psi-\cos^2\psi)]\\ 2\dot\phi\dot\psi(J_{23}-J_{21})\sin\psi\cos\psi\end{bmatrix}\right\} =$$

$$= \dot\phi^2(J_{23}-J_{21})\sin\psi\cos\psi$$

$$h_{23} = -\vec{e}_2\cdot[\vec{r}_{32}\times(\vec{a}_3^{\,o}+\vec{G}_3)+\vec{b}_3^{\,o}] = -\vec{e}_2\cdot\left\{ \begin{bmatrix}-\ell_2\sin\phi-(x+\ell_3^*)\cos\phi\cos\psi\\ \ell_2\cos\phi-(x+\ell_3^*)\sin\phi\cos\psi\\ (x+\ell_3^*)\sin\psi\end{bmatrix} \times \right.$$

$$\times\left(\begin{bmatrix}-\dot\phi^2 m_3[\ell_2\sin\phi+(x+\ell_3^*)\cos\phi\cos\psi]-\dot\psi^2 m_3(x+\ell_3^*)\cos\phi\cos\psi+\\ \quad +2\dot\phi\dot\psi m_3(x+\ell_3^*)\sin\phi\sin\psi-2\dot\phi\dot x m_3\sin\phi\cos\psi-2\dot\psi\dot x m_3\cos\phi\sin\psi\\ \dot\phi^2 m_3[\ell_2\cos\phi-(x+\ell_3^*)\sin\phi\cos\psi]-\dot\psi^2 m_3(x+\ell_3^*)\sin\phi\cos\psi-\\ \quad -2\dot\phi\dot\psi m_3(x+\ell_3^*)\cos\phi\sin\psi+2\dot\phi\dot x m_3\cos\phi\cos\psi-2\dot\psi\dot x m_3\sin\phi\sin\psi\\ \dot\psi^2 m_3(x+\ell_3^*)\sin\psi-2\dot\psi\dot x m_3\cos\psi\end{bmatrix} + \right.$$

$$\left.\left. + \begin{bmatrix}0\\ 0\\ -m_3 g\end{bmatrix}\right) + b_3^o\right\} =$$

$$= -[-\sin\phi,\ \cos\phi,\ 0]\left\{ \begin{bmatrix}-\dot\phi^2 m_3[\ell_2\cos\phi-(x+\ell_3^*)\sin\phi\cos\psi](x+\ell_3^*)\sin\psi\\ -\dot\phi^2 m_3[\ell_2\sin\phi+(x+\ell_3^*)\cos\phi\cos\psi](x+\ell_3^*)\sin\psi\\ 0\end{bmatrix}\right.$$

$$+\begin{bmatrix} +\dot{\psi}^2 m_3\ell_2(x+\ell_3^*)\cos\phi\sin\psi+2\dot{\phi}\dot{\psi}m_3(x+\ell_3^*)^2\cos\phi\sin^2\psi \\ +\dot{\psi}^2 m_3\ell_2(x+\ell_3^*)\sin\phi\sin\psi+2\dot{\phi}\dot{\psi}m_3(x+\ell_3^*)^2\sin\phi\sin^2\psi \\ \dot{\psi}^2 m_3\ell_2(x+\ell_3^*)\cos\psi+2\dot{\phi}\dot{\psi}m_3(x+\ell_3^*)^2\sin\psi\cos\psi \end{bmatrix}+$$

$$+\begin{bmatrix} -2\dot{\phi}\dot{x}m_3(x+\ell_3^*)\cos\phi\sin\psi\cos\psi-2\dot{\psi}\dot{x}m_3[\ell_2\cos\phi\cos\psi-(x+\ell_3^*)\sin\phi] \\ -2\dot{\phi}\dot{x}m_3(x+\ell_3^*)\sin\phi\sin\psi\cos\psi-2\dot{\psi}\dot{x}m_3[\ell_2\sin\phi\cos\psi+(x+\ell_3^*)\cos\phi] \\ -2\dot{\phi}\dot{x}m_3(x+\ell_3^*)\cos^2\psi+2\dot{\psi}\dot{x}m_3\ell_2\sin\psi \end{bmatrix}+$$

$$+\begin{bmatrix} m_3 g[-\ell_2\cos\psi+(x+\ell_3^*)\sin\phi\cos\psi] \\ m_3 g[-\ell_2\sin\phi-(x+\ell_3^*)\cos\phi\cos\psi] \\ 0 \end{bmatrix}+$$

$$+\begin{bmatrix} \dot{\phi}^2(J_{33}-J_{31})\sin\phi\sin\psi\cos\psi+\dot{\phi}\dot{\psi}\cos\phi[J_{32}+(J_{33}-J_{31})(\sin^2\psi-\cos^2\psi)] \\ -\dot{\phi}^2(J_{33}-J_{31})\cos\phi\sin\psi\cos\psi+\dot{\phi}\dot{\psi}\sin\phi[J_{32}+(J_{33}-J_{31})(\sin^2\psi-\cos^2\psi)] \\ 2\dot{\phi}\dot{\psi}(J_{33}-J_{31})\sin\psi\cos\psi \end{bmatrix}\Bigg\} =$$

$$= \dot{\phi}^2 m_3(x+\ell_3^*)^2\sin\psi\cos\psi+2\dot{\psi}\dot{x}m_3(x+\ell_3^*)+m_3 g(x+\ell_3^*)\cos\psi+\phi^2(J_{33}-J_{31})\sin\psi\cos\psi$$

$$\xi_3 = 1 \qquad h^i = -\vec{e}_i\cdot\sum_{j=i}^{n}(\vec{a}_{jo}+\vec{G}_j) \qquad i=3$$

$$h_{33} = -\vec{e}_3\cdot(\vec{a}_3^{\,o}+\vec{G}_3) = -[-\cos\phi\cos\psi,\ -\sin\phi\cos\psi,\ \sin\psi]\{$$

$$\Bigg\{\begin{bmatrix} -\dot{\phi}^2 m_3[\ell_2\sin\phi+(x+\ell_3^*)\cos\phi\cos\psi]-\dot{\psi}^2 m_3(x+\ell_3^*)\cos\phi\cos\psi+ \\ +2\dot{\phi}\dot{\psi}m_3(x+\ell_3^*)\sin\phi\sin\psi-2\dot{\phi}\dot{x}m_3\sin\phi\cos\psi-2\dot{\psi}\dot{x}m_3\cos\phi\sin\psi \\ \dot{\phi}^2 m_3[\ell_2\cos\phi-(x+\ell_3^*)\sin\phi\cos\psi]-\dot{\psi}^2 m_3(x+\ell_3^*)\sin\phi\cos\psi- \\ -2\dot{\phi}\dot{\psi}m_3(x+\ell_3^*)\cos\phi\sin\psi+2\dot{\phi}\dot{x}m_3\cos\phi\cos\psi-2\dot{\psi}\dot{x}m_3\sin\phi\sin\psi \\ \dot{\psi}^2 m_3(x+\ell_3^*)\sin\psi-2\dot{\psi}\dot{x}m_3\cos\psi \end{bmatrix}+$$

$$+\begin{bmatrix}0\\0\\-m_3g\end{bmatrix}\Bigg\} = -\dot{\phi}^2 m_3(x+\ell_3^*)\cos^2\psi-\dot{\psi}^2 m_3(x+\ell_3^*)+m_3g\sin\psi$$

$$h_1 = h_{11}+h_{12}+h_{13} = -\dot{\psi}^2 m_3\ell_2(x+\ell_3^*)\cos\psi-2\dot{\phi}\dot{\psi}[m_3(x+\ell_3^*)^2+J_{23}-J_{21}+$$
$$+ J_{33}-J_{31}]\sin\psi\cos\psi+2\dot{\phi}\dot{x}m_3(x+\ell_3^*)\cos^2\psi-2\dot{\psi}\dot{x}m_3\ell_2\sin\psi$$

$$h_2 = h_{22}+h_{23} = \dot{\phi}^2[m_3(x+\ell_3^*)^2+J_{23}-J_{21}+J_{33}-J_{31}]\sin\psi\cos\psi+$$
$$+ 2\dot{\psi}\dot{x}m_3(x+\ell_3^*)-m_3g(x+\ell_3^*)\cos\psi$$

$$h_3 = h_{33} = -\dot{\phi}^2 m_3(x+\ell_3^*)\cos^2\psi-\dot{\psi}^2 m_3(x+\ell_3^*)-m_3g\sin\psi$$

$$h = \begin{bmatrix} -\dot{\psi}^2 m_3\ell_2(x+\ell_3^*)\cos\psi-2\dot{\phi}\dot{\psi}[m_3(x+\ell_3^*)^2+J_{23}-J_{21}+J_{33}- \\ -J_{31}]\sin\psi\cos\psi+2\dot{\phi}\dot{x}m_3(x+\ell_3^*)\cos^2\psi-2\dot{\psi}\dot{x}m_3\ell_2\sin\psi \\ 2\dot{\psi}\dot{x}m_3(x+\ell_3^*)+m_3g(x+\ell_3^*)\cos\psi+\dot{\phi}^2[m_3(x+\ell_3^*)^2+J_{23}-J_{21}+ \\ +J_{33}-J_{31}]\sin\psi\cos\psi \\ -\dot{\phi}^2 m_3(x+\ell_3^*)\cos^2\psi-\dot{\psi}^2 m_3(x+\ell_3^*)+m_3g\sin\psi \end{bmatrix}$$

$$P = H(q)\ddot{q} + h(q, \dot{q})$$

7. Differential equations of motion

$$P_1^M = [m_2\ell_2^{*2}+m_3(\ell_2^2+(x+\ell_3^*)^2\cos^2\psi)+J_{13}+(J_{21}+J_{31})\sin^2\psi+$$
$$+ (J_{23}+J_{33})\cos^2\psi]\ddot{\phi}-[m_3\ell_2(x+\ell_3^*)\sin\psi]\ddot{\psi}+m_3\ell_2\cos\psi\ddot{x}-$$
$$- \dot{\psi}^2 m_3\ell_2(x+\ell_3^*)\cos\psi-2\dot{\phi}\dot{\psi}[m_3(x+\ell_3^*)^2+J_{23}-J_{21}+J_{33}-J_{31}]\sin\psi\cos\psi+$$
$$+ 2\dot{\phi}\dot{x}m_3(x+\ell_3^*)\cos^2\psi-2\dot{\psi}\dot{x}m_3\ell_2\sin\psi$$

$$P_2^M = -m_3\ell_2(x+\ell_3^*)\sin\psi\ddot{\phi}+[m_3(x+\ell_3^*)^2+J_{22}+J_{32}]\ddot{\psi}+$$

$$+\ \dot{\phi}^2[m_3(x+\ell_3^*)^2+J_{23}-J_{21}+J_{33}-J_{31}]\sin\psi\cos\psi+2\dot{\psi}\dot{x}m_3(x+\ell_3^*)+$$

$$+\ m_3g(x+\ell_3^*)\cos\psi$$

$$P_3^F\ =\ m_3\ell_2\cos\psi\ddot{\phi}+m_3\ddot{x}-\dot{\phi}^2m_3(x+\ell_3^*)\cos^2\psi-\dot{\psi}^2m_3(x+\ell_3^*)+m_3g\sin\psi$$

Appendix 8
Dynamics of "ASEA" Mechanism (Basic Configuration)

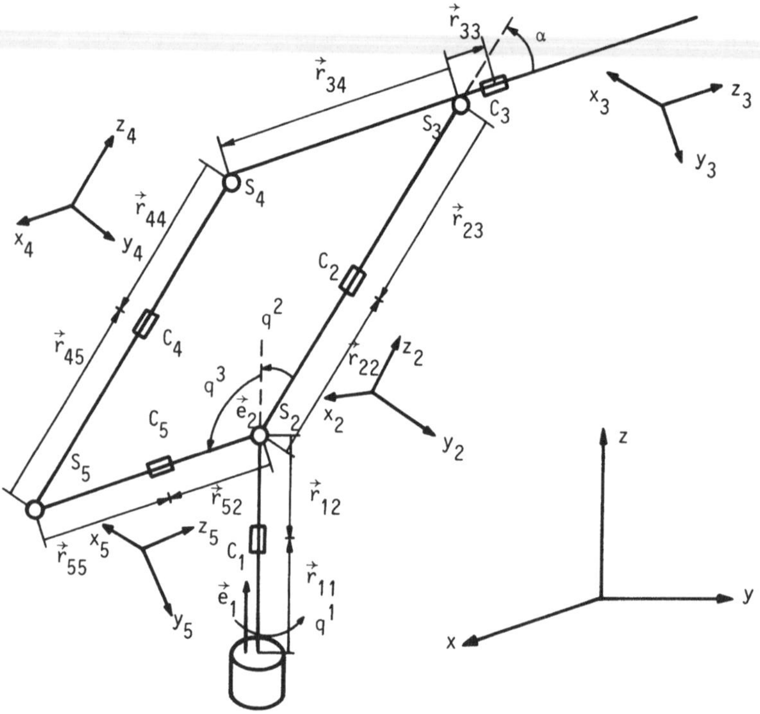

Fig. A.8.1. "ASEA" mechanism configuration

Remarks: - scheme in Fig. A.8.1. is presented in simplified form, but masses and moments of inertia due to movable actuators (shoulder and elbow joint drives) are included,

- only basic (characteristic) robot configuration is considered,

- due to the practical convenience the position vectors $\vec{r}_{34}$ and $\vec{r}_{33}$ in Fig. A.8.1 were adopted different from the conventional case.

The manipulator in Fig. A.8.1. has a configuration with a closed kinematic chain in the form of parallelogram. Because of kinematic constraints introduced by the parallelogram, the manipulator has three degrees of freedom. The powered degrees of freedom are:

- rotation of the base segment around the vertical axis,
- rotation of one segment of the mechanical parallelogram around axis $\vec{e}_2$,
- rotation of the other segment of the mechanical parallelogram around axis $\vec{e}_2$.

Let it be supposed that all segments are of such form, that in the initial position ($q^i=0$) the local coordinate systems either coincide with the absolute system (transformation matrix I) or the transformation matrix is - I. For the links having the property of cane this is satisfied and also for some other types, having the property of axial symmetry. The generalized coordinates are defined as presented in Fig. A.8.1.

For the adopted initial position of mechanism ($q^i=0$, $i=1,2,3$) the positions of the individual local coordinate systems are the following: the systems connected to the segments 1, 2, and 4, coincide with the absolute coordinate system and the systems, connected to segments 3 and 5 are obtained by multiplying the axes of the absolute coordinate system by the matrix -I, i.e. the axes coincide, but are oppositely oriented.

In the initial position, the coordinate system connected to segment 1 coincides with the absolute coordinate system.

KINEMATICS

Segment 1

Let joint S_1 be considered. Vector of joint axis S_1, $\vec{e}_1$ has the coordinates:

$$\vec{e}_1 = (0,\ 0,\ 1)^T$$

Since in the case of S_1 we have the specificity on lower side, instead of $\vec{\tilde{r}}_{11}$ we use $\vec{\tilde{r}}^{\,*}_{11} \perp \vec{\tilde{r}}_{11}$.

$$\vec{r}^{\,*}_{11} = (0,\ r^{*}_{11},\ 0)^T$$

Normalized vector $\vec{a}_1 = \dfrac{-\vec{e}_1\times(\vec{r}^{\,*}_{11}\times\vec{e}_1)}{|\vec{e}_1\times(\vec{r}^{\,*}_{11}\times\vec{e}_1)|} = (0,\ 1,\ 0)^T$

$$\vec{b}_1 = \vec{e}_1\times\vec{a}_1 = (-1,\ 0,\ 0)^T$$

$$\vec{\tilde{a}}_1 = \frac{\vec{\tilde{e}}_1\times(\vec{\tilde{r}}^{\,*}_{11}\times\vec{\tilde{e}}_1)}{|\vec{\tilde{e}}_1\times(\vec{\tilde{r}}^{\,*}_{11}\times\vec{\tilde{e}}_1)|} = (0,\ 1,\ 0)^T$$

$$\vec{\tilde{b}}_1 = \vec{\tilde{e}}_1\times\vec{\tilde{a}}_1 = (-1,\ 0,\ 0)^T$$

Transformation matrix for $q^1=0$ is determined from relation (2.2):

$$Q_1^o = [\vec{e}_1,\ \vec{a}_1,\ \vec{b}_1][\vec{\tilde{e}}_1,\ \vec{\tilde{a}}_1,\ \vec{\tilde{b}}_1]^T$$

$$Q_1^o = \begin{bmatrix} 0 & 0 & -1 \\ 0 & 1 & 0 \\ 1 & 0 & 0 \end{bmatrix} \begin{bmatrix} 0 & 0 & 1 \\ 0 & 1 & 0 \\ -1 & 0 & 0 \end{bmatrix} = \begin{bmatrix} 1 & 0 & 0 \\ 0 & 1 & 0 \\ 0 & 0 & 1 \end{bmatrix}$$

Applying the Rodrigues formula (2.4), $\bar{\xi}_1=1$, after a rotation for angle q^1 about $\vec{e}_1$, the transformation matrix Q_1^o becomes:

$$Q_1 = [\vec{q}_{11},\ \vec{q}_{12},\ \vec{q}_{23}] \text{ where:}$$

$$\vec{q}_{11} = (1,\ 0,\ 0)^T\cdot\cos q^1 + (1-\cos q^1)(0,\ 0,\ 1)\cdot(1,\ 0,\ 0)\vec{e}_1 + \vec{e}_1\times(1,\ 0,\ 0)^T\sin q^1$$

$$\vec{q}_{11} = (\cos q^1,\ \sin q^1,\ 0)^T$$

$$\vec{q}_{12} = (-\sin q^1,\ \cos q^1,\ 0)^T,$$

$$\vec{q}_{13} = (0,\ 0,\ 1)^T$$

so that, finally, the transformation matrix becomes:

$$Q_1 = \begin{bmatrix} \cos q^1 & -\sin q^1 & 0 \\ \sin q^1 & \cos q^1 & 0 \\ 0 & 0 & 1 \end{bmatrix}$$

Velocities and accelerations of the center of mass of segment 1 (relations (2.6) are as follows:

$$\vec{\omega}_1 = \dot{q}^1 \vec{e}_1 = (0,\ 0,\ \dot{q}^1)^T, \qquad \vec{v}_1 = 0$$

$$\vec{\varepsilon}_1 = \ddot{q}^1 \vec{e}_1 = (0,\ 0,\ \ddot{q}^1)^T, \qquad \vec{w}_1 = 0$$

Segment 2

Let $\vec{e}_2$ represent the axis vector of joint S_2. Evidently, the same axis vector is common to joints S_3, S_4, and S_5

$$\tilde{\vec{e}}_2 = (1,\ 0,\ 0)^T, \qquad \tilde{\vec{a}}_2 = \frac{\tilde{\vec{e}}_2 \times (\tilde{\vec{r}}_{22} \times \tilde{\vec{e}}_2)}{|\tilde{\vec{e}}_2 \times (\tilde{\vec{r}}_{22} \times \tilde{\vec{e}}_2)|} = (0,\ 0,\ 1)^T$$

$$\vec{e}_2 = Q_1 \tilde{\vec{e}}_2 = \begin{bmatrix} \cos q^1 & -\sin q^1 & 0 \\ \sin q^1 & \cos q^1 & 0 \\ 0 & 0 & 1 \end{bmatrix} \begin{bmatrix} 1 \\ 0 \\ 0 \end{bmatrix} = \begin{bmatrix} \cos q^1 \\ \sin q^1 \\ 0 \end{bmatrix}$$

$$\vec{r}_{12} = (0,\ 0,\ -r_{12})^T$$

$$\vec{a}_2 = \frac{-\vec{e}_2 \times (\vec{r}_{12} \times \vec{e}_2)}{|\vec{e}_2 \times (\vec{r}_{12} \times \vec{e}_2)|} = (0,\ 0,\ 1)^T$$

$$\tilde{\vec{b}}_2 = \tilde{\vec{e}}_2 \times \tilde{\vec{a}}_2 = (0,\ -1,\ 0)^T, \qquad \vec{b}_2 = \vec{e}_2 \times \vec{a}_2 = (\sin q^1,\ -\cos q^1,\ 0)^T$$

$$Q_2^0 = [\vec{e}_2,\ \vec{a}_2,\ \vec{b}_2][\tilde{\vec{e}}_2,\ \tilde{\vec{a}}_2,\ \tilde{\vec{b}}_2]^T = \begin{bmatrix} \cos q^1 & 0 & \sin q^1 \\ \sin q^1 & 0 & -\cos q^1 \\ 0 & 1 & 0 \end{bmatrix} \begin{bmatrix} 1 & 0 & 0 \\ 0 & 0 & 1 \\ 0 & -1 & 0 \end{bmatrix}$$

$$= \begin{bmatrix} \cos q^1 & -\sin q^1 & 0 \\ \sin q^1 & \cos q^1 & 0 \\ 0 & 0 & 1 \end{bmatrix}$$

A rotation for angle q^2 of segment 2, produces the transformation matrix of the form (applying the Rodrigues' formula):

$$Q_2 = [\vec{q}_{21},\ \vec{q}_{22},\ \vec{q}_{23}], \text{ where:}$$

$$\vec{q}_{21} = (\cos q^1,\ \sin q^1,\ 0)^T \cos q^2 + (\cos q^1,\ \sin q^1,\ 0)^2 \cdot$$

$$\cdot\ (\cos q^1,\ \sin q^1,\ 0)^T (1-\cos q^2)$$

$$\vec{q}_{21} = (\cos q^1,\ \sin q^1,\ 0)^T$$

$$\vec{q}_{22} = (-\sin q^1,\ \cos q^1,\ 0)^T \cos q^2 + (\cos q^1,\ \sin q^1,\ 0)^T \times$$

$$\times\ (-\sin q^1,\ \cos q^1,\ 0)^T \cdot \sin q^2$$

$$\vec{q}_{22} = (-\sin q^1 \cos q^2,\ \cos q^1 \cos q^2,\ \sin q^2)^T$$

$$\vec{q}_{23} = (0,\ 0,\ 1)^T \cos q^2 + (\cos q^1,\ \sin q^1,\ 0)^T \times (0,\ 0,\ 1)^T \sin q^2$$

$$\vec{q}_{23} = (\sin q^1 \sin q^2,\ -\cos q^1 \sin q^2,\ \cos q^2)^T$$

$$Q_2 = \begin{bmatrix} \cos q^1 & -\sin q^1 \cos q^2 & \sin q^1 \sin q^2 \\ \sin q^1 & \cos q^1 \cos q^2 & -\cos q^1 \sin q^2 \\ 0 & \sin q^2 & \cos q^2 \end{bmatrix}$$

Velocities and accelerations of centre of mass of segment 2 (relations (2.6)) are as follows:

$$\vec{\omega}_2 = \vec{\omega}_1 + \dot{q}^2 \vec{e}_2 = (\dot{q}^2 \cos q^1,\ \dot{q}^2 \sin q^1,\ \dot{q}^1)^T,$$

$$\vec{\tilde{r}}_{22} = (0,\ 0,\ 1)^T r_{22}$$

$$\vec{r}_{22} = Q_2 \cdot \vec{\tilde{r}}_{22} = (\sin q^1 \sin q^2,\ -\cos q^1 \sin q^2,\ \cos q^2)^T r_{22}$$

$$\vec{v}_2 = \vec{\omega}_2 \times \vec{r}_{22} = r_{22} (\dot{q}^1 \cos q^1 \sin q^2 + \dot{q}^2 \sin q^1 \cos q^2,\ \dot{q}^1 \sin q^1 \sin q^2 -$$

$$-\ \dot{q}^2 \cos q^1 \cos q^2,\ -\dot{q}^2 \sin q^2)^T$$

$$\vec{\varepsilon}_2 = \vec{\varepsilon}_1 + \ddot{q}^1 \vec{e}_2 + \dot{q}^2 \vec{\omega}_1 \times \vec{e}_2$$

$$\vec{\varepsilon}_2 = (0,\ 0,\ \ddot{q}^1)^T + (\ddot{q}^2 \cos q^1,\ \ddot{q}^2 \sin q^1,\ 0)^T +$$

$$+\ (-\dot{q}^1 \dot{q}^2 \sin q^1,\ \dot{q}^1 \dot{q}^2 \cos q^1,\ 0)^T$$

$$\vec{\varepsilon}_2 = (\ddot{q}^2 \cos q^1 - \dot{q}^1 \dot{q}^2 \sin q^1,\ \ddot{q}^2 \sin q^1 + \dot{q}^1 \dot{q}^2 \cos q^1,\ \ddot{q}^1)^T$$

$$\vec{w}_2 = \vec{\varepsilon}_2 \times \vec{r}_{22} + \vec{\omega}_2 \times (\vec{\omega}_2 \times \vec{r}_{22}) = \vec{\varepsilon}_2 \times \vec{r}_{22} + \vec{\omega}_2 (\vec{\omega}_2 \cdot \vec{r}_{22}) - \vec{r}_{22} (\vec{\omega}_2 \cdot \vec{\omega}_2)$$

$$\vec{w}_2 = (\ddot{q}^2\sin q^1\cos q^2+\dot{q}^1\dot{q}^2\cos q^1\cos q^2+\ddot{q}^1\cos q^1\sin q^2,\ \ddot{q}^1\sin q^1\sin q^2 -$$

$$- \ddot{q}^2\cos q^1\cos q^2+\dot{q}^1\dot{q}^2\sin q^1\cos q^2,\ -\ddot{q}^2\sin q^2)^T r_{22} +$$

$$+ (\dot{q}^1\dot{q}^2\cos q^1\cos q^2,\ \dot{q}^2\dot{q}^1\sin q^1\cos q^2,\ (\dot{q}^1)^2\cos q^2)^T r_{22} -$$

$$- (\sin q^1\sin q^2,\ -\cos q^1\sin q^2,\ \cos q^2)^T r_{22}((\dot{q}^1)^2+(\dot{q}^2)^2)$$

$$\vec{w}_2 = r_{22}(\ddot{q}^2\sin q^1\cos q^2+2\dot{q}^1\dot{q}^2\cos q^1\cos q^2+\ddot{q}^1\cos q^1\sin q^2 -$$

$$- ((\dot{q}^1)^2+(\dot{q}^2)^2)\sin q^1\sin q^2,\ \ddot{q}^1\sin q^1\sin q^2-\ddot{q}^2\cos q^1\cos q^2 +$$

$$+ 2\dot{q}^1\dot{q}^2\sin q^1\cos q^2+((\dot{q}^1)^2+(\dot{q}^2)^2)\cos q^1\sin q^2,\ -\ddot{q}^2\sin q^2 -$$

$$- (\dot{q}^2)^2\cos q^2)^T$$

Segment 3

A slight difficulty arises at segment 3; Namely, rotation angle α is not a generalized coordinate but it can be expressed as $\alpha = \pi - q^2 - q^3$. Therefore, α will be used forthwith though it will be replaced in the final expressions, by its alternate form given above.

From the preceding text, the axis vector of joint S_3 in $\vec{e}_2$. It is further necessary to evaluate vectors $\vec{a}_3$ and $\vec{b}_3$;

$$\vec{a}_3 = \frac{-\vec{e}_2\times(\vec{r}_{23}\times\vec{e}_2)}{|\vec{e}_2\times(\vec{r}_{23}\times\vec{e}_2)|}, \qquad \vec{\vec{r}}_{23} = (0,\ 0,\ -1)^T r_{23}$$

$$\vec{r}_{23} = Q_2\cdot\vec{\vec{r}}_{23} = (-\sin q^1\sin q^2,\ \cos q^1\sin q^2,\ -\cos q^2)^T r_{23}$$

$$\vec{a}_3 = (\sin q^1\sin q^2,\ -\cos q^1\sin q^2,\ \cos q^2)^T$$

$$\vec{\vec{a}}_3 = \frac{\vec{\vec{e}}_2\times(\vec{\vec{r}}_{33}\times\vec{\vec{e}}_2)}{|\vec{\vec{e}}_2\times(\vec{\vec{r}}_{33}\times\vec{\vec{e}}_2)|} = (0,\ 0,\ 1)^T,\ \vec{\vec{b}}_3 = \vec{\vec{e}}_2\times\vec{\vec{a}}_3 = (0,\ -1,\ 0)^T$$

$$\vec{b}_3 = \vec{e}_2\times\vec{a}_3 = (\sin q^1\cos q^2,\ -\cos q^1\cos q^2,\ -\sin q^2)^T$$

Transformation matrix Q_3^o is as follows:

$$Q_3^o = \begin{bmatrix} \cos q^1 & \sin q^1 \sin q^2 & \sin q^1 \cos q^2 \\ \sin q^1 & -\cos q^1 \sin q^2 & -\cos q^1 \cos q^2 \\ 0 & \cos q^2 & -\sin q^2 \end{bmatrix} \begin{bmatrix} 1 & 0 & 0 \\ 0 & 0 & 1 \\ 0 & -1 & 0 \end{bmatrix} =$$

$$= \begin{bmatrix} \cos q^1 & -\sin q^1 \cos q^2 & \sin q^1 \sin q^2 \\ \sin q^1 & \cos q^1 \cos q^2 & -\cos q^1 \sin q^2 \\ 0 & \sin q^2 & \cos q^2 \end{bmatrix}$$

Matrix Q_3^o is clearly equal to matrix Q_2, which is natural considering their respective definitions. It is worth noticing that the position of the coordinate system relating to segment 3 is equal to that relating to segment 2 if the latter was, instead of by q^2, rotated by an angle $q^2+\alpha=\pi-q^3$. Thus, the transformation matrix Q_3 will be obtained by replacing q^2 in the elements of matrix Q_2 by $\pi-q^3$. Therefore:

$$Q_3 = \begin{bmatrix} \cos q^1 & \sin q^1 \cos q^3 & \sin q^1 \sin q^3 \\ \sin q^1 & -\cos q^1 \cos q^3 & -\cos q^1 \sin q^3 \\ 0 & \sin q^3 & -\cos q^3 \end{bmatrix}$$

Velocities and accelerations of the center of mass of segment 3 are;

$$\vec{\omega}_3 = \vec{\omega}_2 + \dot{\alpha}\vec{e}_2 = (-\dot{q}^3 \cos q^1,\ -\dot{q}^3 \sin q^1,\ \dot{q}^1)^T$$

$$\vec{r}_{33} = (\sin q^1 \sin q^3,\ -\cos q^1 \sin q^3,\ -\cos q^3)^T r_{33}$$

$$\vec{v}_3 = (r_{22}+r_{23})(\dot{q}^1 \cos q^1 \sin q^2 + \dot{q}^2 \sin q^1 \cos q^2,\ \dot{q}^1 \sin q^1 \sin q^2 -$$
$$- \dot{q}^2 \cos q^1 \cos q^2,\ -\dot{q}^2 \sin q^2) + r_{33}(\dot{q}^3 \sin q^1 \cos q^3 +$$
$$+ \dot{q}^1 \cos q^1 \sin q^3,\ -\dot{q}^3 \cos q^1 \cos q^3 + \dot{q}^1 \sin q^1 \sin q^3,\ \dot{q}^3 \sin q^3)^T$$

$$\vec{\varepsilon}_3 = \vec{\varepsilon}_2 + \ddot{\alpha}\vec{e}_2 + \dot{\alpha}\vec{\omega}_2 \times \vec{e}_2 = \vec{\varepsilon}_1 + \ddot{q}^2 \vec{e}_2 + \dot{q}^2 \vec{\omega}_1 \times \vec{e}_2 + (-\ddot{q}^2 - \ddot{q}^3)\vec{e}_2 -$$
$$- (\dot{q}^2 + \dot{q}^3)(\vec{\omega}_1 + \dot{q}^2 \vec{e}_2) \times \vec{e}_2$$

$$\vec{\varepsilon}_3 = \vec{\varepsilon}_1 - \ddot{q}^3 \vec{e}_2 - \dot{q}^3 \vec{\omega}_1 \times \vec{e}_2$$

$$\vec{\varepsilon}_3 = (-\ddot{q}^3 \cos q^1 + \dot{q}^1 \dot{q}^3 \sin q^1,\ -\ddot{q}^3 \sin q^1 - \dot{q}^1 \dot{q}^3 \cos q^1,\ \ddot{q}^1)^T$$

$$\vec{w}_3 = \vec{w}_2 - \vec{\varepsilon}_2 \times \vec{r}_{23} - \vec{\omega}_2(\vec{\omega}_2 \cdot \vec{r}_{23}) + \vec{r}_{23}(\vec{\omega}_2 \cdot \vec{\omega}_2) + \vec{\varepsilon}_3 \times \vec{r}_{33} + \vec{\omega}_3(\vec{\omega}_3 \cdot \vec{r}_{33}) - \vec{r}_{33}(\vec{\omega}_3 \cdot \vec{\omega}_3)$$

$$\vec{w}_3 = (r_{22}+r_{23})(\ddot{q}^2\sin q^1\cos q^2+2\dot{q}^1\dot{q}^2\cos q^1\cos q^2+\ddot{q}^1\cos q^1\sin q^2 - ((\dot{q}^1)^2+(\dot{q}^2)^2)\sin q^1\sin q^2,\ \ddot{q}^1\sin q^1\sin q^2-\ddot{q}^2\cos q^1\cos q^2 + 2\dot{q}^1\dot{q}^2\sin q^1\cos q^2+((\dot{q}^1)^2+(\dot{q}^2)^2)\cos q^1\sin q^2,\ -\ddot{q}^2\sin q^2 - (\dot{q}^2)^2\cos q^2)^T+r_{33}(\ddot{q}^3\sin q^1\cos q^3+2\dot{q}^1\dot{q}^3\cos q^1\cos q^3 + \ddot{q}^1\cos q^1\sin q^3-((\dot{q}^1)^2+(\dot{q}^3)^2)\sin q^1\sin q^3,\ -\ddot{q}^3\cos q^1\cos q^3 + 2\dot{q}^1\dot{q}^3\sin q^1\cos q^3+\ddot{q}^1\sin q^1\sin q^3+((\dot{q}^1)^2+(\dot{q}^3)^2)\cos q^1\sin q^3,\ \ddot{q}^3\sin q^3+(\dot{q}^3)^2\cos q^3)^T$$

Segment 5

Since the segments 5 and 3 of the considered mechanism are parallel, it follows:

$$Q_5 = Q_3, \qquad \vec{\varepsilon}_5 = \vec{\varepsilon}_3, \qquad \vec{\omega}_5 = \vec{\omega}_3$$

If we notice that $\vec{r}_{52}||\vec{r}_{33}$ and $\vec{\omega}_5 = \vec{\omega}_3$, it is obvious that $\vec{\omega}_3\times\vec{r}_{33}$ (previously calculated) is parallel to $\vec{\omega}_5\times\vec{r}_{52}$, in other words,

$$\vec{\omega}_5\times\vec{r}_{52} = -\frac{|\vec{r}_{52}|}{|\vec{r}_{33}|}(\vec{\omega}_3\times\vec{r}_{33}).$$

Similar conclusion can be deduced for accelerations $\vec{w}_5$ and $\vec{w}_3$, because $\vec{r}_{52}$ and $\vec{r}_{33}$ are parallel and $\vec{\varepsilon}_3$ equal to $\vec{\varepsilon}_5$, both terms in the expression for $\vec{w}_5$ are already calculated and it is therefore only required, to replace $\vec{r}_{33}$ by $-\vec{r}_{52}$.

Therefore, the linear velocity and acceleration are:

$$\vec{v}_5 = \vec{\omega}_5\times\vec{r}_{52} = -r_{52}(\dot{q}^3\sin q^1\cos q^3+\dot{q}^1\cos q^1\sin q^3,\ -\dot{q}^3\cos q^1\cos q^3+\dot{q}^1\sin q^1\sin q^3,\ \dot{q}^3\sin q^3)^T$$

$$\vec{w}_5 = \vec{\varepsilon}_5\times\vec{r}_{52}+\vec{\omega}_5\times(\vec{\omega}_5\times\vec{r}_{52}) = -r_{52}(\ddot{q}^3\sin q^1\cos q^3+2\dot{q}^1\dot{q}^3\cos q^1\cos q^3 + \ddot{q}^1\cos q^1\sin q^3-((\dot{q}^1)^2+(\dot{q}^3)^2)\sin q^1\sin q^3,\ -\ddot{q}^3\cos q^1\cos q^3 + 2\dot{q}^1\dot{q}^3\sin q^1\cos q^3+\ddot{q}^1\sin q^1\sin q^3+((\dot{q}^1)^2+(\dot{q}^3)^2)\cos q^1\sin q^3,\ \ddot{q}^3\sin q^3+(\dot{q}^3)^2\cos q^3)^T$$

Segment 4

Because segments 2 and 4 are parallel, it follows:

$$Q_4 = Q_2, \qquad \vec{\varepsilon}_4 = \vec{\varepsilon}_2, \qquad \vec{\omega}_4 = \vec{\omega}_2$$

The fact that $(\vec{r}_{52}-\vec{r}_{55})||\vec{r}_{52}$, $\vec{r}_{45}||\vec{r}_{22}$, $\vec{\omega}_4=\vec{\omega}_2$ and $\vec{\varepsilon}_4=\vec{\varepsilon}_2$, here is utilized and the procedure for segment 5 repeated.

Linear velocity and acceleration are:

$$\vec{v}_4 = \vec{v}_5-\vec{\omega}_5\times\vec{r}_{55}+\vec{\omega}_4\times\vec{r}_{45} = \vec{\omega}_5\times(\vec{r}_{52}-\vec{r}_{55})+\vec{\omega}_2\times\vec{r}_{22} \cdot \frac{|\vec{r}_{45}|}{|\vec{r}_{22}|}$$

$$\vec{v}_4 = -(\dot{q}^3\sin q^1\cos q^3+\dot{q}^1\cos q^1\sin q^3,\ -\dot{q}^3\cos q^1\cos q^3 +$$
$$+\ \dot{q}^1\sin q^1\sin q^3,\ \dot{q}^3\sin q^3)^T(r_{52}+r_{55})+r_{45}(\dot{q}^1\cos q^1\sin q^2 +$$
$$+\ \dot{q}^2\sin q^1\cos q^2,\ \dot{q}^1\sin q^1\sin q^2-\dot{q}^2\cos q^1\cos q^2,\ -\dot{q}^2\sin q^2)^T$$

$$\vec{w}_4 = \vec{w}_5-\vec{\varepsilon}_5\times\vec{r}_{55}-\vec{\omega}_5\times(\vec{\omega}_5\times\vec{r}_{55})+\vec{\varepsilon}_4\times\vec{r}_{45}+\vec{\omega}_4\times(\vec{\omega}_4\times\vec{r}_{45})$$

$$\vec{w}_4 = \vec{\varepsilon}_5\times(\vec{r}_{52}-\vec{r}_{55})+\vec{\omega}_5\times(\vec{\omega}_5\times(\vec{r}_{52}-\vec{r}_{55}))+[\vec{\varepsilon}_2\times\vec{r}_{22}+\vec{\omega}_2\times(\vec{\omega}_2\times\vec{r}_{22})]\cdot\frac{|\vec{r}_{45}|}{|\vec{r}_{22}|}$$

$$\vec{w}_4 = -(r_{52}+r_{55})(\ddot{q}^3\sin q^1\cos q^3+2\dot{q}^1\dot{q}^3\cos q^1\cos q^3+\ddot{q}^1\cos q^1\sin q^3 -$$
$$-\ ((\dot{q}^1)^2+(\dot{q}^3)^2)\sin q^1\sin q^3,\ -\ddot{q}^3\cos q^1\cos q^3+2\dot{q}^1\dot{q}^3\sin q^1\cos q^3 +$$
$$+\ \ddot{q}^1\sin q^1\sin q^3+((\dot{q}^1)^2+(\dot{q}^3)^2)\cos q^1\sin q^3,\ \ddot{q}^3\sin q^3+(\dot{q}^3)^2\cos q^3)^T+$$
$$+\ r_{45}(\ddot{q}^2\sin q^1\cos q^2+2\dot{q}^1\dot{q}^2\cos q^1\cos q^2+\ddot{q}^1\cos q^1\sin q^2 -$$
$$-\ ((\dot{q}^1)^2+(\dot{q}^2)^2)\sin q^1\sin q^2,\ \ddot{q}^1\sin q^1\sin q^2-\ddot{q}^2\cos q^1\cos q^2 +$$
$$+\ 2\dot{q}^1\dot{q}^2\sin q^1\cos q^2+((\dot{q}^1)^2+(\dot{q}^2)^2)\cos q^1\sin q^2,$$
$$-\ddot{q}^2\sin q^2-(\dot{q}^2)^2\cos q^2)^T$$

DYNAMICS

The peculiarity of this manipulator is in that it contains a closed kinematic chain. In order to apply the procedure for open kinematic structures, the manipulator is fictitiously decoupled into two open kinematic chains, whereby the mutual reaction forces acting between the chains are treated as external forces (Fig. A.8.2).

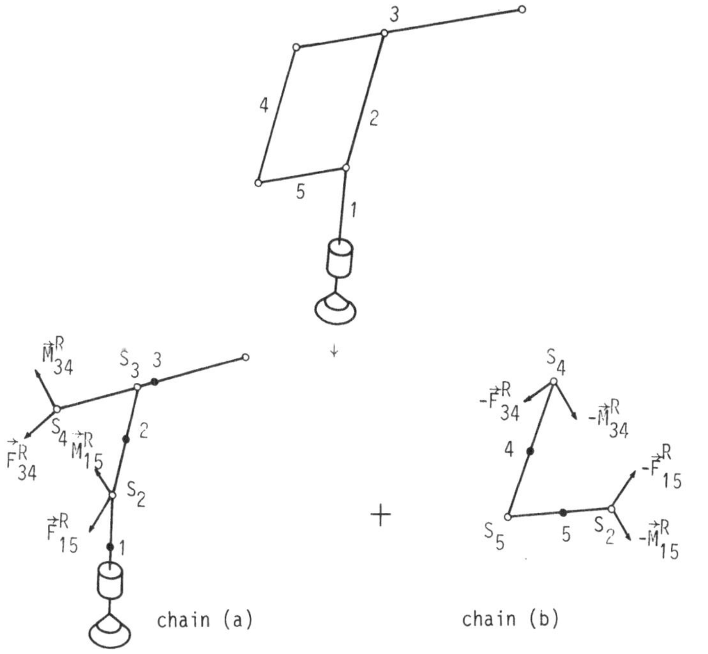

Fig. A.8.2. Decomposition of the "ASEA" mechanism

Chain (a)

If we fictitiously break the chain (a) at joint S_3 and write the equation of moments balance with respect to the point of break, the next relation holds:

$$m_3\vec{r}_{33}\times(\vec{w}_3-\vec{g}) + \vec{M}_3 = \vec{r}_{34}\times\vec{F}^R_{34}+\vec{M}^R_{34}+\vec{M}^R_3 \tag{A.8.1}$$

where $\vec{M}^R_3$ represents the reaction moment in joint S_3, and $\vec{M}_3$ denotes the moment of inertial forces of the third segment.

Scalar multiplication of (A.8.1) and $\vec{e}_2$ gives:

$$(m_3\vec{r}_{33}\times(\vec{w}_3-\vec{g})+\vec{M}_3)\cdot\vec{e}_2 = \vec{e}_2\cdot(\vec{r}_{34}\times\vec{F}^R_{34}) \tag{A.8.2}$$

Repeating the procedure for joint S_2:

$$(m_2\vec{r}_{22}\times(\vec{w}_2-\vec{g})+\vec{M}_2+m_3(\vec{r}_{22}-\vec{r}_{23}+\vec{r}_{33})\times(\vec{w}_3-\vec{g})+\vec{M}_3)\cdot\vec{e}_2 =$$
$$=\vec{e}_2\cdot((\vec{r}_{22}-\vec{r}_{23}+\vec{r}_{34})\times\vec{F}^R_{34})+P_2 \tag{A.8.3}$$

Chain (b)

If we fictitiously break the chain (b) at joint S_5 and write the equation of moments balance with respect to the point of break:

$$(m_4\vec{r}_{45}\times(\vec{w}_4-\vec{g})+\vec{M}_4)\cdot\vec{e}_2 = -\vec{e}_2\cdot((\vec{r}_{45}-\vec{r}_{44})\times\vec{F}^R_{34}) \tag{A.8.4}$$

The balanced moment equation with respect to the point S_2 is:

$$(m_5\vec{r}_{52}\times(\vec{w}_5-\vec{g})+\vec{M}_5+m_4(\vec{r}_{52}-\vec{r}_{55}+\vec{r}_{45})\times(\vec{w}_4-\vec{g})+\vec{M}_4)\cdot\vec{e}_2 =$$
$$-\vec{e}_2\cdot((\vec{r}_{52}-\vec{r}_{55}+\vec{r}_{45}-\vec{r}_{44})\times\vec{F}^R_{34})+P_3 \tag{A.8.5}$$

If we notice from Fig. A.8.1. that $\vec{r}_{22}-\vec{r}_{23}+\vec{r}_{34}=\vec{r}_{52}-\vec{r}_{55}+\vec{r}_{45}-\vec{r}_{44}$, and add equations (A.8.3) and (A.8.5), we obtain:

$$(m_2\vec{r}_{22}\times(\vec{w}_2-\vec{g})+\vec{M}_2+m_3(\vec{r}_{22}-\vec{r}_{23}+\vec{r}_{33})\times(\vec{w}_3-\vec{g})+\vec{M}_3+m_5\vec{r}_{52}\times(\vec{w}_5-\vec{g})+$$
$$+\vec{M}_5+m_4(\vec{r}_{52}-\vec{r}_{55}+\vec{r}_{45})(\vec{w}_4-\vec{g})+\vec{M}_4)\cdot\vec{e}_2=P_2+P_3 \tag{A.8.6}$$

If (A.8.2) and (A.8.5) are added and (A.8.4) subtracted from the sum, the right hand side of the obtained equation will, contain the product of $(\vec{r}_{34}-\vec{r}_{52}+\vec{r}_{55}-\vec{r}_{45}+\vec{r}_{44}+\vec{r}_{45}-\vec{r}_{44})\equiv 0$ and $\vec{F}^R_{34}$. Therefore we get:

$$(m_3\vec{r}_{33}\times(\vec{w}_3-\vec{g})+\vec{M}_3-m_4\vec{r}_{45}\times(\vec{w}_4-\vec{g})-\vec{M}_4+m_5\vec{r}_{52}\times(\vec{w}_5-\vec{g})+\vec{M}_5+$$
$$+m_4(\vec{r}_{52}-\vec{r}_{55}+\vec{r}_{45})\times(\vec{w}_4-\vec{g})+\vec{M}_4)\cdot\vec{e}_2 = P_3 \tag{A.8.7}$$

or, in another form:

$$(m_3\vec{r}_{33}\times(\vec{w}_3-\vec{g})+\vec{M}_3+m_5\vec{r}_{52}\times(\vec{w}_5-\vec{g})+\vec{M}_5+m_4(\vec{r}_{52}-\vec{r}_{55})\times(\vec{w}_4-\vec{g}))\cdot\vec{e}_2=P_3 \tag{A.8.8}$$

Now we can join the chains (a) and (b) and introduce a fictitious break at joint S_1. The equation of moments balance at this mechanism point of break is:

$$(\vec{M}_1+m_2(\vec{r}_{11}-\vec{r}_{12}+\vec{r}_{22})\times(\vec{w}_2-\vec{g})+\vec{M}_2+m_3(\vec{r}_{11}-\vec{r}_{12}+\vec{r}_{22}-\vec{r}_{23}+\vec{r}_{33})\times$$
$$\times(\vec{w}_3-\vec{g})+\vec{M}_3+m_4(\vec{r}_{11}-\vec{r}_{12}+\vec{r}_{52}-\vec{r}_{55}+\vec{r}_{45})\times \tag{A.8.9}$$
$$\times(\vec{w}_4-\vec{g})+\vec{M}_4+m_5(\vec{r}_{11}-\vec{r}_{12}+\vec{r}_{52})\times(\vec{w}_5-\vec{g})+\vec{M}_5)\cdot\vec{e}_1=P_1$$

Equations (A.8.6), (A.8.8) and (A.8.9) represent the required equations of motions of the manipulator in vector form.

The remaining quantities to be determined within equations (A.8.6), (A.8.8) and (A.8.9) are $\vec{M}_i$, the moments of the inertial forces resulting from the motion about respective center of mass. Applying equations (2.14), (2.15), (2.16) gives the required moments.

For segment (1):

$$T_1 = \begin{bmatrix} J_{11}\cos^2 q^1 + J_{12}\sin^2 q^1 & (J_{11}-J_{12})\sin q^1 \cos q^1 & 0 \\ (J_{11}-J_{12})\sin q^1 \cos q^1 & J_{11}\sin^2 q^1 + J_{12}\cos^2 q^1 & 0 \\ 0 & 0 & J_{13} \end{bmatrix}$$

Directly from Eulers' equations (2.11) it is evident that, because of the form of $\vec{\omega}_1$ the term $\vec{\lambda}_1 = 0$. Therefore,

$$\vec{M}_1 = -T_1\vec{\varepsilon}_1 = -(0,\ 0,\ J_{13}\ddot{q}^1)^T$$

For segment (2):

$$T_2 = \begin{bmatrix} J_{21}\cos^2 q^1 + J_{22}\sin^2 q^1 \cos^2 q^2 + J_{23}\sin^2 q^1 \sin^2 q^2,\ (J_{21}-J_{22}\cos^2 q^2 - \\ -J_{23}\sin^2 q^2)\sin q^1 \cos q^1,\ (J_{23}-J_{22})\sin q^1 \sin q^2 \cos q^2 \\ \\ (J_{21}-J_{22}\cos^2 q^2 - J_{23}\sin^2 q^2)\sin q^1 \cos q^1,\ J_{21}\sin^2 q^1 + (J_{22}\cos^2 q^2 + \\ +J_{23}\sin^2 q^2)\cos^2 q^1,\ (J_{22}-J_{23})\cos q^1 \sin q^2 \cos q^2 \\ \\ (J_{23}-J_{22})\sin q^1 \sin q^2 \cos q^2,\ (J_{22}-J_{23})\cos q^1 \sin q^2 \cos q^2, \\ J_{22}\sin^2 q^2 + J_{23}\cos^2 q^2 \end{bmatrix}$$

First term of the expression (2.15) is $-T_2\vec{\varepsilon}_2 =$

$$= \begin{bmatrix} (J_{21}\cos^2 q^1 + J_{22}\sin^2 q^1 \cos^2 q^2 + J_{23}\sin^2 q^1 \sin^2 q^2)(\ddot{q}^2 \cos q^1 - \dot{q}^1 \dot{q}^2 \sin q^1) + \\ +(J_{21}-J_{22}\cos^2 q^2 - J_{23}\sin^2 q^2)\sin q^1 \cos q^1 (\ddot{q}^2 \sin q^1 + \\ +\dot{q}^1 \dot{q}^2 \cos q^1) + \ddot{q}^1 (J_{23}-J_{22})\sin q^1 \sin q^2 \cos q^2 \end{bmatrix}$$

$$= \begin{bmatrix} (J_{21}-J_{22}\cos^2 q^2-J_{23}\sin^2 q^2)\sin q^1\cos q^1\cdot(\ddot{q}^2\cos q^1-\dot{q}^1\dot{q}^2\sin q^1)+ \\ \quad +[J_{21}\sin^2 q^1+(J_{22}\cos^2 q^2+J_{23}\sin^2 q^2)\cos^2 q^1](\ddot{q}^2\sin q^1+ \\ \quad +\dot{q}^1\dot{q}^2\cos q^1)+\ddot{q}^1(J_{22}-J_{23})\cos q^1\sin q^2\cos q^2 \\ \\ (J_{23}-J_{22})\sin q^1\sin q^2\cos q^2(\ddot{q}^2\cos q^1-\dot{q}^1\dot{q}^2\sin q^1)+ \\ \quad +(J_{22}-J_{23})\cos q^1\sin q^2\cos q^2(\ddot{q}^2\sin q^1+\dot{q}^1\dot{q}^2\cos q^1)+ \\ \quad +\ddot{q}^1(J^{22}\sin^2 q^2+J_{23}\cos^2 q^2) \end{bmatrix}$$

Calculating the second term of expression (2.15), $\vec{\lambda}_2$:

$$\vec{\lambda}_2 = Q_2 \underbrace{\begin{bmatrix} (\vec{\omega}_2\cdot\vec{q}^{22})(\vec{\omega}_2\cdot\vec{q}^{23})(J_{22}-J_{23}) \\ (\vec{\omega}_2\cdot\vec{q}^{21})(\vec{\omega}_2\cdot\vec{q}^{23})(J_{23}-J_{21}) \\ (\vec{\omega}_2\cdot\vec{q}^{21})(\vec{\omega}_2\cdot\vec{q}^{22})(J_{21}-J_{22}) \end{bmatrix}}_{\vec{r}}$$

The vector which multiplies Q_2 is:

$$\vec{r} = \begin{bmatrix} (\dot{q}^1)^2\sin q^2\cos q^2(J_{22}-J_{23}) \\ \dot{q}^2\dot{q}^1\cos q^2(J_{23}-J_{21}) \\ \dot{q}^2\dot{q}^1\sin q^2(J_{21}-J_{22}) \end{bmatrix}$$

$$\vec{\lambda}_2 = \begin{bmatrix} (\dot{q}^1)^2(J_{22}-J_{23})\cos q^1\sin q^2\cos q^2-\dot{q}^2\dot{q}^1(J_{23}\cos^2 q^2\sin q^1+ \\ \quad +J_{22}\sin^2 q^2\sin q^1-J_{21}\sin q^1) \\ \\ (\dot{q}^1)^2(J_{22}-J_{23})\sin q^1\sin q^2\cos q^2+\dot{q}^1\dot{q}^2(J_{23}\cos^2 q^2\cos q^1+ \\ \quad +J_{22}\sin^2 q^2\cos q^1-J_{21}\cos q^1) \\ \\ \dot{q}^1\dot{q}^2(J_{23}-J_{22})\sin q^2\cos q^2 \end{bmatrix}$$

Therefore, after rearranging, vector $\vec{M}_2$ can be represented as:

$$\vec{M}_2 = \begin{bmatrix} \ddot{q}^1(J_{23}-J_{22})\sin q^1 \sin q^2 \cos q^2 + \ddot{q}^2 J_{21}\cos q^1 + (\dot{q}^1)^2 (J_{22}- \\ -J_{23})\cos q^1 \sin q^2 \cos q^2 + \dot{q}^1\dot{q}^2 (J_{21}-J_{22}-J_{23})\sin q^1 \\ \ddot{q}^1(J_{22}-J_{23})\cos q^1 \sin q^2 \cos q^2 + \ddot{q}^2 J_{21}\sin q^1 + (q^1)^2 (J_{22}- \\ -J_{23})\sin q^1 \sin q^2 \cos q^2 - \dot{q}^1\dot{q}^2 (J_{21}-J_{22}-J_{23})\cos q^1 \\ \ddot{q}^1 (J_{22}\sin^2 q^2 + J_{23}\cos^2 q^2) \end{bmatrix}$$

With reference to discussions relating to the evaluation of matrix Q_3, similar reasoning can be adopted here since $\vec{M}_3$ depends on $\vec{\omega}_3$, Q_3, $\vec{\varepsilon}_3$, all of which are obtained from $\vec{\omega}_2$, Q_2, $\vec{\varepsilon}_2$ with the substitution of q^2 by $\pi-q^3$. Therefore, $\vec{M}_3$ can be obtained from the expression for $\vec{M}_2$ by applying the mentioned substitution bearing in mind that in this case J_{2i} becomes J_{3i}

$$\cos(\Pi-q^3) = -\cos q^3, \qquad \sin(\Pi-q^3) = \sin q^3$$

$$\vec{M}_3 = \begin{bmatrix} -\ddot{q}^1(J_{33}-J_{32})\sin q^1 \sin q^3 \cos q^3 - \ddot{q}^3 J_{31}\cos q^1 - (\dot{q}^1)^2 (J_{32}- \\ -J_{33})\cos q^1 \sin q^3 \cos q^3 - \dot{q}^1\dot{q}^3 (J_{31}-J_{32}-J_{33})\sin q^1 \\ -\ddot{q}^1(J_{32}-J_{33})\cos q^1 \sin q^3 \cos q^3 - \ddot{q}^3 J_{31}\sin q^1 - (\dot{q}^1)^2 (J_{32}- \\ -J_{33})\sin q^1 \sin q^3 \cos q^3 + \dot{q}^1\dot{q}^3 (J_{31}-J_{32}-J_{33})\cos q^1 \\ \ddot{q}^1 (J_{32}\sin^2 q^3 + J_{33}\cos^2 q^3 \end{bmatrix}$$

Vectors $\vec{M}_4$ and $\vec{M}_5$ will not be evaluated, because the expression for $\vec{M}_4$ is equal to that for $\vec{M}_2$ (since $\vec{\omega}_4=\vec{\omega}_2$, $\vec{\varepsilon}_4=\vec{\varepsilon}_2$, $Q_4=Q_2$), with the exception that inertia tensor J_2 be replaced by J_4, in other words, J_{2i} should be replaced by J_{4i} $(i=1,2,3)$. The same is valid for $\vec{M}_3$ and $\vec{M}_5$.

SOURCE FILE FOR THE COMPUTATION OF

"ASEA" MECHANISM DRIVING TORQUES

```
C-----------------------------------------------------------------------
C       PROGRAM : ASEA
C-----------------------------------------------------------------------

C       FUNCTION : SUBROUTINE FOR CALCULATION OF KINEMATICS AND DYNAMICS
C                  OF ASEA-MECHANISM , CLOSED LOOP MECHANISM WITH FIVE
C                  REVOLUTE JOINTS
C-----------------------------------------------------------------------
C       INPUT VARIABLES :
C                  Rij     DISTANCE BETWEEN J-TH JOINT AND THE CENTER OF
C                          MASS OF THE I-TH LINK
C                  Qi      GENERALIZED COORDINATE AT THE I-TH JOINT
C                  QiD     GENERALIZED VELOCITY AT THE I-TH JOINT
C                  QiDD    GENERALIZED ACCELERATION AT THE I-TH JOINT
C                  Mi      MASS OF THE I-TH LINK
C             Ji1,Ji2,Ji3  VECTOR OF MOMENT OF INERTIA OF THE I-TH LINK
C-----------------------------------------------------------------------
C       OUTPUT VARIABLES :
C                  Pi      DRIVING TORQUE AT I-TH JOINT
C-----------------------------------------------------------------------
        SUBROUTINE ASEA
        REAL MM1X,MM1Y,MM1Z,MM2X,MM2Y,MM2Z,MM3X,MM3Y,MM3Z,MM4X,MM4Y,MM4Z
        REAL MM5X,MM5Y,MM5Z,M1,M2,M3,M4,M5,J11,J12,J13,J21,J22,J23
        REAL J31,J32,J33,J41,J42,J43,J51,J52,J53
        COMMON/DIMENSIONS/R11,R12,R22,R23,R33,R34,R44,R45,R55,R52
        COMMON/STATE/Q1,Q1D,Q1DD,Q2,Q2D,Q2DD,Q3,Q3D,Q3DD
        COMMON/DYNAMIC/M1,M2,M3,M4,M5,J11,J12,J13,J21,J22,J23,J31,J32,J33
     *,J41,J42,J43,J51,J52,J53
C-----------------------------------------------------------------------
C    Calculating of kinematics
C-----------------------------------------------------------------------
        OPEN(UNIT=1,FILE='REZ.DAT',STATUS='OLD')
        S1=SIN(Q1)
        S2=SIN(Q2)
        S3=SIN(Q3)
        C1=COS(Q1)
        C2=COS(Q2)
        C3=COS(Q3)
C-----------------------------------------------------------------------

C    link 1
C-----------------------------------------------------------------------
        E1X=Ø.
        E1Y=Ø.
        E1Z=1.
        R11X=Ø.
        R11Y=Ø.
        R11Z=R11
        R12X=Ø.
        R12Y=Ø.
        R12Z=-R1
        OM1X=Ø.
        OM1Y=Ø.
        OM1Z=Q1D
        V1X=Ø.
        V1Y=Ø.
        V1Z=Ø.
```

```
      EP1X=Ø.
      EP1Y=Ø.
      EP1Z=Q1DD
      W1X=Ø.
      W1Y=Ø.
      W1Z=Ø.
C-----------------------------------------------------------------------
C    link 2
C-----------------------------------------------------------------------
      E2X=C1
      E2Y=S1
      E2Z=Ø.
      R22X=S1*S2*R22
      R22Y=-C1*S2*R22
      R22Z=C2*R22
      R23X=-S1*S2*R23
      R23Y=C1*S2*R23
      R23Z=-C2*R23
      OM2X=Q2D*C1
      OM2Y=Q2D*S1
      OM2Z=Q1D
      V2X=R22*(Q1D*C1*S2+Q2D*S1*C2)
      V2Y=R22*(Q1D*S1*S2-Q2D*C1*C2)
      V2Z=R22*(-Q2D*S2)
      EP2X=Q2DD*C1-Q1D*Q2D*S1
      EP2Y=Q2DD*S1+Q1D*Q2D*C1
      EP2Z=Q1DD
      W2X=R22*(Q2DD*S1*C2+2.*Q1D*Q2D*C1*C2+
     *Q1DD*C1*S2-(Q1D**2+Q2D**2)*S1*S2)
      W2Y=R22*(Q1DD*S1*S2-Q2DD*C1*C2+
     *2.*Q1D*Q2D*S1*C2+(Q1D**2+Q2D**2)*C1*S2)
      W2Z=R22*(-Q2DD*S2-Q2D**2*C2)
C-----------------------------------------------------------------------
C    link 3
C-----------------------------------------------------------------------
      R33X=R33*S1*S3
      R33Y=R33*(-C1*S3)
      R33Z=R33*(-C3)
      OM3X=-Q3D*C1
      OM3Y=-Q3D*S1
      OM3Z=Q1D
      V3X=(R22+R23)*(Q1D*C1*S2+Q2D*S1*C2)+
     *R33*(Q3D*S1*C3+Q1D*C1*S3)
      V3Y=(R22+R23)*(Q1D*S1*S2-Q2D*C1*C2)+
     *R33*(Q1D*S1*S3-Q3D*C1*C3)
      V3Z=(R22+R23)*(-Q2D*S2)+R33*Q3D*S3
      EP3X=-Q3DD*C1+Q1D*Q3D*S1
      EP3Y=-Q3DD*S1-Q1D*Q3D*C1
      EP3Z=Q1DD
      W3X=(R22+R23)*(Q2DD*S1*C2+2.*Q1D*Q2D*C1*C2+
     *Q1DD*C1*S2-(Q1D**2+Q2D**2)*S1*S2)+
     *R33*(Q3DD*S1*C3+2.*Q1D*Q3D*C1*C3+
     *Q1DD*C1*S3-(Q1D**2+Q3D**2)*S1*S3)
      W3Y=(R22+R23)*(Q1DD*S1*S2-Q2DD*C1*C2+
     *2.*Q1D*Q2D*S1*C2+(Q1D**2+Q2D**2)*C1*S2)+
```

```
     *R33*(-Q3DD*C1*C3+2.*Q1D*Q3D*S1*C3+
     *Q1DD*S1*S3+(Q1D**2+Q3D**2)*C1*S3)
      W3Z=(R22+R23)*(-Q2DD*S2-Q2D**2*C2)+R33*(Q3DD*S3+Q3D**2*C3)
C-----------------------------------------------------------------
C    link 5
C-----------------------------------------------------------------
      R52X=R52*(-S1*S3)
      R52Y=R52*C1*S3
      R52Z=R52*C3
      R55X=R55*S1*S3
      R55Y=R55*(-C1*S3)
      R55Z=R55*(-C3)
      OM5X=OM3X
      OM5Y=OM3Y
      OM5Z=OM3Z
      V5X=-R52*(Q3D*S1*C3+Q1D*C1*S3)
      V5Y=-R52*(Q1D*S1*S3-Q3D*C1*C3)
      V5Z=-R52*Q3D*S3
      EP5X=EP3X
      EP5Y=EP3Y
      EP5Z=EP3Z
      W5X=-R52*(Q3DD*S1*C3+2.*Q1D*Q3D*C1*C3+
     *Q1DD*C1*S3-(Q1D**2+Q3D**2)*S1*S3)
      W5Y=-R52*(-Q3DD*C1*C3+2.*Q1D*Q3D*S1*C3+
     *Q1DD*S1*S3+(Q1D**2+Q3D**2)*C1*S3)
      W5Z=-R52*(Q3DD*S3+Q3D**2*C3)
C-----------------------------------------------------------------
C    link 4
C-----------------------------------------------------------------
      R45X=R45*S1*S2
      R45Y=R45*(-C1*S2)
      R45Z=R45*C2
      OM4X=OM2X
      OM4Y=OM2Y
      OM4Z=OM2Z
      V4X=-(R52+R55)*(Q3D*S1*C3+Q1D*C1*S3)+
     *R45*(Q1D*C1*S2+Q2D*S1*C2)
      V4Y=-(R52+R55)*(-Q3D*C1*C3+Q1D*S1*S3)+
     *R45*(Q1D*S1*S2-Q2D*C1*C2)
      V4Z=-(R52+R55)*Q3D*S3-R45*Q2D*S2
      EP4X=EP2X
      EP4Y=EP2Y
      EP4Z=EP2Z
      W4X=-(R52+R55)*(Q3DD*S1*C3+2.*Q1D*Q3D*C1*C3+Q1DD*C1*S3-
     *(Q1D**2+Q3D**2)*S1*S3)+R45*(Q2DD*S1*C2+Q1DD*C1*S2+
     *2.*Q1D*Q2D*C1*C2-(Q1D**2+Q2D**2)*S1*S2)
      W4Y=-(R52+R55)*(-Q3DD*C1*C3+2.*Q1D*Q3D*S1*C3+Q1DD*S1*S3+
     *(Q1D**2+Q3D**2)*C1*S3)+R45*(Q1DD*S1*S2-Q2DD*C1*C2+
     *2.*Q1D*Q2D*S1*C2+(Q1D**2+Q2D**2)*C1*S2)
      W4Z=-(R52+R55)*(Q3DD*S3+Q3D**2*C3)+R45*(-Q2DD*S2-Q2D**2*C2)
C-----------------------------------------------------------------
C    Calculation of dynamics
C-----------------------------------------------------------------
      MM1X=Ø.
      MM1Y=Ø.
```

```
      MM1Z=-J13*Q1DD
      MM2X=Q1DD*(J23-J22)*S1*S2*C2+Q2DD*J21*C1+Q1D**2*(J22-J23)*C1*S2*
     *C2+Q1D*Q2D*(J21-J22-J23)*S2
      MM2Y=Q1DD*(J22-J23)*C1*S2*C2+Q2DD*J21*S1+Q1D**2*(J22-J23)*S1*S2*
     *C2-Q1D*Q2D*(J21-J22-J23)*C1
      MM2Z=Q1DD*(J22*S2**2+J23*C2**2)
      MM3X=-Q1DD*(J33-J32)*S1*S3*C3-Q3DD*J31*C1-Q1D**2*(J32-J33)*
     *C1*S3*C3-Q1D*Q3D*(J31-J32-J33)*S1
      MM3Y=-Q1DD*(J33-J32)*C1*S3*C3-Q3DD*J31*S1-Q1D**2*(J32-J33)*
     *S1*S3*C3+Q1D*Q3D*(J31-J32-J33)*C1
      MM3Z=Q1DD*(J32*S3**2+J33*C3**2)
      MM4X=Q1DD*(J43-J42)*S1*S2*C2+Q2DD*J21*C1+Q1D**2*(J42-J43)*C1*S2*
     *C2+Q1D*Q2D*(J41-J42-J43)*S2
      MM4Y=Q1DD*(J42-J43)*C1*S2*C2+Q2DD*J21*S1+Q1D**2*(J42-J43)*S1*S2*
     *C2-Q1D*Q2D*(J41-J42-J43)*C1
      MM4Z=Q1DD*(J42*S2**2+J43*C2**2)
      MM5X=-Q1DD*(J53-J52)*S1*S3*C3-Q3DD*J31*C1-Q1D**2*(J52-J53)*
     *C1*S3*C3-Q1D*Q3D*(J51-J52-J53)*S1
      MM5Y=-Q1DD*(J53-J52)*C1*S3*C3-Q3DD*J31*S1-Q1D**2*(J52-J53)*
     *S1*S3*C3+Q1D*Q3D*(J51-J52-J53)*C1
      MM5Z=Q1DD*(J52*S3**2+J53*C3**2)
      GX=Ø.
      GZ=9.81
      GY=Ø.
      P1=E1X*(MM1X+MM2X+MM3X+MM4X+MM5X+M2*((R11Y-R12Y+R22Y)*(W2Z-GZ)-
     *(R11Z-R12Z+R22Z)*(W2Y-GY))+M3*((R11Y-R12Y+R22Y-R23Y+R33Y)*
     *(W3Z-GZ)-(R11Z-R12Z+R22Z-R23Z+R33Z)*(W3Y-GY))+M4*((R11Y-R12Y+
     *R52Y-R55Y+R45Y)*(W4Z-GZ)-(R11Z-R12Z+R52Z-R55Z+R45Z)*(W4Y-GY))+
     *M5*((R11Y-R12Y+R52Y)*(W5Z-GZ)-(R11Z-R12Z+R52Z)*(W5Y-GY)))
     *+E1Y*(MM1Y+MM2Y+MM3Y+MM4Y+MM5Y+M2*(-(R11X-R12X+R22X)*(W2Z-GZ)+
     *(R11Z-R12Z+R22Z)*(W2X-GX))+M3*(-(R11X-R12X+R22X-R23X+R33X)*
     *(W3Z-GZ)+(R11Z-R12Z+R22Z-R23Z+R33Z)*(W3X-GX))+M4*(-(R11X-R12X+
     *R52X-R55X+R45X)*(W4Z-GZ)+(R11Z-R12Z+R52Z-R55Z+R45Z)*(W4X-GX))+
     *M5*(-(R11X-R12X+R52X)*(W5Z-GZ)+(R11Z-R12Z+R52Z)*(W5X-GX)))+
     *E1Z*(MM1Z+MM2Z+MM3Z+MM4Z+MM5Z+M2*(-(R11Y-R12Y+R22Y)*(W2X-GX)+
     *(R11X-R12X+R22X)*(W2Y-GY))+M3*(-(R11Y-R12Y+R22Y-R23Y+R33Y)*
     *(W3X-GX)+(R11X-R12X+R22X-R23X+R33X)*(W3Y-GY))+M4*(-(R11Y-R12Y+
     *R52Y-R55Y+R45Y)*(W4X-GX)+(R11X-R12X+R52X-R55X+R45X)*(W4Y-GY))+
     *M5*(-(R11Y-R12Y+R52Y)*(W5X-GX)+(R11X-R12X+R52X)*(W5Y-GY)))
      P3=E2X*(MM3X+MM5X+M3*(R33Y*(W3Z-GZ)-R33Z*(W3Y-GY))+M5*(R52Y*
     *(W5Z-GZ)-R52Z*(W5Y-GY))+M4*((R52Y-R55Y)*(W4Z-GZ)-
     *(R52Z-R55Z)*(W4Y-GY)))+
     *E2Y*(MM3Y+MM5Y+M3*(-R33X*(W3Z-GZ)+R33Z*(W3X-GX))+M5*(-R52X*
     *(W5Z-GZ)+R52Z*(W5X-GX))+M4*(-(R52X-R55X)*(W4Z-GZ)+
     *(R52Z-R55Z)*(W4X-GX)))+
     *E2Z*(MM3Z+MM5Z+M3*(-R33Y*(W3X-GX)+R33X*(W3Y-GY))+M5*(-R52Y*
     *(W5X-GX)+R52X*(W5Y-GY))+M4*(-(R52Y-R55Y)*(W4X-GX)+
     *(R52X-R55X)*(W4Y-GY)))
      P2=-P3+E2X*(MM2X+MM3X+MM5X+MM4X+M2*(R22Y*(W2Z-GZ)-R22Z*(W2Y-GY))
     *+M3*((R22Y-R23Y+R33Y)*(W3Z-GZ)-(R22Z-R23Z+R33Z)*(W3Y-GY))+
     *M5*(R52Y*(W5Z-GZ)-R52Z*(W5Y-GY))+M4*((R52Y-R55Y+R45Y)*(W4Z-
     *GZ)-(R52Z-R55Z+R45Z)*(W4Y-GY)))
     *+E2Y*(MM2Y+MM3Y+MM5Y+MM4Y+M2*(-R22X*(W2Z-GZ)+R22Z*(W2X-GX))
     *+M3*(-(R22X-R23X+R33X)*(W3Z-GZ)+(R22Z-R23Z+R33Z)*(W3X-GX))+
     *M5*(-R52X*(W5Z-GZ)+R52Z*(W5X-GX))+M4*(-(R52X-R55X+R45X)*(W4Z-
```

```
     *GZ)+(R52Z-R55Z+R45Z)*(W4X-GX)))
     *+E2Z*(MM2Z+MM3Z+MM5Z+MM4Z+M2*(-R22Y*(W2X-GX)+
     *R22X*(W2Y-GY))+
     *+M3*(-(R22Y-R23Y+R33Y)*(W3X-GX)+(R22X-R23X+R33X)*(W3Y-GY))+
     *M5*(-R52Y*(W5X-GX)+R52X*(W5Y-GY))+M4*(-(R52Y-R55Y+R45Y)*(W4X-
     *GX)+(R52X-R55X+R45X)*(W4Y-GY)))
C
      WRITE(6,100)
100   FORMAT(5X,'DRIVING TORQUES AT JOINTS 1 ; 2 ; 3',/)
      WRITE(6,20)P1,P2,P3
20    FORMAT(3(E15.5))
      WRITE(1,101)
101   FORMAT(5X,'DRIVING TORQUES AT JOINTS 1 ; 2 ; 3',/)
      WRITE(1,20)P1,P2,P3
      CLOSE(UNIT=1)
      RETURN
      END
```

Appendix 9
Programme Support for Dynamics Modelling of Manipulation Robots

The program NOMDYN calculates nominal dynamics of manipulation robots based on Newton-Euler's equations, described in Paragraph 2.1. This program is written in a programming language FORTRAN-77 and it can be used on arbitrary computer system with FORTRAN compiler.

The presented algorithm provides an interactive work between the user and computer. Flow chart of main program NOMDYN is presented in Fig. A.9.1.

First, the user has to set the name of the robot for which he wants to calculate a nominal dynamics. For generating the dynamic model of manipulation robots it is necessary to form special formatted files, i.e. special data files with kinematic and dynamic parameters of the robot. All input - output data files obtain their names according to the following convention:

<NAME OF THE ROBOT>.EXT

where: NAME OF THE ROBOT is string with five characters (the user chooses the name of the robot), EXT denotes a character of the data file.

The user has to prepare the following input data files:

<NAME OF THE ROBOT>.CNF	- data about geometric, structure and kinematic of the robot,
<NAME OF THE ROBOT>.DNM	- dynamic parameters of the robot,
<NAME OF THE ROBOT>.TRA	- data about desired trajectories of the robot with respect to internal coordinate system.

As output from data file appear the following data

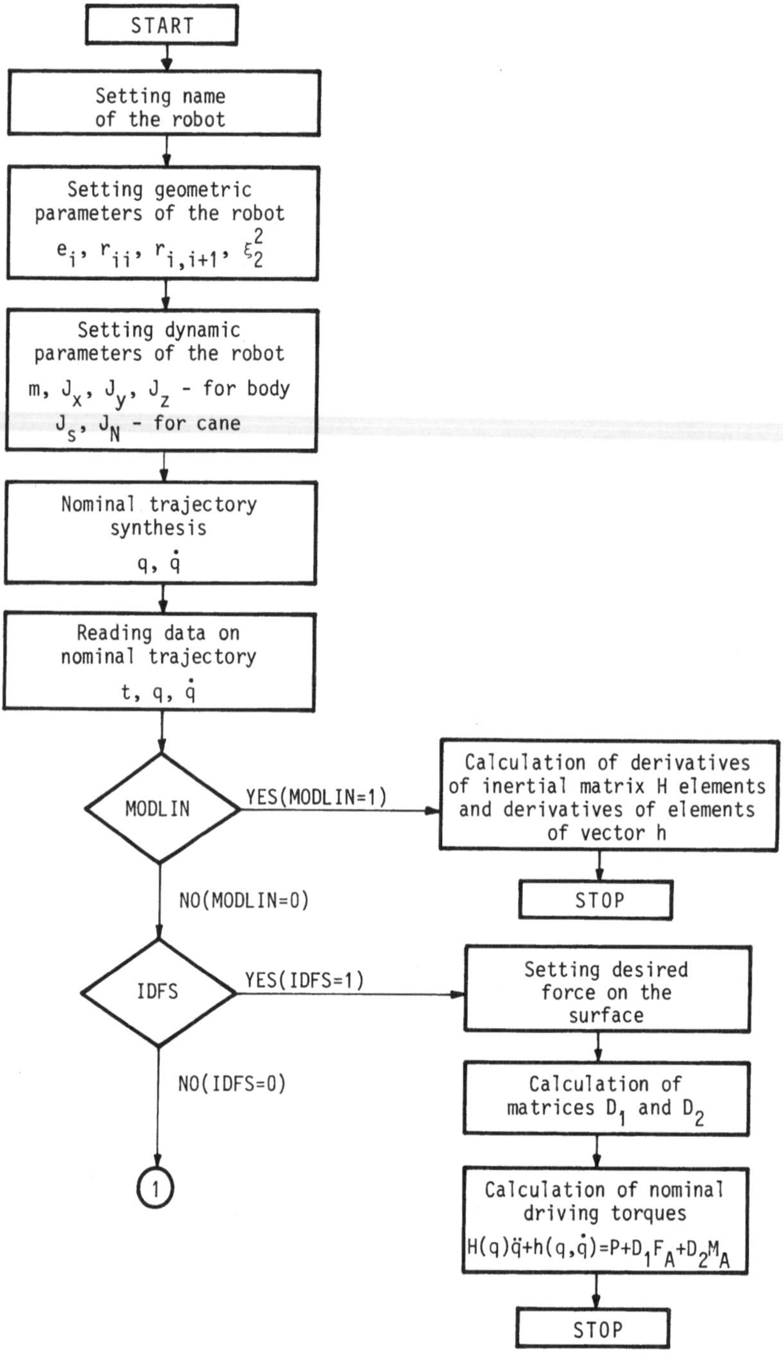
START
Setting name of the robot
Setting geometric parameters of the robot
e_i, r_{ii}, $r_{i,i+1}$, ξ_2^2
Setting dynamic parameters of the robot
m, J_x, J_y, J_z - for body
J_s, J_N - for cane
Nominal trajectory synthesis
q, $\dot{q}$
Reading data on nominal trajectory
t, q, $\dot{q}$
MODLIN
YES(MODLIN=1)
Calculation of derivatives of inertial matrix H elements and derivatives of elements of vector h
STOP
NO(MODLIN=0)
IDFS
YES(IDFS=1)
Setting desired force on the surface
Calculation of matrices D_1 and D_2
Calculation of nominal driving torques
$H(q)\ddot{q}+h(q,\dot{q})=P+D_1F_A+D_2M_A$
STOP
NO(IDFS=0)
1

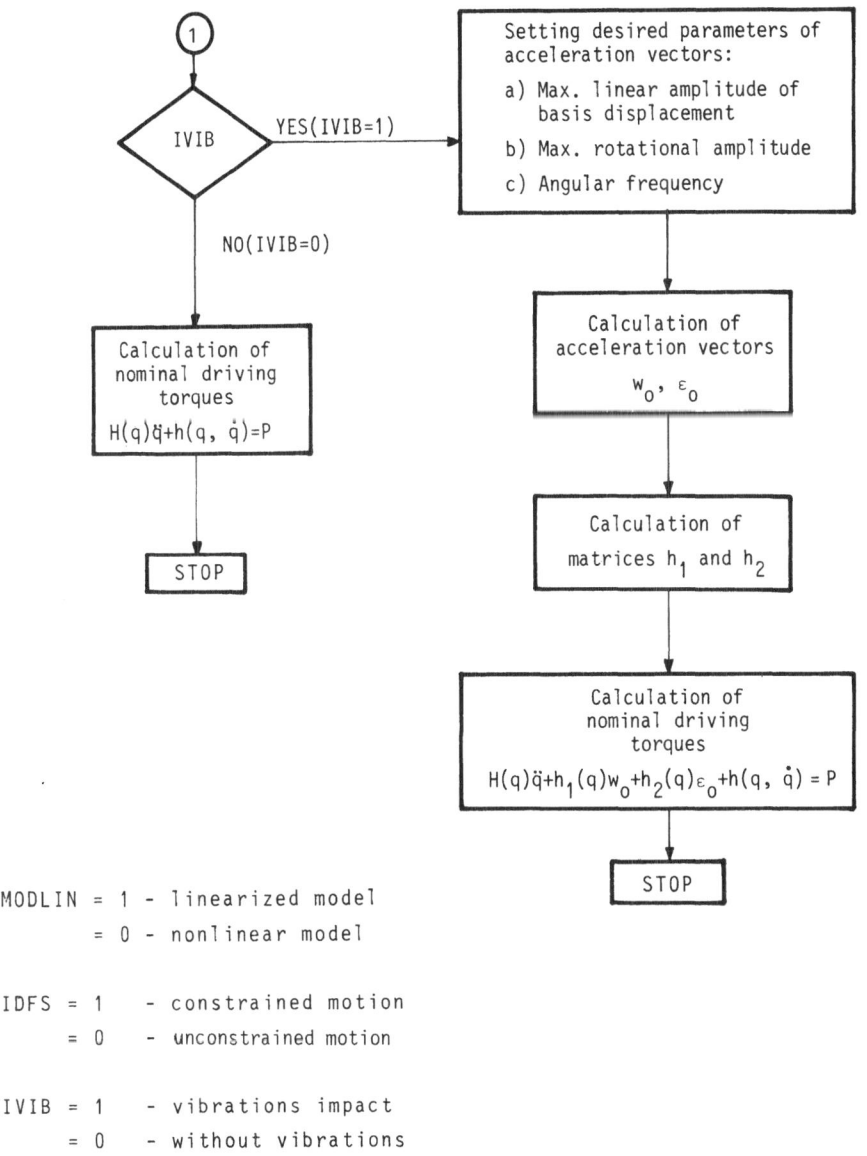

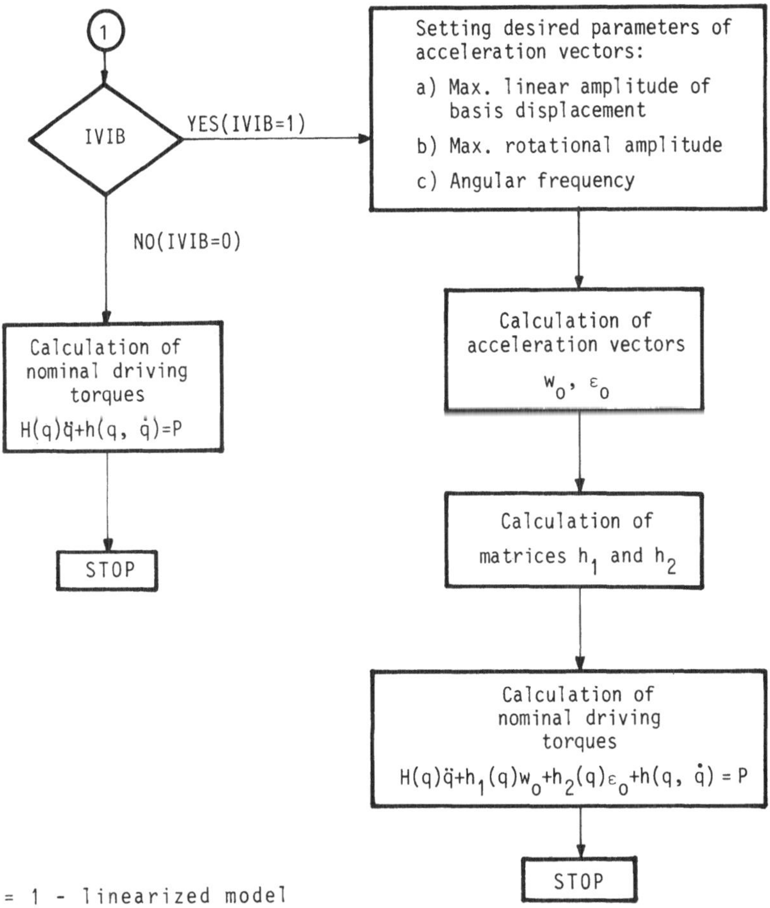

MODLIN = 1 - linearized model
       = 0 - nonlinear model

IDFS = 1 - constrained motion
     = 0 - unconstrained motion

IVIB = 1 - vibrations impact
     = 0 - without vibrations

Fig. A.9.1. Flow chart of main program NOMDYN

<NAME OF THE ROBOT>.ANG - nominal trajectories in time instants.

(The name of the robot is formed according to the previous convention).

Data file *****. CNF consists of parameters which defines a configuration of the robot. In the following text, we shall present the parameters which have to be set:

1) Number of degrees of freedom (format 1X, I1),

2) First joint type indicator (0 - rotational, 1 - linear) (format 1X, F15.6),

3) X coordinate of unit vector of first joint axis with respect to internal coordinate system connected to the first segment $\vec{e}_{11}$ (format 1X, F15.6),

4) Y coordinate of unit vector of first joint axis with respect to first internal coordinate system $\vec{e}_{11}$ (format 1X, F15.6),

5) Z coordinate of unit vector of first joint axis with respect to first internal coordinate system $\vec{e}_{11}$ (format 1X, F15.6),

6) X coordinate of unit vector of second joint axis with respect to first internal coordinate system $\vec{e}_{12}$ (format 1X, F15.6),

7) Y coordinate of unit vector of second joint axis with respect to first internal coordinate system $\vec{e}_{12}$ (format 1X, F15.6),

8) Z coordinate of unit vector of second joint axis with respect to first internal coordinate system $\vec{e}_{12}$ (format 1X, F15.6),

9) X coordinate of vector from the first joint to the gravity center of the first segment with respect to first internal coordinate system $\vec{r}_{11}$ (format 1X, F15.6),

10) Y coordinate of vector from the first joint to the gravity center of the first segment with respect to first internal coordinate system $\vec{r}_{11}$ (format 1X, F15.6),

11) Z coordinate of vector from the first joint to the gravity center of the first segment with respect to first internal coordinate system $\vec{r}_{11}$ (format 1X, F15.6),

12) X coordinate of vector from the second joint to the gravity center of the first segment with respect to first internal coordinate system $\vec{r}_{12}$ (format 1X, F15.6),

13) Y coordinate of vector from the second joint to the gravity center of the first segment with respect to first internal coordinate system $\vec{r}_{12}$ (format 1X, F15.6)

14) Z coordinate of vector from the second joint to the gravity center of the first segment with respect to first internal coordinate system $\vec{r}_{12}$ (format 1X, F15.6).

If the i-th joint is linear, then the user has to define two vectors $\vec{r}^*_{ii}$ and $\vec{r}^*_{i-1,i}$. Here, we schall breafly, explain the sence of introducing of these vectors.

When the i-th joint is linear, it follows $\vec{r}_{ii} || \vec{e}_i$ or $\vec{r}_{i-1,i} || \vec{e}_i$. If $\vec{r}_{i-1,i} || \vec{e}_i$ we call it the "specificity" of the (i-1)-th segment on the upper end.

Then we introduce a unit vector $\vec{r}^*_{i-1,i}$ perpendicular to $\vec{e}_i (\vec{r}^*_{i-1,i} \perp \vec{e}_i)$ (Fig. A.9.2a). Further, the vector $\vec{r}^*_{i-1,i}$ is used instead of $\vec{r}_{i-1,i}$ for determing the generalized coordinate q_i. If $\vec{r}_{ii} || \vec{e}_i$ we call it the "specificity" of the i-th segment on the lower end. Then we introduce a unit vector $\vec{r}^*_{ii}$ perpendicular to $\vec{e}_i (\vec{r}_{ii} \perp \vec{e}_i)$ (Fig. A.9.2b) and use it instead of $\vec{r}_{ii}$.

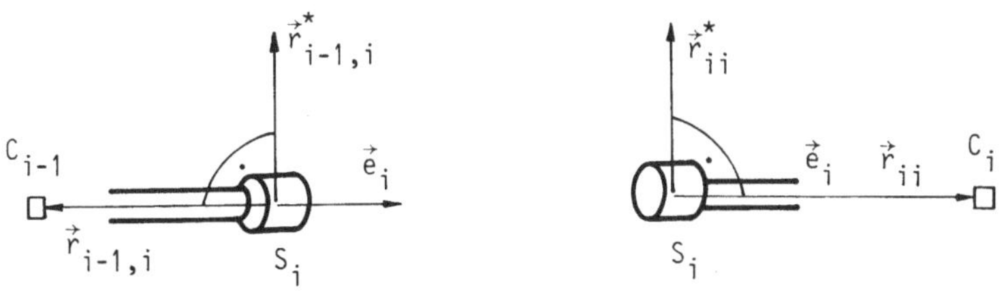

Fig. A.9.2(a) "Specificity" of (i-1)-th segment on the upper end (b) "Specificity" of i-th segment on the lower end

The definition of generalized coordinate in the case of "specificity" is shown in Fig. A.9.3.

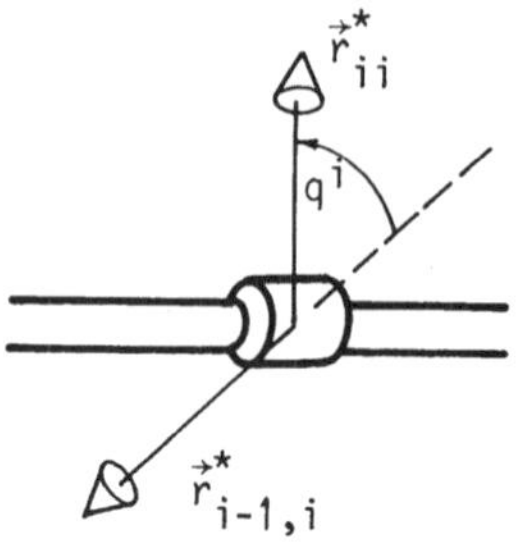

Fig. A.9.3. Definition of the generalized coordinate in the case of "specificity"

If the i-th joint is linear, then the user, as we said before, has to define two vectors r^*_{ii} and $\vec{r}^*_{i-1,i}$:

15. X coordinate of vector $\vec{r}^*_{i-1,i}$ (format 1X, F15.6),

16. Y coordinate of vector $\vec{r}^*_{i-1,i}$ (format 1X, F15.6),

17. Z coordinate of vector $\vec{r}^*_{i-1,i}$ (format 1X, F15.6),

18. X coordinate of vector $\vec{r}^*_{ii}$ (format 1X, F15.6),

19. Y coordinate of vector $\vec{r}^*_{ii}$ (format 1X, F15.6),

20. Z coordinate of vector $\vec{r}^*_{ii}$ (format 1X, F15.6)

In the same way, the user has to set parameters for each segment.

At the end of this data file, the user has to define unit vector of the first joint axis with respect to external coordinate system:

X coordinate of unit vector of the first joint axis with respect to external coordinate system (format 1X, F15.6),

Y coordinate of unit vector of the first joint axis with respect to external coordinate system (format 1X, F15.6),

Z coordinate of unit vector of first joint axis with respect to external coordinate system (format 1X, F15.6).

Data file *****. DNM consists of dynamic parameters which have to be set:

1) Type of segment (1-rod, 0-body) (format 1X, F15.6),

2) Mass of segment [kg] (format 1X, F15.6),

3) Moment of inertia J_{xx} or J_s depending on type of joint [kgm^2] (format 1X, F15.6),

4) Moment of inertia J_{yy} or J_N depending on type of joint [kgm^2] (format 1X, F15.6),

5) Moment of inertia J_{zz} depending on type of joint [kgm^2] (format 1X, F15.6).

In the same way, the user has to set parameters for each segment.

Data file *****.TRA consists of data which are used for generating robot trajectory. These data which have to be set are:

1) Q0 - initial position vector (joint coordinates vector in the initial point on the trajectory) ([m] - for prismatic joint, [rad] - for revolute joint) (format 1X, 6F10.5),

2) QF - final position vector (joint coordinates vector in the final point on the trajectory) ([m] - for prismatic joint, [rad] - for revolute joint) format 1X, 6F10.5),

3) H - integration step [s] (format 1X, 6F10.5)

4) T - duration time of the movement [s] (format 1X, 6F10.5).

Data file *****. ANG consists of calculated data about kinematics of the robot (trajectory, velocities, accelerations in particular time instants). These data can be presented as follows:

1) TE - current time [s] (format 1X, 6E13.5),

2) Q - joint coordinates vector in time instant ([m] - for prismatic joint, [rad] - for revolute joint) (format 1X, 6E13.5),

3) DQ - joint velocities vector in time instant ([m/s] - for prismatic joint, [rad/s] - for revolute joint) (form 1X, 6E13.5),

4) DDQ - joint accelerations vector in time instant ($[m/s^2]$ - for prismatic joint, $[rad/s^2]$ - for revolute joint) (format 1X, 6E13.5).

For nominal trajectory synthesis the algorithm uses the subroutine TRAJEK. This subroutine is programed on the basis of theoretical considerations presented in Paragraph 2.5 of this book.

For calculation of inertial matrix H(q) and culomn vector $h(q, \dot{q})$ is used software module MODEL. As mentioned in the begining of this Appendix, this module is based on Newton-Euler equations, described in Paragraph 2.1 of this book.

There are two options: linearized (MODLIN=1) or nonlinear (MODLIN=0) model. The linearized dynamic model of manipulation robots is based on the algorithm presented in Fig. 3.3 in Chapter 3 of this book. In this case derivatives of inertial matrix H elements and derivatives of elements of vector h are calculated.

If the user wants to work with nonlinear model, then MODLIN=0.

The variable MOD defines which of the dynamic effects are taken into account (MOD=0 - all effects are taken, MOD=1 - inertial and gravity terms are taken, MOD=2 - only gravity terms are taken, MOD=3 - calculation of Jacobian matrix). The variable ITIP defines a calculation of inertial matrix H(q) of dynamic model (ITIP=0) or a calculation of driving torques/forces (ITIP=1).

Using the program NOMDYN the user can calculate the driving torques/ /forces for previous defined manipulation task, in two cases:

1) when the manipulator gripper motion is constrained (surface type constraint),

2) in conditions of mechanical vibrations impact.

In the first case the user has to define the external force F_A which he wants to realize during motion (in general case the user has to define the external moment M_A). For calculation of matrices D_1 and D_2, joined to the external force F_A and external moment M_A, respectively, is used the subroutine MATD.

In the second case, in conditions of mechanical vibrations impact, the user has to define desired parameters of acceleration vectors as follows:

a) - Max. linear amplitude of basis displacement,

b) - Max. rotational amplitude,

c) - Angular frequency.

On the basis of these data, the algorithm calculates the translational acceleration vector w_o and angular acceleration vector ε_o. For calculation matrices h_1 and h_2 joined to these acceleration vectors, w_o and ε_o, subroutines MATH1 and MATH2, respectively, are used.

At the end of these considerations it should be said, that the presented programs and soubroutines are limited to the case of manipulator with sixth degrees of freedom.

After the test example, in the text to follow, the hierarchical structure of the task NOMDYN, the list of all programs with their functions and the listings of all subroutines of the task NOMDYN are presented.

TEST EXAMPLE

Let us consider the manipulation robot during grinding of the working object (Fig. A.9.4). During the period T_1 the manipulator moves freely, i.e. the manipulator with a cutting tool moves towards the working object. Let us assume that the contact between the gripper (cutting tool) and the working object is impactless and that the end of period T_1 is, at the same time the start of the interval T_2 (point A_1 in Fig. A.9.4).

The motion from A_1 to A_2 is imposed, i.e. the manipulator is considered as a closed chain. This is a case of the surface-type constraint, considered in Paragraph 2.4. During the motion from A_1 to A_2 the reaction force (cutting force) $\vec{F}_A$ appears.

On the basis of the cutting theory, it is known that, during grinding, two perpendicular components appear, radial F_r and tangential F_t. It is also known that their ratio is 1.5÷3. Due to simplicity, the radial force component F_r perpendicular to the surface A_1A_2O is taken into account only.

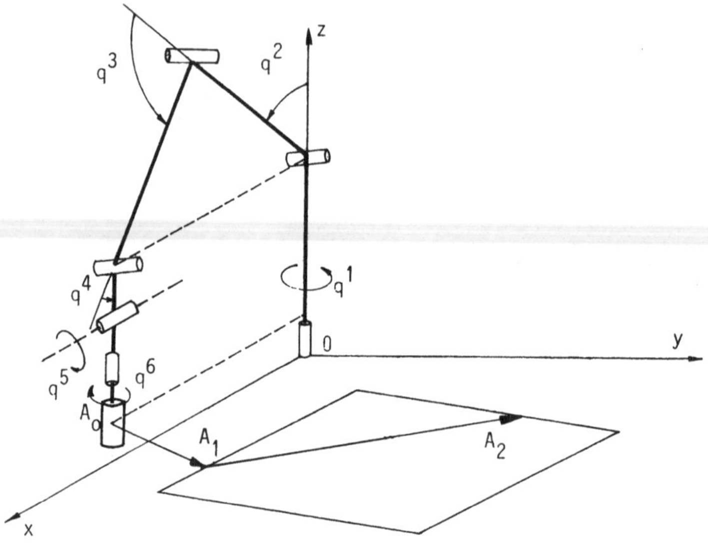

Fig. A.9.4. Manipulation robot

In data files ROBOT.CNF and ROBOT.DNM the geometric and kinematic parameters, as well as dynamic parameters of manipulator are stored, respectively.

In data file ROBOT.TRA the internal manipulator coordinates in initial time instant (point A_o), end time instant (point A_1), integration time and movement duration time are stored.

The internal manipulator coordinates for the complete movement ($A_o \rightarrow A_1 \rightarrow A_2$) are given in Table A.9.1. It is evidently that the user has to form data file ROBOT.TRA for each particular trajectory part. The presented data file relates to the movement part $A_o \rightarrow A_1$, only.

```
         ROBOT.CNF
6          ;   number of degrees of freedom
0.000000   ;    type of joint                                   1
0.000000   ;    joint unit axis- eii (x)                        1
0.000000   ;    joint unit axis- eii (y)                        1
1.000000   ;    joint unit axis- eii (z)                        1
1.000000   ;    i+1 joint unit axis- ei,i+1 (x)                 1
0.000000   ;    i+1 joint unit axis- ei,i+1 (y)                 1
0.000000   ;    i+1 joint unit axis- ei,i+1 (z)                 1
0.000000   ;    link vectors-Rii (x)                            1
0.000000   ;    link vectors- Rii (y)                           1
0.400000   ;    link vectors- Rii (z)                           1
0.000000   ;    link vectors- Ri,i+1 (x)                        1
0.000000   ;    link vectors- Ri,i+1 (y)                        1
-0.400000  ;    link vectors- Ri,i+1 (z)                        1
1.000000   ;    spec. vector for lin.j.- uii (x)                1
0.000000   ;    spec. vector for lin.j.- uii (y)                1
0.000000   ;    spec. vector for lin.j.- uii (z)                1
1.000000   ;    spec. vector for lin.j.- ui,i+1 (x)             1
0.000000   ;    spec. vector for lin.j.- ui,i+1 (y)             1
0.000000   ;    spec. vector for lin.j.- ui,i+1 (z)             1
0.000000   ;    type of joint                                   2
1.000000   ;    joint unit axis- eii (x)                        2
0.000000   ;    joint unit axis- eii (y)                        2
0.000000   ;    joint unit axis- eii (z)                        2
1.000000   ;    i+1 joint unit axis- ei,i+1 (x)                 2
0.000000   ;    i+1 joint unit axis- ei,i+1 (y)                 2
0.000000   ;    i+1 joint unit axis- ei,i+1 (z)                 2
0.000000   ;    link vectors-Rii (x)                            2
0.400000   ;    link vectors- Rii (y)                           2
0.000000   ;    link vectors- Rii (z)                           2
0.000000   ;    link vectors- Ri,i+1 (x)                        2
-0.400000  ;    link vectors- Ri,i+1 (y)                        2
0.000000   ;    link vectors- Ri,i+1 (z)                        2
1.000000   ;    spec. vector for lin.j.- uii (x)                2
0.000000   ;    spec. vector for lin.j.- uii (y)                2
0.000000   ;    spec. vector for lin.j.- uii (z)                2
1.000000   ;    spec. vector for lin.j.- ui,i+1 (x)             2
0.000000   ;    spec. vector for lin.j.- ui,i+1 (y)             2
0.000000   ;    spec. vector for lin.j.- ui,i+1 (z)             2
0.000000   ;    type of joint                                   3
1.000000   ;    joint unit axis- eii (x)                        3
0.000000   ;    joint unit axis- eii (y)                        3
0.000000   ;    joint unit axis- eii (z)                        3
1.000000   ;    i+1 joint unit axis- ei,i+1 (x)                 3
0.000000   ;    i+1 joint unit axis- ei,i+1 (y)                 3
0.000000   ;    i+1 joint unit axis- ei,i+1 (z)                 3
0.000000   ;    link vectors-Rii (x)                            3
0.000000   ;    link vectors- Rii (y)                           3
-0.400000  ;    link vectors- Rii (z)                           3
0.000000   ;    link vectors- Ri,i+1 (x)                        3
0.000000   ;    link vectors- Ri,i+1 (y)                        3
0.400000   ;    link vectors- Ri,i+1 (z)                        3
0.000000   ;    spec. vector for lin.j.- uii (x)                3
0.000000   ;    spec. vector for lin.j.- uii (y)                3
0.000000   ;    spec. vector for lin.j.- uii (z)                3
0.000000   ;    spec. vector for lin.j.- ui,i+1 (x)             3
0.000000   ;    spec. vector for lin.j.- ui,i+1 (y)             3
0.000000   ;    spec. vector for lin.j.- ui,i+1 (z)             3
0.000000   ;    type of joint                                   4
```

```
 1.000000   ;   joint unit axis- eii (x)                    4
 0.000000   ;   joint unit axis- eii (y)                    4
 0.000000   ;   joint unit axis- eii (z)                    4
 0.000000   ;   i+1 joint unit axis- ei,i+1 (x)             4
 0.000000   ;   i+1 joint unit axis- ei,i+1 (y)             4
 1.000000   ;   i+1 joint unit axis- ei,i+1 (z)             4
 0.000000   ;   link vectors-Rii (x)                        4
 0.075000   ;   link vectors- Rii (y)                       4
 0.000000   ;   link vectors- Rii (z)                       4
 0.000000   ;   link vectors- Ri,i+1 (x)                    4
-0.075000   ;   link vectors- Ri,i+1 (y)                    4
 0.000000   ;   link vectors- Ri,i+1 (z)                    4
 0.000000   ;   spec. vektor for lin.j.- uii (x)            4
 0.000000   ;   spec. vector for lin.j.- uii (y)            4
 0.000000   ;   spec. vector for lin.j.- uii (z)            4
 0.000000   ;   spec. vector for lin.j.- ui,i+1 (x)         4
 0.000000   ;   spec. vector for lin.j.- ui,i+1 (y)         4
 0.000000   ;   spec. vector for lin.j.- ui,i+1 (z)         4
 0.000000   ;   type of joint                               5
 0.000000   ;   joint unit axis- eii (x)                    5
 0.000000   ;   joint unit axis- eii (y)                    5
 1.000000   ;   joint unit axis- eii (z)                    5
 0.000000   ;   i+1 joint unit axis- ei,i+1 (x)             5
 1.000000   ;   i+1 joint unit axis- ei,i+1 (y)             5
 0.000000   ;   i+1 joint unit axis- ei,i+1 (z)             5
 0.000000   ;   link vectors-Rii (x)                        5
 0.075000   ;   link vectors- Rii (y)                       5
 0.000000   ;   link vectors- Rii (z)                       5
 0.000000   ;   link vectors- Ri,i+1 (x)                    5
-0.075000   ;   link vectors- Ri,i+1 (y)                    5
 0.000000   ;   link vectors- Ri,i+1 (z)                    5
 0.000000   ;   spec. vektor for lin.j.- uii (x)            5
 0.000000   ;   spec. vector for lin.j.- uii (y)            5
 0.000000   ;   spec. vector for lin.j.- uii (z)            5
 0.000000   ;   spec. vector for lin.j.- ui,i+1 (x)         5
 0.000000   ;   spec. vector for lin.j.- ui,i+1 (y)         5
 0.000000   ;   spec. vector for lin.j.- ui,i+1 (z)         5
 0.000000   ;   type of joint                               6
 0.000000   ;   joint unit axis- eii (x)                    6
 1.000000   ;   joint unit axis- eii (y)                    6
 0.000000   ;   joint unit axis- eii (z)                    6
 1.000000   ;   i+1 joint unit axis- ei,i+1 (x)             6
 0.000000   ;   i+1 joint unit axis- ei,i+1 (y)             6
 1.000000   ;   i+1 joint unit axis- ei,i+1 (z)             6
 0.000000   ;   link vectors-Rii (x)                        6
 0.150000   ;   link vectors- Rii (y)                       6
 0.000000   ;   link vectors- Rii (z)                       6
 0.000000   ;   link vectors- Ri,i+1 (x)                    6
-0.150000   ;   link vectors- Ri,i+1 (y)                    6
 0.000000   ;   link vectors- Ri,i+1 (z)                    6
 1.000000   ;   spec. vector for lin.j.- uii (x)            6
 0.000000   ;   spec. vector for lin.j.- uii (y)            6
 0.000000   ;   spec. vector for lin.j.- uii (z)            6
 1.000000   ;   spec. vector for lin.j.- ui,i+1 (x)         6
 0.000000   ;   spec. vector for lin.j.- ui,i+1 (y)         6
 0.000000   ;   spec. vector for lin.j.- ui,i+1 (z)         6
 0.000000   ;   First joint axis-ext. (x)
 0.000000   ;   First joint axis-ext. (y)
 1.000000   ;   First joint axis-ext. (z)
```

```
          ROBOT.DNM
0.000000   ;type of link                                    1
0.000000   ;mass of links                                   1
0.000000   ;moment of inertia Jxx/Js                        1
0.000000   ;moment of inertia Jyy/Jn                        1
0.200000   ;moment of inertia Jzz                           1
0.000000   ;type of link                                    2
5.000000   ;mass of links                                   2
0.250000   ;moment of inertia Jxx/Js                        2
0.010000   ;moment of inertia Jyy/Jn                        2
0.250000   ;moment of inertia Jzz                           2
0.000000   ;type of link                                    3
5.000000   ;mass of links                                   3
0.250000   ;moment of inertia Jxx/Js                        3
0.250000   ;moment of inertia Jyy/Jn                        3
0.010000   ;moment of inertia Jzz                           3
0.000000   ;type of link                                    4
1.000000   ;mass of links                                   4
0.002000   ;moment of inertia Jxx/Js                        4
0.002000   ;moment of inertia Jyy/Jn                        4
0.002000   ;moment of inertia Jzz                           4
0.000000   ;type of link                                    5
1.000000   ;mass of links                                   5
0.002000   ;moment of inertia Jxx/Js                        5
0.002000   ;moment of inertia Jyy/Jn                        5
0.002000   ;moment of inertia Jzz                           5
0.000000   ;type of link                                    6
2.000000   ;mass of links                                   6
0.010000   ;moment of inertia Jxx/Js                        6
0.002000   ;moment of inertia Jyy/Jn                        6
0.010000   ;moment of inertia Jzz                           6
```

```
              ROBOT.TRA
-1.57080  -0.52360  -2.09439  -0.52360  0.        0.
-1.05441  -0.66327  -2.18166  -0.57596  -0.05236  0.
0.01      1.
```

Point on trajectory	Internal coordinates q^i[rad]					
	q^1	q^2	q^3	q^4	q^5	q^6
A_o	-1.57080	-0.52360	-2.09439	-0.52360	0.	0.
A_1	-1.05441	-0.66327	-2.18166	-0.57596	-0.05236	0.
A_2	-0.52360	-0.80260	-2.09439	-0.48870	-0.10470	0.

Table A.9.1. Internal coordinates

In Table A.9.2. the external coordinates A_o, A_1 and A_2 and duration times are given.

Point on trajectory	External coordinates (Descartes coordinates) [m]			Movement	Duration time [s]
	x	y	z		
A_o	0.8	0.	0.2	$A_o \to A_1$	1
A_1	0.718	0.404	-0.029		
A_2	0.425	0.714	-0.001	$A_1 \to A_2$	1

Table A.9.2. External coordinates

Using the program NOMDYN the nominal driving torques for the free motion case ($A_o \to A_1$) and the constrained motion case ($A_1 \to A_2$) are calculated.

For the movement $A_o \to A_1$ the driving torques in mechanism joints in conditions of mechanical vibrations impact, which are transmitted through the basis on the carrying structure are calculated.

a) Free motion ($A_o \to A_1$)

The motion between the initial and final position (Table A.9.1) is performed over the straight line segment while adopting a triangular velocity profile.

Using program TRAJEK, the time history of internal coordinates and velocities can be calculated according to the following expressions:

$$q(t) = q^O + \lambda(t)(q^F - q^O), \quad \dot{q}(t) = \dot{\lambda}(t)(q^F - q^O), \quad 0 \le t \le T$$

where: q^O - initial position, q^F - final position, T - duration time, $\lambda(t) \in [0, 1]$ - scalar parameter which defines the velocity rate of the internal coordinates

$$\dot{\lambda}(t) = \begin{cases} \dfrac{4t}{T^2}, & 0 \le t \le T/2 \\ \dfrac{4(T-t)}{T^2}, & T/2 \le t \le T \end{cases}$$

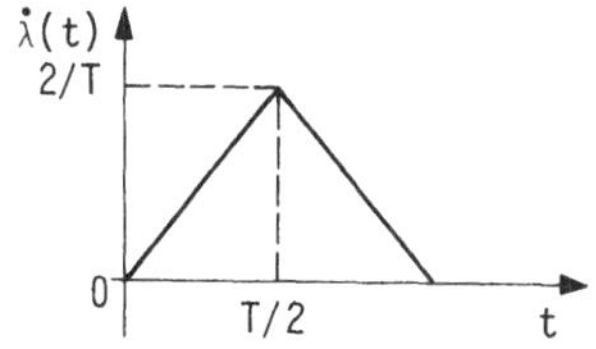

The simulation results (time history of internal coordinates) are shown in Fig. A.9.5.

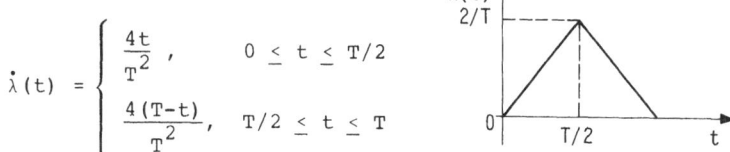

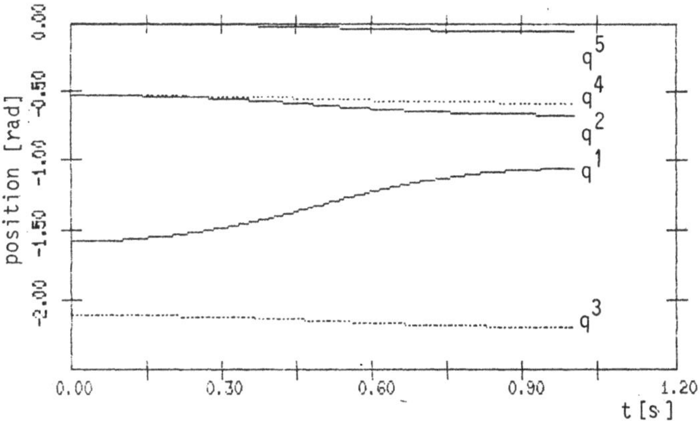

Fig. A.9.5. Time history of internal coordinates

Nominal driving torques are calculated using the following matrix equation:

$$H(q)\ddot{q} + h(q, \dot{q}) = P$$

where: $H(q)$ is the inertial n×n matrix (n is the number of degrees of freedom), $\ddot{q}$ is the acceleration vector (n×1), $h(q, \dot{q})$ is the (n×1) column vector including gravity, centrifugal and Coriolis' effects, P is the (n×1) driving torque vector in the mechanism joints.

In Fig. A.9.6. variations of the driving torques in joints 2 and 3 are presented.

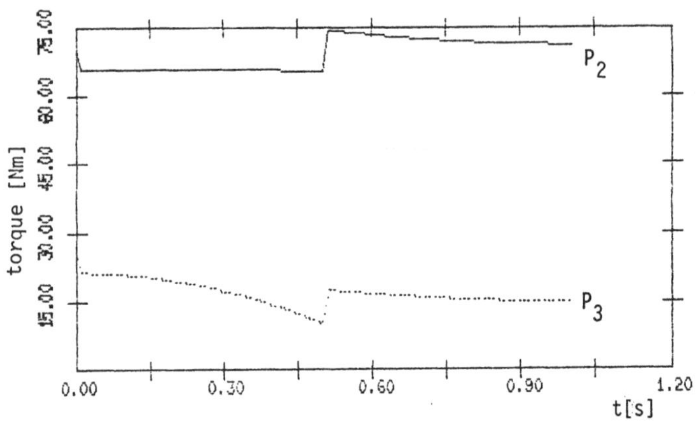

Fig. A.9.6. Nominal driving torques during $A_o \to A_1$ motion

b) Simulation of motion in conditions of mechanical vibrations impact.

Mechanical vibrations are presented by vectors of the translational $\vec{w}_o$ and angular $\vec{\varepsilon}_o$ acceleration. It has been supposed that on the manipulator translational acceleration is acting in the vertical direction w_{oz} and the angular acceleration ε_{oz} around that direction. The accelerations are presented in the form of simple sine function:

$$w_{oz} = -e_\ell \Omega^2 \sin\Omega t, \qquad \varepsilon_{oz} = -e_r \Omega^2 \sin\Omega t$$

This form corresponds to the vibrations, appearing due to centrifugal forces with machines with unbalanced rotor.

In the previous expressions meaning of the particular values is:

e_ℓ - the linear amplitude of basis deplacement,

e_r - the angular amplitude,

Ω - the angular frequency.

The driving torques were calculated using the following matrix equations:

$$H(q)\ddot{q} + h_1(q)w_o + h_2(q)\varepsilon_o + \xi(q, \dot{q}) = P$$

where: $h_1(q)$ is the matrix adjoint to the translational acceleration vector w_o of dimension (n×3), $h_2(q)$ is the matrix adjoint to the angular acceleration vector ε_o of dimension (n×3), $\zeta(q, \dot{q})$ in the column vector, identical to the vector $h(q, \dot{q})$ in the case of free manipulator motion.

Adopting that: e_ℓ = 100 μm, $e_r = 0{,}1 \cdot 10^{-3}$ rad, $\Omega = 100\ s^{-1}$, the driving torques of the robotic mechanism individual degrees of freedom were calculated. In Fig. A.9.7. variation of the driving torques of the mechanism degrees of freedom 2 and 3 were given. By comparing the diagrams in Fig. A.9.6. and A.9.7. it is easily noted, that due to vibrations, oscillating variations of the torque (Fig. A.9.7) occur about the its nominal value presented in the Fig. A.9.6.

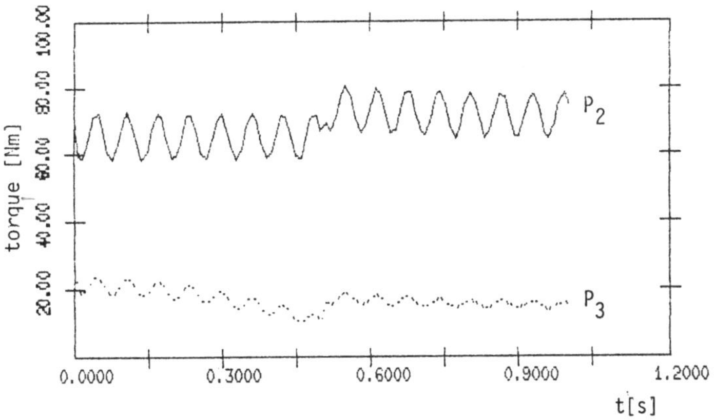

Fig. A.9.7. Driving torques under mechanical vibrations impact

c) As with free motion ($A_o \to A_1$), adopting the initial and final values of the internal coordinates (Table A.9.1), triangular velocity profile and movement time, the time history of the internal coordinates q^i were presented in Fig. A.9.8.

From Table A.9.2, based on the value of the Cartesian coordinates of points A_1 and A_2, it is evident that the processed surface is inclined with respect to the Oxy plane.

Equation of the plane in normal form can be presented as:

$$-0.012120x - 0.011607y + 0.631025z = 0.$$

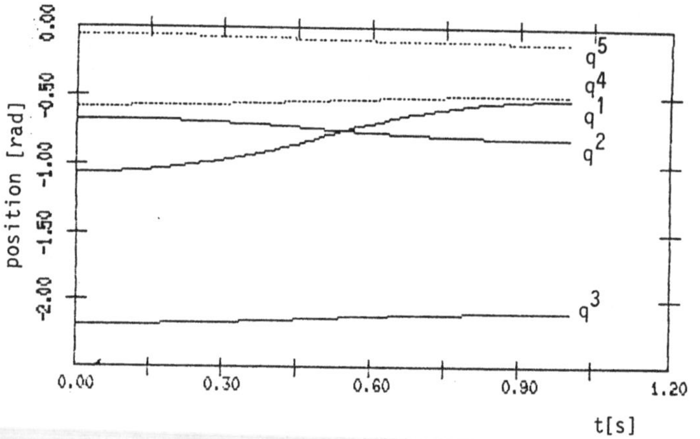

Fig. A.9.8. Internal coordinates during motion ($A_1 \rightarrow A_2$)

From the previous equation the projections of unit vector of the normal to the plane are obtained:

$$\cos\alpha = -0.0192, \quad \cos\beta = -0.0184, \quad \cos\gamma = 0.996.$$

If it is adopted that the reaction force of the support (cutting force) $|\vec{F}_r| = 30\text{N}$, then, under the supposition that it is acting in the last segment center of gravity, its projections are:

$$F_{rx} = |\vec{F}_r|\cos\alpha = -0.576 \text{ N}$$

$$F_{ry} = |\vec{F}_r|\cos\beta = -0.552 \text{ N}$$

$$F_{rz} = |\vec{F}_r|\cos\gamma = 29.880 \text{ N}$$

The nominal driving torques are calculated using the following matrix equations (in the considered case $F_A = F_r$):

$$H(q)\ddot{q} + h(q, \dot{q}) = P + D_1 F_A + D_2 M_A$$

where: D_1 is the matrix adjoint to the external force F_A (reaction force) of dimension (n×3), D_2 is the matrix adjoint to the external moment M_A (in the considered case, constraint of the surface type $M_A = 0$).

In Fig. A.9.9. is presented the time history of the driving torques of the second and third mechanism degree of freedom.

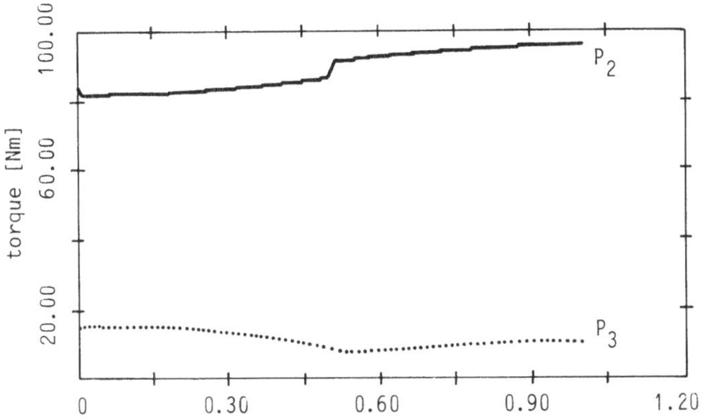

Fig. A.9.9. Driving torques during constrained motion ($A_1 \rightarrow A_2$)

R E M A R K

In this test-example a redundant manipulator configuration with respect to the tool positioning task for grinding was adopted. Therefrom different values of supplementary angular coordinates (q^4 and q^5) appear in points of contact of the tool and inclined working surface. The values of the corresponding internal coordinates of the mechanism in the characteristic working points represent one of the possible cases of the redundant mechanism configuration related to the considered industrial task.

Let it be noted, that in scope of the programme in this Appendix the subroutine for mapping of the external to internal coordinate was not included. The user should calculate himself the internal coordinates of the mechanism at the basis of the known external coordinates (see e.g. (2.117-2.121) and (2.123-2.125)).

Here can be mentioned, too, that the programme support in Appendix is of educational and research character, while the user-oriented PC software based on symbolic models, developed also in the Robotics Laboratory of the "Mihailo Pupin" Institute, is valuable for real-time application.

```
C-----------------------------------------------------------------------
C        HIERARCHICAL STRUCTURE OF THE TASK N O M D Y N
C-----------------------------------------------------------------------
C
C  LEVEL: I          II            III         IV         V          VI
C
C        NOMDYN
C                    INPUT_KIN     COLIAS
C                                  MODEL
C                                              ASSEM
C                                                         CROSS2
C                                                                    DOT1
C                                                         CROSS0
C                                                         PRMM
C                                              ZAKRET
C                                                         CROSS
C                                                         DOT
C                                              DPOS
C                                                         CROSS
C                                                         CROSS0
C                                                         CROSS6
C                                                         ILF1
C                                              TOTR
C                                                         CROSS4
C                                              DTOTR
C                                              BRZUB
C                                                         CROSS0
C                                                         CROSS3
C                                                         CROSS2
C                                                                    DOT1
C                                                         CROSS1
C                                              DOMEG
C                                              DALFA
C                                              DALFA0
C                                                         CROSS0
C                                                         CROSS1
C                                              DDOMEG
C                                                         CROSS0
C                                              DBETAI
C                                                         CROSS1
C                                              DBETA
C                                                         CROSS1
C                                                         CROSS3
C                                              DBETA0
C                                                         CROSS1
C                                                         CROSS0
C                                              DDBE0
C                                                         CROSS0
C                                                         CROSS1
C                                              SILAR
C                                              DINAM
C                                              MOM11
C                                              MOM21
C                                              MOM2
C                                              DINAMC
C                                                         CROSS0
C                                              DBIC
C                                                         CROSS7
```

```
C                                                  DOT2
C                                        DTAUC
C                                                  CROSS
C                                                  CROSS1
C                                                  DOT1
C                                                  DOT2
C                                        DBIOC
C                                                  CROSS7
C                                                  DOT2
C                                        DDTAUC
C                                        DDBIOC
C                                        MOM0
C                                                  DOT0
C                                                  PRMV
C                                        MOM10
C                                        DTSB
C                                        DB1SB
C                                        DVSB
C                                        DDB0SB
C                                        DTRANS
C                                        MATH
C                                                  CROSS5
C                                        DHMOD
C                                                  CROSS1
C                                        VEKH
C                                                  CROSS4
C                                                  CROSS1
C                                                  DOT1
C                                        DH1MOD
C                                                  CROSS1
C                                        MOMUK
C                                                  CROSS1
C                                                  CROSS4
C                                                  DOT1
C                              MODCHK
C                    INPUT_DIN
C                    TRAJEK
C                    MODEL
C                    MATD
C                    MATH1
C                    MATH2
C                              PP1
C                              PP2
C-------------------------------------------------------------
```

```
C-----------------------------------------------------------------------
C                LIST ALL OF PROGRAMS WITH THEIR FUNCTIONS
C-----------------------------------------------------------------------
C
C   Task NOMDYN includes the following subroutines:
C
C  INPUT_KIN          setting kinematic parameters
C
C     COLIAS          generation of assembling vectors if vectors
C                     (Rii,ei) or (Ri,i+1,ei+1) are colinear
C
C     MODEL            main program for robot mathematical model
C                      computation based on Newton - Euler's method,
C                      described in Paragraph 2.1 of this book
C
C        ASSEM         assembling of the mechanism
C
C           CROSS2     double cross product a x (a x b), where
C                      a and b are 3x1 vectors
C
C              DOT1    scalar product of two vectors (dot product)
C
C           CROSS0     cross product of two vectors (first vector is
C                      3x1 and the second is the i-th row of the matrix
C                      6x3)
C
C           PRMM       product of two square matrix (nxn)
C
C        ZAKRET        calculation of transformation matrix Q01 (Ai)
C                      and ei=Qi.ei(int), rii=Qi.rii(int), ri,i+1=
C                      Qi.ri,i+1(int)
C
C           CROSS      cross product of two vectors (vector product)
C                      (one vector is the i-th row of 6x3 matrix and
C                      the other is the j-th column of 3x3 matrix)
C
C           DOT        scalar product of two vectors (one vector is
C                      the i-th row of 6x3 matrix and the other is the
C                      j-th column of 3x3 matrix)
C
C        DPOS          calculation of expresions dl(q(i,j)), dl(rii),
C                      dl(ri,i+1), dl(ei+1)
C
C           CROSS      cross product of two vectors (vector product)
C                      (one vector is the i-th row of 6x3 matrix and
C                      the other is the j-th column of 3x3 matrix)
C
C           CROSS0     cross product of two vectors (first vector is
C                      3x1 and the second is the i-th row of the matrix
C                      6x3)
C
C           CROSS6     auxiliary subroutine
C
C           ILF1       auxiliary subroutine
C
C        TOTR          calculation of position vectors rij,j=1,...,i-1
C
C           CROSS4     cross product of the 1-th row and i-th column
```

```
C                of the 6x6x3 array as the first vector and the
C                j-th row of the 6x3 matrix as the second vector
C
C    DTOTR       calculation of d1(r(i,j))
C
C    BRZUB       calculation of coefficients accompanying angular
C                and linear accelerations beta-ij, j=1,...,i-1
C                and alfa -ij, j=1,...,i-1 and coefficients
C                beta0-i, alfa0-i dependent on velocities
C
C       CROSS0   cross product of two vectors (first vector is
C                3x1 and the second is the i-th row of the matrix
C                6x3)
C
C       CROSS3   cross product of two vectors (first vector is the
C                l-th row and the i-th column of the 6x6x3 array
C                and the second is 3x1 vector)
C
C       CROSS2   double cross product a x (a x b), where
C                a and b are 3x1 vectors
C
C          DOT1  scalar product of two vectors (dot product)
C
C       CROSS1   cross product of two vectors
C
C
C    DOMEG       calculation of d1OMEG(i) -1- derivative of OMEG(i)
C
C    DALFA       calculation of d1ALFA(i,j)
C
C    DALFA0      calculation of d1ALFA0(i)
C
C       CROSS0   cross product of two vectors (first vector is
C                3x1 and the second is the i-th row of the matrix
C                6x3)
C
C       CROSS1   cross product of two vectors
C
C    DDOMEG      calculation of derivatives of ALFA0(i) wrt
C                SIDOT(i)
C
C       CROSS0   cross product of two vectors (first vector is
C                3x1 and the second is the i-th row of the matrix
C                6x3)
C
C    DBETAI      calculation of 1- derivative of BETAii
C
C       CROSS1   cross product of two vectors
C
C    DBETA       calculation of 1- derivative of BETA(i,j)
C
C       CROSS1   cross product of two vectors
C
C       CROSS3   cross product of two vectors (first vector is the
C                l-th row and the i-th column of the 6x6x3 array
C                and the second is 3x1 vector)
C
C    DBETA0      calculation of d1BETA0(i)
C
C       CROSS1   cross product of two vectors
```

```
C         CROSS0    cross product of two vectors (first vector is
C                   3x1 and the second is the i-th row of the matrix
C                   6x3)
C
C      DDBE0        calculation of derivatives of BETA0(i)
C                   wrt SIDOT(i)
C
C         CROSS0    cross product of two vectors (first vector is
C                   3x1 and the second is the i-th row of the matrix
C                   6x3)
C
C         CROSS1    cross product of two vectorsC
C
C      SILAR        calculation of total force (relevant for
C                   ITIP=1)
C                   calculation of coefficients aij accompanying
C                   accelerations and a0 coefficient dependent
C                   on velocity (relevant for ITIP=0)
C
C      DINAM        calculation of 1-derivatives of dynamic coeffi-
C                   cients a(i,j) and a0(i), and the derivatives
C                   of a0(i) wrt SIDOT(i)
C
C      MOM11        calculation of the equivalent acceleration
C                   for the bar
C
C      MOM21        calculation of coefficients alfa-ij, alfa-i0
C                   for the bar
C
C      MOM2         calculation of coefficients bij,j=1,...,i and
C                   bi0 for the link considered as the bar (cane)
C
C      DINAMC       calculation of derivative of Si wrt S1(1)
C
C         CROSS0    cross product of two vectors (first vector is
C                   3x1 and the second is the i-th row of the matrix
C                   6x3)
C
C      DBIC         calculation of derivatives of coefficients
C                   b(i,j) wrt SI(1)
C
C         CROSS7    auxiliary subroutine
C
C         DOT2      auxiliary subroutine
C
C      DTAUC        calculations of derivatives of dynamic
C                   coefficients TAU(i) wrt S1(1)
C
C         CROSS     cross product of two vectors (vector product)
C                   (one vector is the i-th row of 6x3 matrix and
C                   the other is the j-th column of 3x3 matrix)
C
C         CROSS1    cross product of two vectors
C
C         DOT1      scalar product of two vectors (dot product)
C
C         DOT2      auxiliary subroutine
C
C      DBI0C        calculation of derivatives of dynamic
C
```

```
C                 coefficients b0(i) wrt SI(1)
C
C      CROSS7     auxiliary subroutine
C
C      DOT2       auxiliary subroutine
C
C    DDTAUC       calculation of derivatives of dynamic
C                 coeficients TAU(i) wrt SIDOT(i)
C
C    DDBI0C       calculation of derivatives of dynamic
C                 coefficients b0(i) wrt SIDOT(i)
C
C    MOM0         calculation of transformation matrix for
C                 moments of inertia wrt fixed coordinate
C                 system and calculation of velocity dependent
C                 component lamda (for the body)
C
C      DOT0       scalar product (dot product) of two vectors
C                 (one vector is the three component array and
C                 the other is the j-th column of the 3x3 matrix)
C
C      PRMV       nxn matrix multiplied by nx1 vector
C
C
C    MOM10        calculation of coefficient bij, bi0 for the body
C
C    DTSB         calculation of derivatives of terms of T(i)
C                 matrix wrt SI(1)
C
C    DBISB        calculation of derivatives of dynamic coeffi-
C                 cients b(i,j) wrt SI(1)
C
C    DVSB         calculation of derivatives of dynamic vector
C                 (Euler's equation) v(i)
C
C    DDB0SB       calculation of derivatives of LAMD(i) and
C                 b0(i) wrt SI(1), and b0(i) wrt SIDOT(1)
C
C    DTRANS       saving of vectors d1(r(i-1,i)), d1(r(i,i)),
C                 d1(OMEG(i)) and d1(ALFA0(i))
C
C    MATH         calculation of inertial matrix H
C
C      CROSS5     cross product of the j-th row and i-th column
C                 of the 6x6x3 array as the first vector and the
C                 j-th row and the 1-th column of the 6x6x3 array
C                 as the second vector
C
C    DHMOD        calculation of derivatives of inertial matrix
C                 elements wrt joint coordinates
C      CROSS1     cross product of two vectors
C
C    VEKH         calculation of total vector h0 (except for
C                 frictions)
C
C      CROSS4     cross product of the 1-th row and i-th column
C                 of the 6x6x3 array as the first vector and the
C                 j-th row of the 6x3 matrix as the second vector
C
C      CROSS1     cross product of two vectors
```

```
C
C           DOT1      scalar product of two vectors (dot product)
C
C        DH1MOD       calculation of derivatives of elements of
C                     vector h(q,q') wrt joint coordinates q(1)
C                     and joint velocities q'(1)
C
C           CROSS1    cross product of two vectors
C
C        MOMUK        calculation of total driving torques/forces
C                     (P=P(q,q',q''))
C           CROSS1    cross product of two vectors
C
C           CROSS4    cross product of the 1-th row and i-th column
C                     of the 6x6x3 array as the first vector and the
C                     j-th row of the 6x3 matrix as the second vector
C           DOT1      scalar product of two vectors (dot product)
C
C     MODCHK          testing of successful assembling (types
C                     transformation matrix An and characteristic
C                     kinematic parameters)
C
C   INPUT_DIN         setting dynamic parameters
C
C   TRAJEK            calculation of nominal trajectory of the robot
C
C   MODEL             main program for robot mathematical model
C                     computation based on Newton - Euler's method,
C                     described in Paragraph 2.1 of this book
C
C   MATD              calculation of matrices D1 and D2 adjointed
C                     to the external force FA and external moment
C                     MA, respectively
C
C   MATH1             calculation of matrix h1 adjointed to the
C                     linear acceleration vector w0 of the first
C                     (basis) segment of the robot
C
C   MATH2             calculation of matrix h2 adjointed to the
C                     rotational acceleration vector e0 of the first
C                     (basis) segment of the robot
C
C     PP1             cross product of two vectors (vector product)
C
C     PP2             final calculation of matrix h2
C
C--------------------------------------------------------------------
```

```
C*********************************************************************
C                                                                    *
C     MAIN PROGRAM:         N   O   M   D   Y   N                    *
C                                                                    *
C*********************************************************************
C                                                                    *
C     FUNCTION:                                                      *
C                                                                    *
C     Main program for calculation nominal dynamics (nominal         *
C     driving torques) of manipulation robots based on               *
C     Newton-Euler' method                                           *
C                                                                    *
C--------------------------------------------------------------------*
C                                                                    *
C     The dynamic model of manipulation robots can be presented      *
C     in the form                                                    *
C                                                                    *
C             H(q)q" + h(q,q') = P                                   *
C                                                                    *
C     where H(q) is the inertial matrix, dimension, nxn(n-degrees    *
C     of freedom), q" is internal acceleration vector, h(q,q') is    *
C     the culomn vector which includes inertial, centrifugal and     *
C     Coriolis effects, dimension (nxl), P is the driving torques    *
C     (forces) vector in the manipulator joints, dimension (nxl).    *
C                                                                    *
C     If the manipulator gripper motion is constrained, the dynamic  *
C     model can be presented in following form                       *
C                                                                    *
C             H(q)q" + h(q,q') = P + D1Fa + D2Ma                     *
C                                                                    *
C     where D1 and D2 are matrices adjointed to the external force   *
C     Fa and external moment Ma, respectively.                       *
C                                                                    *
C     If we want to calculate the driving torques in conditions of   *
C     mechanical vibrations impact which are transmitting trough     *
C     the basis to the carrying structure of manipulator, the        *
C     dynamic model is                                               *
C                                                                    *
C             H(q)q" + h1(q)w0 + h2(q)e0 + h'(q,q') = P              *
C                                                                    *
C      where h1(q) and h2(q) are matrices adjointed to the linear    *
C      acceleration vector w0 and rotational acceleration vector e0,*
C      respectively, h'(q,q') is the culomn vector which includes    *
C      inertial, centrifugal and Coriolis effects.                   *
C                                                                    *
C--------------------------------------------------------------------*
C                                                                    *
C     INPUT VARIABLES:                                               *
C                                                                    *
C             Q01(N)    nominal angles of the joints                 *
C             QT01(N)   nominal velocities of the joints             *
C             VR2,VR1   time instants on nominal trajectory          *
C             N         number of d.o.f.                             *
C             HH(N,N)   inertia matrix                               *
C             H1(N)     vector of gravity,centrifugal and            *
C                       Coriolis moments (forces)                    *
C             MODLIN    =1 - linearized model                        *
C                       =0 - nonlinear model                         *
C             IDFS      =1 - motion is constrained                   *
```

```
C                    =0 - gripper moves freely                        *
C                         (without constraints)                       *
C         IVIB       =1 - simulation of motion in conditions of       *
C                         mechanical vibrations impact                *
C                    =0 - without vibrations                          *
C         W0(3)      linear acceleration                              *
C         E0(3)      rotational acceleration                          *
C         DFS(3)     desired force on the surface                     *
C                                                                     *
C---------------------------------------------------------------------*
C                                                                     *
C  OUTPUT VARIABLES:                                                  *
C                                                                     *
C         QU0(6)     accelerations of the joints                      *
C         P0(6)      nominal driving torques                          *
C                                                                     *
C---------------------------------------------------------------------*
C                                                                     *
C  SUBROUTINES REQUIRED:                                              *
C                                                                     *
C         INPUT_KIN  setting geometric parameters                     *
C         INPUT_DIN  setting dynamic parameters                       *
C         TRAJEK     subroutine calculates nominal trajectory         *
C         MODEL      module that computes matrices of the             *
C                    mechanical part of the robot                     *
C         MATD       subroutine calculates matrices D1 and D2         *
C                    adjointed to the desired external force Fa       *
C                    and external moment Ma,respectively,             *
C                    accting  on the manipulator's gripper            *
C         MATH1      subroutine calculates matrix h1 adjointed        *
C                    to the linear acceleration vector w0             *
C         MATH2      subroutine calculates matrix h2 adjointed        *
C                    to the rotational acceleration vector e0         *
C                                                                     *
C---------------------------------------------------------------------*
C
      DIMENSION R(6,6,3)
      DIMENSION XPOM(6)
      DIMENSION QU0(6)
      DIMENSION POP(6)
      DIMENSION PX0(6)
      DIMENSION HPOM(6,6)
      DIMENSION DFS(3)
      DIMENSION W0(3),E0(3)
      DIMENSION H10(6,3),H20(6,3)
C
      COMMON/SOPC/ NST,INOM,IPOM1,IPOM2,IPOM3,INOM1
      COMMON/NPOM1/ Q00(6),QT00(6),QU00(6),P00(6)
      COMMON/NPOM2/ Q01(6),QT01(6),QU01(6),P01(6)
      COMMON/SUBSYS/ A(3,3,6),B(3,6),F(3,6)
      COMMON/SUBSHM/ ASHM(3,3,6),BSHM(3,6),FSHM(3,6)
      COMMON/NKOORD/ P0(6),X0(18),U0(6)
      COMMON/SKOORD/ DX0(18)
      COMMON/SKOORA/ XU0(6)
      COMMON/MODSW/KOD
      COMMON/SUR/ D1(6,3),D2(6,3),VECP(3)
      COMMON/PRSTA/ PINTPS(6)
      COMMON/VIBRO/ R
C
      INCLUDE 'MODELM.MOD'
```

```
      INCLUDE 'CONFIG.MOD'
C
      COMMON/MODOPC/ MOD,ITIP,MODLIN
C
      COMMON/ORDGLB/ NUK,ISLOBODE
      COMMON/FILE/FILE(5)
      BYTE FILE
      COMMON/ZZTSK1/INDMEH
      CHARACTER*1 OPC,FILEP*5
C
C------------------------------------------------------------------
C     Setting name of the robot
C------------------------------------------------------------------
C
      TYPE 137
137   FORMAT(1X,'Name of the robot [5a1]:',$)
      ACCEPT 1981,FILE
1981  FORMAT (5A1)
C
C------------------------------------------------------------------
C     Setting geometric parameters of the robot
C------------------------------------------------------------------
C
      call input_kin
C
C------------------------------------------------------------------
C     Setting dynamic parameters of the robot
C------------------------------------------------------------------
C
      call input_din
C
C------------------------------------------------------------------
C     Nominal trajectory synthesis
C------------------------------------------------------------------
C
      CALL TRAJEK
C
C------------------------------------------------------------------
C     Reading data on nominal trajectory
C------------------------------------------------------------------
C
      FILEP=CHAR(FILE(1))//CHAR(FILE(2))//CHAR(FILE(3))//
     *      CHAR(FILE(4))//CHAR(FILE(5))
      OPEN(UNIT=4,FILE=FILEP//'.ANG',STATUS='OLD')
C
      READ(4,100,END=3100) VR2
      READ(4,100)(Q01(I),I=1,N)
      READ(4,100)(QT01(I),I=1,N)
 100  FORMAT(6E13.5)
C
      DO 1 I=1,N
      QU0(I)=QU01(I)
      SI(I)=Q01(I)
 1    SIDOT(I)=QT01(I)
C
C------------------------------------------------------------------
C     Dynamic model of the robot
C------------------------------------------------------------------
C
C------------------------------------------------------------------
```

```
C       Linearized or nonlinear model
C------------------------------------------------------------
C
        WRITE (6,1783)
 1783   FORMAT (1X,'Want you linearized model of the robot (Y/N)?:',$)
        READ  (5,1235) OPC
        IF (OPC.EQ.'Y') THEN
           MODLIN=1
        ELSE
           MODLIN=0
        END IF
C
        CALL MODEL
C
        IF(MODLIN.EQ.1) GO TO 300
C
C------------------------------------------------------------
C       Constrained gripper motion
C------------------------------------------------------------
C
        WRITE (6,1234)
1234    FORMAT (1X,'Want you constrained gripper motion?(Y/N):',$)
        READ (5,1235) OPC
1235    FORMAT (A1)
        IF (OPC.EQ.'Y') THEN
           TYPE *,'Desired force on the surface which you
     *want to realise during motion in [N]:'
           IDFS=1
           WRITE(6,62)
           READ (5,6001) DFS(1)
           WRITE(6,63)
           READ (5,6001) DFS(2)
           WRITE(6,64)
           READ (5,6001) DFS(3)
 62        FORMAT(1X,'Along x-axis:',$)
 63        FORMAT(1X,'Along y-axis:',$)
 64        FORMAT(1X,'Along z-axis:',$)
6001       FORMAT(F10.5)
        ELSE
           IDFS=0
        END IF
C
C------------------------------------------------------------
C       Mechanical vibrations impact
C------------------------------------------------------------
C
        WRITE(6,1236)
1236    FORMAT(1X,'Want you simulation in conditions
     * of mechanical vibrations impact?(Y/N):',$)
        READ(5,1237) OPC
1237    FORMAT(A1)
        IF(OPC.EQ.'Y') THEN
           TYPE *,'Desired parameters of acceleration vectors:'
           TYPE *,'a) Max. basis linear movement amplitude in [m]:'
           IVIB=1
           READ(5,65)EL
           TYPE *,'b) Max. rotational amplitude in [rad]:'
           READ(5,66)ER
           TYPE *,'c) Angular frequency in [1/s]:'
           READ(5,67)OMEGA
```

```
 65        FORMAT(F10.5)
 66        FORMAT(F10.5)
 67        FORMAT(F10.5)
 6002      FORMAT(F10.5)
         ELSE
           IVIB=0
         END IF
C
         POM=OMEGA*VR2
         W0(3)=-EL*OMEGA**2*SIN(POM)
         E0(3)=-ER*OMEGA**2*SIN(POM)
C
C--------------------------------------------------------------
C        We assume that there is just linear acceleration in
C        vertical direction and angular acceleration about this
C        direction
C--------------------------------------------------------------
C
         OPEN(UNIT=7,FILE='ROBOT.TRQ',STATUS='NEW')
C
         DO 77 I=1,N
         XPOM(I)=0.
         DO 77 J=1,N
         XPOM(I)=XPOM(I)+HH(I,J)*QU0(J)
 77      HPOM(I,J)=HH(I,J)
C
         DO 12 I=1,N
         PX0(I)=XPOM(I)+H1(I)
C
C--------------------------------------------------------------
C        Calculation matrix D1 and D2 adjointed to the
C        external force FA and moment MA, respectivelly
C--------------------------------------------------------------
C
         CALL MATD(E)
C
C--------------------------------------------------------------
C        Calculation matrix h1 and h2 adjointed to the
C        linear acceleration vector w0  and angular
C        acceleration vector e0
C--------------------------------------------------------------
C
         CALL MATH1(E,R,H10)
         CALL MATH2(E,T,R,H20)
C
C--------------------------------------------------------------
C        Calculation of nominal driving torques
C--------------------------------------------------------------
C
         DO 149 K=1,3
         PX0(I)=PX0(I)-D1(I,K)*DFS(K)*IDFS
 149     PX0(I)=PX0(I)+(H10(I,K)*W0(K)+H20(I,K)*E0(K))*IVIB
C
         WRITE (66,*) VR2,PX0(2),PX0(3)
C

 12      POP(I)=PX0(I)
C        PRINT *,'DRIVING TORQUES',(PX0(IPOM),IPOM=1,N)
 300     CONTINUE
 3       CONTINUE
```

```
      DO 32 I=1,N
 32   P0(I)=PX0(I)
      WRITE(7,100) VR2
      WRITE(7,100)(PX0(I),I=1,N)
      VR1=VR2
      DO 4 I=1,N
      Q00(I)=Q01(I)
      QT00(I)=QT01(I)
      QU00(I)=QU01(I)
  4   CONTINUE
C
C-----------------------------------------------------------------
C     Reading data on nominal trajectory
C-----------------------------------------------------------------
C
      READ(4,100, END=3100) VR2
      READ(4,100)(Q01(I),I=1,N)
      READ(4,100)(QT01(I),I=1,N)
      DO 6 I=1,N
      SI(I)=Q01(I)
      SIDOT(I)=QT01(I)
      QU0(I)=(QT01(I)-QT00(I))/(VR2-VR1)
  6   CONTINUE
C
      CALL MODEL
C
      IF(MODLIN.EQ.1) GO TO 4000
      DO 88 I=1,N
      XPOM(I)=0.
      DO 88 J=1,N
      XPOM(I)=XPOM(I)+HH(I,J)*QU0(J)
 88   HPOM(I,J)=HH(I,J)
C
      POM=OMEGA*VR2
      W0(3)=-EL*OMEGA**2*SIN(POM)
      E0(3)=-ER*OMEGA**2*SIN(POM)
C
C
      DO 7 I=1,N
      PX0(I)=XPOM(I)+H1(I)
C
      CALL MATD(E)
      CALL MATH1(E,R,H10)
      CALL MATH2(E,T,R,H20)
C
      DO 18 K=1,3
      PX0(I)=PX0(I)-D1(I,K)*DFS(K)*IDFS
 18   PX0(I)=PX0(I)+(H10(I,K)*W0(K)+H20(I,K)*E0(K))*IVIB
C
C
  7   POP(I)=PX0(I)
C     PRINT *,'DRIVING TORQUES',(PX0(IPOM),IPOM=1,N)
      WRITE (66,*) VR2,PX0(2),PX0(3)
C
C-----------------------------------------------------------------
C     Writing of driving torques
C-----------------------------------------------------------------
C
      WRITE(7,100) VR2
      WRITE(7,100)(PX0(I),I=1,N)
```

```
1000    FORMAT(4(2X,E13.5))
        WRITE(6,100) VR2
        WRITE(6,100)(PX0(I),I=1,N)
4000    CONTINUE
        GO TO 3
3100    CLOSE(UNIT=4)
        CLOSE(UNIT=7)
        STOP
        END
```

```
      SUBROUTINE INPUT_KIN
***********************************************************************
* SUBROUTINE:  Input_kin                                              *
*.....................................................................*
* FUNCTION:  Input program for setting kinematic parameters           *
*.....................................................................*
* SUBROUTINE CALLED: Input_kin                                        *
*.....................................................................*
* INPUT VARIABLES:  ISLOBODE= variable which defines number of d.o.f. *
*                   I0= index of d.o.f.                               *
* OUTPUT VARIABLES: Kinematic parameters                              *
*.....................................................................*
* LOCAL VARIABLES: INEW- indicator which denots whether data with     *
*                        manipulator's parameters exists (1-yes;0-no) *
***********************************************************************
      INCLUDE 'CONFIG.MOD'
      INCLUDE 'MODELM.MOD'
      COMMON /MODSW/KOD
      COMMON/LINSPE/ UD(6,3),UG(6,3)
      COMMON/ORDGLB/ NUK,ISLOBODE
      COMMON/FILE/ FILE(5)
      BYTE file
*
      COMMON /INTR/ SIMB(19,6)
      DIMENSION EPOC1(3)
      CHARACTER*45 HELP(19),HXX0,hepoc(3)
      CHARACTER*2 A$(6),OPC1
      CHARACTER*1 POM,PP,A1*79,A$COM*15,POM1*4
      CHARACTER*9 FILEP,FILEP2,FILEP4,KNZ(13),KNZ1(13)
*
      DATA HELP/'type of joint','joint unit axis- eii (x)',
     >          'joint unit axis- eii (y)','joint unit axis- eii (z)',
     >          'i+1 joint unit axis- ei,i+1 (x)',
     >          'i+1 joint unit axis- ei,i+1 (y)',
     >          'i+1 joint unit axis- ei,i+1 (z)',
     >          'link vectors- Rii (x)',
     >          'link vectors- Rii (y)',
     >          'link vectors- Rii (z)',
     >          'link vectors- Ri,i+1 (x)',
     >          'link vectors- Ri,i+1 (y)',
     >          'link vectors- Ri,i+1 (z)',
     >          'spec. vector for lin.j.- uii (x)',
     >          'spec. vector for lin.j.- uii (y)',
     >          'spec. vector for lin.j.- uii (z)',
```

```
>            'spec. vector for lin.j.- ui,i+l (x)',
>            'spec. vector for lin.j.- ui,i+l (y)',
>            'spec. vector for lin.j.- ui,i+l (z)'/
   DATA KNZ/'KSI2','eii (x)','eii (y)','eii (z)',
>           'ei,i+l:x','ei,i+l:y','ei,i+l:z',
>           'rii (x)','rii (y)','rii (z)',
>           'ri,i+l:x','ri,i+l:y','ri,i+l:z'/
   DATA KNZ1/'KSI2','eii (x)','eii (y)','eii (z)',
>           '%','%','%',
>           'rii (x)','rii (y)','rii (z)',
>           'ri,i+l:x','ri,i+l:y','ri,i+l:z'/
   DATA HEPOC/'First joint axis-ext. (x)',
>             'First joint axis-ext. (y)',
>             'First joint axis-ext. (z)'/

      FILEP=CHAR(FILE(1))//CHAR(FILE(2))//CHAR(FILE(3))//
    >      CHAR(FILE(4))//CHAR(FILE(5))//'.CNF'
*.......................................................................
*     READING DATA
*.......................................................................
      INEW=0             ! INDICATOR FOR DATA- 1:OLD; 0:NEW
*
      DESNI=1            ! INITIALIZATION OF COORDINATE SYSTEM
*
      OPEN (55,FILE=FILEP,STATUS='OLD',ERR=11)
      INEW=1
      WRITE (6,14) FILEP
14    FORMAT (1X,' WARNING*** File:',a10,'  EXISTS',/)
      read (55,'(1X,i1)') islobode
      DO IPOM=1,ISLOBODE
         DO JPOM=1,19
            READ (55,'(1X,F15.6)') SIMB(JPOM,IPOM)
         END DO
      END DO
      DO IPOM=1,3
         READ (55,'(1X,F15.6)') EPOC(IPOM)
      END DO
      CLOSE (55)
      if (inew.eq.1) GO TO 12
11    WRITE (6,13) FILEP
13    FORMAT (1X,'MESSAGE*** File ',a10,' does not exist')
      STOP
*.......................................................................
*     REMOVING OF VARIABLES
*.......................................................................
12          do ipom=1,islobode
               ksi2(ipom)=simb(1,ipom)
               do kpom=1,3
                  eu(ipom,kpom)=simb(kpom+1,ipom)
                  eul(ipom,kpom)=simb(kpom+4,ipom)
                  r0u(ipom,kpom)=simb(kpom+7,ipom)
                  if (ipom.eq.islobode) then
                     rt(kpom)=simb(kpom+10,ipom)
                  else
                     r0u(ipom+islobode,kpom)=simb(kpom+10,ipom)
                  end if
               end do
            end do
*.......................................................................
```

```
*       TESTING OF VARIABLES
*...........................................................
              do ipom=1,islobode
*
*                vector ei must be unit vector
*
                 eum=sqrt(eu(ipom,1)*eu(ipom,1)+
     >               eu(ipom,2)*eu(ipom,2)+eu(ipom,3)*eu(ipom,3))
                 if (eum.gt.1.01.OR.eum.lt.0.99) then
                      write (6,'(''+DIAG*** Vector ei'',i2,
     >                    '' is not unit vector'')') ipom
                      return
                 end if
              end do
              do ipom=1,islobode
*

*            vector ei+1 must be junit vector,
*            but vector ei,i+1 in last robot joint is
*            not unit
*
*
             if (ipom.lt.islobode) then
                eulm=sqrt(eul(ipom,1)*eul(ipom,1)+
   >                eul(ipom,2)*eul(ipom,2)+eul(ipom,3)*eul(ipom,3))
                if (eulm.gt.1.01.OR.eulm.lt.0.99) then
                     write (6,'(''+DIAG*** Vector ei+1'',i2,
   >                     '' is not unit vector'')') ipom
                     return
                end if
             end if
          end do
*
*         testing whether exists one linear joint
*
          do ipom=1,islobode
             if (ksi2(ipom).eq.1) if10=1
          end do
          if (if10.eq.1) then                  !IFL0=1;exists
                                               !
             do ipom=1,islobode                !     =0;dos't exist
                if (ksi2(ipom).eq.1) then
                   do kpom=1,3
                      ud(ipom,kpom)=simb(kpom+13,ipom)
                      ug(ipom,kpom)=simb(kpom+16,ipom)
                   end do
                else
                   do kpom=1,3
                      ud(ipom,kpom)=0
                      ug(ipom,kpom)=0
                   end do
                end if
             end do
*.................................................
*   testing variables
*.................................................
             do ipom=1,islobode
*
*   chacking whether uii,ui,i+1 are unit vectors
*
```

```
*
*       vector uii must be unit vector
*
                    if (ksi2(ipom).eq.1) then
                       udm=sqrt(ud(ipom,1)*ud(ipom,1)+
     >                          ud(ipom,2)*ud(ipom,2)+
     >                          ud(ipom,3)*ud(ipom,3))
                       if (udm.gt.1.01.or.udm.lt.0.99) then
                          write (6,'('' DIAG*** Vector uii'',i2,
     >                              ''is not unit vector'')') ipom
                          return
                       end if
                    end if
                 end do
*
*       vector ui,i+1 must be unit vector
*
                 do ipom=1,islobode
                    if (ksi2(ipom).eq.1) then
                       ugm=sqrt(ug(ipom,1)*ug(ipom,1)+
     >                          ug(ipom,2)*ug(ipom,2)+
     >                          ug(ipom,3)*ug(ipom,3))
                       if (ugm.gt.1.01.or.ugm.lt.0.99) then
                          write (6,'('' DIAG*** Vector ui,i+1'',i2,
     >                              '' is not unit vector'')') ipom
                          return
                       end if
                    end if
                 end do
              else
                 epocm=sqrt(epoc(1)*epoc(1)+epoc(2)*epoc(2)+
     >                   epoc(3)*epoc(3))
                 if (epocm.lt..99.or.epocm.gt.1.01) then
                    return
                 end if
              end if
*.....................................................................
*             chacking of colinearity and assembling
*.....................................................................
              n=islobode
              do ipom=1,islobode
                 call colias(ipom)
              end do
              kod=1
              call model
              call modchk(imod,ino)
              if (imod.lt.0) then
                 write (6,'('' FATAL*** Assembling: UNSUCCESSFUL'')')
                 return
              else
                 write (6,'('' MESSAGE*** Assembling: SUCCESSFUL'')')
              end if
              return
          END
```

```
C---------------------------------------------------------------------|
C    SUBROUTINE:     COLIAS                                           |
C---------------------------------------------------------------------|
C    FUNCTION:       Generation of assembling vectors if vectors      |
C                    (Rii,ei or Ri,i+1,ei+1)are colinear              |
C---------------------------------------------------------------------|
C    INPUT VARIABLES:                                                 |
C            KSI2    type of joint                                    |
C            EPOC    unit vector of first joint axis in external      |
C                    coordinate system                                |
C            UD,UG   unit vectors which are connected for             |
C                    i-th internal coordinate system in the case      |
C                    when joint i is linear and "special"             |
C            ROU     link vectors(Rii and Ri,i+1)                     |
C            EU      unit vector of i-th joint axis in the i-th       |
C                    internal coordinate system                       |
C            EU1     unit vector of i-th joint axis                   |
C                    internal coordinate system                       |
C---------------------------------------------------------------------|
C    OUTPUT VARIABLES:                                                |
C            IAS1    speciality of link with lower side(Ei||Rii)      |
C                    =0 link is not special                           |
C                    =1 link is special                               |
C            IAS2   speciality of link with upper side(Ei,i+1||Ri,i+1|
C                    =0 link is not special                           |
C                    =1 link is special                               |
C            RAS    assembling vectors if link is special             |
C---------------------------------------------------------------------|
C    SUBROUTINES CALLED:                                              |
C---------------------------------------------------------------------|
C
      SUBROUTINE  COLIAS(I)
C
      INCLUDE 'IN2:CONFIG.MOD'
      COMMON/LINSPE/ UD(6,3),UG(6,3)
      REAL IDEL(3)
      DIMENSION IAS(6)
C
      IAS1(I)=0
      IAS(1)=0
      IAS2(I)=0
C
      IF(I.EQ.1.AND.KSI2(I).EQ.1)GO TO 567
      ROPOC(1)=-EPOC(3)
      ROPOC(2)=-EPOC(1)
      ROPOC(3)=-EPOC(2)
      GO TO 4000
567   DO 403 K=1,3
      ROPOC(K)=UD(I,K)
403   CONTINUE
4000  IND=0
      M=I
      IF(KSI2(I).EQ.1)GO TO 444
      DO 15 K=1,3
      IMUL1=0
      IF(ROU(M,K).EQ.0.)IMUL1=1
      IMUL2=0
      IF(EU(I,K).EQ.0.)IMUL2=1
```

```
      IF(IMUL1.NE.IMUL2)GO TO 200
      IF(IMUL1.EQ.0.AND.IMUL2.EQ.0)GO TO 56
      IDEL(K)=0.
      GO TO 15
56    IDEL(K)=ROU(M,K)/EU(I,K)
15    CONTINUE
C
300   IF(IDEL(1).EQ.0.)GO TO 101
      IF(IDEL(2).EQ.0.)GO TO 102
      IF(IDEL(3).EQ.0.)GO TO 103
      IF(IDEL(1).EQ.IDEL(2).AND.IDEL(2).EQ.IDEL(3))GO TO 104
      IAS(I)=0
      GO TO 200
104   IAS(I)=1
      GO TO 201
103   IF(IDEL(1).EQ.IDEL(2))GO TO 104
      IAS(I)=0
      GO TO 200
102   IF(IDEL(3).EQ.0.)GO TO 105
      IF(IDEL(1).EQ.IDEL(3))GO TO 104
      IAS(I)=0
      GO TO 200
105   IAS(I)=1
      GO TO 202
101   IF(IDEL(2).EQ.0.)GO TO 106
      IF(IDEL(3).EQ.0.)GO TO 105
      IF(IDEL(2).EQ.IDEL(3))GO TO 104
      IAS(I)=0
      GO TO 200
106   IF(IDEL(3).EQ.0.)GO TO 107
      GO TO 105
107   TYPE 1016
1016  FORMAT(' DIAG***** Joint unit axis vector and link vector are
     *vectors*******')                                              zero-
      RETURN
200   RAS(M,1)=0.
      RAS(M,2)=0.
      RAS(M,3)=0.
      GO TO 204
201   RAS(M,1)=1.
      RAS(M,2)=0.
      RAS(M,3)=0.
      GO TO 204
202   IF(ROU(M,1).EQ.0.)GO TO 599
      RAS(M,1)=0.
      RAS(M,2)=1.
      IF(IND.EQ.1)RAS(M,2)=-1.
      RAS(M,3)=0.
      GO TO 204
599   RAS(M,1)=1.
      IF(IND.EQ.1)RAS(M,1)=-1.
      RAS(M,2)=0.
      RAS(M,3)=0.
      GO TO 204
444   IAS1(I)=1
      DO 446 K=1,3
      RAS(M,K)=UG(I,K)
446   CONTINUE
      GO TO 666
445   IAS2(I)=1
```

```
      DO 447 K=1,3
      RAS(M,K)=-UD(I+1,K)
447   CONTINUE
      GO TO 900
204   IF(TND.EQ.1)GO TO 500
      IAS1(I)=IAS(I)
666   M=I+N
      IF(KSI2(I+1).EQ.1)GO TO 445
      IF(I.EQ.N)GO TO 505
      DO 25   K=1,3
      IMUL1=0
      IF(ROU(M,K).EQ.0.)IMUL1=1
      IMUL2=0
      IF(EU1(I,K).EQ.0.)IMUL2=1
      IF(IMUL1.NE.IMUL2)GO TO 600
      IF(IMUL1.EQ.0.AND.IMUL2.EQ.0)GO TO 79
      IDEL(K)=0.
      GO TO 25
79    IDEL(K)=ROU(M,K)/EU1(I,K)
25    CONTINUE
      IND=1
      GO TO 300
500   IAS2(I)=IAS(I)
900   RETURN
505   IAS2(I)=0
      RETURN
600   IAS2(I)=0
      RAS(M,K)=0.
      RAS(M,K)=0.
      RAS(M,K)=0.
      RETURN
      END
```

```
C***********************************************************************
C       MODULE:          M  O  D  E  L                                  *
C***********************************************************************
C                                                                       *
C       FUNCTION:                                                       *
C                                                                       *
C       Main program for robot mathematical model                       *
C       computation based on Newton - Euler's method.                   *
C                                                                       *
C       For given joint coordinates and velocities, and                 *
C       parameters of links, this program calculates                    *
C       dynamic model matrices. The dynamic model is                    *
C                                                                       *
C               P = H(q)q" + h(q,q')                                    *
C                                                                       *
C       where H(q) is the inertia matrix which is n x n                 *
C       dimensional, symetric and positive definite. This               *
C       matrix depends on joint coordinate vector q.                    *
C       h(q,q') is the n dimensional vector which depends               *
C       on joint coordinates and velocities. P is the                   *
C       driving torque/force vector.                                    *
C                                                                       *
C       This program also calcualtes the driving torque/force           *
C       vector, if it is pointed out by an input pointer (ITIP)         *
C                                                                       *
C-----------------------------------------------------------------------*
C
C       INPUT VARIABLES:                                                *
C                                                                       *
C               NDIM     -   number of joints                           *
C                                                                       *
C               SI       -   joint coordinate vector                    *
C               SIDOT    -   joint velocities vector                    *
C               SIDDOT   -   joint acceleration vector                  *
C                                                                       *
C                            (SI,SIDOT and SIDDOT comes through         *
C                            common region MODELM.MOD)                  *
C               ITIP     =   0 - evaluation of model matrices           *
C                        =   1 - evaluation of driving torque/force     *
C                             vector                                    *
C                                                                       *
C               EU(6,3) -    joint axes in local coordinate systems     *
C                                                                       *
C                            (EU(i,1), EU(i,2), EU(i,3)) - unit         *
C                            vector of joint axis i in local coo-       *
C                            rdinate system of link i                   *
C                                                                       *
C               EU1(6,3)-    unit vectors of joint axes "i" given       *
C                            in local coordinate systems of links       *
C                            "i+1"                                      *
C                                                                       *
C                            (EU1(i,1),EU1(i,1),EU1(i,3)) - unit        *
C                            vector of joint axis i in local coo-       *
C                            rdinate system of link i+1                 *
C                                                                       *
C               ROU(11,3) - vectors R(i,i), R(i,i+1) given with         *
C                            respect to local coordinate system i,      *
C                            for i = 1,2,...,6                          *
```

```
C                                                                          *
C                          (ROU(i,1), ROU(i,2), ROU(i,3)) -               *
C                          vector Rii                                     *
C                          (ROU(6+i,1),ROU(6+i,2),ROU(6+i,3))-            *
C                          -vector Ri,i+1                                 *
C                                                                          *
C                                                                          *
C       Characteristic vectors of LINK i :                                *
C                                                                          *
C       ---------------------------------------------------               *
C       |                                                 |               *
C       |       ^           ^              ^              |               *
C       |       |Ei         |Zi            |Ei+1          |               *
C       |       |    Rii    |      Ri,i+1  |              |               *
C       |       *---------->0<-------------*              |               *
C       |   joint i        / \         joint i+1          |               *
C       |                 /   \                           |               *
C       |               Xi     Yi                         |               *
C       |             local coordinate                    |               *
C       |             system of link i                    |               *
C       ---------------------------------------------------               *
C                                                                          *
C-------------------------------------------------------------------------*
C                                                                          *
C       OUTPUT VARIABLES:                                                 *
C                                                                          *
C               H       Inertia matrix H(q) (if ITIP=0)                   *
C               H0      Vector h(q,q') (if ITIP=0)                        *
C               UMOM    Driving torque/force vector (if ITIP=1)           *
C                                                                          *
C-------------------------------------------------------------------------*
C
      SUBROUTINE MODEL
      REAL MI(6,3),LAMD(3)
      COMMON /MOM/ MI
      COMMON/SILA/ FI(6,3)
      DIMENSION R0(3)
      DIMENSION RI(3)
      INCLUDE 'CONFIG.MOD'
      INCLUDE 'MODELM.MOD'
      DIMENSION BETI(6,6,3),BETI0(6,3),AI(6,6,3),AI0(6,3)
      DIMENSION R00(3)
      DIMENSION BII(6,6,3),BI0(6,3)
      DIMENSION OMEGI(3),OMEG0(3),ALP(6,6,3),ALP0(6,3)
      DIMENSION R(6,6,3)
      DIMENSION R00P(3),R0P(3),RIP(3)
      DIMENSION ALSII(6,6,3),ALNII(6,6,3),T(3,3),S(3)
      DIMENSION ALS0(6,3),ALN0(6,3)
      DIMENSION VL(3),BB(6,6,3),B1(6,3)
C
      DIMENSION SPOM(6),Q0(3,3)
C
C
      COMMON /MODOPC/ MOD,ITIP,MODLIN
      COMMON /MODSW / KOD
      COMMON /VIBRO/ R
C-----------------------------------------------------------------------
      KOD=1
C-----------------------------------------------------------------------
C       Initialization
```

```
C-----------------------------------------------------------------
      DO 5 I=1,N
      DO 5 K=1,3
      BETI0(I,K)=0.
      AI0(I,K)=0.
      BI0(I,K)=0.
      MI(I,K)=0.
      FI(I,K)=0.
      ALP0(I,K)=0.
5     OMEG0(K)=0.
C-----------------------------------------------------------------
C     Main loop I = 1,....N
C-----------------------------------------------------------------
      DO 10 I=1,N
      DO 15 K=1,3
      RI(K)=ROU(I,K)
      RIP(K)=RAS(I,K)
      IF(IAS1(I).EQ.1.AND.I.EQ.1) R0P(K)=R0POC(K)
      IF(I.NE.1) GO TO 15
      E(1,K)=EPOC(K)
      R0(K)=R0POC(K)
15    CONTINUE
C-----------------------------------------------------------------
C     Asembling process
C-----------------------------------------------------------------
      IF(IAS1(I).EQ.1) GO TO 300
      IF(KSI2(I).EQ.1) GO TO 300
      CALL ASSEM(R0,RI,E,EU,I,Q)
      GO TO 301
300   CALL ASSEM(R0P,RIP,E,EU,I,Q)
301   CONTINUE
C-----------------------------------------------------------------
C     Calculation of transformation matrices
C-----------------------------------------------------------------
      CALL ZAKRET(E,Q,I,ROU,RI,R0,EU,R00,R0P,RIP,R00P,RAS)
C-----------------------------------------------------------------
C     Calculation of position vectors wrt inertial frame
C-----------------------------------------------------------------
      IF(MODLIN.NE.1) GO TO 5000
      CALL DPOS(I,E,RI,R0)
 5000 CONTINUE
C
      CALL TOTR(R,RI,R00,E,I)
C
      IF(MODLIN.NE.1) GO TO 5001
      CALL DTOTR(I)
 5001 CONTINUE
      IF(MOD.EQ.3) GO TO 10
      IF(MOD.EQ.2) GO TO 10
C-----------------------------------------------------------------
C     Calculation of coefficients for angular and
C     linear velocities
C-----------------------------------------------------------------
      CALL BRZUB(OMEGI,OMEG0,E,ALP,ALP0,BETI,BETI0,
     1R00,RI,I)
C
      IF(MODLIN.NE.1) GO TO 5002
      CALL DOMEG(I)
      CALL DALFA(I)
      CALL DALFA0(I,E,OMEG0)
```

```
      CALL DDOMEG(I,E,OMEGO)
      CALL DBETAI(I,RI,E)
      CALL DBETA(I,RI,R00,ALP)
      CALL DBETA0(I,E,RI,R00,OMEGI,OMEGO,ALP0)
      CALL DDBE0(I,E,RI,R00,OMEGO,OMEGI)
 5002 CONTINUE
C-----------------------------------------------------------------
C     Calculation of coefficients of acceleration
C-----------------------------------------------------------------
      CALL SILAR(BETI,BETI0,AI,AI0,I)
C
      IF(MODLIN.NE.1) GO TO 5003
C
      CALL DINAM(I)
 5003 CONTINUE
      IF(KSI1(I).EQ.0) GO TO 50
C-----------------------------------------------------------------
C     Calcultion of acceleration coefficients for
C     rod - approximation
C-----------------------------------------------------------------
      CALL MOM11(I,RI,OMEGI,S)
      CALL MOM21(I,ALP,ALP0,ALSII,ALNII,ALS0,ALN0,S)
      CALL MOM2(I,BII,BI0,ALSII,ALS0,ALNII,ALN0)
C
      IF(MODLIN.NE.1) GO TO 5004
C
      CALL DINAMC(I,E,S)
      CALL DBIC(I,ALP,S)
      CALL DTAUC(I,S,OMEGI)
      CALL DBI0C(I,S,ALP0)
      CALL DDTAUC(I,E,OMEGI,S)
      CALL DDBI0C(I,S)
 5004 CONTINUE
      GO TO 60
   50 CONTINUE
C-----------------------------------------------------------------
C     Acceleration coefficients for body case
C-----------------------------------------------------------------
      CALL MOM0(OMEGI,Q,I,T,LAMD,VL)
      CALL MOM10(BII,BI0,T,ALP,ALP0,LAMD,I)
C
      IF(MODLIN.NE.1) GO TO 5005
C
C
      CALL DTSB(I)
      CALL DBISB(I,T,ALP)
      CALL DVSB(I,E,OMEGI)
      CALL DDB0SB(I,VL,T,ALP0)
 5005 CONTINUE
   60 CONTINUE
C-----------------------------------------------------------------
C     Calculation of moments of inertial forces(ITIP=1)
C-----------------------------------------------------------------
      IF(ITIP.EQ.1) CALL MOMI(BII,BI0,I)
C
      DO 20 K=1,3
      OMEGO(K)=OMEGI(K)
   20 CONTINUE
      IF(MODLIN.NE.1) GO TO 5006
      CALL DTRANS(I)
```

```
 5006   CONTINUE
C-----------------------------------------------------------------
C       End of main loop
C-----------------------------------------------------------------
   10   CONTINUE
C       IF(MOD.EQ.3) GO TO 1
        IF(MOD.EQ.3)RETURN
        IF(ITIP.EQ.1) GO TO 65
C-----------------------------------------------------------------
C       Calculation of inertia matrix H and vector h (ITIP=0)
C-----------------------------------------------------------------
        CALL MATH(R,AI,BII,E,HH,BB)
C
        IF(MODLIN.NE.1) GO TO 5007
C
        CALL DHMOD(E,BB,AI,R)
C
 5007   CONTINUE
        CALL VEKH(R,AI0,BI0,E,H1,B1)
        IF(MODLIN.NE.1) GO TO 5008
        CALL DH1MOD(B1,E,AI0,R)
 5008   CONTINUE
        RETURN
C-----------------------------------------------------------------
C       Calculation of mechanical torques
C       UMOM  (ITIP=1)
C-----------------------------------------------------------------
65      CALL MOMUK(R,E,UMOM)
        RETURN
        END
```

```
C-----------------------------------------------------------------
C   MODULE:          MODEL
C-----------------------------------------------------------------
C   SUBROUTINE:      ASSEM
C-----------------------------------------------------------------
C   FUNCTION:        Assembling of the mechanism
C-----------------------------------------------------------------
C   INPUT VARIABLES:
C           I        index of joint to be assembled
C           ROA      vector rii already assembled
C           ROU      vector rii to be assembled
C           EA       vector ei already assembled
C           EU       vector ei to be assembled
C-----------------------------------------------------------------
C   OUTPUT:
C           A4       Transformation matrix Qi, (Ai) after
C                    assembling
C-----------------------------------------------------------------
C   SUBROUTINE CALLED:
C           CROSS0   cross product
C           CROSS2   cross product
C           PRMM     product of two matrices
C-----------------------------------------------------------------
   SUBROUTINE ASSEM(ROA,ROU,EA,EU,I,A4)
   DIMENSION ROA(3),ROU(3),EA(6,3),EU(6,3),A2(3,3)
   DIMENSION TEMP0(3),TEMP1(3),TEMP2(3),TEMP3(3)
   DIMENSION E1(3),E3(3)   ,A3(3,3),A4(3,3)
     COMMON /POMOC/ E1,E3,TEMP0,TEMP1,TEMP2,TEMP3,A2,A3
C-----------------------------------------------------------------
   DO 10 K=1,3                      ! prepare vectors for assembling
   E1(K)=EA(I,K)
   E3(K)=EU(I,K)
10 CONTINUE
   CALL CROSS2(E1,ROA,TEMP1)        ! a=ei X (ei X rii)
   CALL CROSS2(E3,ROU,TEMP3)        ! a=ei X (ei X rii) link coord.s.
   T1=0.
   T3=0.
     DO 15 K=1,3                           ! prepare for |a|
   T1=T1+TEMP1(K)*TEMP1(K)
   T3=T3+TEMP3(K)*TEMP3(K)
15 CONTINUE
   T1=1./SQRT(T1)
   T3=1./SQRT(T3)
   CALL CROSS0(TEMP1,EA,TEMP0,I)           ! b=a X ei
   CALL CROSS0(TEMP3,EU,TEMP2,I)           ! b=a X ei link coord.s.
   DO 20 K=1,3                      ! forming Q matrix components
   A2(K,1)=TEMP1(K)*T1                     ! aa=a/|a| (fixed)
   A2(K,2)=EA(I,K)                         ! ei
   A2(K,3)=TEMP0(K)*T1                     ! b
   A3(1,K)=-TEMP3(K)*T3                    ! aa=-a/|a| (internal)
   A3(2,K)=EU(I,K)                         ! ei
   A3(3,K)=-TEMP2(K)*T3                    ! -b
20 CONTINUE
     CALL PRMM(A2,A3,A4,3,3)               ! Q=[a2][a3]
   RETURN
   END
```

```
C----------------------------------------------------------------
C       SUBROUTINE:      CROSS2
C----------------------------------------------------------------
C       FUNCTION:        Double cross product a x (a x b)
C                        a and b are 3x1 vectors
C----------------------------------------------------------------
C       INPUT VARIABLES:
C                OMEGO   vector 3x1
C                RO      vector 3x1
C----------------------------------------------------------------
C       OUTPUT VARIABLES:
C                TEMP4   result vector for double cross  product
C                        OMEGO(x) X (OMEGO(x) X RO(x))
C----------------------------------------------------------------
C       SUBROUTINE CALLED:
C                DOT1    dot product of two 3x1 vectors
C----------------------------------------------------------------
        SUBROUTINE CROSS2(OMEGO,RO,TEMP4)
        DIMENSION OMEGO(3),RO(3),TEMP4(3)
         TEMP=DOT1(OMEGO,RO)
        TEMP6=DOT1(OMEGO,OMEGO)
         DO 211 K=1,3
        TEMP4(K)=TEMP*OMEGO(K)-TEMP6*RO(K)
  211   CONTINUE
        RETURN
        END
```

```
C----------------------------------------------------------------
C       SUBROUTINE:      DOT1
C----------------------------------------------------------------
C       FUNCTION:        Scalar product of two vectors (Dot product)
C----------------------------------------------------------------
C       INPUT VARIABLES:
C                OMEGO   vector (3x1)
C                RO      vector (3x1)
C----------------------------------------------------------------
C       OUTPUT VARIABLES:
C                DOT1    Dot product OMEGO.RO
C----------------------------------------------------------------
C
        FUNCTION DOT1(OMEGO,RO)
        DIMENSION OMEGO(3),RO(3)
        DOT1=OMEGO(1)*RO(1)+OMEGO(2)*RO(2)+OMEGO(3)*RO(3)
        RETURN
        END
```

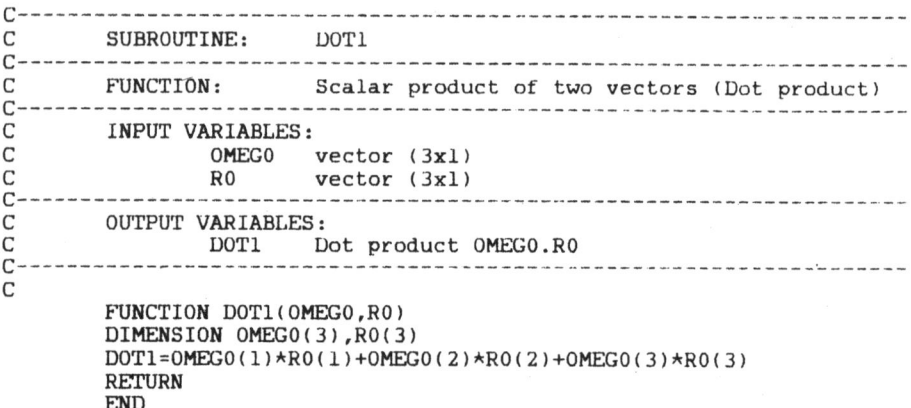

```
C----------------------------------------------------------------------
C       SUBROUTINE:      CROSSO
C----------------------------------------------------------------------
C       FUNCTION:        Cross product of two vectors
C                        First vector is 3x1 and the second is the i-th
C                        row of the matrix 6x3
C----------------------------------------------------------------------
C       INPUT VARIABLES:
C               OMEGO    vector 3x1
C               E        matrix 6x3
C               I        index of the i-the row of the matrix E
C----------------------------------------------------------------------
C       OUTPUT VARIABLES:
C               TEMP1    cross product (3x1 vector) OMEGO(x) X E(I,x)
C----------------------------------------------------------------------
      SUBROUTINE CROSSO(OMEGO,E,TEMP1,I)
     DIMENSION OMEGO(3),E(6,3)
      DIMENSION TEMP1(3)
      TEMP1(1)=OMEGO(2)*E(I,3)-OMEGO(3)*E(I,2)
      TEMP1(2)=OMEGO(3)*E(I,1)-OMEGO(1)*E(I,3)
      TEMP1(3)=OMEGO(1)*E(I,2)-OMEGO(2)*E(I,1)
      RETURN
      END
```

```
C----------------------------------------------------------------------
C       SUBROUTINE:      PRMM
C----------------------------------------------------------------------
C       FUNCTION:        Product of two square matrices (nxn)
C----------------------------------------------------------------------
C       INPUT VARIABLES:
C               AM       Matrix nxn
C               BM       Matrix nxn
C               N        Dimension n
C----------------------------------------------------------------------
C       OUTPUT VARIABLES:
C               RM       Result matrix nxn
C----------------------------------------------------------------------
      SUBROUTINE PRMM(AM,BM,RM,N)
      DIMENSION AM(N,N),BM(N,N),RM(N,N)
      DO 10 J=1,N
      DO 10 K=1,N
      RM(J,K)=0.
      DO 10 L=1,N
10    RM(J,K)=RM(J,K)+AM(J,L)*BM(L,K)
      RETURN
      END
```

```
C-----------------------------------------------------------------------
C     MODULE:      MODEL
C-----------------------------------------------------------------------
C     SUBROUTINE: ZAKRET
C-----------------------------------------------------------------------
C     FUNCTION:         Calculation of transformation matrix Q0i (Ai)
C                       and ei=Qi.ei(int), rii=Qi.rii(int),ri,i+1=Qi.
C                       ri,i+1(int)
C-----------------------------------------------------------------------
C     INPUT VARIABLES:
C              N        no. of d.o.f
C              I        index of joint/link being considered
C              KSI2     type of joint
C              SI       internal coordinate qi
C              E        joint axes before change in qi
C              Q        transformation matrix before change in qi
C              EU       joint axes wrt link coord. system
C              KSI3     speciality of joint/link
C              DESNI    orientation of coord. systems
C              ROU      vectors rii and ri,i+1 wrt link coord.sys.
C              RAS      special vectors rii and ri,i+1 wrt link coord.
C              RT       vector rn,n+1
C              RI       vector rii      before change in qi
C              R0       vector ri-1,i before change in qi
C              RIP      special vector rii before change in qi
C              R0P      special vector ri-1,i before change in qi
C-----------------------------------------------------------------------
C     OUTPUT VARIABLES:
C              Q        transformation matrix Q-i (A-i)
C              RI       vectore rii after change in qi
C              R0       vector ri,i+1
C              RIP      special vector rii
C              R0P      special vector ri,i+1
C              E        joint axes after change in qi (ei,ei+1)
C              RT0      vector rn,n+1
C-----------------------------------------------------------------------
C     SUBROUTINE CALLED:
C              PRMV     matrix multiplied by vector
C              CROSS    cross product
C              DOT      dot product
C-----------------------------------------------------------------------
      SUBROUTINE ZAKRET(E,Q,I,ROU,RI,R0,EU,R00,R0P,RIP,R00P,RAS)
        INTEGER DESNI
      DIMENSION R00(3)         ,R0P(3),R00P(3),RIP(3),RAS(11,3)
      DIMENSION E(6,3),EU(6,3),Q(3,3),RI(3),R0(3)
      DIMENSION ROU(11,3)      ,POM(3)
      DIMENSION TEMP0(3),TEMP1(3),TEMP2(3),TEMP3(3),TEMP4(3)
        DIMENSION TEMP5(3),TEMP6(3)
      COMMON /TIPL/ N,KSI1(6),KSI2(6)    ,KSI3(6)
        COMMON /BRZL/ RT(3),RP(6,6,3),RT0(3)
      COMMON /UG/ SI(6),SIDOT(6)
        COMMON /SISTEM/ DESNI
C
        COMMON/ZZZCE/IAS1(6),IAS2(6)
        COMMON/ZZZCD/EU1(6,3)
C
        COMMON /POMOC/ TEMP0,TEMP1,TEMP2,TEMP3,TEMP4,TEMP5
```

```
        1       ,TEMP6,POM,SND,CSD
C
C -----------------------------------------------------------------
c       Initial calculations for index i
C       and preparation of all vectors
C -----------------------------------------------------------------
        IF(KSI2(I).EQ.1) GO TO 5
      SN=SIN(SI(I))
      CS=COS(SI(I))
        SND=SN*DESNI
5       CONTINUE
      DO 10 K=1,3
      TEMP0(K)=EU(I,K)
      TEMP2(K)=ROU(I,K)
        IF(IAS1(I).EQ.1) TEMP6(K)=RAS(I,K)
        IF(I.EQ.N) GO TO 100
      TEMP1(K)=EU1(I,K)
      TEMP3(K)=ROU(I+N,K)
        IF(IAS2(I).EQ.1) TEMP5(K)=RAS(I+N,K)
        GO TO 10
100     RT0(K)=RT(K)
        TEMP3(K)=RT(K)
   10 CONTINUE
C -----------------------------------------------------------------
C       Calculation of ei,ri-1,i before change in qi,
C       calculation of special vector ri-1,i
C -----------------------------------------------------------------
        CALL PRMV(Q,TEMP0,TEMP4,3)
      DO 11 K=1,3
      R00(K)=R0(K)
        IF(IAS2(I).EQ.1) R0OP(K)=R0P(K)
   11 E(I,K)=TEMP4(K)
        IF(KSI2(I).EQ.1) GO TO 50
C -----------------------------------------------------------------
C       RODRIGUES FORMULA
C       Calculation of transformation matrix Qi (Ai) after
C       change of coord. qi
C -----------------------------------------------------------------
      DO 15 J=1,3
      TEMP=DOT(E,Q,I,J)
        CSD=(1.-CS)*TEMP
      CALL CROSS(E,Q,TEMP0,I,J)
      DO 15 K=1,3
      Q(K,J)=Q(K,J)*CS+CSD*E(I,K)+SND*TEMP0(K)
   15 CONTINUE
50      CONTINUE
C -----------------------------------------------------------------
C       Calculation of ei+1, rii, ri,r+1 after change in qi
C       calculation of special vectors rii,ri,i+1
C -----------------------------------------------------------------
        CALL PRMV(Q,TEMP2,RI,3)
        IF(IAS1(I).EQ.1) CALL PRMV(Q,TEMP6,RIP,3)
        IF(IAS2(I).EQ.1) CALL PRMV(Q,TEMP5,R0P,3)
40      CONTINUE
      IF(I.EQ.N) GO TO 200
        CALL PRMV(Q,TEMP3,R0,3)
        CALL PRMV(Q,TEMP1,TEMP3,3)
      DO 20 K=1,3
      E(I+1,K)=TEMP3(K)
        IF(KSI2(I).EQ.1) RI(K)=RI(K)+E(I,K)*SI(I)
```

```
   20 CONTINUE
      RETURN
C _________________________________________________
C        Last link rn,n+1
C _________________________________________________
200      CALL PRMV(Q,TEMP3,POM,3)
         DO 21 K=1,3
         IF(KSI2(I).EQ.1) RI(K)=RI(K)+E(I,K)*SI(I)
21       RT0(K)=POM(K)
         RETURN
      END


C-----------------------------------------------------------------
C       SUBROUTINE:      CROSS
C-----------------------------------------------------------------
C       FUNCTION:        Cross product of two vectors (Vector product)
C                        One vector is the i-th row of 6x3 matrix and
C                        the other is the j-th column of 3x3 matrix
C-----------------------------------------------------------------
C       INPUT VARIABLES:
C               E        matrix 6x3
C               Q        matrix 3x3
C               I        index of the i-the row of matrix E
C               J        index of the j-th column of matrix Q
C-----------------------------------------------------------------
C       OPUTUT VARIABLES:
C               TEMP0    cross product E(i,x)XQ(x,j)
C-----------------------------------------------------------------
      SUBROUTINE CROSS(E,Q,TEMP0,I,J)
     DIMENSION E(6,3),Q(3,3),TEMP0(3)
      TEMP0(1)=E(I,2)*Q(3,J)-E(I,3)*Q (2,J)
      TEMP0(2)=E(I,3)*Q(1,J)-E(I,1)*Q(3,J)
      TEMP0(3)=E(I,1)*Q(2,J)-E(I,2)*Q(1,J)
      RETURN
      END


C-----------------------------------------------------------------
C       SUBROUTINE:      DOT
C-----------------------------------------------------------------
C       FUNCTION:        Scalar product of two vectors
C                        One vector is the i-th row of 6x3 matrix, the
C                        other is the j-th column of 3x3 matrix)
C-----------------------------------------------------------------
C       INPUT VARIABLES:
C               A        matrix 6x3
C               B        matrix 3x3
C               I        index of the i-th row
C               J        index of the j-th column
C-----------------------------------------------------------------
C       OUTPUT VARIABLES:
C               DOT      scalar product A(i,x).B(x,j)
C-----------------------------------------------------------------
C
      FUNCTION DOT(A,B,I,J)
      DIMENSION A(6,3),B(3,3)
      DOT=A(I,1)*B(1,J)+A(I,2)*B(2,J)+A(I,3)*B(3,J)
      RETURN
      END
```

```
C*********************************************************************
C    MODULE:           MODEL, linearisation
C--------------------------------------------------------------------
C    SUBROUTINE:       DPOS
C-------------------------------------------------------------------|
C    FUNCTION:         Calculation of the expresions:               |
C                      dl(q(i,j))=(e(l)xq(i,j))*NKSI2(l)*ETA(l,i)   |
c                      where:                                        |
c                      NKSI2(l)=1-KSI2(l)=1 for rotational joint else 0|
c                      ETA(l,i)=1 for l=<i else 0                    |
c                                                                    |
c                      dl(rii)=(elxrii)*NKSI2(l)*ETA(l,i)+           |
c                              el*KSI2(i)*DELTA(l,i)                 |
c                      (for i=1,N,  l=1,i)                           |
c                      where:                                        |
c                      DELTA(l,i)=1 for l=i else 0                   |
c                                                                    |
c                      dl(ri,i+1)=(elx(ri,i+1))*NKSI2(l)*ETA(l,i)    |
c                      (for i=1,N-1,  l=1,N)                         |
c                                                                    |
c                      dl(ei+1)=(elxei+1)*NKSI2(l)*ETA(l,i)          |
c                      (for i=1,N-1)                                 |
c                                                                    |
c                                                                    |
C-------------------------------------------------------------------|
C    INPUT VARIABLES:                                                |
c                                                                    |
c            I        - degree of freedom i                          |
c            E(6,3)   - joint unit vectors wrt fixed coord. syst.    |
c            RI(3)    - vector rii                                   |
c            R0(3)    - vector ri,i+1                                |
c            Q(3,3)   - transformation matrix from local to fixed    |
c                       coord. syst.                                 |
c            KSI2(6)  - joint type (0 if rotational else 1)          |
c                                                                    |
C-------------------------------------------------------------------|
C    OUTPUT VARIABLES:                                               |
c                                                                    |
c            DQ(3,3,6)=DQ(k,j,l) - l-derivative of the k-th component|
c                                  of q(i,j)                         |
c            DRI(3,6)=DRI(k,l)=dl(rii) - l-derivative of rii         |
c            DR0(3,6)=DR0(k,l)=dl(ri,i+1) - l-derivative of ri,i+1   |
c            DE(3,15)=DE(k,IL)=dl(ei) - l-derivative of ei           |
c            DRIP(3,21) - matrix whith rows equal to dl(rii)         |
c                                                                    |
C-------------------------------------------------------------------|
C    SUBROUTINES CALLED:                                             |
c            CROSS,CROSS0,CROSS6,ILF1                                |
C*******************************************************************|
C
      SUBROUTINE DPOS(I,E,RI,R0)
      DIMENSION E(6,3),RI(3),R0(3)
C
      INCLUDE 'DMOD.MOD'
      INCLUDE 'DCONF.MOD'
      INCLUDE 'MODLM1.MOD'
C
C
```

```
      DO 10 L=1,I
      DO 10 J=1,3
      IF(KSI2(L).EQ.1) GO TO 11
      CALL CROSS(E,Q,TEMP0,L,J)
      DO 12 K=1,3
   12 DQ(K,J,L)=TEMP0(K)
      GO TO 10
   11 DO 14 K=1,3
   14 DQ(K,J,L)=0.
   10 CONTINUE
C
      I1=I+1
      DO 20 L=1,I
      IF(KSI2(L).EQ.1) GO TO 21
      CALL CROSS0(RI,E,TEMP0,L)
      IF(I.EQ.N) GO TO 24
      CALL CROSS0(R0,E,TEMP1,L)
      CALL CROSS6(E,TEMP2,L,I1)
   24 CONTINUE
      DO 22 K=1,3
      DRI(K,L)=-TEMP0(K)
      IF(I.EQ.N) GO TO 22
      DR0(K,L)=-TEMP1(K)
      I1L=ILF(I1,L)
      DE(K,I1L)=TEMP2(K)
   22 CONTINUE
      GO TO 28
   21 IF(L.LT.I) GO TO 29
      DO 23 K=1,3
   23 DRI(K,L)=E(I,K)
      GO TO 28
   29 DO 33 K=1,3
   33 DRI(K,L)=0.
   28 IL=ILF1(I,L)
      DO 25 K=1,3
   25 DRIP(K,IL)=DRI(K,L)
   20 CONTINUE
      RETURN
      END


C----------------------------------------------------------------------
C       SUBROUTINE:     CROSS6
C----------------------------------------------------------------------
C       FUNCTION: Auxiliary subroutine
C----------------------------------------------------------------------
C
        SUBROUTINE CROSS6(E,TEMP0,L,I)
        DIMENSION E(6,3),TEMP0(3)
        TEMP0(1)=E(L,2)*E(I,3)-E(L,3)*E(I,2)
        TEMP0(2)=E(L,3)*E(I,1)-E(L,1)*E(I,3)
        TEMP0(3)=E(L,1)*E(I,2)-E(L,2)*E(I,1)
        RETURN
        END
```

```
C-----------------------------------------------------
C       SUBROUTINE:    ILF1
C-----------------------------------------------------
C       FUNCTION:      Auxilary subroutine
C-----------------------------------------------------
        FUNCTION ILF1(I,L)
        IF(I.EQ.1) GO TO 20
        ILF1=0
        DO 10 K=1,I-1
   10   ILF1=ILF1+K
        ILF1=ILF1+L
        RETURN
   20   ILF1=1
        RETURN
        END
```

```
C---------------------------------------------------------------------
C       MODULE:            MODEL
C---------------------------------------------------------------------
C       SUBROUTINE:        TOTR
C---------------------------------------------------------------------
C       FUNCTION:          Calculates position vectors rij,j=1,...i-1
C---------------------------------------------------------------------
C       INPUT VARIABLES:
C               N          No. of d.o.f
C               DESNI      orientation of the system
C                          DESNI=1 (right or. default)
C               I          index of link/joint being considered
C               KSI2       type of joint
C               R          array of vectors ri-1,j,j=1,...i-1
C               E          array of joints axes
C               RI         vector rii
C               R0         vector ri-1,i
C               RT0        last segment- vector rn,n+1
C               MOD        type of calculation
C                          MOD=0 or 1 or 3 only R ,RP  are calculated
C                          MOD=3 R, RP and DEL are calculated
C---------------------------------------------------------------------
C       OUTPUT VARIABLES:
C               R          array of position vectors rij
C                          link centers wrt joints centers
C               RP         array of position vectors rij ,where for the
C                          last segment the tip is considered ( vector
C                          rn,n+1 taken into account)
C               DEL        array of Jacobian components
C                          (only calculated if MOD=3)
C---------------------------------------------------------------------
C       SUBROUTINE CALLED:
C               CROSS4  cross product
C---------------------------------------------------------------------
      SUBROUTINE TOTR(R,RI,R0,E,I)
      DIMENSION RI(3),R0(3),R(6,6,3)
        DIMENSION E(6,3)
        DIMENSION TEMP(3),TEMP1(3),TEMP2(3)
        INTEGER DESNI
```

```
      COMMON /SISTEM/ DESNI
      COMMON /BRZL/ RT(3),RP(6,6,3),RT0(3)
      COMMON /BRZUG/ DEL(6,6,3)
      COMMON /TIPL/ N,KSI1(6),KSI2(6)
      COMMON /POMOC/ TEMP,TEMP1,TEMP2
      COMMON /MODOPC/ MOD
      COMMON /ZZZCF/ RTT(6,6,3)
C ______________________________________________________________
C     POSITION VECTORS
C ______________________________________________________________
      DO 5 K=1,3                              ! initialization
      TEMP2(K)=0.
      TEMP1(K)=0.
5     TEMP(K)=RI(K)-R0(K)
    DO 15 J=1,I                               ! position vectors rij and rp(ij)
150   CONTINUE
    DO 15 K=1,3
    IF(J-I) 12,20,12

   20 R(J,I,K)=RI(K)
        RP(J,I,K)=R(J,I,K)
        IF(I.EQ.N) RP(I,I,K)=R(I,I,K)-RT0(K)
        IF(N.GT.3.AND.I.EQ.4) GO TO 444
        GO TO 15
444     RTT(I,I,K)=RP(I,I,K)
        RP(I,I,K)=0.
        GO TO 15
12    R(I,J,K)=R(I-1,J,K)+TEMP(K)
        RP(I,J,K)=R(I,J,K)
        IF(N.GT.3.AND.I.EQ.4) RP(I,J,K)=R(I,J,K)-RI(K)
        IF(I.EQ.N) RP(I,J,K)=R(I,J,K)-RT0(K)
   15 CONTINUE
C ______________________________________________________________
C       COMPONENTS OF THE JACOBIAN MATRIX
C ______________________________________________________________
        IF(MOD.NE.3) RETURN                   ! only if mod=3
        DO 25 J=1,I
        KSI=1-KSI2(J)
        IF(KSI2(J).NE.1) CALL CROSS4(RP,E,TEMP1,I,J,J)
        DO 25 K=1,3
        DEL(I,J,K)=-TEMP1(K)*KSI+E(J,K)*KSI2(J)
        DEL(I,J,K)=DEL(I,J,K)*DESNI
25      CONTINUE
      RETURN
        END
```

```
C-------------------------------------------------------------------
C    SUBROUTINE:     CROSS4
C-------------------------------------------------------------------
C    FUNCTION:       Cross product of the l-th row and i-th column
C                    of the 6x6x3 array as the first vector and the
C                    j-the row of the 6x3 matrix as the second vector
C-------------------------------------------------------------------
C    INPUT VARIABLES:
C           R        array 6x6x3
C           AI       matrix 6x3
C           L0       index of the l-th row of the array R
C           I        index of the i-th column of the array R
C           J        index of the j-th row of the matrix AI
C-------------------------------------------------------------------
C    OUTPUT VARIABLES:
C           TEMP0    result 3xl vector:
C                    R(l,i,x) X AI(j,x)
C-------------------------------------------------------------------
      SUBROUTINE CROSS4(R,AI,TEMP0,L0,I,J)
     DIMENSION R(6,6,3),AI(6,3),TEMP0(3)
      TEMP0(1)=R(L0,I,2)*AI(J,3)-R(L0,I,3)*AI(J,2)
      TEMP0(2)=R(L0,I,3)*AI(J,1)-R(L0,I,1)*AI(J,3)
      TEMP0(3)=R(L0,I,1)*AI(J,2)-R(L0,I,2)*AI(J,1)
       RETURN
      END
```

```
C*******************************************************************|
C   MODULE:  MODEL, linearisation                                   |
C-------------------------------------------------------------------|
C   SUBROUTINE:     DTOTR                                           |
C-------------------------------------------------------------------|
C   FUNCTION: Claculation of dl(r(i,j))                             |
C             dl(r(i,j))=dl(r(i-l,j))+dl(r(i,i))-dl(r(i-l,i)        |
C             (for j=l,i-l,  l=l,i)                                 |
C-------------------------------------------------------------------|
C   INPUT VARIABLES:                                                |
c                                                                   |
c           DRI(3,6)=DRI(k,l)=dl(rii)    l-derivative fo rii        |
c           DR00(3,6)=DR00(k,l)=dl(r(i-l,i) l-derivative of r(i-l,i)|
c           DR(3,50)=dl(r(i,j)) - l-derivative of r(i,j)            |
c           (for i=2,N, j=l,i-l, l=l,i-l), organized as:            |
c                                                                   |
c           i | 2 3 3 3 3       |                                   |
c           j | l l l 2 2       |                                   |
c           l | l l 2 l 2       |                                   |
c                                                                   |
C-------------------------------------------------------------------|
C   OUTPUT VARIABLES:                                               |
c                                                                   |
c           DR(3,50)=dl(r(i,j)) - l-derivative of r(i,j)            |
C-------------------------------------------------------------------|
C   SUBROUTINES CALLED:                                             |
C*******************************************************************|
```

```
C
      SUBROUTINE DTOTR(I)
      INCLUDE 'DMOD.MOD'
C
      IF(I.EQ.1) RETURN
      I1=I-1
      I2=I-2
  100 FORMAT(3F16.5)
      DO 10 L=1,I1
      DO 20 K=1,3
   20 TEMP0(K)=DRI(K,L)-DR00(K,L)
      DO 10 J=1,I1
      IJL=IJLF2(I,J,L)
      IJL1=IJLF2(I1,J,L)
      IF(J.LE.I2) GO TO 30
   40 CONTINUE
      DO 50 K=1,3
   50 DR(K,IJL)=DRP(K,L)+TEMP0(K)
      GO TO 10
   30 IF(L.GT.I2) GO TO 40
      DO 60 K=1,3
   60 DR(K,IJL)=DR(K,IJL1)+TEMP0(K)
   10 CONTINUE
      RETURN
      END
```

```
C--------------------------------------------------------------------
C       MODUL:             MODEL
C--------------------------------------------------------------------
C       SUBROUTINE:        BRZUB
C--------------------------------------------------------------------
C       FUNCTION:          Calculation of coefficients accompanying
C                          angular and linear accelerations beta-ij,
C                          j=1,...,i-1 and alfa-ij,j=1,...,i-1
C                          and coefficients beta0-i,alfa0-i dependent
C                          on velocities
C--------------------------------------------------------------------
C       INPUT VARIABLES:
C                  N       No. d.o.f
C                  I       index of the link/joint currently processed
C                  KSI2    type of joint
C                  SIDOT   internal velocity q'-i
C                  OMEG0   angular velocity omega(i-1)
C                  E       joints axes (ei)
C                  R00     vector ri-1,i
C                  RI      vector rii
C                  MOD     type of calculation
C                          =0 all effects
C                          =1 inertial and gravitational
C                          =2 only gravitational
C--------------------------------------------------------------------
C       OUTPUT VARIABLES:
C                  OMEGI   angular velocity omega(i)
C                  ALP     coefficients alfa(i,j),j=1,...i
C                  ALP0    coefficient alfa0(i)
C                  BETI    coefficients beta(i,j),j=1,...,i
C                  BETI0   coefficient beta0(i)
C--------------------------------------------------------------------
```

```
C       SUBROUTINES CALLED:
C               CROSS0  cross product of two vectors
C               CROSS1  cross product of two vectors
C               CROSS2  double cross product
C               CROSS3  cross product of two vectors
C----------------------------------------------------------------------
      SUBROUTINE BRZUB(OMEGI,OMEGO,E,ALP,ALP0,BETI,BETI0,
        1R00,RI,I)
C
      DIMENSION OMEGI(3),OMEGO(3),E(6,3),RI(3),R00(3)
      DIMENSION ALP(6,6,3),ALP0(6,3),BETI(6,6,3),BETI0(6,3)
C
      DIMENSION TEMP0(3),TEMP1(3),TEMP2(3),TEMP3(3),TEMP4(3),TEMP5(3),
        1TEMP6(3),TEMP7(3),TEMP8(3),TEMP9(3),TEMP10(3)
C
      COMMON /BRZUG/ DEL(6,6,3)
      COMMON /UG/ SI(6),SIDOT(6)
        COMMON /TIPL/ N,KSI1(6),KSI2(6),KSI3(6)
        COMMON /BRZL/ RT(3),RP(6,6,3),RT0(3)
C
        COMMON /POMOC/ TEMP0,TEMP1,TEMP2,TEMP3,TEMP4,TEMP5
        1       ,TEMP6,TEMP7,TEMP8,TEMP9,TEMP10
C
        COMMON /MODOPC/ MOD
C
C--------------------------------------------------------------------
C       Initialization
C--------------------------------------------------------------------
C
        DO 1000 K=1,3
        TEMP1(K)=0.
1000    CONTINUE
        KSI=1-KSI2(I)
      I2=I-1
C--------------------------------------------------------------------
C       Forming of omega(i-1) X ei
C       and q' . (omega(i-1) X ei)
C--------------------------------------------------------------------
        IF(MOD.NE.1) CALL CROSS0(OMEGO,E,TEMP0,I)
      DO 16 K=1,3
        IF(MOD.NE.1) TEMP0(K)=SIDOT(I)*TEMP0(K)
      IF(I2.EQ.0) GO TO 15
C--------------------------------------------------------------------
C       Forming the coefficients alfa and alfa0
C       angular velocity omega-i
C--------------------------------------------------------------------
      TEMP6(K)=RI(K)-R00(K)
        IF(MOD.NE.1.AND.KSI.EQ.0) ALP0(I,K)=ALP0(I-1,K)

        IF(MOD.NE.1.AND.KSI.NE.0) ALP0(I,K)=ALP0(I-1,K)+TEMP0(K)
   15 CONTINUE
        IF(KSI2(I).EQ.1) GO TO 160
      OMEGI(K)=OMEGO(K)+SIDOT(I)*E(I,K)
      ALP(I,I,K)=E(I,K)
        GO TO 16
 160    OMEGI(K)=OMEGO(K)
        ALP(I,I,K)=0.
  16  CONTINUE
```

```
C-----------------------------------------------------------------
C       Final calculation of coefficients alfa and
C       getting ready for coefficients beta(i,j)
C-----------------------------------------------------------------
      IF(I2.EQ.0) GO TO 5
      DO 17 J=1,I2
      CALL CROSS3(ALP,TEMP6,TEMP1,I2,J)
      DO 18 K=1,3
      ALP(I,J,K)=ALP(I-1,J,K)
18    BETI(I,J,K)=BETI(I-1,J,K)+TEMP1(K)
17    CONTINUE
C-----------------------------------------------------------------
C       Final calculation of coefficients beta (i,j),
C       j=1,...,i
C       Calculation of coefficient beta0(i)
C-----------------------------------------------------------------
 5    CONTINUE
        IF(KSI2(I).NE.1) CALL CROSS0(RI,E,TEMP2,I)
        IF(MOD.EQ.1) GO TO 40
        IF(KSI2(I).NE.1) CALL CROSS1(TEMP0,RI,TEMP3)
      CALL CROSS2(OMEGI,RI,TEMP4)
      CALL CROSS2(OMEG0,R00,TEMP5)
      IF(I.NE.1) CALL CROSS0(TEMP6,ALP0,TEMP1,I2)
      DO 30 K=1,3
        IF(KSI2(I).EQ.1) GO TO 33
        BETI0(I,K)=-TEMP1(K)+TEMP3(K)+TEMP4(K)-TEMP5(K)
        BETI(I,I,K)=-TEMP2(K)

       GO TO 30
33     BETI0(I,K)=-TEMP1(K)+TEMP4(K)-TEMP5(K)+2.*TEMP0(K)
       BETI(I,I,K)=E(I,K)
30     IF(I.NE.1) BETI0(I,K)=BETI0(I-1,K)+BETI0(I,K)
      RETURN
40     CONTINUE
       DO 41 K=1,3
       IF(KSI.NE.0) BETI(I,I,K)=-TEMP2(K)
       IF(KSI.EQ.0) BETI(I,I,K)=E(I,K)
41     CONTINUE
       RETURN
       END
```

```
C-----------------------------------------------------------------
C    SUBROUTINE:     CROSS3
C-----------------------------------------------------------------
C    FUNCTION:       Cross product of two vectors
C                    First vector is the 1-th row and the i-th
C                    column of the 6x6x3 array and the second is 3x1
C                    vector
C-----------------------------------------------------------------
C    INPUT VARIABLES:
C            R       array 6x6x3
C            F       vector 3x1
C            L0      index of the 1-th row of the array R
C            I       Index of the j-th column of the array R
C-----------------------------------------------------------------
C    OUTPUT VARIABLES:
C            TEMP0   result vector 3x1: R(1,i,x) X F(x)
C-----------------------------------------------------------------
      SUBROUTINE CROSS3(R,F,TEMP0,L0,I)
     DIMENSION R(6,6,3),F(3),TEMP0(3)
      TEMP0(1)=R(L0,I,2)*F(3)-R(L0,I,3)*F(2)
      TEMP0(2)=R(L0,I,3)*F(1)-R(L0,I,1)*F(3)
      TEMP0(3)=R(L0,I,1)*F(2)-R(L0,I,2)*F(1)
      RETURN
      END
```

```
C-----------------------------------------------------------------
C       SUBROUTINE:      CROSS1
C-----------------------------------------------------------------
C       FUNCTION:        Cross product of two vectors
C-----------------------------------------------------------------
C       INPUT VARIABLES:
C               E0       vector 3x1
C               R0       vector 3x1
C-----------------------------------------------------------------
C       OUTPUT VARIABLES:
C               TEMP2    result vector 3x1
C                        E0xR0
C-----------------------------------------------------------------
        SUBROUTINE CROSS1(E0,R0,TEMP2)
        DIMENSION E0(3),R0(3),TEMP2(3)
        TEMP2(1)=E0(2)*R0(3)-E0(3)*R0(2)
        TEMP2(2)=E0(3)*R0(1)-E0(1)*R0(3)
        TEMP2(3)=E0(1)*R0(2)-E0(2)*R0(1)
        RETURN
        END
```

```
C***********************************************************************
C      MODULE: MODEL, linearisation
C-----------------------------------------------------------------------
C      SUBROUTINE:     DOMEG
C-----------------------------------------------------------------------
C      FUNCTION:  Calculation of dlOMEG(i) - l- derivative of OMEG(i)
c                 dlOMEG(i)=dlOMEG(i-1)+SIDOT(i)*dl(ei)*NKSI2(I)
C                 (for i=2,N;  l=1,i-1)
C-----------------------------------------------------------------------
C      INPUT VARIABLES:
c
c              DOMEGO(3,4)=DOMEGO(k,l)=dlOMEG(k)=dlOMEG(i-1)
c                          - l-derivative of OMEG(i-1)
c              (for i=3,N;  l=1,i-2)
c              DE(3,15)=DE(k,IL)=dl(ei) - l-derivative of ei
c              SIDOT(i) - i-th joint velocity
C-----------------------------------------------------------------------
C      OUTPUT VARIABLES:
c
c              DOMEGI(3,5)=DOMEGI(k,l)=dlOMEG(i) - l-derivative of
C                                                  OMEG(i)
C-----------------------------------------------------------------------
C      SUBROUTINES CALLED:
C***********************************************************************
      SUBROUTINE DOMEG(I)
      INCLUDE 'DMOD.MOD'
      INCLUDE 'DCONF.MOD'
      INCLUDE 'MODLM1.MOD'
      I1=I-1
      IF(I1.EQ.0) RETURN
      DO 10 L=1,I1
      IL =ILF(I,L)
      IF(L.EQ.I1) GO TO 14
      DO 11 K=1,3
   11 DOMEGI(K,L)=DOMEGO(K,L)
      IF(KSI2(I).EQ.1) GO TO 10
      DO 12 K=1,3
   12 DOMEGI(K,L)=DOMEGI(K,L)+SIDOT(I)*DE(K,IL)
      GO TO 10
   14 IF(KSI2(I).EQ.1) GO TO 15
      DO 16 K=1,3
   16 DOMEGI(K,L)=SIDOT(I)*DE(K,IL)
      GO TO 10
   15 DO 17 K=1,3
   17 DOMEGI(K,L)=0.
   10 CONTINUE
      RETURN
      END
```

```
C*******************************************************************
C       MODULE:            MODEL,linearization
C-------------------------------------------------------------------
C       SUBROUTINE:        DALFA
C-------------------------------------------------------------------
C       FUNCTION:          Calculation of d1ALFA(i,j)
c
c                          d1ALFA(i,j)=d1ALFA(j,j)=d1(ej)*NKSI2(j)
C-------------------------------------------------------------------
C       INPUT VARIABLES:
C
C               DE(3,15)=DE(k,IL)=d1(ei) - 1-derivative of unit
c                                          joint axis vector ei
C-------------------------------------------------------------------
C       OUTPUT VARIABLES:
C
C               DALP(3,15)=DALP(k,IL)=d1ALFA(i,i)
c               (for i=2,N;  1=1,i-1)
C-------------------------------------------------------------------
C       SUBROUTINES CALLED:
C*******************************************************************
C
        SUBROUTINE DALFA(I)
        INCLUDE 'DMOD.MOD'
        INCLUDE 'MODLM1.MOD'
        INCLUDE 'DCONF.MOD'
        I1=I-1
        IF(I.EQ.1) RETURN
        DO 10 L=1,I1
        IL=ILF(I,L)
        IF(KSI2(I).EQ.1) GO TO 11
        DO 12 K=1,3
   12   TEMP0(K)=DE(K,IL)
        GO TO 13
   11   DO 14 K=1,3
   14   TEMP0(K)=0.
   13   CONTINUE
        DO 20 K=1,3
   20   DALP(K,IL)=TEMP0(K)
C       IF(I.EQ.2.AND.L.EQ.1) WRITE(1,100) TEMP0
C 100   FORMAT(3F10.6)
   10   CONTINUE
        RETURN
        END
```

```
C*******************************************************************
C       MODULE:          MODEL, linearization
C-------------------------------------------------------------------
C       SUBROUTINE:      DALFA0
C-------------------------------------------------------------------
C       FUNCTION:        Calculation of dlALFA0(i)
c
c                        dlALFA0(i)=dlALFA0(i-1)+SIDOT(i)*
c                        *(dlOMEG(i-1)xei+OMEG(i-1)*dl(ei))*NKSI2(i)
C-------------------------------------------------------------------
C       INPUT VARIABLES:
C
C               DALP00(3,4)=DALP00(k,l)=dlALFA0(i-1) - l-derivative of
c                                                       ALFA0(i-1)
c               (for i=3,N;  l=1,i-2)
c
c               DOMEG0(3,4)=DOMEG0(k,l)=dlOMEG(i-1) - l-derivative of
C                                                      OMEG(i-1)
c               (for i=3,N;  l=1,i-2)
c
c               OMEG(i-1) - angular velocity of (i-1)st joint
c               E(3,6)    - matrix of ei
c               DE(3,15)=dl(ei) - l-derivative of ei
c               (for i=2,N;     l=1,i-1)
C-------------------------------------------------------------------
C       OUTPUT VARIABLES:
C
C               DALP0(3,5)=dlALFA0(i)
c               (for i=2,N;     l=1,i-1)
C-------------------------------------------------------------------
C       SUBROUTINES CALLED:
c
c               CROSS0,CROSS1
C*******************************************************************
C
      SUBROUTINE DALFA0(I,E,OMEG0)
      INCLUDE 'DMOD.MOD'
      INCLUDE 'DCONF.MOD'
      INCLUDE 'MODLM1.MOD'
      DIMENSION E(6,3),OMEG0(3)
      IF(I.EQ.1) RETURN
      I1=I-1
      DO 10 L=1,I1
      IL=ILF(I,L)
      IF(L.EQ.I1) GO TO 11
      DO 12 K=1,3
   12 TEMP0(K)=DOMEG0(K,L)
      CALL CROSS0(TEMP0,E,TEMP2,I)
   11 CONTINUE
      DO 13 K=1,3
   13 TEMP1(K)=DE(K,IL)
      CALL CROSS1(OMEG0,TEMP1,TEMP3)
      IF(L.EQ.I1) GO TO 14
      DO 15 K=1,3
   15 DOMEGE(K,L)=TEMP2(K)+TEMP3(K)
      GO TO 10
   14 DO 16 K=1,3
   16 DOMEGE(K,L)=TEMP3(K)
```

```
10    CONTINUE
      IF(KSI2(I).EQ.1) GO TO 20
      DO 31 L=1,I1
      IF(L.EQ.I1) GO TO 32
      DO 33 K=1,3
33    DALP(K,L)=DALP00(K,L)+SIDOT(I)*DOMEGE(K,L)
      GO TO 31
32    DO 34 K=1,3
34    DALP0(K,L)=SIDOT(I)*DOMEGE(K,L)
31    CONTINUE
      GO TO 21
20    DO 21 L=1,I1
      IF(L.EQ.I1) GO TO 22
      DO 23 K=1,3
23    DALP0(K,L)=DALP00(K,L)
      GO TO 21
22    DO 24 K=1,3
24    DALP0(K,I1)=0.
21    CONTINUE
      RETURN
      END
```

```
C*************************************************************************
C   MODULE:          MODEL, linearization
C-------------------------------------------------------------------------
C   SUBROUTINE:      DDOMEG
C-------------------------------------------------------------------------
C   FUNCTION:        Calculation of derivatives of ALFA0(I)
c                    wrt SIDOT(1)
c
c                   ddlALP0(i)=ddlALP0(I-1)+[(OMEG(i-1)xei)DELTA(1,i)
c                    +SIDOT(i)*(el*NKSI2(1)xei)]*NKSI2(i)
c
c                    (for i=1,N;   l=1,i)
c
C-------------------------------------------------------------------------
C   INPUT VARIABLES:
C
C           OMEG(3)=OMEG(i-1) - anagular velocity of link i-1
c           E(6,3)            - matrix of ei
c           DE(3,15)=dl(ei)   - l-derivative of ei
c           DDALP0(3,21)=ddlALP0(i-1) - derivative of ALP0(i-1)
c                                      wrt SIDOT(1)
C-------------------------------------------------------------------------
C   OUTPUT VARIABLES:
C
C           DDALP0(3,21)=ddlALP0(i) - derivative of ALP0(i)
c                                    wrt SIDOT(1)
C-------------------------------------------------------------------------
C   SUBROUTINES CALLED:    CROSS0
C*************************************************************************
```

```
C
      SUBROUTINE DDOMEG(I,E,OMEGO)
      INCLUDE 'DMOD.MOD'
      INCLUDE 'DCONF.MOD'
      INCLUDE 'MODLM1.MOD'
      DIMENSION E(6,3),OMEGO(3)
      I1=I-1
      IF(KSI2(I).EQ.1) GO TO 10
      IF(I.EQ.1) GO TO 25
      DO 21 L=1,I1
      IL=ILF1(I,L)
      IL1=ILF1(I1,L)
      ILE=ILF(I,L)
      DO 22 K=1,3
   22 DDALPO(K,IL)=DDALPO(K,IL1)+SIDOT(I)*DE(K,ILE)
   21 CONTINUE
      GO TO 24
   25 DO 26 K=1,3
   26 TEMP1(K)=0.
      GO TO 27
   24 CALL CROSSO(OMEGO,E,TEMP1,I)
   27 IL=ILF1(I,I)
      DO 23 K=1,3
   23 DDALPO(K,IL)=TEMP1(K)
      GO TO 11
   10 IF(I.EQ.1) GO TO 12
      DO 11 L=1,I
      IL=ILF1(I,L)
      IF(L.EQ.I) GO TO 12
      IL1=ILF1(I1,L)
      DO 13 K=1,3
   13 DDALPO(K,IL)=DDALPO(K,IL1)
      GO TO 11
   12 DO 14 K=1,3
   14 DDALPO(K,IL)=0.
   11 CONTINUE
      RETURN
      END
```

```
C***********************************************************************
C      MODULE:          MODEL, linearization
C-----------------------------------------------------------------------
C      SUBROUTINE:      DBETAI
C-----------------------------------------------------------------------
C      FUNCTION:        Calculation of l-derivative of BETAii
c
c                       dlBETAii=(dl(ei)xrii+eixdl(rii))*NKSI2(i)
c                       +dl(ei)*KSI2(i),     for l<i
c                       dlBETAii=(eixdi(rii))*NKSI2(i),  for l=i
C-----------------------------------------------------------------------
C       INPUT VARIABLES:
C
C               RI(3)=rii - vector from joint i to mass-center of link i
c               E(6,3)     - matrix of ei
c               DRI(3,6)=dl(rii)=DRI(k,l) - l-derivative of rii
c               DE(3,15)=dl(ei) - l-derivative of ei
C-----------------------------------------------------------------------
C       OUTPUT VARIABLES:
C
C               DBETAI(3,91)=dlBETAii - l-derivative of BATAii
C-----------------------------------------------------------------------
C       SUBROUTINES CALLED: CROSS1
C***********************************************************************
       SUBROUTINE DBETAI(I,RI,E)
       INCLUDE 'DMOD.MOD'
       INCLUDE 'MODLM1.MOD'
       INCLUDE 'DCONF.MOD'
       DIMENSION RI(3),E(6,3)
       IF(KSI2(I).EQ.1) GO TO 20
       DO 10 L=1,I
       IF(L.EQ.I) GO TO 11
       IL=ILF(I,L)
       DO 12 K=1,3
  12   TEMP3(K)=DE(K,IL)
       CALL CROSS1(TEMP3,RI,TEMP0)
       GO TO 13
  11   DO 14 K=1,3
  14   TEMP0(K)=0.
  13   DO 15 K=1,3
       TEMP2(K)=E(I,K)
  15   TEMP4(K)=DRI(K,L)
       CALL CROSS1(TEMP2,TEMP4,TEMP1)
       IIL=IJLF1(I,I,L)
       DO 16 K=1,3
  16   DBETI(K,IIL)=TEMP0(K)+TEMP1(K)
  10   CONTINUE
       GO TO 21
  20   CONTINUE
       DO 21 L=1,I
       IIL=IJLF1(I,I,L)
       IF(L.EQ.I) GO TO 22
       IL=ILF(I,L)
       DO 23 K=1,3
  23   DBETI(K,IIL)=DE(K,IL)
       GO TO 21
  22   DO 24 K=1,3
  24   DBETI(K,IIL)=0.
  21   CONTINUE
       RETURN
       END
```

```
C***********************************************************************|
C   MODULE:          MODEL, linearization                               |
C-----------------------------------------------------------------------|
C   SUBROUTINE:      DBETA                                              |
C-----------------------------------------------------------------------|
C   FUNCTION:        Calculation of l-derivative of BETA(i,j)           |
c                                                                       |
c                    dlBETA(i,j)=dlBETA(i-l,j)+                         |
c                    dlALFA(i-l,j)x(rii-r(i-l,i))+ALFA(i-l,j)x          |
c                    x(dl(rii)-dl(r(i-l,i))                             |
C-----------------------------------------------------------------------|
C   INPUT VARIABLES:                                                    |
C                                                                       |
C          RI(3)=rii - vector from joint i to mass-center of link i|
c          R00(3)=r(i-l,i) - vector from joint i to mass-center of |
c                            link i-l                                   |
c          DRI(3,6)=DRI(k,l)=dl(rii) - l-derivative of rii              |
c          DR00(3,6)=dl(r(i-l,i)) - l-derivative of r(i-l,i)            |
c                                                                       |
c          DE(3,15)=dl(ei) - l-derivative of ei                         |
c          dlBETA(i-l,j) - l-derivative of BETA(i-l,l)                  |
c                                                                       |
C-----------------------------------------------------------------------|
C   OUTPUT VARIABLES:                                                   |
C                                                                       |
C          DBETI(3,91)=dlBETA(i,j) - l-derivative of BETA(i,j)          |
c                                                                       |
c          organized as:                                                |
c                                                                       |
c                    i | 1 2 2 2 2      |                               |
c                    j | 1 1 1 2 2      |                               |
c                    l | 1 1 2 1 2      |                               |
c                                                                       |
c                                                                       |
C-----------------------------------------------------------------------|
C   SUBROUTINES CALLED:                                                 |
C***********************************************************************|
      SUBROUTINE DBETA(I,RI,R00,ALP)
      INCLUDE 'DMOD.MOD'
      INCLUDE 'MODLM1.MOD'
      INCLUDE 'DCONF.MOD'
      DIMENSION RI(3),R00(3),ALP(6,6,3)
C
      IF(I.EQ.1) RETURN
      I1=I-1
      DO 11 K=1,3
C100  FORMAT(3F14.6)
   11 TEMP0(K)=RI(K)-R00(K)
      DO 10 J=1,I1
      DO 30 L=1,I1
      IF(KSI2(J).EQ.1) GO TO 31
      IF(J.EQ.1) GO TO 31
      IF(L.GE.J) GO TO 31
      JL=ILF(J,L)
      DO 32 K=1,3
   32 TEMP4(K)=DE(K,JL)
      CALL CROSS1(TEMP4,TEMP0,TEMP1)
   31 CONTINUE
```

```
      DO 33 K=1,3
   33 TEMP2(K)=DRI(K,L)-DR00(K,L)
      CALL CROSS3(ALP,TEMP2,TEMP3,J,J)
      IJL=IJLF1(I,J,L)
      IJL0=IJLF1(I1,J,L)
      DO 34 K=1,3
   34 DBETI(K,IJL)=DBETI(K,IJL0)+TEMP3(K)
      IF(KSI2(J).EQ.1) GO TO 30
      IF(J.EQ.1) GO TO 30
      IF(L.GE.J) GO TO 30
      DO 35 K=1,3
   35 DBETI(K,IJL)=DBETI(K,IJL)+TEMP1(K)
   30 CONTINUE
      L=I
      IJL=IJLF1(I,J,L)
      DO 36 K=1,3
   36 TEMP1(K)=DRI(K,L)
      CALL CROSS3(ALP,TEMP1,TEMP2,J,J)
      DO 37 K=1,3
   37 DBETI(K,IJL)=TEMP2(K)
   10 CONTINUE
      IF(I.NE.6) RETURN
      IPP=IJLF1(6,4,1)
      DO 1001 K=1,3
 1001 TEMP1(K)=DBETI(K,IPP)
      RETURN
      END
```

```
C*********************************************************************
C   MODULE:            MODEL, linearization
C---------------------------------------------------------------------
C   SUBROUTINE:        DBETA0
C---------------------------------------------------------------------
C   FUNCTION:          Calculation of dlBETA0(i)
c
c           dlBETA0(i)=dlBETA0(i-1)+dlALFA0(i-1)x(rii-r(i-1,i))+
c           ALFA0(i-1)xdl(rii-r(i-1,i))+SIDOT(i)*
C           [dl(OMEG(i-1)xei)xrii*NKSI2(i)+
c           (OMEG(i-1)xeixdl(rii*NKSI2(i)+
c           2*dl(OMEG(i-lxei)*KSI2(i)]+dlGAMA(i,i)-dlGAMA(i-1,i)
c
c           where:
c
c           dlGAMA(i,i)=dlOMEG(I)X(OMEG(i)xrii)+
c                       OMEG(i)x(dlOMEG(i)xrii)+
c                       OMEG(i)x(OMEG(i)xdl(rii)
c
c           dlGAMA(i-1,i)=dl(omeg(i-1)x(OMEG(i-1)xr(i-1,i)+
c                         OMEG(i-1)x(dlOMEG(i-1)xr(i-1,i))+
C                         OMEG(i-1)x(OMEG(i-1)xdl(r(i-1,i)
C---------------------------------------------------------------------
C   INPUT VARIABLES:
C
C           BETI0(i,k) - coefficient BETA0(i)
c           DBETI0(3,21)=dlBETI0(i) - l-derivative of BETA0(i)
c           ALP0(6,3)  -    coefficient dlALFA0(i-1)
c           DALP00(3,4)-    coefficient dlALFA0(i-1)
c           SIDOT(3)=SIDOT(i) - internal velocity of joint i
c           OMEG(3)=OMEG(i)   - angular velocity of link i
c           DOMEGI(3,5)=dlOMEG(i) - l-derivative of OMEG(i)
c           OMEG0(3)=OMEG(i-1) - angular velocity of link i-1
c           DOMEG0(3,4)=DLOMEG(i-1) - l-derivative of OMEG(i-1)
c           DBETI0(3,21)=dlBETA(i-1) - l-derivative of BETA(i-1)
c
C---------------------------------------------------------------------
C   OUTPUT VARIABLES:
C
C           DBETI0(3,21)=dlBETA(i) - l-derivative of BETA(i)
c
C---------------------------------------------------------------------
C   SUBROUTINES CALLED:     CROSS0,CROSS1
C*********************************************************************
      SUBROUTINE DBETA0(I,E,R1,R00,OMEGI,OMEG0,ALP0)
      INCLUDE 'DMOD.MOD'
      INCLUDE 'MODLM1.MOD'
      INCLUDE 'DCONF.MOD'
      DIMENSION RI(3),R00(3),OMEGI(3),OMEG0(3),ALP0(6,3),E(6,3)
      I1=I-1
      IF(I.GT.1) GO TO 20
      DO 10 K=1,3
 10   TEMP0(K)=DRI(K,1)
      CALL CROSS1(OMEGI,TEMP0,TEMP1)
      CALL CROSS1(OMEGI,TEMP1,TEMP0)
      DO 11 K=1,3
 11   DBETI0(K,1)=TEMP0(K)
      GO TO 21
```

```
 20    DO 21 L=1,I
       IL=ILF1(I,L)
       IL0=ILF1(I-1,L)
       DO 22 K=1,3
       TEMP0(K)=ALP0(I1,K)
 22    TEMP1(K)=DRI(K,L)
       CALL CROSS1(TEMP0,TEMP1,TEMP2)
       CALL CROSS1(OMEGI,TEMP1,TEMP3)
       CALL CROSS1(OMEGI,TEMP3,TEMP4)
       IF(L.EQ.I) GO TO 100
       DO 23 K=1,3
 23    DBETI0(K,IL)=DBETI0(K,IL0)+TEMP2(K)+TEMP4(K)
       GO TO 110
100    DO 120 K=1,3
120    DBETI0(K,IL)=TEMP2(K)+TEMP4(K)
110    CONTINUE
       IF(KSI2(I).EQ.1) GO TO 24
       CALL CROSS0(OMEG0,E,TEMP2,I)
       CALL CROSS1(TEMP2,TEMP1,TEMP3)
       DO 255 K=1,3
255    DBETI0(K,IL)=DBETI0(K,IL)+SIDOT(I)*TEMP3(K)
 24    IF(L.EQ.I) GO TO 21
       DO 25 K=1,3
       TEMP2(K)=DR00(K,L)
 25    TEMP3(K)=DOMEGI(K,L)
       CALL CROSS1(TEMP0,TEMP2,TEMP4)
       CALL CROSS1(OMEGI,RI,TEMP5)
       CALL CROSS1(TEMP3,TEMP5,TEMP6)
       CALL CROSS1(TEMP3,RI,POM)
       CALL CROSS1(OMEGI,POM,TEMP1)
       CALL CROSS1(OMEG0,DR00,TEMP5)
       CALL CROSS1(OMEG0,TEMP5,POM)
       DO 26 K=1,3
 26    DBETI0(K,IL)=DBETI0(K,IL)-TEMP4(K)+TEMP6(K)+TEMP1(K)-POM(K)
       IF(KSI2(I).EQ.1) GO TO 27
       DO 28 K=1,3
 28    TEMP0(K)=DOMEGE(K,L)
       CALL CROSS1(TEMP0,RI,TEMP1)
       DO 29 K=1,3
 29    DBETI0(K,IL)=DBETI0(K,IL)+SIDOT(I)*TEMP1(K)
       GO TO 30
 27    DO 31 K=1,3
 31    DBETI0(K,IL)=DBETI0(K,IL)+SIDOT(I)*(DOMEGE(K,L)+DOMEGE(K,L))
 30    CONTINUE
       IF(L.EQ.I1) GO TO 21
       DO 32 K=1,3
       TEMP0(K)=RI(K)-R00(K)
       TEMP3(K)=DOMEG0(K,L)
 32    TEMP2(K)=DALP00(K,L)
       CALL CROSS1(TEMP2,TEMP0,TEMP1)
       CALL CROSS1(OMEG0,R00,TEMP0)
       CALL CROSS1(TEMP3,TEMP0,TEMP2)
       CALL CROSS1(TEMP3,R00,TEMP0)
       CALL CROSS1(OMEG0,TEMP0,TEMP4)
       DO 33 K=1,3
 33    DBETI0(K,IL)=DBETI0(K,IL)+TEMP1(K)-TEMP2(K)-TEMP4(K)
 21    CONTINUE
       RETURN
       END
```

```
C***********************************************************************
C      MODULE:            MODEL, linearization
C-----------------------------------------------------------------------
C      SUBROUTINE:        DDBE0
C-----------------------------------------------------------------------
C      FUNCTION:          Calculation of derivatives of BETA0(i)
c                         wrt SIDOT(i)
c
c                         dd1BETA0(i)=dd1BETA0(i-1)+dd1ALFA0(i-1)x
c                         (rii-r(i-1,i))+[(OMEG(i-1)xei)xrii*NKSI2(i)+
c                         +2*(OMEG(i-1)xei)*KSI2(i)*DELTA(1,i)+
c                         +SIDOT(i)*[(dd1OMEG(i-1)xei)xrii*NKSI2(i)+
c                         +2*(dd1OMEG(i-1)xei)*KSI2(i)]*ETA(1,i-1)+
c                         dd1OMEG(i)x(OMEG(i)xrii)+OMEG(i)x
c                         dd1OMEG(I)XRII)-dd1OMEG(i-1)x(OMEG(i-1)x
c                         r(i-1,i))-OMEG(i-1)x(dd1OMEG(i-1)xr(i-1,i)
c
c                         where:
c
c                         dd1OMEG(k)=ek*NKSI2(k)
c
c
C-----------------------------------------------------------------------
C      INPUT VARIABLES:
C
C              DDBET0(3,21) -- dd1BETA(i-1) - derivative of BETA(i-1)
c                                             wrt SIDOT(1)
c              DDALP0(3,21) =  dd1ALFA0(i)  - derivative fo ALFA0(i)
c                                             wrt SIDOT(1)
c              RI(3)=rii - vector from joint i to mass-center of link i
c              R00(3)=r(i-1,i) - vector from joint i to masscenter of
c                                link i-1
c              OMEG(i) - angular velocity of link i
c              E(6,3)  - matrix of ei
c
C-----------------------------------------------------------------------
C      OUTPUT VARIABLES:
C
C              DDBET0(3,21)=dd1BETA(i) - derivative of BETA(i)
c                                        wrt SIDOT(i)
c
C-----------------------------------------------------------------------
C      SUBROUTINES CALLED:     CROSS0, CROSS1
C***********************************************************************
      SUBROUTINE DDBE0(I,E,RI,R00,OMEGO,OMEGI)
      INCLUDE 'DMOD.MOD'
      INCLUDE 'MODLM1.MOD'
      INCLUDE 'DCONF.MOD'
      DIMENSION E(6,3),RI(3),R00(3),OMEGO(3),OMEGI(3)
      DO 10 L=1,I
      IL=ILF1(I,L)
      IF(I.EQ.1) GO TO 30
      IF(L.EQ.I) GO TO 11
      IL1=ILF1(I-1,L)
      DO 12 K=1,3
 12   DDBET0(K,IL)=DDBET0(K,IL1)
      DO 13 K=1,3
      TEMP0(K)=RI(K)-R00(K)
```

```
 13   TEMP1(K)=DDALP0(K,IL1)
      CALL CROSS1(TEMP1,TEMP0,TEMP2)
      DO 14 K=1,3
 14   DDBET0(K,IL)=DDBET0(K,IL)+TEMP2(K)
 11   CONTINUE
      IF(L.NE.I) GO TO 20
      CALL CROSS0(OMEGO,E,TEMP0,I)
      IF(KSI2(I).EQ.1) GO TO 15
      CALL CROSS1(TEMP0,RI,TEMP1)
      DO 16 K=1,3
 16   DDBET0(K,IL)=TEMP1(K)
      GO TO 20
 15   DO 17 K=1,3
 17   DDBET0(K,IL)=TEMP0(K)+TEMP0(K)
 20   CONTINUE
      IF(KSI2(L).EQ.1) GO TO 10
      IF(L.EQ.I) GO TO 30
      DO 21 K=1,3
 21   TEMP0(K)=E(L,K)*SIDOT(I)
      CALL CROSS0(TEMP0,E,TEMP1,I)
      IF(KSI2(I).EQ.1) GO TO 22
      CALL CROSS1(TEMP1,RI,TEMP2)
      DO 23 K=1,3
 23   DDBET0(K,IL)=DDBET0(K,IL)+TEMP2(K)
      GO TO 250
 22   DO 24 K=1,3
 24   DDBET0(K,IL)=DDBET0(K,IL)+TEMP1(K)+TEMP1(K)
250   CONTINUE
      DO 25 K=1,3
 25   TEMP0(K)=E(L,K)
      CALL CROSS1(OMEGO,R00,TEMP1)
      CALL CROSS1(TEMP0,TEMP1,TEMP2)
      CALL CROSS1(TEMP0,R00,TEMP3)
      CALL CROSS1(OMEGO,TEMP3,TEMP4)
      DO 26 K=1,3
 26   DDBET0(K,IL)=DDBET0(K,IL)-TEMP2(K)-TEMP4(K)
 30   CONTINUE
      IF(KSI2(L).EQ.1) GO TO 10
      DO 31 K=1,3
 31   TEMP0(K)=E(L,K)
      CALL CROSS1(OMEGI,RI,TEMP1)
      CALL CROSS1(TEMP0,TEMP1,TEMP2)
      CALL CROSS1(TEMP0,RI,TEMP1)
      CALL CROSS1(OMEGI,TEMP1,TEMP3)
      DO 32 K=1,3
 32   DDBET0(K,IL)=DDBET0(K,IL)+TEMP2(K)+TEMP3(K)
 10   CONTINUE
      RETURN
      END
```

```
C-----------------------------------------------------------------
C   MODULE:           MODEL
C------------------------------------------------------------------
C   SUBROUTINE:       SILAR
C------------------------------------------------------------------
C   FUNCTION:         Calculation of total force ( relevant for ITIP=1)
C                     calculation of coefficients aij
C                     accompanying accelerations and a0 coefficient
C                     dependent on velocity (relevant for ITIP=0)
C------------------------------------------------------------------
C   INPUT VARIABLES:
C           I         index of joint/link being considered
C           MM        mass of the link mi
C           BETI0     coefficient beta0-i
C           BETI      coefficients beta-ij, j=1,...,i-1
C           SIDDOT    internal acceleration q''-i
C           MOD       type of calculations
C                     =0 all effects
C                     =1 inertial and gravitational
C                     =2 only gravitational
C- ----------------------------------------------------------------
C   OUTPUT VARIABLES:
C           AI        coefficients of inertial forces aij,j=1,...,i-1
C                     a=-m.beta
C           AI0       coefficient dependent of velocity a0-i
C                     a0=-m.beta0
C           FI        total force (without weight) F-i
C- ----------------------------------------------------------------
    SUBROUTINE SILAR(BETI,BETI0,AI,AI0,I)
    REAL L(6),LZ(6),MM(6),JS(6),JN(6),JX(6),JY(6),JZ(6)
    COMMON /MIN/ MM,JS,JN,JX,JY,JZ
    DIMENSION BETI(6,6,3),BETI0(6,3)
    DIMENSION AI(6,6,3),AI0(6,3)
C
    COMMON /SILA/ FI(6,3)
    COMMON /UG1/ SIDDOT(6)
C
    COMMON/MODOPC/ MOD
C
C   ---------------------------------------------------------------
C   calculation of coefficients aij,a0
C   calculation of force Fi
C   ---------------------------------------------------------------
    DO 10 K=1,3
    IF(MOD.EQ.0) AI0(I,K)=-MM(I)*BETI0(I,K)
    FI(I,K)=AI0(I,K)
    IF(MOD.EQ.2) GO TO 10
    DO 10 J=1,I
    AI(I,J,K)=-MM(I)*BETI(I,J,K)
    FI(I,K)=FI(I,K)+AI(I,J,K)*SIDDOT(J)
10  CONTINUE
    END
```

```
C*****************************************************************
C    MODULE:           MODEL, linearization
C-----------------------------------------------------------------
C    SUBROUTINE:       DINAM
C-----------------------------------------------------------------
C    FUNCTION:         Calculation of l-derivatives of dynamic
c                      coefficients a(i,j) and a0(i), and
c                      the derivatives of a0(i) wrt SIDOT(1)
c
c                      dl(a(i,j))=-m(i)*dlBETA(i)
c                      dl(a0(i)) =-m(i)*dlBETA0(i)
c                      ddl(a0(i))=-m(i)*ddlBETA0(i)
C-----------------------------------------------------------------
C    INPUT VARIABLES:
C
C          DBETI(3,91)=dlBETA(i) - l-derivative of BETA(i)
c          DBETI0(3,21)=dlBETA0(i)-l-derivative of BETA0(i)
C          DDBET0(3,21)=dlBETA0(i)-l-derivative of BETA0(i)
C-----------------------------------------------------------------
C    OUTPUT VARIABLES:
C
C          DAI(3,91)=dl(a(i)) - l-derivative of a(i)
c          DAI0(3,21)=dl(a0(i)- l-derivative of a0(i)
c          DDAI0(3,21)=ddl(a0(i)) -derivative of a0(i) wrt SIDOT(1)
C-----------------------------------------------------------------
C    SUBROUTINES CALLED:     no
C*****************************************************************
      SUBROUTINE DINAM(I)
      INCLUDE 'DMOD.MOD'
      INCLUDE 'MODELM.MOD'
      INCLUDE 'DCONF.MOD'
      DO 10 J=1,I
      DO 10 L=1,I
      IJL=IJLF1(I,J,L)
      DO 11 K=1,3
 11   DAI(K,IJL)=-MM(I)*DBETI(K,IJL)
 10   CONTINUE
      DO 12 L=1,I
      IL=ILF1(1,L)
      DO 13 K=1,3
      DAI0(K,IL)=-MM(I)*DBETI0(K,IL)
 13   DDAI0(K,IL)=-MM(I)*DDBET0(K,IL)
 12   CONTINUE
      RETURN
      END
```

```
C-----------------------------------------------------------------
C   MODULE:          MODEL
C-----------------------------------------------------------------
C   SUBROUTINE:      MOM11
C-----------------------------------------------------------------
C   FUNCTION:        Calculation of the equivalent acceleration
C                    for the bar
C-----------------------------------------------------------------
C   INPUT VARIABLES:
C           RI       rii vector for the link (bar)
C           OMEGI    angular velocity wrt fixed coord. s. for the link
C           MOD      type of calculation
C                    =0 all effects
C                    =1 no velocity terms considered
C           I        index of the link being processed
C-----------------------------------------------------------------
C   OUTPUT VARIABLES:
C           S        unit vector for the bar
C           TAU      equivalent acceleration for the bar
C-----------------------------------------------------------------
C   SUBROUTINE CALLED
C           CROSS1   cross product
C           DOT1     dot product
C-----------------------------------------------------------------
      SUBROUTINE MOM11(I,RI,OMEGI,S)
      DIMENSION RI(3),OMEGI(3)
      DIMENSION TEMP(3)
      DIMENSION S(3)
      COMMON /UBRZ/ TAU(6,3)
C
        COMMON /POMOC/ TEMP,TEMP1,RAS
C
        COMMON/MODOPC/ MOD
C
      RAS=0.
      DO 10 K=1,3                         ! unit vector for the bar
   10 RAS=RAS+RI(K)*RI(K)
        IF(RAS.LT.0.00001) RAS=0.00001
      RAS=1./SQRT(RAS)
      DO 15 K=1,3
   15 S(K)=RAS*RI(K)
        IF(MOD.EQ.1) RETURN               ! no velocity dependance
      CALL CROSS1(S,OMEGI,TEMP)           ! equivalent acceler.
      TEMP1=DOT1(OMEGI,S)
      DO 20 K=1,3
   20 TAU(I,K)=TEMP1*TEMP(K)
      RETURN
      END
```

```
C-----------------------------------------------------------------
C       MODULE: MODEL
C-----------------------------------------------------------------
C       SUBROUTINE:     MOM21
C-----------------------------------------------------------------
C       FUNCTION:       Calculation of coefficients alfa-ij, alfa-i0
C                       for the bar
C-----------------------------------------------------------------
C       INPUT VARIABLES:
C               I       index of the joint/link considered
C               ALP     coefficients alfa-ij
C               ALP0    coefficient alfa-i0
C               S       link unit vector wrt fixed coord. s.
C               MOD     type of calculations:
C                       =0 all effects
C                       =1 no velocity dependance
C-----------------------------------------------------------------
C       OUTPUT VARIABLES:
C               ALSII   'parallel' components of alfa-ij
C               ALNII   'normal' components of alfa-ij
C               ALS0    'parallel' component of alfa-i0
C               ALN0    'normal' component of alfa-i0
C-----------------------------------------------------------------
C       SUBROUTINE CALLED:
C               MOM12   normal and parallel projections of vector
C-----------------------------------------------------------------
C
      SUBROUTINE MOM21(I,ALP,ALP0,ALSII,ALNII,ALS0,ALN0,S)
      DIMENSION ALP(6,6,3),ALP0(6,3),ALSII(6,6,3),ALNII(6,6,3)
      DIMENSION ALS0(6,3),ALN0(6,3)
      DIMENSION EPS(3),EPSS(3),EPSN(3)
      DIMENSION S(3)
        COMMON /MODOPC/ MOD
C
      DO 15 J=1,I                    ! coefficients alsii,alnii
      DO 10 K=1,3
   10 EPS(K)=ALP(I,J,K)
      CALL MOM12(EPS,EPSS,EPSN,S)
      DO 11 K=1,3
      ALSII(I,J,K)=EPSS(K)
   11 ALNII(I,J,K)=EPSN(K)
   15 CONTINUE
        IF(MOD.EQ.1) RETURN          !. exit if no velocity dependance
      DO 16 K=1,3                    ! coefficients als0,aln0
   16 EPS(K)=ALP0(I,K)
      CALL MOM12(EPS,EPSS,EPSN,S)
      DO 17 K=1,3
      ALS0(I,K)=EPSS(K)
   17 ALN0(I,K)=EPSN(K)
      RETURN
      END
```

```
C-----------------------------------------------------------------
C     MODULE:          MODEL
C-----------------------------------------------------------------
C     SUBROUTINE:      MOM2
C-----------------------------------------------------------------
C     FUNCTION:        Calculation of coefficients bij,j=1,...,i and
C                      bi0 for the link considered as the bar (cane)
C-----------------------------------------------------------------
C     INPUT VARIABLES:
C             MOD      type of calculations:
C                      =0 all effects are considered
C                      =1 inertial and gravitational
C                      =2 only gravitational
C             JS       Js component of the moment of inertia
C             JN       Jn component of the moment of inertia
C             ALSII    coefficients alfa--ij for the "parallel" comp.
C             ALNII    coefficients alfa--ij for the "normal" comp.
C             ALS0     coeff. alfa-i0 for the "parallel" comp.
C             ALN0     coeff. alfa-i0 for the "normal" comp.
C             TAU      equivalent acceleration for the bar
C             I        index of the link being processed
C-----------------------------------------------------------------
C     OUTPUT VARIABLES:
C             BII      coefficients b-ij
C             BI0      coefficients b-i0
C-----------------------------------------------------------------
    SUBROUTINE MOM2(I,BII,BI0,ALSII,ALS0,ALNII,ALN0)
    REAL L(6),LZ(6),MM(6),JS(6),JN(6),JX(6),JY(6),JZ(6)
    DIMENSION BII(6,6,3),ALSII(6,6,3),ALNII(6,6,3)
    DIMENSION BI0(6,3),ALS0(6,3),ALN0(6,3)
    COMMON /UBRZ/ TAU(6,3)
    COMMON /MIN/ MM,JS,JN,JX,JY,JZ
C
      COMMON /MODOPC/ MOD
C
      COMMON /POMOC/ POM1,POM2
C
      POM1=0.
      POM2=0.
    DO 10 J=1,I                 ! coefficient bij
    DO 10 K=1,3
      IF(JS(I).NE.0.) POM1=JS(I)*ALSII(I,J,K)
      IF(JN(I).NE.0.) POM2=JN(I)*ALNII(I,J,K)
      BII(I,J,K)=-(POM1+POM2)
10    CONTINUE
C
      IF(MOD.EQ.1) RETURN       ! no velocity dependance considered
      POM1=0.
      POM2=0.
C
      DO 15 K=1,3               ! coefficient bi0
      IF(JS(I).NE.0.) POM1=JS(I)*ALS0(I,K)
      IF(JN(I).NE.0.) POM2=JN(I)*(ALN0(I,K)+TAU(I,K))
      BI0(I,K)=-(POM1+POM2)
15    CONTINUE
      RETURN
      END
```

```
C***********************************************************************
C       MODULE:           MODEL, linearization
C-----------------------------------------------------------------------
C       SUBROUTINE:       DINAMC
C-----------------------------------------------------------------------
C       FUNCTION:         Calculation of derivative of Si wrt SI(1)
c
c                         dl(Si)=(elxSi)NKSI2(1)*ETA(1,i)
C-----------------------------------------------------------------------
C       INPUT VARIABLES:
C
C               E(6,3) - matrix of ei
C               S(3)   - vector Si
C-----------------------------------------------------------------------
C       OUTPUT VARIABLES:
C
C               DS(3,6) - l-derivative of Si
C-----------------------------------------------------------------------
C       SUBROUTINES CALLED:      CROSS0
C***********************************************************************
      SUBROUTINE DINAMC(I,E,S)
      INCLUDE 'DMOD.MOD'
      INCLUDE 'MODLM1.MOD'
      INCLUDE 'DCONF.MOD'
      DIMENSION E(6,3),S(3)
      DO 16 L=1,I
      IF(KSI2(L).EQ.1) GO TO 20
      CALL CROSS0(S,E,TEMP0,L)
      DO 17 K=1,3
   17 DS(K,L)=-TEMP0(K)
      GO TO 16
   20 DO 21 K=1,3
   21 DS(K,L)=0.
   16 CONTINUE
      RETURN
      END
```

```
C***********************************************************************
C       MODULE:           MODEL, linearization
C-----------------------------------------------------------------------
C       SUBROUTINE:       DBIC
C-----------------------------------------------------------------------
C       FUNCTION:         Calculation of derivatives of coefficients
c                         b(i,j) wrt SI(1)
c
C                         dl(b(i,j))=-JN(i)*
c                         [(dl(S(i))xALFA(i,j))xS(i)+
c                         (S(i)xdlALFA(i,j))xS(i)+
c                         (S(i)xALFA(i,j))xdl(S(i))]
c                         -JS(i)*
c                         [(dlS(i)).ALFA(i,j))S(i)+
c                         (S(i).dlALFA(i,j))S(i)+
c                         (S(i).ALFA(i,j))dl(S(i))]
c
```

```
C-----------------------------------------------------------------
C   INPUT VARIABLES:
C
C           JS(6)=JS(i) - longitudeinal moments of inertia of links
c           JN(6)=JS(i) - transversal moments of inertia of link
c           S(6)=S(i)   - unit vector along longitudinal axes of
c                         links
c           DS(3,6)=dl(S(i)) - l-derivative of S(i)
c           ALP(6,6,3)=ALFA(i,j,k) - alfa coefficients
c           DALP(3,15)=dl(ALFA(i,j)) - l-derivative of ALFA coeff.
c
C-----------------------------------------------------------------
C   OUTPUT VARIABLES:
C
C           DBII(3,91)=dl(b(i,j)) - l-derivative of b(i,j)
C-----------------------------------------------------------------
C   SUBROUTINES CALLED:     CROSS7, DOT2
C*****************************************************************
C
      SUBROUTINE DBIC(I,ALP,S)
      INCLUDE 'DMOD.MOD'
      INCLUDE 'MODLM1.MOD'
      INCLUDE 'DCONF.MOD'
      DIMENSION ALP(6,6,3),S(3)
C
      DO 10 J=1,I
      DO 11 K=1,3
   11 TEMP0(K)=ALP(I,J,K)
      DO 10 L=1,I
      DO 12 K=1,3
   12 TEMP1(K)=DS(K,L)
      CALL CROSS7(TEMP1,TEMP0,S,TEMP3)
      CALL DOT2(TEMP1,TEMP0,S,TEMP6)
      IF(I.EQ.1) GO TO 14
      IF(L.EQ.I) GO TO 14
      JL=ILF(J,L)
      DO 13 K=1,3
   13 TEMP2(K)=DALP(K,JL)
      CALL CROSS7(S,TEMP2,S,TEMP4)
      CALL DOT2(S,TEMP2,S,POM)
      GO TO 15
   14 DO 16 K=1,3
   16 TEMP4(K)=0.
   15 CALL CROSS7(S,TEMP0,TEMP1,TEMP5)
      CALL DOT2(S,TEMP0,TEMP1,TEMP2)
      IJL=IJLF1(I,J,L)
      DO 17 K=1,3
   17 DBII(K,IJL)=-JN(I)*(TEMP3(K)+TEMP4(K)+TEMP5(K))
     1             -JS(I)*(TEMP6(K)+POM(K)+TEMP2(K))
   10 CONTINUE
      RETURN
      END
```

```
C-------------------------------------------------------------------
C       MODULE:          MODEL
C-------------------------------------------------------------------
C       SUBROUTINE:      CROSS7
C-------------------------------------------------------------------
C       FUNCTION: Auxiliary subroutine
C-------------------------------------------------------------------
C
        SUBROUTINE CROSS7(A,B,C,TEMP0)
        DIMENSION A(3),B(3),C(3),TEMP0(3)
        TEMP=DOT1(A,C)
        TEMP6=DOT1(B,C)
        DO 10 K=1,3
   10   TEMP0(K)=TEMP*B(K)-TEMP6*A(K)
        RETURN
        END
```

```
C-------------------------------------------------------------------
C       MODULE:          MODEL
C-------------------------------------------------------------------
C       SUBROUTINE:      DOT2
C-------------------------------------------------------------------
C       FUNCTION:        Auxiliary subroutine
C-------------------------------------------------------------------
C
        SUBROUTINE DOT2(A,B,C,TEMP0)
        DIMENSION A(3),B(3),C(3),TEMP0(3)
        TEMP=DOT1(A,B)
        DO 10 K=1,3
   10   TEMP0(K)=TEMP*C(K)
        RETURN
        END
```

```
C*********************************************************************
C   TASK:              MODEL, linearization
C---------------------------------------------------------------------
C   MODULE:            DTAUC
C---------------------------------------------------------------------
C   FUNCTION:          CAlculation of derivatives of dynamic
c                      coefficents TAU(i) wrt SI(1)
c
c          d1TAU(i)=[d1OMEG(i).S(i)+OMEG(i).d1(S(i)](S(i)xOMEG(i))+
c          (OMEG(i).S(i))[(d1(S(I)xOMEG(i))+(S(i))xd1(OMEG(i))]
C---------------------------------------------------------------------
C   INPUT VARIABLES:
C
C          OMEGI(3) - angular velocity of link i
c          S(3)     - unit vector along longitudinal axis of link i
c          DOMEGI(3,5)=d1(OMEG(i)) - 1-derivative of OMEG(i)
c          DS(3,6)=d1(S(i)) - 1-derivative of S(i)
C---------------------------------------------------------------------
C   OUTPUT VARIABLES:
C
C          DTAU(3,6)=d1(TAU(i)) - 1-derivative of TAU(i)
C---------------------------------------------------------------------
C   SUBROUTINES CALLED:      CROSS,CROSS1,DOT1,DOT2
C*********************************************************************
      SUBROUTINE DTAUC(I,S,OMEGI)
      INCLUDE 'DMOD.MOD'
      INCLUDE 'MODLM1.MOD'
      INCLUDE 'DCONF.MOD'
      DIMENSION S(3),OMEGI(3)
C
      CALL CROSS1(S,OMEGI,TEMP0)
      OMS=DOT1(OMEGI,S)
      DO 10 L=1,I
      DO 12 K=1,3
      TEMP1(K)=DOMEGI(K,L)
   12 TEMP2(K)=DS(K,L)
      IF(I.EQ.1) GO TO 11
      IF(L.EQ.I) GO TO 11
      CALL DOT2(TEMP1,S,TEMP0,TEMP3)
      CALL CROSS1(S,TEMP1,TEMP6)
      GO TO 14
   11 DO 13 K=1,3
      TEMP3(K)=0.
   13 TEMP6(K)=0.
   14 CALL DOT2(OMEGI,TEMP2,TEMP0,TEMP4)
      CALL CROSS1(TEMP2,OMEGI,TEMP5)
      DO 15 K=1,3
   15 DTAU(K,L)=TEMP3(K)+TEMP4(K)+OMS*(TEMP5(K)+TEMP6(K))
   10 CONTINUE
      RETURN
      END
```

```
C*****************************************************************
C       MODULE:           MODEL, linearization
C-----------------------------------------------------------------
C       SUBROUTINE:       DBIOC
C-----------------------------------------------------------------
C       FUNCTION:         Calculation of derivatives of dynamic
c                         coefficients b0(i) wrt SI(1)
c
c                         dl(b0(i))=-JN(i)*
c                         [(dl(S(i))xALFA0(i))xS(i)+
c                          (S(i)xdl(ALFA0(i))xS(i)+
c                          (S(i)xALFA0(i))xdl(S(i))]-
c                         -JS(i)*
c                         [(dl(S(i)).ALFA0(i))S(i)+
c                          (S(i).dl(ALFA0(i))S(i)+
c                          (S(i).ALFA0(i))dl(S(i))]
C-----------------------------------------------------------------
C       INPUT VARIABLES:
C
C               JN(6) - transversal moments of inertia of links
c               JS(6) - longitudinal moments of inertia of links
c               S(3)  - unit vector along longitudinal axis of link i
c               ALP0(6,3) - cofficients ALFA0(i)
c               DS(3,6)    - 1-derivatives of S(i)
c               DALP0(3,5)- 1-derivatives of ALFA0(i)
c
C-----------------------------------------------------------------
C       OUTPUT VARIABLES:
C
C               DBI0(3,21)=dl(b0(i)) - 1-derivatives of b0(i)
C-----------------------------------------------------------------
C       SUBROUTINES CALLED:     CROSS7,DOT2
C*****************************************************************
C
      SUBROUTINE DBIOC(I,S,ALP0)
      INCLUDE 'DMOD.MOD'
      INCLUDE 'MODLM1.MOD'
      INCLUDE 'DCONF.MOD'
      DIMENSION S(3),ALP0(6,3)
C
      DO 9 K=1,3
    9 TEMP0(K)=ALP0(I,K)
      DO 10 L=1,I
      DO 12 K=1,3
   12 TEMP1(K)=DS(K,L)
      CALL CROSS7(TEMP1,TEMP0,S,TEMP2)
      CALL CROSS7(S,TEMP0,TEMP1,TEMP3)
      CALL DOT2(TEMP1,TEMP0,S,TEMP4)
      CALL DOT2(S,TEMP0,TEMP1,TEMP5)
      IF(I.EQ.1) GO TO 13
      IF(L.EQ.I) GO TO 13
      DO 14 K=1,3
   14 TEMP1(K)=DALP0(K,L)
      CALL CROSS7(S,TEMP1,S,TEMP6)
      CALL DOT2(S,TEMP1,S,POM)
      GO TO 15
   13 DO 16 K=1,3
      TEMP6(K)=0.
```

```
 16    POM(K)=0.
 15    CONTINUE
       IL=ILF1(I,L)
       DO 17 K=1,3
 17    DBI0(K,IL)=-JN(I)*(TEMP2(K)+TEMP3(K)+TEMP6(K)+DTAU(K,L))
  1               -JS(I)*(TEMP4(K)+TEMP5(K)+POM(K))
 10    CONTINUE
       RETURN
       END
```

```
C-------------------------------------------------------------
C       SUBROUTINE:    DDTAUC
C-------------------------------------------------------------
C       FUNCTION:      calculation of derivatives od dynamic
C                      coeficients TAU(i) wrt SIDOT(i)
C-------------------------------------------------------------
        SUBROUTINE DDTAUC(I,E,OMEGI,S)
        INCLUDE 'DMOD.MOD'
        INCLUDE 'MODLM1.MOD'
        INCLUDE 'DCONF.MOD'
        DIMENSION E(6,3),OMEGI(3),S(3)
C
        CALL CROSS1(S,OMEGI,TEMP1)
        C2=DOT1(OMEGI,S)
        DO 10 L=1,I
        IF(KSI2(L).EQ.1) GO TO 11
        DO 12 K=1,3
   12   TEMP0(K)=E(L,K)
        CL1=DOT1(TEMP0,S)
        CALL CROSS1(S,TEMP0,TEMP2)
        DO 13 K=1,3
   13   DDTAU(K,L)=CL1*TEMP1(K)+C2*TEMP2(K)
        GO TO 10
   11   DO 15 K=1,3
   15   DDTAU(K,L)=0.
   10   CONTINUE
        RETURN
        END
```

```
C***********************************************************************
C       MODULE:           MODEL, linearization
C-----------------------------------------------------------------------
C       SUBROUTINE:       DDBIOC
C-----------------------------------------------------------------------
C       FUNCTION:         Calculation of derivatives of dynamic
c                         coefficients b0(i) wrt SIDOT(i)
c
c                         ddl(b0(i))=-JN(i)*
c                         [(S(i)xddl(ALFA0(i))xS(i)+ddl(TAU(i)]-
c                         -JN(i)*
c                         [(S(i).ddl(ALFA0(i))S(i)]
C-----------------------------------------------------------------------
C       INPUT VARIABLES:
C
C                 JN(6) - transverzal moments of inertia of links
c                 JS(6) - longitudinal moments of inertia of links
c                 S(3)  - unit vector along longitudinal axis of link i
c                 DDALP0(3,21)=ddl(ALFA0(i)) - deriavives of ALFA0(i)
c                                              wrt SIDOT(i)
c                 DDTAU(3,6)=ddl(TAU(i)) - derivatives of TAU(i)
c                                          wrt SIDOT(1)
c
C-----------------------------------------------------------------------
C       OUTPUT VARIABLES:
C
C                 DDBI0(3,21)=ddl(b0(i)) - derivatives of b0(i)
c                                          wrt SIDOT(I)
C
C-----------------------------------------------------------------------
C       SUBROUTINES CALLED:     CROSS7, DOT2
C***********************************************************************
        SUBROUTINE DDBIOC(I,S)
        INCLUDE 'DMOD.MOD'
        INCLUDE 'MODLM1.MOD'
        INCLUDE 'DCONF.MOD'
        DIMENSION S(3)
C
        DO 10 L=1,I
        IL=ILF1(I,L)
        DO 11 K=1,3
   11   TEMP0(K)=DDALP0(K,IL)
        CALL CROSS7(S,TEMP0,S,TEMP1)
        CALL DOT2(S,TEMP0,S,TEMP2)
        DO 12 K=1,3
   12   DDBI0(K,IL)=-JN(I)*(TEMP1(K)+DDTAU(K,L))-JS(I)*TEMP2(K)
   10   CONTINUE
        RETURN
        END
```

```
C-----------------------------------------------------------------------
C   MODULE: MODEL
C-----------------------------------------------------------------------
C   SUBROUTINE:     MOMO
C-----------------------------------------------------------------------
C   FUNCTION:       Calculation of transformation matrix for moments
C                   of inertia wrt fixed coord. system and calcula-
C                   tion of velocity dependent component lamda
C                   ( for the body)
C-----------------------------------------------------------------------
C   INPUT VARIABLES:
C           I       index of the joint/link being considered
C           KSI2    type of the joint
C           OMEGI   angular velocity omega-i
C           JX      Jxx component of the main inertia tensor
C           JY      Jyy component
C           JZ      Jzz component
C           Q       Transformation matrix Q0i(A-i)
C-----------------------------------------------------------------------
C   OUTPUT VARIABLES:
C           T       moment of inertia wrt fixed coord. system
C           LAMD    component lamda
C           VL      component  lamda before transformation wrt
C                   fixed coordinate system
C-----------------------------------------------------------------------
C   SUBROUTINES CALLED:
C           DOTO    dot product
C           PRMV    matrix multiplied by a vector
C-----------------------------------------------------------------------
      SUBROUTINE MOMO(OMEGI,Q,I,T,LAMD,VL)
C
      REAL LAMD(3)
      REAL L(6),LZ(6),MM(6),JS(6),JN(6),JX(6),JY(6),JZ(6)
      DIMENSION VL(3),TEMP9(3,3)
      DIMENSION OMEGI(3),Q(3,3),T(3,3)
      COMMON /MIN/ MM,JS,JN,JX,JY,JZ
        COMMON /TIPL/ N,KSI1(6),KSI2(6),KSI3(6)
C
        COMMON /POMOC/ TEMP9
C
C-----------------------------------------------------------------------
C   Component lambda
C-----------------------------------------------------------------------
        IF(KSI2(I).NE.0) GO TO 15
      OMEG1=DOTO(OMEGI,Q,1)              ! transformation of omega-i
      OMEG2=DOTO(OMEGI,Q,2)
      OMEG3=DOTO(OMEGI,Q,3)
      VL(1)=(JY(I)-JZ(I))*OMEG2*OMEG3 ! comp.of lamda before final transf.
      VL(2)=(JZ(I)-JX(I))*OMEG1*OMEG3
      VL(3)=(JX(I)-JY(I))*OMEG1*OMEG2
        CALL PRMV(Q,VL,LAMD,3)            ! final transformation for lamda
C-----------------------------------------------------------------------
C  Transformation T for the tensor of inertia wrt fixed coord. system
C-----------------------------------------------------------------------
      DO 12 J=1,3
      TEMP9(J,1)=JX(I)*Q(J,1)             ! components of the matrix T
      TEMP9(J,2)=JY(I)*Q(J,2)
      TEMP9(J,3)=JZ(I)*Q(J,3)
```

```
   12 CONTINUE
15       CONTINUE
      DO 13 J=1,3                   ! final calculations of matrix T
      DO 13 K=1,3
      T(J,K)=0.
        IF(KSI2(I).NE.0) GO TO 13
      DO 13 LO=1,3
      T(J,K)=T(J,K)+TEMP9(J,LO)*Q(K,LO)
   13 CONTINUE
      RETURN
      END


C------------------------------------------------------------------
C   MODULE:          MODEL
C------------------------------------------------------------------
C   SUBROUTINE:      DOTO
C------------------------------------------------------------------
C   FUNCTION:        Scalar product (dot product) of two vectors
C                    One vector is the three component array and the
C                    other is the j-th column of the 3x3 matrix
C------------------------------------------------------------------
C   INPUT VARIABLES:
C          OMEGI    vector (3x1), first component of the dot product
C          Q        matrix 3x3
C          J        index of the j-th column of matrix Q
C------------------------------------------------------------------
C   OUTPUT VARIABLES:
C          DOTO     dot product OMEGI(x).Q(x,j)
C------------------------------------------------------------------
C
   FUNCTION DOTO(OMEGI,Q,J)
   DIMENSION OMEGI(3),Q(3,3)
   DOTO=OMEGI(1)*Q(1,J)+OMEGI(2)*Q(2,J)+OMEGI(3)*Q(3,J)
   RETURN
   END


C------------------------------------------------------------------
C       MODULE:          MODEL
C------------------------------------------------------------------
C       SUBROUTINE:      PRMV
C------------------------------------------------------------------
C       FUNCTION:        nxn matrix multiplied by nx1 vector
C------------------------------------------------------------------
C       INPUT VARIABLES:
C              AM        Matrix nxn
C              V         Vector nx1
C              N         dimension n
C------------------------------------------------------------------
C       OUTPUT VARIABLES:
C              R         Result vector nx1
C------------------------------------------------------------------
        SUBROUTINE PRMV(AM,V,R,N)
        DIMENSION AM(N,N),V(N),R(N)
        DO 10 I=1,N
        R(I)=0.
        DO 10 J=1,N
10      R(I)=R(I)+AM(I,J)*V(J)
        RETURN
        END
```

```
C-----------------------------------------------------------------------
C     MODULE:           MODEL
C-----------------------------------------------------------------------
C     SUBROUTINE:       MOM10
C-----------------------------------------------------------------------
C     FUNCTION:         Calculation of coefficient bij,bi0 for the body
C-----------------------------------------------------------------------
C     INPUT VARIABLES:
C             I         index of the link being processed
C             KSI2      type of the link (0=rot.,1=lin.)
C             MOD       type of calculation
C                       =0 all effects
C                       =1 no velocity dependance
C             T         moment of inertia wrt fixed coord.s.
C             ALP       coefficients alfa-ij for the body
C             ALP0      coefficients alfa-i0 for the body
C             LAMD      component dependant on velocity for the body
C-----------------------------------------------------------------------
C     OUTPUT VARIABLES:
C             BII       coefficients bij for the body
C             BI0       coefficient bi0 for the body
C-----------------------------------------------------------------------
      SUBROUTINE MOM10(BII,BI0,T,ALP,ALP0,LAMD,I)
      REAL LAMD(3)
      DIMENSION BII(6,6,3),BI0(6,3),T(3,3)
      DIMENSION ALP(6,6,3),ALP0(6,3)
        COMMON /TIPL/ N,KSI1(6),KSI2(6),KSI3(6)
        COMMON /MODOPC/ MOD
C
        DO 5 K=1,3                          ! initialization
5       BI0(I,K)=0.
      DO 10 J=1,I                           ! coefficients bij
        DO 10 K=1,3
      BII(I,J,K)=0.
        IF(KSI2(I).NE.0) GO TO 10
      DO 10 L0=1,3
      BII(I,J,K)=BII(I,J,K)-T(K,L0)*ALP(I,J,L0)
   10 CONTINUE
        IF(KSI2(I).NE.0) RETURN             ! for linear joints exit
        IF(MOD.EQ.1) RETURN                 ! if no vel. dependance exit
      DO 11 K=1,3                           ! coefficient bi0
      BI0(I,K)=BI0(I,K)+LAMD(K)
      DO 11 L0=1,3
      BI0(I,K)=BI0(I,K)-T(K,L0)*ALP0(I,L0)
11      CONTINUE
        RETURN
        END
```

```
C***********************************************************************|
C    MODULE:           MODEL, linearization                             |
C-----------------------------------------------------------------------|
C    SUBROUTINE:       DTSB                                             |
C-----------------------------------------------------------------------|
C    FUNCTION:         Calculation of derivatives of terms of T(i)      |
c                      matrix wrt SI(l)                                 |
c                                                                       |
c                      dl(T(i,j,k))=sum/m=1,3/[(dl(q(j,i,m))q(k,i,m)+   |
c                              q(j,i,m)(dl(q(k,i,m))]Ji,m               |
C-----------------------------------------------------------------------|
C    INPUT VARIABLES:                                                   |
C                                                                       |
C          Q(3,3)  - matrix (3x3) which contains elements qi(k,m)       |
c                    - unit vectors of local frame attached to          |
c                    link i                                             |
c          DQ(3,3,6)=DQ(j,m,l) - l-derivatives of q(j,i,m)              |
c          JX(6),JY(6),JZ(6) - moments of inertia of links              |
c                                                                       |
C-----------------------------------------------------------------------|
C    OUTPUT VARIABLES:                                                  |
C                                                                       |
C          DT(3,3,6) - dl(Ti(j,k)) - l-derivatives of Ti(j,k)           |
c                                                                       |
C-----------------------------------------------------------------------|
C    SUBROUTINES CALLED:     no                                         |
C***********************************************************************|
      SUBROUTINE DTSB(I)
      INCLUDE 'DMOD.MOD'
      INCLUDE 'MODLM1.MOD'
      INCLUDE 'DCONF.MOD'
C
      TEMP0(1)=JX(I)
      TEMP0(2)=JY(I)
      TEMP0(3)=JZ(I)
      DO 10 J=1,3
      DO 10 K=1,3
      DO 10 L=1,I
      DT(J,K,L)=0.
      DO 11 M=1,3
   11 DT(J,K,L)=DT(J,K,L)+(DQ(J,M,L)*Q(K,M)+Q(J,M)*DQ(K,M,L))*TEMP0(M)
   10 CONTINUE
      RETURN
      END
```

```
C***********************************************************************
C  MODULE:           MODEL, linearization
C-----------------------------------------------------------------------
C  SUBROUTINE:       DBISB
C-----------------------------------------------------------------------
C  FUNCTION:         Calculation of derivatives of dynamic
C                    coefficients b(i,j) wrt SI(l)
C
C                    dl(b(i,j))=-dl(T(i))ALFA(i,j)-T(i)(dl(ALFA(i,j))
C-----------------------------------------------------------------------
C  INPUT VARIABLES:
C
C           ALP(6,6,3)=alfa(i,j,k) - dynamic coeffic. ALFA(i,j)
C           T(3,3) - dynamic matrix T
C           DT(3,3,6) - l-derivative of T(i)
C           DALP(3,15)=dl(ALFA(i,j)) - l-derivative of ALFA(i,j)
C
C-----------------------------------------------------------------------
C  OUTPUT VARIABLES:
C
C           DBII(3,91)=dl(b(i,j)) - l-derivative of b(i,j)
C
C-----------------------------------------------------------------------
C  SUBROUTINES CALLED:     no
C***********************************************************************
      SUBROUTINE DBISB(I,T,ALP)
      INCLUDE 'DMOD.MOD'
      INCLUDE 'MODLM1.MOD'
      INCLUDE 'DCONF.MOD'
      DIMENSION T(3,3),ALP(6,6,3)
C
      DO 10 J=1,I
      DO 10 L=1,I
      IJL=IJLF1(I,J,L)
      DO 13 K=1,3
      DBII(K,IJL)=0.
      DO 12 M=1,3
   12 DBII(K,IJL)=DBII(K,IJL)-DT(K,M,L)*ALP(I,J,M)
   13 CONTINUE
      IF(I.EQ.1) GO TO 10
      IF(L.EQ.I) GO TO 10
      IF(KSI2(J).EQ.1) GO TO 10
      JL=ILF(J,L)
      DO 14 K=1,3
      DO 15 M=1,3
   15 DBII(K,IJL)=DBII(K,IJL)-T(K,M)*DE(M,JL)
   14 CONTINUE
   10 CONTINUE
      RETURN
      END
```

```
C*******************************************************************
C       MODULE:           MODEL, linearization
C-------------------------------------------------------------------
C       SUBROUTINE:       DVSB
C-------------------------------------------------------------------
C       FUNCTION:         Calculation of derivatives of dynamic
c                         vector (Euler-s equation) v(i)
c
c                         |{[dl(OMEG(i).q(i,2)](OMEG(i).q(i,3)+      |
c                         |       +(OMEG(i).q(i,2))dl(OMEG(i).q(i,3))}|
c               dl(v(i)=|              ...          *(Ji2-Ji3)    |
c                         |              ...                       |
c
c
C-------------------------------------------------------------------
C       INPUT VARIABLES:
C
C               Q(3,3) - matrix of unit vectors of local coordinate
c                        system axes of link i
c
c               OMEG(3)=OMEG(i) - angular velocity of link i
c
C-------------------------------------------------------------------
C       OUTPUT VARIABLES:
C
C               DV(3,6)=dl(v(i)) - 1-derivative of v(i)
c
C-------------------------------------------------------------------
C       SUBROUTINES CALLED:     no
C*******************************************************************
        SUBROUTINE DVSB(I,E,OMEGI)
        INCLUDE 'DMOD.MOD'
        INCLUDE 'MODLM1.MOD'
        INCLUDE 'DCONF.MOD'
        DIMENSION E(6,3),OMEGI(3)
C
        DO 9 K=1,3
    9   TEMP1(K)=DOTO(OMEGI,Q,K)
        DO 10 L=1,I
        DO 11 J=1,3
        TEMP2(J)=0.
        DO 12 M=1,3
   12   TEMP2(J)=TEMP2(J)+OMEGI(M)*DQ(M,J,L)
        IF(I.EQ.1) GO TO 11
        IF(L.EQ.I) GO TO 11
        DO 14 M=1,3
   14   TEMP2(J)=TEMP2(J)+DOMEGI(M,L)*Q(M,J)
   11   CONTINUE
        DV(1,L)=(TEMP2(2)*TEMP1(3)+TEMP1(2)*TEMP2(3))*(JY(I)-JZ(I))
        DV(2,L)=(TEMP2(3)*TEMP1(1)+TEMP1(3)*TEMP2(1))*(JZ(I)-JX(I))
        DV(3,L)=(TEMP2(1)*TEMP1(2)+TEMP1(1)*TEMP2(2))*(JX(I)-JY(I))
   10   CONTINUE
C
        DO 29 K=1,3
   29   TEMP1(K)=DOTO(OMEGI,Q,K)
        DO 20 L=1,I
        IF(KSI2(L).EQ.1) GO TO 22
        DO 23 J=1,3
```

```
 23    TEMP2(J)=DOT(E,Q,L,J)
       DDV(1,L)=(TEMP2(2)*TEMP1(3)+TEMP1(2)*TEMP2(3))*(JY(I)-JZ(I))
       DDV(2,L)=(TEMP2(3)*TEMP1(1)+TEMP1(3)*TEMP2(1))*(JZ(I)-JX(I))
       DDV(3,L)=(TEMP2(1)*TEMP1(2)+TEMP1(1)*TEMP2(2))*(JX(I)-JY(I))
       GO TO 20
 22    DO 25 K=1,3
 25    DDV(K,L)=0.
 20    CONTINUE
       RETURN
       END
```

```
C*****************************************************************
C       MODULE:            MODEL, linearization
C-----------------------------------------------------------------
C       SUBROUTINE:        DDB0SB
C-----------------------------------------------------------------
C       FUNCTION:          Calculation of derivatives of LAMD(i)
c                          and b0(i) wrt SI(1), and
c                          b0(i) wrt SIDOT(1)
c
c                          dl(LAMD(i))=(dl(Qi)vi+Qi(dl(vi))
c
c                          dl(b0(i))=-(dl(Ti))ALFA0(i)-Ti(dl(ALFA0(i)+
C                          dl(LAMD(i))
c
c                          ddl(b0(i))=-Ti(ddl(ALFA0(i))+Qi(ddl(vi))
c
C-----------------------------------------------------------------
C       INPUT VARIABLES:
C
C                DQ(3,3,6)=dl(q(i,j,k)) - l-derivative of qi(j,k)
c
c                VL(3) - vector v(i)
c                DT(3,3,6)=dl(T(i)) - l-derivatives of T(i)
c                ALP0(6,3) - dynamic coefficients ALFA0(i,k)
c                T(3,3) - matrix T(i)
c                DALP0(3,5)=dl(ALFA0(i)) - l- derivatives of ALFA0(i)
c                DLAMD(3,6)=dl(LAMD(i))  - l- derivatives of LAMD(i)
c
c                DDALP0(3,21)=ddl(ALFA0(i)) - derivative of ALFA0(i)
c                                             wrt SIDOT(i)
c                DDV(3,6)=ddl(v(i)) - derivative of v(i) wrt SIDOT(1)
c
```

```
C------------------------------------------------------------------------
C    OUTPUT VARIABLES:
C
C           DBI0(3,21)=dl(b0(i)) - l-derivatives of b0(i)
c          DDBI0(3,21)=ddl(b0(i)) - derivative of b0(i) wrt SIDOT(i)
c
C------------------------------------------------------------------------
C      SUBROUTINES CALLED:     no
C************************************************************************
      SUBROUTINE DDB0SB(I,VL,T,ALP0)
      INCLUDE 'DMOD.MOD'
      INCLUDE 'MODLM1.MOD'
      INCLUDE 'DCONF.MOD'
      DIMENSION VL(3),T(3,3),ALP0(6,3)
C
      DO 10 L=1,I
      DO 10 K=1,3
      DLAMD(K,L)=0.
      DO 12 M=1,3
   12 DLAMD(K,L)=DLAMD(K,L)+DQ(K,M,L)*VL(M)+Q(K,M)*DV(M,L)
   10 CONTINUE
C
      DO 20 L=1,I
      IL=ILF1(I,L)
      DO 20 K=1,3
      DBI0(K,IL)=0.

      DO 22 M=1,3
   22 DBI0(K,IL)=DBI0(K,IL)-DT(K,M,L)*ALP0(I,M)
      DBI0(K,IL)=DBI0(K,IL)+DLAMD(K,L)
      IF(I.EQ.1) GO TO 20
      IF(L.EQ.I) GO TO 20
      DO 24 M=1,3
   24 DBI0(K,IL)=DBI0(K,IL)-T(K,M)*DALP0(M,L)
   20 CONTINUE
C
      DO 30 L=1,I
      IL=ILF1(I,L)
      DO 30 K=1,3
      DDBI0(K,IL)=0.
      DO 31 M=1,3
   31 DDBI0(K,IL)=DDBI0(K,IL)-T(K,M)*DDALP0(M,IL)+Q(K,M)*DDV(M,L)
   30 CONTINUE
      RETURN
      END
```

```
C***********************************************************************
C       MODULE:          MODEL, linearization
C-----------------------------------------------------------------------
C       SUBROUTINE:      DTRANS
C-----------------------------------------------------------------------
C       FUNCTION:        Saving of vectors dl(r(i-1,i)), dl(r(i,i)),
c                        dl(OMEG(i)) and dl(ALFA0(i))
C-----------------------------------------------------------------------
C       INPUT VARIABLES:
C
c               DR0(3,6)=dl(r(i-1,i)) - 1-derivative of r(i-1.i)
c               DRI(3,6)=dl(r(i,i))   - 1-derivative of r(i,i)
C               DOMEGI(3,6)=dl(OMEG(i)) - 1-derivative of OMEG(i)
c               DALP0(3,6)=dl(ALFA0(i)) - 1-derivative of ALFA0(i)
c
C-----------------------------------------------------------------------
C       OUTPUT VARIABLES:
C
C               DR00(3,6) saves DR0(3,6)
C               DRP(3,6)  saves DRI(3,6)
C               DOMEGO(3,6) saves DOMEGI(3,6)
c               DALP00(3,6) saves DALP0(3,6)
C
C-----------------------------------------------------------------------
C       SUBROUTINES CALLED:     no
C***********************************************************************
C
      SUBROUTINE DTRANS(I)
      INCLUDE 'DMOD.MOD'
      INCLUDE 'DCONF.MOD'
      DO 20 J=1,I
      DO 20 K=1,3
      DR00(K,J)=DR0(K,J)
   20 DRP(K,J)=DRI(K,J)
      IF(I.EQ.1.AND.I.EQ.N) GO TO 31
      DO 32 J=1,I-1
      DO 32 K=1,3
      DOMEGO(K,J)=DOMEGI(K,J)
   32 DALP00(K,J)=DALP0(K,J)
   31 CONTINUE
      RETURN
      END
```

```
C------------------------------------------------------------------
C       MODULE:          MODEL
C------------------------------------------------------------------
C       SUBROUTINE:      MATH
C------------------------------------------------------------------
C       FUNCTION:        Calculation of inertial matrix H
C------------------------------------------------------------------
C       INPUT VARIABLES:
C               N        No. of d.o.f
C               R        Position vectors Rij, i,j=1,...,n
C               E        joints axes
C               AI       inertial forces coefficients aij
C               BII      inertial moments coefficients bij
C------------------------------------------------------------------
C       OUTPUT VARIABLES:
C               H        inertial matrix H
C               B1       array of complete inertial matrix before
C                        dot product with ei
C------------------------------------------------------------------
C       SUBROUTINE CALLED:
C               DOT1     dot product
C               CROSS5   cross product
C
C------------------------------------------------------------------
      SUBROUTINE MATH(R,AI,BII,E,H,B1)
C------------------------------------------------------------------
      DIMENSION R(6,6,3),AI(6,6,3),BII(6,6,3),E(6,3),H(6,6)
        DIMENSION TEMP1(3),B(6,6,3),B1(6,6,3)
      DIMENSION TEMPO(3)
C------------------------------------------------------------------
      COMMON /TIPL/ N,KSI1(6),KSI2(6)
C------------------------------------------------------------------
        COMMON /POMOC/ TEMPO,TEMP1
C------------------------------------------------------------------
C------------------------------------------------------------------
C       Recursive forward calculations
C------------------------------------------------------------------
      DO 10 I=1,N
        DO 11 K=1,3
11      TEMP1(K)=E(I,K)
      DO 10 J=I,N
        IF(KSI2(I).EQ.1) GO TO 18
      DO 15 LO=I,N
      CALL CROSS5(R,AI,TEMPO,J,I,LO)             ! rij X aij
      DO 15 K=1,3
15    B(LO,J,K)=BII(LO,J,K)+TEMPO(K)             ! total bij
      DO 16 K=1,3
      TEMPO(K)=0.
      DO 16 JO=I,N
  16   TEMPO(K)=TEMPO(K)+B(JO,J,K)
        GO TO 101
18      DO 19 K=1,3
        TEMPO(K)=0.
        DO 19 JO=I,N
19      TEMPO(K)=TEMPO(K)+AI(JO,J,K)             ! total aij
  101   CONTINUE
        DO 102 K=1,3              ! store complete H(i,j)
  102   B1(I,J,K)=TEMPO(K)
```

```
 10     H(I,J)=-DOT1(TEMP1,TEMP0)        ! calculate elements H(i,j)
        DO 20 I=2,N               ! final forming of complete matrix H
        DO 20 J=1,I-1
 20     H(I,J)=H(J,I)
      RETURN
      END
```

```
C--------------------------------------------------------------------
C     MODULE:           MODEL
C--------------------------------------------------------------------
C     SUBROUTINE:       CROSS5
C--------------------------------------------------------------------
C     FUNCTION:         Cross product of the j-th row and i-th column of
C                       the 6x6x3 array as the first vector and the j-th
C                       row and the l-th column of the 6x6x3 array as the
C                       second vector
C--------------------------------------------------------------------
C     INPUT VARIABLES:
C             R         array 6x6x3
C             AI        array 6x6x3
C             L0        index of the l-th  column of the AI array
C             I         index of the i-th column of the R array
C             J         index of the j-th row of the R and AI array
C--------------------------------------------------------------------
C     OUTPUT VARIABLES:
C             TEMP0     result 3x1 vector:
C                       R(j,i,x) X AI(j,l,x)
C--------------------------------------------------------------------
      SUBROUTINE CROSS5(R,AI,TEMP0,L0,I,J)
      DIMENSION R(6,6,3),AI(6,6,3),TEMP0(3)
      TEMP0(1)=R(J,I,2)*AI(J,L0,3)-R(J,I,3)*AI(J,L0,2)
      TEMP0(2)=R(J,I,3)*AI(J,L0,1)-R(J,I,1)*AI(J,L0,3)
      TEMP0(3)=R(J,I,1)*AI(J,L0,2)-R(J,I,2)*AI(J,L0,1)
      RETURN
      END
```

```
C*********************************************************************
C    MODULE:          MODEL, linearization
C---------------------------------------------------------------------
C    SUBROUTINE:      DHMOD
C---------------------------------------------------------------------
C    FUNCTION:        Calculation of derivatives of
c                     INERTIAL MATRIX elements wrt joint coordinates
c
c            dl(H(i,j))=-dl(ei).
c                          sum/k=max(i,j) TO n/[b(k,j)+r(k,j)xa(k,j)]
c                          -ei.
c                          sum/k=max(i,j) TO n/[dl(b(k,j))+
c                               dl(r(k,i))xa(k,j)+r(k,i)xdl(a(k,j))]
c
C---------------------------------------------------------------------
C    INPUT VARIABLES:
C
C            E(6,3)           - matrix which rows are ei
c            DE(3,15)         - 1-derivatives of vectors ei
c            B(6,6,3)         - vector which has been formed in
C                               subroutine MATH :
C
C                  =sum/k=(max(i,j) TO n/[b9k,j)+r(k,i)xa(k,j)]
c
c            DBII(3,91)       - 1-derivatives of b(k,j)
c            DR(3,55)         - 1-derivatives of r(k,i)
c            AI(6,6,3)        - dynamic coefficients a(k,j)
c            DAI(3,91)        - 1-derivatives of a(k,j)
c
C---------------------------------------------------------------------
C    OUTPUT VARIABLES:
C
C            DH(6,6,6)        - 1-derivatives of inertial matrix
c
C---------------------------------------------------------------------
C    SUBROUTINES CALLED:      CROSS1
C*********************************************************************
      SUBROUTINE DHMOD(E,B1,AI,R)
      INCLUDE 'DMOD.MOD'
      INCLUDE 'MODLM1.MOD'
      INCLUDE 'DCONF.MOD'
      COMMON/DHCOM/ DH(6,6,6),DH1(6,6),DDH1(6,6)
      DIMENSION B1(6,6,3),E(6,3),AI(6,6,3),R(6,6,3)
C
      DO 10 I=1,N
      IND=1
      IF(I.EQ.4) IND=0
      DO 10 J=I,N
      L00=I
      IF(J.GT.I) L00=J
      DO 10 L=1,N
      DH(I,J,L)=0.
      IF(I.EQ.1) GO TO 11
      IF(L.GE.I) GO TO 11
      IL=ILF(I,L)
      DO 12 M=1,3
   12 DH(I,J,L)=DH(I,J,L)-DE(M,IL)*B1(I,J,M)
   11 CONTINUE
```

```
      DO 13 K=1,3
   13 TEMP0(K)=0.
      DO 14 L0=L00,N
      IF(L.GT.L0) GO TO 14
      IJL1=IJLF1(L0,J,L)
      IF(KSI2(I).EQ.1) GO TO 27
      IJL2=IJLF2(L0,I,L)
      IL1=ILF1(L0,L)
      IF(L0.EQ.I) GO TO 15
      IF(L.EQ.L0) GO TO 31
      DO 16 K=1,3
   16 TEMP1(K)=DR(K,IJL2)
      GO TO 17
   31 IL11=ILF1(L0,L0)
      DO 32 K=1,3
   32 TEMP1(K)=DRIP(K,IL11)
      GO TO 17
   15 DO 18 K=1,3
   18 TEMP1(K)=DRIP(K,IL1)
   17 CONTINUE
      DO 19 K=1,3
   19 TEMP2(K)=AI(L0,J,K)
      CALL CROSS1(TEMP1,TEMP2,TEMP3)
      DO 20 K=1,3
      TEMP1(K)=R(L0,I,K)
   20 TEMP2(K)=DAI(K,IJL1)
      CALL CROSS1(TEMP1,TEMP2,TEMP4)
      DO 1000 K=1,3
 1000 TEMP5(K)=TEMP3(K)+TEMP4(K)
      DO 21 K=1,3
   21 TEMP0(K)=TEMP0(K)+DBII(K,IJL1)+TEMP3(K)+TEMP4(K)
      GO TO 14
   27 DO 28 K=1,3
   28 TEMP0(K)=TEMP0(K)+DAI(K,IJL1)
   14 CONTINUE
      DO 25 M=1,3
   25 DH(I,J,L)=DH(I,J,L)-E(I,M)*TEMP0(M)
   10 CONTINUE
      DO 26 I=2,N
      DO 26 J=1,I-1
      DO 26 L=1,N
   26 DH(I,J,L)=DH(J,I,L)
C
      RETURN
      END
```

```
C----------------------------------------------------------------------
C     MODULE:          MODEL
C----------------------------------------------------------------------
C     SUBROUTINE:      VEKH
C----------------------------------------------------------------------
C     FUNCTION:        Calculation of total vector h0
C                      (except for frictions)
C----------------------------------------------------------------------
C     INPUT VARIABLES:
C             N        No. of d.o.f
C             KSI2     (1-n) types of joints
C             R        array of positions vectors rij,i,j=1,...,n
C             AI0      array of coefficients a0,i=1,..,n
C             G        array of links wights G,i=1,...,n
C             BI0      array of coefficients b0,i=1,...,n
C             E        array of joints axes, i=1,...,n
C             MOD      type of calculations:
C                      =0 all effects
C                      =1 inertial and gravitational
C                      =2 only gravitational
C----------------------------------------------------------------------
C     OUTPUT VARIABLES:
C             UMOM     total vector h0
C             B1       array of complete vectors h0 before reduction
C                      on driving components (dot product with ei)
C----------------------------------------------------------------------
C     SUBROUTINE CALLED:
C             CROSS1   cross product
C             CROSS4   cross product
C             DOT1     dot product
C----------------------------------------------------------------------
      SUBROUTINE VEKH(R,AI0,BI0,E,UMOM,B1)
C
      DIMENSION AI0(6,3),BI0(6,3)
      DIMENSION UMOM(6)    ,B1(6,3)
      DIMENSION R(6,6,3)
      DIMENSION E(6,3)
      DIMENSION TEMP0(3),TEMP1(3),TEMP2(3),TEMP4(6,3),TEMP3(3)
      DIMENSION FI1(3),FI0(3),RMOM(3),PMOM(3),TEMP5(3)
C
      COMMON /SPMOM/ GM(6,3),G(6)
      COMMON /TIPL/ N,KSI1(6),KSI2(6),KSI3(6)
C
      COMMON /POMOC/ TEMP0,TEMP1,TEMP2,TEMP3,TEMP4,FI1,FI0,RMOM,PMOM
C
      COMMON /MODOPC/ MOD
C
C     _______________________________________________________
C     Recursive backward calculations
C     _______________________________________________________
      DO 15 I=N,1,-1
      DO 6 K=1,3                            ! initialization
      TEMP3(K)=0.
      TEMP5(K)=0.
      IF(I.NE.N) TEMP5(K)=R(I+1,I,K)-R(I+1,I+1,K)
      IF(I.NE.N) GO TO 5
      FI1(K)=0.
      PMOM(K)=0.
```

```
        TEMP3(K)=0.
5       CONTINUE
        TEMP2(K)=E(I,K)
        IF(MOD.NE.0) GO TO 56              ! prepare forces
        TEMP4(I,K)=AI0(I,K)
        IF(K.EQ.3) TEMP4(I,K)=AI0(I,K)+G(I)
        GO TO 6
56      TEMP4(I,K)=0.
        IF(K.EQ.3) TEMP4(I,K)=G(I)
6       CONTINUE
c ------------------------------------
C       calculation torques due to total force
C       (except for the gravitational)
c ------------------------------------
        CALL CROSS4(R,TEMP4,TEMP0,I,I,I)
        IF(I.NE.N) CALL CROSS1(TEMP5,FI1,TEMP3)
        DO 12 K=1,3
        RMOM(K)=TEMP0(K)+PMOM(K)+TEMP3(K)
        IF(MOD.EQ.0) RMOM(K)=RMOM(K)+BI0(I,K)
12      PMOM(K)=RMOM(K)
10      DO 13 K=1,3                        ! update total force
        FI0(K)=TEMP4(I,K)+FI1(K)
13      FI1(K)=FI0(K)
!       final forming of driving component of h0
        IF(KSI2(I).NE.1) UMOM(I)=-DOT1(TEMP2,PMOM)       ! rotational
        IF(KSI2(I).EQ.1) UMOM(I)=-DOT1(TEMP2,FI1)        ! linear
!       storing complete h0 ( before dot product with ei)
        IF(KSI2(I).EQ.1) GO TO 100
        DO 101 K=1,3
  101   B1(I,K)=PMOM(K)
        GO TO 200
  100   DO 102 K=1,3
  102   B1(I,K)=FI1(K)
  200   CONTINUE
15      CONTINUE
        RETURN
        END


C*****************************************************************
C   MODULE:          MODEL, linearization
C----------------------------------------------------------------
C   SUBROUTINE:      DH1MOD
C----------------------------------------------------------------
C   FUNCTION:        Calculation of derivatives of
c                    elements of vector h(q,q') wrt
c                    joint coordinates q(l) and
c                    joint velocities q'(l)
c
c          dl(h(i))=-dl(ei).
c           sum/k=i TO n/[(r(k,i)x(a0(k)+G(k))+b0(k))*NKSI2(i)+
c                        +(a0(k)+G(k)KSI2(i)]
c                   -ei.
c           sum/k=i TO n/[dl(r(k,i))x(a0(k)+G(k))+r(k,i)xdl(a0(k)+
c                          dl(b0(k)]*NKSI2(i)+dl(a0(k))*KSI2(i)]
c
c          ddl(h(i))=-ei.
c           sum/k=i TO n/[r(k,i)xddl(a0(k))+ddl(b0(k))*NKS2(i)+
c                         +ddl(a0(k))*KSI2(i)]
c
C----------------------------------------------------------------
```

```
      -----------------------------------------------------------
C     INPUT VARIABLES:
C
C              DE(3,15)          - 1-derivative of ei
c              B1(6,3)           - vector formed in nonlinear model part
c
c              =sum/k=i TO n/[r(k,i)x(a0(k)+G(k))+b0(k))*NKSI2(i)+
c                                  (a0(k)+G(k))*KSI2(i)]
c              E(6,3)            - martix which rows are ei
c              DR(3,55)          - 1-derivatives of r(k,i)
c              AI0(6,3)          - dynamic coefficients a0(i)
c              G(6)              - z-components of gravity vectors
c              R(6,6,3)          - vectors r(k,i)
c              DAI0(3,21)        - 1-derivatives of a0(i)
c              DBI0(3,21)        - 1-derivatives of b0(i)
c
c              DDAI0(3,21)       - derivatives of a0(i) wrt SIDOT(1)
c              DDBI0(3,21)       - derivatives of b0(i) wrt SIDOT(1)
c
C-----------------------------------------------------------------------
C     OUTPUT VARIABLES:
C
C              DH1(6,6)          - 1-derivative of h(i)
c
C-----------------------------------------------------------------------
C     SUBROUTINES CALLED:      CROSS1
C***********************************************************************
      SUBROUTINE DH1MOD(B1,E,AI0,R)
      INCLUDE 'DMOD.MOD'
      INCLUDE 'DCONF.MOD'
C **********************************************************************
C     COMMON: INPUT/OUTPUT VARIABLES (FROM/TO OTHER TASKS)
C ----------------------------------------------------------------------
      COMMON /UG/ SI(6),SIDOT(6)
      COMMON /UG1/ SIDDOT(6)
      COMMON /DINAM/ HH(6,6),H1(6),UMOM(6)
      COMMON /BRZL/ RT(3),RP(6,6,3),RT0(3)
      COMMON /BRZUG/DEL(6,6,3)
      COMMON /MATAN/ Q(3,3),IAX
C-----------------------------------------------------------------------
      DIMENSION B1(6,3),E(6,3),AI0(6,3),R(6,6,3)
      COMMON/DHCOM/ DH(6,6,6),DH1(6,6),DDH1(6,6)
      COMMON/FRICT/CFRICT(6)
C
      DO 10 I=1,N
      DO 10 L=1,N
      DH1(I,L)=0.
      IF(I.EQ.1) GO TO 11
      IF(L.GE.I) GO TO 11
      IL=ILF(I,L)
      DO 12 M=1,3
   12 DH1(I,L)=DH1(I,L)-DE(M,IL)*B1(I,M)
   11 CONTINUE
      DO 13 K=1,3
   13 TEMP0(K)=0.
      DO 14 L0=I,N
      IF(L.GT.L0) GO TO 14
      IL1=ILF1(L0,L)
      IF(KSI2(I).EQ.1) GO TO 27
      IJL2=IJLF2(L0,I,L)
      IL1=ILF1(L0,L)
```

```
      IF(LO.EQ.I) GO TO 15
      IF(L.EQ.LO) GO TO 31
      DO 16 K=1,3
   16 TEMP1(K)=DR(K,IJL2)
      GO TO 17
   31 IL11=ILF1(LO,LO)
      DO 32 K=1,3
   32 TEMP1(K)=DRIP(K,IL11)
      GO TO 17
   15 DO 18 K=1,3
   18 TEMP1(K)=DRIP(K,IL1)
   17 CONTINUE
      DO 19 K=1,3
      TEMP2(K)=AI0(LO,K)
      IF(K.EQ.3) TEMP2(K)=TEMP2(K)+G(LO)
   19 CONTINUE
      CALL CROSS1(TEMP1,TEMP2,TEMP3)
      DO 20 K=1,3
      TEMP1(K)=R(LO,I,K)
   20 TEMP2(K)=DAI0(K,IL1)
      CALL CROSS1(TEMP1,TEMP2,TEMP4)
      DO 21 K=1,3
   21 TEMP0(K)=TEMP0(K)+DBI0(K,IL1)+TEMP3(K)+TEMP4(K)
      GO TO 14
   27 DO 28 K=1,3
   28 TEMP0(K)=TEMP0(K)+DAI0(K,IL1)
   14 CONTINUE
      DO 25 M=1,3
   25 DH1(I,L)=DH1(I,L)-E(I,M)*TEMP0(M)
   10 CONTINUE
C
C
      DO 100 I=1,N
      DO 100 L=I,N
      DDH1(I,L)=0.
      DO 110 K=1,3
  110 TEMP0(K)=0.
      DO 140 LO=I,N
      IF(L.GT.LO) GO TO 140
      IL1=ILF1(LO,L)
      IF(KSI2(I).EQ.1) GO TO 270
      DO 160 K=1,3
  160 TEMP1(K)=R(LO,I,K)
      DO 200 K=1,3
  200 TEMP2(K)=DDAI0(K,IL1)
      CALL CROSS1(TEMP1,TEMP2,TEMP3)
      DO 210 K=1,3
  210 TEMP0(K)=TEMP0(K)+TEMP3(K)+DDBI0(K,IL1)
      GO TO 140
  270 DO 280 K=1,3
  280 TEMP0(K)=TEMP0(K)+DDAI0(K,IL1)
  140 CONTINUE
      DO 150 M=1,3
  150 DDH1(I,L)=DDH1(I,L)-E(I,M)*TEMP0(M)
  100 CONTINUE
      DO 1000 I=1,N
 1000 DDH1(I,I)=DDH1(I,I)-CFRICT(I)
      RETURN
      END
```

```
C-----------------------------------------------------------------
C   MODULE:          MODEL
C-----------------------------------------------------------------
C   SUBROUTINE:      MOMUK
C-----------------------------------------------------------------
C   FUNCTION:        Calculation of total driving torques/forces
C                    (P=P(q,q',q''))
C-----------------------------------------------------------------
C   INPUT VARIABLES:
C           R        Position vectors of link centers wrt fixed
C                    coord.s. and position vectors of links centers
C                    wrt previous joints centers
C           E        Joints axes wrt fixed coord. s.
C           MOD      Type of calculations:
C                    =0 all effects are considered
C                    =1 inertial and gravitational
C                    =2 only gravitational
C                    =3 only kinematics
C           N        number of d.o.f.
C           FI       Total forces except for the gravitational forces
C           G        Links weights
C           MI       Moments of inertial forces wrt fixed coord. s.
C           KSI2     Type of joints
C                    KSI2(i)=0 rotational
C                    KSI2(i)=1 linear joint
C-----------------------------------------------------------------
C   OUTPUT VARIABLES:
C           UMOM     Vector of total driving torques/forces
C                    (except for the friction)
C-----------------------------------------------------------------
C   SUBROUTINES CALLED:
C           CROSS1   Cross product
C           CROSS4   cross product
C           DOT1     dot product
C-----------------------------------------------------------------
C
      SUBROUTINE MOMUK(R,E,UMOM)
C
      REAL MI(6,3)
      DIMENSION UMOM(6)
      DIMENSION R(6,6,3)
      DIMENSION E(6,3)
      DIMENSION TEMP0(3),TEMP1(3),TEMP2(3),TEMP4(6,3),TEMP3(3)
      DIMENSION FI1(3),FI0(3),RMOM(3),PMOM(3),TEMP5(3)
C
      COMMON /SILA/ FI(6,3)
      COMMON /MOM/ MI
      COMMON /SPMOM/ GM(6,3),G(6)
      COMMON /TIPL/ N,KSI1(6),KSI2(6),KSI3(6)
C
      COMMON /POMOC/ TEMP0,TEMP1,TEMP2,TEMP3,TEMP4,FI1,FI0,RMOM,PMOM
C
      COMMON /MODOPC/ MOD
C
C-----------------------------------------------------------------
C  Backward recursive calculations (I=N,...,1)
C-----------------------------------------------------------------
C
```

```
      DO 15 I=N,1,-1
      DO 6 K=1,3              ! initialization of local variables
      TEMP3(K)=0.
      IF(I.NE.N) TEMP5(K)=R(I+1,I,K)-R(I+1,I+1,K)
      IF(I.NE.N) GO TO 5
      FI1(K)=0.
      PMOM(K)=0.
      TEMP3(K)=0.
5     CONTINUE
      TEMP2(K)=E(I,K)
      IF(MOD.EQ.2) GO TO 56            ! only gravitational forces
      TEMP4(I,K)=FI(I,K)               ! total force
      IF(K.EQ.3) TEMP4(I,K)=FI(I,K)+G(I)
      GO TO 6
56    TEMP4(I,K)=0.
      IF(K.EQ.3) TEMP4(I,K)=G(I)       ! only gravitational forces
6     CONTINUE                         ! torque due to total force
      CALL CROSS4(R,TEMP4,TEMP0,I,I,I)
      IF(I.NE.N) CALL CROSS1(TEMP5,FI1,TEMP3)
      DO 12 K=1,3                      ! total torque
      RMOM(K)=TEMP0(K)+PMOM(K)+TEMP3(K)
      IF(MOD.NE.2) RMOM(K)=RMOM(K)+MI(I,K)
12    PMOM(K)=RMOM(K)
10    DO 13 K=1,3                      ! updating total forces
      FI0(K)=TEMP4(I,K)+FI1(K)
13    FI1(K)=FI0(K)
C
C---------------------------------------------------------------------
C     If rotational joint calculate driving torque
C---------------------------------------------------------------------
C
      IF(KSI2(I).NE.1) UMOM(I)=-DOT1(TEMP2,PMOM)
C
C---------------------------------------------------------------------
C     If linear joint calculate driving force
C---------------------------------------------------------------------
C
      IF(KSI2(I).EQ.1) UMOM(I)=-DOT1(TEMP2,FI1)
15    CONTINUE
      RETURN
      END
```

```
C----------------------------------------------------------------------
C     SUBROUTINE:     MODCHK
C----------------------------------------------------------------------
C     FUNCTION:  Testing of successful assembling
C                Types transformation matrix An and characteristic
C                kinematic parameters
C----------------------------------------------------------------------
C     INPUT VARIABLES:
C              Q       (3,3) matrix An (after assembling)
C              IAX     indicator of the end-effector orientation
C              INO     flag for existing of dialogue
C                      =0 dialogue is not existing
C                      =1 dialogue is existing
C              SI      joint coordinates
C              E       unit joint axes in external coor.system
C              RP
C              RT      position vectors in external coordinate system
C              RTT
C     OUTPUT VARIABLES:
C              ICHK    indicator if successful assembling
C                      ICHK=0  assembling successful
C                      ICHK=-1 assembling unsuccessful
C----------------------------------------------------------------------
C     SUBROUTINES CALLED:
C----------------------------------------------------------------------
C
      SUBROUTINE MODCHK(ICHK,INO)
C
      INCLUDE 'IN2:CONFIG.MOD'
      INCLUDE 'IN2:MODELM.MOD'
      DIMENSION RP1(6,6,3),RP2(6,6,3),RP3(6,6,3),RP4(6,6,3),RP5(6,6,3)
C
      ICHK=0
      DO 1 I=1,3
      LL=0
      DO 2 K=1,3
      IF(Q(I,K).NE.0) LL=LL+1
 2    CONTINUE
      IF(LL.EQ.0) ICHK=-1
 1    CONTINUE
      IF(INO.EQ.1)RETURN
      TYPE *,' Characteristic kinematic variables'
      TYPE *,' -----------------------------------'
      TYPE *,'          Joint coordinates: '
      TYPE 1099,(SI(I),I=1,N)
1099     FORMAT(6(E13.5))
         TYPE *,' Transformation matrix An:'
         TYPE 100,((Q(I,J),J=1,3),I=1,3)
100      FORMAT(1X,3F10.5)
         TYPE 77
77       FORMAT(/,'-------------------------------------------------
     *   '-----------------------')
         TYPE *,' Joint axis in external coordinate system
         DO 202 I=1,N
         TYPE 333,I,(E(I,K),K=1,3)
333      FORMAT(/,'  E(',I1,')= ',3(F10.5))
202      CONTINUE
         TYPE 77
```

```
      TYPE *,' Position vectors '
      DO 303 I=1,N
      IF(I.EQ.N) GO TO 505
      J=I+1
      IF(I.EQ.3) GO TO 997
      IF(I.EQ.4) GO TO 996
      DO 339 K=1,3
      RP4(I,J,K)=RP(I,I,K)+RP(J,J,K)-RP(J,I,K)
339   CONTINUE
      TYPE 444,I,I,(RP(I,I,K),K=1,3),I,J,(RP4(I,J,K),K=1,3)
444   FORMAT(/,' R(',I1,',',I1,')=',X,3F10.5,
     *  ' R(',I1,',',I1,')=',X,3F10.5)
      GO TO 303
997   DO 887 K=1,3
      RP1(I,J,K)=RP(I,I,K)-RP(J,I,K)
887   CONTINUE
      TYPE 444,I,I,(RP(I,I,K),K=1,3),I,J,(RP1(I,J,K),K=1,3)
      GO TO 303
996   DO 774 K=1,3
      RP2(I,I,K)=RTT(I,I,K)
      RP3(I,J,K)=RTT(I,I,K)+RP(J,J,K)-RP(J,I,K)
774   CONTINUE
      TYPE 444,I,I,(RP2(I,I,K),K=1,3),I,J,(RP3(I,J,K),K=1,3)
      GO TO 303
505   DO 677 K=1,3
677   RP5(I,I,K)=RP(I,I,K)+RT0(K)
      TYPE 666,I,I,(RP5(I,I,K),K=1,3),(RT0(K),K=1,3)
666   FORMAT(/,' R(',I1,',',I1,')=',X,3F10.5,' RT0= ',3F10.5)
303   CONTINUE
      TYPE 77
      RETURN
      END
```

```
      SUBROUTINE INPUT_DIN
*.....................................................................
* SUBROUTINE: Input_din
*.....................................................................
* FUNCTION: Input program for setting dynamic parameters
*.....................................................................
* SUBROUTINE CALLED: Input_din
*.....................................................................
* INPUT VARIABLES:    ISLOBODE=variable which defines number of d.o.f
*                     IO=index of d.o.f.
* OUTPUT VARIABLES:   dynamic parameters
*.....................................................................
      real l(6),lz(6),mm(6),js(6),jn(6),jx(6),jy(6),jz(6)
      common/tipl/ nx,ksil(6),ksi2(6),ksi3(6)
      COMMON/spmom/gm(6,3),g(6)
      COMMON/MIN/ MM,JS,JN,JX,JY,JZ
      COMMON/ORDGLB/ NUK,ISLOBODE
      COMMON/FILE/ FILE(5)
      BYTE file
*
      CHARACTER*70 HELP(5)
      CHARACTER*15 A$COM
      CHARACTER*2 A$(6),opcl*1
      CHARACTER*1 POM,PP,A1*80
      CHARACTER*6 MAT$(5)
      DIMENSION SIMB(17,6)
      CHARACTER*9 FILEP,FILEP2
      DATA HELP/'type of link','mass of links',
     >           'moment of inertia Jxx/Js',
     >           'moment of inertia Jyy/Jn',
     >           'moment of inertia Jzz'/
      DATA MAT$/'KSI2','MM','Jxx/Js','Jyy/Jn','Jzz'/
      FILEP=CHAR(FILE(1))//CHAR(FILE(2))//CHAR(FILE(3))//
     >       CHAR(FILE(4))//CHAR(FILE(5))//'.DNM'
*.....................................................................
*     Formats
*.....................................................................
14    FORMAT (1X,' WARNING*** File:',a10,'  exists',/)
13    FORMAT (1X,'MESSAGE*** File:',a10,' does not exist')
2     FORMAT ('+',a6,':',F15.6)
20    format ('+WARNING*** Unallowed parameter
     > value',i2,i2)
4575  format (a2)
*.....................................................................
*     Reading data
*.....................................................................
      OPEN (55,FILE=FILEP,STATUS='OLD',ERR=11)
      INEW=1
      WRITE (6,14) FILEP
      DO IPOM=1,ISLOBODE
         DO JPOM=1,5
            READ (55,'(1X,F15.6)') SIMB(JPOM,IPOM)
         END DO
      END DO
      close (55)
      if (inew.eq.1) GO TO 12
11    inew=0
      WRITE (6,13) FILEP
```

```
      CLOSE (55)
      STOP
*...............................................................
*     Data exists
*...............................................................
*
*     Chacking of parametar KSI1, if isn't 0 or 1 then
*     setting 0
*
12          do ipom=1,islobode
               if (simb(1,ipom).ne.0.and.simb(1,ipom).ne.1) then
                  simb(1,ipom)=0
               end if
            end do
*................................................................
*     Removing of variables
*................................................................
            do ipom=1,islobode
               ksil(ipom)=simb(1,ipom)
               mm(ipom)=simb(2,ipom)
               g(ipom)=-mm(ipom)*9.81
               if (ksil(ipom).eq.0) then
                  jx(ipom)=simb(3,ipom)
                  jy(ipom)=simb(4,ipom)
                  jz(ipom)=simb(5,ipom)
               else if (ksil(ipom).eq.1) then
                  js(ipom)=simb(3,ipom)
                  jn(ipom)=simb(4,ipom)
               end if
            end do
            RETURN
      END
```

```
C----------------------------------------------------------------
C    SUBROUTINE:      TRAJEK
C------------------------------------------------------------------
C    FUNCTION:        Calculation of nominal trajectory of the robot
C------------------------------------------------------------------
C    INPUT VARIABLES:
C            N        number of d.o.f.
C            Q0(I)    initial values of internal coordinates
C            QF(I)    final values of internal coordinates
C            H        time interval - step
C            T        movement duration time
C            BETA     parameter which defines trapesoidal velocity
C                     profile
C
C                   ^ DLMB
C                   |
C                   |
C                   |
C   1./(T*(1-BETA)) +     .--------------.
C                   |    .                .
C                   |   .                  .
C                   |  .                    .
C                   | .                      .
C                 --|----+--------------+------+---------------->
C                   0    T1             T2     T              TE
C
C
C            PROFIL   character of variable for choosing desired
C                     velocity profile
C            IND(I)   indicator of joint type (rotational or linear)
C            KSI2(I)  indicator of joint (linear/rotational)
C                               =1   linear
C                               =0   rotational
C    OUTPUT VARIABLES:
C            TE       currently time
C            Q(I,J)   internal coordinates
C            DQ(I,J)  internal velocities
C            DDQ(I,J) internal accelerations
C
C    LOCAL VARIABLES:
C            T1       time instant when velocity profile is changed
C            T2       second time instant
C            M        number of time steps
C            LMB      scalar parameter which defines velocity profile
C            DLMB     first derivative
C            DDLMB    second derivative
C            DLTQ(I)  deference between final and initial values
C
C----------------------------------------------------------------
C
      SUBROUTINE TRAJEK
C
C----------------------------------------------------------------
      CHARACTER *2 PROFIL,FILEP*9,STAMPA*1,FILET*9
      CHARACTER *4 IND(6)
      BYTE FTOT(14),FILE(5)
      COMMON/FILE/FILE
      COMMON/DINT_TR/ H
```

```
      COMMON/ORDGLB/ NUK,N
      COMMON/TIPL/NPOM,KSI1(6),KSI2(6),KSI3(6)

      DIMENSION Q0(6),QF(6),Q(6),DQ(6),DDQ(6),DLTQ(6)
      REAL LMB
C
C-----------------------------------------------------------------
C     Reading initial data from <name of the robot>.tra
C-----------------------------------------------------------------
C
      FILET=CHAR(FILE(1))//CHAR(FILE(2))//CHAR(FILE(3))//
     *      CHAR(FILE(4))//CHAR(FILE(5))//'.TRA'

      write (6,199)
199   format (1x,/,' Trajectory synthesis for all joints'//)
      open(unit=7,file=filet,status='old',err=1001)
      read(7,210)(q0(i),i=1,n)
      read(7,210)(qf(i),i=1,n)
      read(7,210)h,t
200   format(1x,i3)
210   format(1x,6f10.5)
      close(unit=7)



      DATA FTOT(1)/'I'/,FTOT(2)/'N'/,FTOT(3)/'1'/,FTOT(4)/':'/,
     1     FTOT(10)/'.'/,
     2     FTOT(14)/' '/
      FTOT(11)='A'
      FTOT(12)='N'
      FTOT(13)='G'
      FTOT(5)=FILE(1)
      FTOT(6)=FILE(2)
      FTOT(7)=FILE(3)
      FTOT(8)=FILE(4)
      FTOT(9)=FILE(5)

      TE=0.
      M=T/H
C
C-----------------------------------------------------------
C     Joint type initialization(TRA/ROT)
C-----------------------------------------------------------
C

      DO I=1,N
         IF(KSI2(I).EQ.1)THEN
            IND(I)='TRA'
         ELSE
            IND(I)='ROT.'
         END IF
      END DO

      TYPE40
40    FORMAT($,' Want you print of results on the screen? [Y/N]')
      READ(5,45)STAMPA
45    FORMAT(A1)

      FILEP=CHAR(FILE(1))//CHAR(FILE(2))//CHAR(FILE(3))//
     *      CHAR(FILE(4))//CHAR(FILE(5))//'.ANG'
```

```
      OPEN(UNIT=4,FILE=FILEP,STATUS='NEW')

      TYPE 10
10    FORMAT($,' Choos velocity profile - trapesoidal or ',
     *        'triangular!![TP/TG]: ')
      READ(5,20)PROFIL
20    FORMAT(A2)


      IF(PROFIL.EQ.'TG')GO TO 70

25    TYPE 30
30    FORMAT($,' Define trapesoidal profile -
     * with parameter beta [0.0-0.5]')
      READ (5,35) BETA
35    FORMAT (F10.5)
      IF((BETA.LT.0.0).OR.(BETA.GE.0.5)) GO TO 25

      T1=BETA*T
      T2=(1.-BETA)*T

C-------------------------------------------------------------
C     Calculation of scalar parameter of velocity profile-LMB
C     and its second derivative DDLMB
C-------------------------------------------------------------
C     Constants of velocity parameter
C-------------------------------------------------------------
      P=1./((BETA*T**2)*(1.-BETA))
      P1=1./(T*(1.-BETA))
C------------------------------------------------------------
C     After integration constant of velocity parameter
C-------------------------------------------------------------
      CONST=0.5*BETA/(1.-BETA)

      IF (STAMPA.EQ.'N') GO TO 50
      WRITE(6,22)(IND(I),I=1,N)
22    FORMAT(6(4X,A4,4X))

50    CONTINUE
      IF(TE.GT.T)GO TO 65
      IF(TE.LT.T1) THEN
              LMB=0.5*P*TE**2
              DLMB=P*TE
              DDLMB=P
      ELSE IF(TE.LT.T2)THEN
              LMB=P1*TE-CONST
              DLMB=P1
              DDLMB=0.
      ELSE IF(TE.LE.T) THEN
              LMB=-0.5*P*(T-TE)**2+1.
              DLMB=P*(T-TE)
              DDLMB=-P
      END IF


C----------------------------------------------------------
C     Position, velocity and acceleration calculation
C----------------------------------------------------------

      DO I=1,N
```

```
      DLTQ(I)=QF(I)-Q0(I)
      Q(I)=Q0(I)+LMB*DLTQ(I)
      DQ(I)=DLMB*DLTQ(I)
      DDQ(I)=DDLMB*DLTQ(I)
      END DO

      WRITE(4,100)TE
      WRITE(4,100)(Q(I),I=1,N)
      WRITE(4,100)(DQ(I),I=1,N)
C     WRITE(4,100)(DDQ(I),I=1,N)

      IF(STAMPA.EQ.'N')GO TO 110
      WRITE(6,100)TE
      WRITE(6,100)(Q(I),I=1,N)
      WRITE(6,100)(DQ(I),I=1,N)
C     WRITE(6,100)(DDQ(I),I=1,N)
100   FORMAT(6E13.5)

110   TE=TE+H
      GO TO 50
65    CONTINUE

      CLOSE(UNIT=4)

      RETURN

70    CONTINUE
      T1=T/2.
C
C-------------------------------------------------------------------
C     Calculation of scalar parameter of velocity profile - lambda and
C     its second derivative
C     Constant of velocity parameter
C-------------------------------------------------------------------
C

      P=4./(T**2)


150   CONTINUE
      IF(TE.GT.T) GO TO 160

      IF(TE.LT.T1)THEN
              LMB=0.5*P*TE**2
              DLMB=P*TE
              DDLMB=P
      ELSE IF(TE.LE.T)THEN
              LMB=-0.5*P*(T-TE)**2+1.
              DLMB=P*(T-TE)
              DDLMB=-P
      END IF

      DO I=1,N
      DLTQ(I)=QF(I)-Q0(I)
      Q(I)=Q0(I)+LMB*DLTQ(I)
      DQ(I)=DLMB*DLTQ(I)
      DDQ(I)=DDLMB*DLTQ(I)
      END DO

      WRITE(4,100)TE
```

```
      WRITE(4,100)(Q(I),I=1,N)
      WRITE(4,100)(DQ(I),I=1,N)
C     WRITE(4,100)(DDQ(I),I=1,N)

      IF(STAMPA.EQ.'N')GO TO 90
      WRITE(6,100)TE
      WRITE(6,100)(Q(I),I=1,N)
      WRITE(6,100)(DQ(I),I=1,N)
C     WRITE(6,100)(DDQ(I),I=1,N)

90    TE=TE+H
      GO TO 150
160   CONTINUE

      CLOSE(UNIT=4)

      RETURN

1001  WRITE (6,1002) FILET
1002  FORMAT (1X,'File ',a10,1X,'does not exist')
      STOP
      END
```

```
C-------------------------------------------------------------
C    SUBROUTINE:     MATD
C-------------------------------------------------------------
C    FUNCTION:       subroutine calculates the matrices D1 and D2
C                    adjointed to the external force FA and external
C                    moment MA,respectively
C-------------------------------------------------------------
C    INPUT VARIABLES:
C          E(I,K)    unit vector of axis of the joint i
C          RAI(I,K)  vector from the center of the i-th joint
C                    to the point of contact between manipulator
C                    and constraint
C          N         number of joints
C          KSI2(I)   type of joint
C          VECP(I)   vector from the center of the last segment
C                    to the contact point between manipulator and
C                    constraint
C-------------------------------------------------------------
C    OUTPUT VARIABLES:
C          D1(I,K)   matrix D1
C          D2(I,K)   matrix D2
C-------------------------------------------------------------
      SUBROUTINE MATD(E)
      INTEGER DESNI
      DIMENSION E(6,3),RAI(6,3)
      COMMON /TIPL/ N,KSI1(6),KSI2(6),KSI3(6)
      COMMON /BRZL/ RT(3),RP(6,6,3)
      COMMON /SUR/ D1(6,3),D2(6,3),VECP(3)
      DO 10 J=1,N
      DO 20 K=1,3
      RAI(J,K)=RP(N,J,K)+VECP(K)
20    CONTINUE
10    CONTINUE
      DO 30 I=1,N
      IF(KSI2(I).EQ.1) GO TO 40
      D1(I,1)=E(I,2)*RAI(I,3)-E(I,3)*RAI(I,2)
      D1(I,2)=E(I,3)*RAI(I,1)-E(I,1)*RAI(I,3)
      D1(I,3)=E(I,1)*RAI(I,2)-E(I,2)*RAI(I,1)
      GO TO 30
40    D1(I,1)=E(I,1)
      D1(I,2)=E(I,2)
      D1(I,3)=E(I,3)
30    CONTINUE
      DO 60 I=1,N
      IF(KSI2(I).EQ.1) GO TO 70
      D2(I,1)=E(I,1)
      D2(I,2)=E(I,2)
      D2(I,3)=E(I,3)
      GO TO 60
70    D2(I,1)=0.
      D2(I,2)=0.
      D2(I,3)=0.
60    CONTINUE
      RETURN
      END
```

```
C---------------------------------------------------------------------
C SUBROUTINE:   MATH1
C---------------------------------------------------------------------
C FUNCTION:     Calculation of matrix h1 adjointed to the vector of
C               linear acceleration w0 of the first(basis) segment of
C               the robot
C---------------------------------------------------------------------
C INPUT VARIABLES:
C     E(I,K)    unit vector of axis of the i-th joint
C     MM(J)     mass of the j-th segment
C     R(J,I,K)  vector from the center of the i-th joint to the
C               gravity center of the j-th segment
C     N         number of joint
C     KSI2(I)   type of joint(for rotacional KSI2(I)=0, for
C               linear KSI2(I)=1)
C---------------------------------------------------------------------
C OUTPUT VARIABLES:
C     H10(I,K)  matrix h1
C---------------------------------------------------------------------
        SUBROUTINE MATH1(E,R,H10)
        REAL MM(6)
        DIMENSION E(6,3),R(6,6,3),R1(6,3),H10(6,3)
        COMMON /TIPL/ N,KSI1(6),KSI2(6),KSI3(6)
        COMMON /MIN/ MM
        DO 15 I=1,N
        DO 70 K=1,3
        IF(KSI2(I).EQ.0) R1(I,K)=0.
        GO TO 40
40      H10(I,K)=0.
        DO 20 J=I,N
        IF(KSI2(I).EQ.0) GO TO 50
        GO TO 60
50      R1(I,K)=R1(I,K)+MM(J)*R(J,I,K)
        GO TO 20
60      H10(I,K)=H10(I,K)+MM(J)
20      CONTINUE
        IF(KSI2(I).EQ.0) GO TO 70
        H10(I,K)=H10(I,K)*E(I,K)
70      CONTINUE
        IF(KSI2(I).EQ.1) GO TO 80
        H10(I,1)=E(I,2)*R1(I,3)-E(I,3)*R1(I,2)
        H10(I,2)=E(I,3)*R1(I,1)-E(I,1)*R1(I,3)
        H10(I,3)=E(I,1)*R1(I,2)-E(I,2)*R1(I,1)
80      CONTINUE
C       WRITE(5,90) (H10(I,L),L=1,3)
C       FORMAT(1X,'H10(I,L):',3E15.5)
15      CONTINUE
        RETURN
        END
```

```
C-----------------------------------------------------------------------
C SUBROUTINE:   MATH2
C-----------------------------------------------------------------------
C FUNCTION:     Calculation of matrix h2 adjointed to the vector of
C               rotational acceleration e0 of the first(basis) segment
C               of the robot
C-----------------------------------------------------------------------
C INPUT VARIABLES:
C     E(I,K)      unit vector of axis of the i-th joint
C     T(J)        tenzor of inertia of the j-th segment
C     MM(J)       mass of the j-th segment
C     R(J,I,K)    vector from the center of the i-th joint to the
C                 gravity center of the j-th segment
C     N           number of joints
C     KSI2(I)     type of joint(for rotacional KSI2(I)=0,for linear
C                 KSI2(I)=1)
C-----------------------------------------------------------------------
C OUTPUT VARIABLES:
C     H20         matrix h2
C-----------------------------------------------------------------------
      SUBROUTINE MATH2(E,T,R,H20)
      REAL MM(6)
      DIMENSION E(6,3),T(3,3),R(6,6,3),R2(3),H20(6,3)
      COMMON /TIPL/ N,KSI1(6),KSI2(6),KSI3(6)
      COMMON /MIN/ MM
      DO 50 I=1,N
      IF(KSI2(I).EQ.0) GO TO 35
C
C----------------------------------------------------------------
C     LINEAR JOINT
C----------------------------------------------------------------
C
      DO 20 K=1,3
      R2(K)=0.
      DO 20 J=I,N
      R2(K)=R2(K)+MM(J)*R(J,1,K)
20    CONTINUE
      CALL PP1(R2,E,I,H20)
      GO TO 50
C
C----------------------------------------------------------------
C     ROTACIONAL JOINT
C----------------------------------------------------------------
C
35    DO 40 J=I,N
      DO 40 K=1,3
      R2(K)=MM(J)*R(J,1,K)
      CALL PP2(E,R2,R,I,K,J,H20)
40    CONTINUE
50    CONTINUE
      RETURN
      END
```

```
C-----------------------------------------------------------------
C   SUBROUTINE:  PP1
C-----------------------------------------------------------------
C   FUNCTION:    Cross product of two vectors(Vector product)
C-----------------------------------------------------------------
C   INPUT VARIABLES:
C       R2         vector 3x1
C       E          matrix 6x3
C       I          index of the i-th row of matrix E
C-----------------------------------------------------------------
C   OUTPUT VARIABLES:
C       H20        cross product
C-----------------------------------------------------------------
        SUBROUTINE PP1(R2,E,I,H20)
        DIMENSION R2(3),E(6,3),H20(6,3)
        H20(I,1)=R2(2)*E(I,3)-R2(3)*E(I,2)
        H20(1,2)=R2(3)*E(I,1)-R2(1)*E(I,3)
        H20(I,3)=R2(1)*E(I,2)-R2(2)*E(I,1)
        RETURN
        END
```

```
C-----------------------------------------------------------------
C  SUBROUTINE:   PP2
C-----------------------------------------------------------------
C  FUNCTION:     Final calculation of matrix h2
C-----------------------------------------------------------------
C  INPUT VARIABLES:
C      R2         vector 3X1
C      E          matrix 6x3
C      R          matrix 6x6x3
C-----------------------------------------------------------------
C  OUTPUT VARIABLES:
C      H20     matrix h2
C-----------------------------------------------------------------
    SUBROUTINE PP2(E,R2,R,I,K,J,H20)
    DIMENSION E(6,3),R2(3),R(6,6,3),H20(6,3)
    H20(I,K)=H20(I,K)+(R2(K)*R(J,I,K))*E(I,K)-(R2(K)*E(I,K))*R(J,I,K)
    RETURN
    END
```

Subject Index

M. Vukobratović (Ed.)

Introduction to Robotics

With contributions by M. Djurović, D. Hristić, B. Karan, M. Kirćanski, N. Kirćanski, D. Stokić, D. Vuijić, M. Vukobratović

1989. 228 figures. XIV, 301 pp.
Hardcover ISBN 3-540-17452-4

Contents: Preface. - General Introduction to Robotics. - Manipulator Kinematic Model. - Dynamics and Dynamic Analysis of Manipulation Robots. - Hierarchical Control of Robots. - Microprocessor Implementation of Control Algorithms. - Industrial Robot Programming Systems. - Sensors in Robotics. - Elements, Structures and Application of Industrial Robots. - Robotics and Flexible Automation Systems. - Appendix.

This book provides a general introduction to robot technology with an emphasis on robot mechanisms and kinematics. It is conceived as a reference book for students in the field of robotics.

Springer-Verlag
Berlin Heidelberg New York
London Paris
Tokyo Hong Kong

Springer